Dominik G. Rabus, Karsten Rebner, Cinzia Sada
**Optofluidics**

# Also of interest

*Optofluidics Systems Technology*
Dominik G. Rabus, 2014
ISBN 978-3-11-033602-3, e-ISBN 978-3-11-035021-0

*Optical Measurement Mechanics*
Kaifu Wang, 2018
ISBN 978-3-11-057304-6, e-ISBN 978-3-11-057305-3

*Plasma and Plasmonics*
Kushai Shah, 2018
ISBN 978-3-11-056994-0, e-ISBN: 978-3-11-057003-8

*Multiphoton Microscopy and Fluorescence Lifetime Imaging*
*Applications in Biology and Medicine*
Karsten König, 2017
ISBN 978-3-11-043898-7, e-ISBN 978-3-11-042998-5

*Microfluidics*
*Theory and Practice for Beginners*
Sebastian Seiffert, Julian Thiele, 2018
ISBN 978-3-11-048777-0, e-ISBN 978-3-11-048770-1

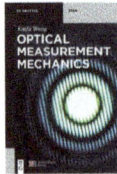

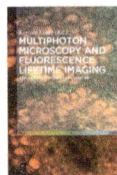

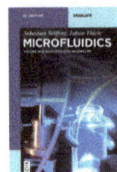

Dominik G. Rabus, Karsten Rebner,
Cinzia Sada

# Optofluidics

Process Analytical Technology

**DE GRUYTER**

**Authors**
Prof. Dominik G. Rabus
RABUS.TECH
Ahornstr. 47
74670 Forchtenberg
Germany
dominik@optofluidics.info

Prof. Karsten Rebner
Hochschule Reutlingen
Process Analysis and Technology
Alteburgstr. 150
72762 Reutlingen
Germany
karsten.rebner@reutlingen-university.de

Prof. Cinzia Sada
University of Padova
Physics and Astronomy Department
Via Marzolo 8
35131 Padova
Italy
cinzia@optofluidics.info

ISBN 978-3-11-054614-9
e-ISBN (E-BOOK) 978-3-11-054615-6
e-ISBN (EPUB) 978-3-11-054622-4

**Library of Congress Cataloging-in-Publication Data**
Names: Rabus, D. G. (Dominik G.), author. | Sada, Cinzia, author. | Rebner,
    Karsten, author.
Title: Optofluidics : process analytical technology / Dominik G. Rabus,
    Cinzia Sada, Karsten Rebner.
Description: Boston : De Gruyter, [2018] | Series: De Gruyter textbook |
    Includes bibliographical references and index.
Identifiers: LCCN 2018018089 (print) | LCCN 2018020956 (ebook) | ISBN
    9783110546156 (electronic Portable Document Format (pdf)) | ISBN
    9783110546149 (alk. paper) | ISBN 9783110546156 (e-book pdf) | ISBN
    9783110546224 (e-book epub)
Subjects: LCSH: Optofluidics.
Classification: LCC TJ853.4.O68 (ebook) | LCC TJ853.4.O68 R325 2018 (print) |
    DDC 621.36—dc23
LC record available at https://lccn.loc.gov/2018018089

**Bibliographic information published by the Deutsche Nationalbibliothek**
The Deutsche Nationalbibliothek lists this publication in the Deutsche Nationalbibliografie; detailed
bibliographic data are available on the Internet at http://dnb.dnb.de.

© 2019 Walter de Gruyter GmbH, Berlin/Boston
Typesetting: Integra Software Services Pvt. Ltd.
Printing and binding: CPI books GmbH, Leck
Cover image: © Nanoscribe

www.degruyter.com

To Anouk, Tizian, Olivia, Linus, Quentin, Leila and Dave

# Foreword

Optofluidics is a truly interdisciplinary field which needs a dedicated team of experts with cross discipline and in-depth knowledge to be able to create novel devices and systems. The nature of any collaboration is the principle in understanding each other and to be able to speak a common language. This book is the perfect example of this interdisciplinary collaboration. Optofluidics requires material science, photonics, fluidics, biology combined with engineering.

We would like this book to be the starting point of this fascinating topic and to provide a hands-on guide to start the process of a common cross discipline way of understanding, the source of innovation.

> Life is wonderful. There are moments when you think you'd like to die. But then something new happens, and you think you´re in heaven.
>
> Edith Piaf

https://doi.org/10.1515/9783110546156-201

# Contents

# List of Abbreviations

| | |
|---|---|
| $\Delta p$ | pressure drop through valve in MPa |
| A | cross section |
| ANSC | American National Standards Committee |
| ANSI | American National Standards Institute (www.ansi.org) |
| ASE | Amplified spontaneous emission |
| ASIC | Application Specific Integrated Circuit |
| ASTM | American Society for Testing and Materials |
| BANSAI | Biomedical analysis system with laser light |
| BCG | Bromcresol green |
| BOD | Biological oxygen demand |
| C | carbon |
| CA | Controlled atmosphere |
| CFD | Computational fluid dynamics |
| CIE | The International Commission on Illumination (Commission Internationale de l´Eclairage) |
| $C_J$ | junction capacitance |
| CTE | coefficients of thermal expansion |
| DPSS | diode-pumped solid-state |
| DUT | device under test |
| DUV | deep ultraviolet |
| ECL | external cavity laser |
| ELSA | electron stretcher accelerator |
| EMF | electromotive force |
| eV | electron volts |
| FRET | Fluorescence Resonance Energy Transfer |
| FTIR | fourier transform infrared spectroscopy |
| GRIN | graded index |
| h | thickness of the material |
| HA | Human albumin |
| HeNe | helium neon |
| $I_D$ | diode current |
| $I_L$ | current generated by incident light |
| IONAS | Institute for Optofluidics and Nanophotonics |
| IRS | integrated reverse symmetry |
| $I_S$ | photodiode reverse saturation current |
| $I_{SC}$ | short circuit current |
| J | Joule |
| k | extinction coefficient/wavenumber/Boltzmann constant |
| K | Kelvin |
| $k_v$ | defined volume flow of water in $m^3/h$ |
| LED | light-emitting diode |
| M | molecular weight/mass |
| M´ | mass flow in kg/h |
| MID | magnetic inductive flow |
| MMI | multimode interference |
| n | refractive index |
| $N_A$ | Avogadros number |

https://doi.org/10.1515/9783110546156-202

| | |
|---|---|
| NA | numerical aperture |
| NC | normally closed |
| $n_{eff}$ | effective refractive index |
| $n_f$ | refractive index of the film layer of a waveguide |
| NO | normally open |
| NTC | negative temperature coefficient |
| OLED | organic light-emitting diode |
| p1 | absolute pressure at valve inlet in MPa |
| p2 | absolute pressure at valve outlet in MPa |
| PC | polycarbonate |
| PDMS | polydimethylesiloxane |
| PIC | photonic integrated circuit |
| PLC | polarization controller |
| PMMA | polymethylmethacrylate |
| PSD | position sensitive device |
| psi | pounds per square inch |
| PTC | positive temperature coefficient |
| PVA | polyvenylalcohol |
| q | electron charge |
| QCM | quartz micro gravimetry |
| QD | quantum dot |
| R | molar refractivity |
| RI | refractive index |
| $R_m$ | molar refraction |
| $R_S$ | series resistance |
| SMA | shape memory actuator |
| SMP | shape memory polymer |
| SPD | spectral power distribution |
| T | absolute temperature in Kelvin |
| V´ | volume flow in m³/h (at 0.1 MPa and 20 °C) |
| $v_k$ | specific volume at p ½ in m³/kg |
| $V_{OC}$ | open circuit voltage |
| vph | phase velocity |
| WKB | Wentzel-Kramers-Brillouin Approximation |
| ZLMT | Zentrum für Labormedizin, Mikrobiologie und Transfusionsmedizin |
| α | flow coefficient |
| $α_e$ | polarizability of electrons |
| $ε_0$ | dielectric constant |
| λ | wavelength |
| ρ | density in kg/m³ (at 0.1 MPa and 20 °C) |
| $ρ_1$ | density of the gas upstream of valve |
| ψ | outflow function (function of pressure ratio p2/p1) |

# 1 Introduction

Over the past decade, various field of studies such as biology, microfluidics and photonics has begun to merge with each other. The result was visible in the advent of optofluidics and biophotonics, toward sensing new applications. As biophotonics has helped to open new doors to fundamental research and to rethink the way "biology" has functioned before, similar breakthroughs have been recently reported on the newly established merger of photonics and fluidics, that is, optofluidics that has again brought new ideas to life.

This book takes the merging of technologies a step further in showing the way of integrating fluidics and optics, with some insight dedicated to sensors in optofluidic systems ranging from standard physical measurement principles to analytical sensors to spectroscopy as well as some applications in biology. In this book, the merger of technologies is key and program at the same time because new ideas that result into devices that are only possible with the combination of appropriate technologies. Optics, microfluidics, sensing and photonics are vast research areas and to cover all aspects of these three key technologies would go beyond the scope of this work.

Therefore to narrow down the spectrum of possible research topics to be covered in this book, the focus lies on the integration and system aspect of future micro- and nanosystems. When dealing with micro- and nanosystems, the next question is on which material. Depending on the final application, the choice of the best material can span from polymers to glasses, from semiconductors to insulators or "hybrid" integration in smart solutions. As a matter of fact, although polymers have been widely explored substrates for biophotonic devices because of their biocompatibility, the fabrication flexibility and low cost [58, 147], they still have some limits of applicability, especially when long-term stability is a requirement. This is why polymers have been recently coupled and integrated to other materials in the frame of realizing smart multifunctional platforms for chemical and bio-chemical sensors [161].

One of the key topics addressed in this book describes the materials and the technologies to modify their chemical and physical properties and how to integrate them together and with other devices to provide a smart multifunctional optofluidic system. As a key study to show some concrete high technological achievements in optofluidics, in particular, this book presents how polymers can be exploited in biophotonics by merging the potentialities of microfluidics, integrated optics and relative technologies. In this framework, some highlights on culturing and patterning of live single cells for a sufficient period of time will be described, enabling the integration of optical evanescent field-enhanced photonic devices, and providing means to integrate electronic readout and other photonic components such as photodiodes and light sources. This is realized by a combination of engineering

https://doi.org/10.1515/9783110546156-001

nanofabrication and biological methods forming a cross-discipline approach, which is the building block of the optofluidics concept.

Over the past few years, there has been an increasing demand and interest both scientifically and technologically for small and portable analytical tools, the so-called Lab-on-Chip (LoC) systems, for present and future applications in areas such as medicine, food, environment and public security [66, 125]. The ultimate goal has been to integrate as many analytical functions as possible into a single device, and it has been undertaken by universities, research institutes and companies all over the world [10]. At the heart of this effort are the noninvasive analysis and the possibility to carry out real-time investigations and monitoring of the system under study, including sensing and its analytical applications down to microliters to nanoliters scale of reference for the analyzed volumes.

Microfluidic chips are ideal platforms for housing and handling small volumes. A large variety of microfluidic systems is also available for biology, such as cell analysis: the cell handling procedure in microfluidic chips follows cell sampling, cell trapping and sorting, cell treatment and cell analysis. Very often, however, large and expensive diagnostic tools are required for physical and chemical analysis of the investigated sample with a detrimental effect of lack of the device portability. The strong demand of a new concept of low-cost and portable platform, which is easy to handle and use, has pushed the research and the industrial efforts toward the delivery of an integrated system where both the "optical tools" are coupled and assembled together with the fluidics/microfluidics and are perfectly matched. In this frame of reference, the integrated optics has been definitively addressed as the "new horizon" to be settled into a fluidic/microfluidics device: photonic waveguides and devices such as splitters, couplers and complex photonic-integrated circuits play a key role. Consequently, the fabrication of optical waveguides on substrates has the potential to solve a major integration challenge. The focus on processing technologies provides the necessary platform in order to combine biology, fluidics and photonics because of the fact that all three technologies are well established and eventually needs only a new approach to be simply integrated. For example, using well-known polymers such as methacrylate-based polymers opens up the possibility to use readily available micro-nano fabrication tools such as lithography, hot embossing or etching technologies. Polymers, for example, are available in solid and liquid form, which enable the manufacturing of hybrid polymer devices, for example, polymer photonic devices on silicon. Polymers can not only serve as the basis of the three mentioned technologies, but can also serve merely as a carrier for other devices on non-polymer substrates such as silicon photodiodes or InP laser sources. This fact is again a major advantage for creating a truly merging material platform.

Optical waveguides are the basic and the most important optical elements that must be integrated into LoC microsystems [157]. So far, there have been many bioanalytical systems that use optical waveguides as their main analytical tool

[67, 144]. The most common methods for waveguide fabrication are conventional deposition techniques such as chemical vapor deposition, flame hydrolysis deposition and ion exchange on glass substrates.

Various concepts for biosensors based on evanescent field sensing [172] have been demonstrated [25, 34, 82]. Recently, it has been shown that live animal cell adhesion and spreading can be monitored online and quantitatively via the interaction of cells with the evanescent electromagnetic field present on the surface of an optical waveguide [193]. The idea to study living cells in combination with optical waveguides has also been used in context with optical waveguide light mode spectroscopy (OWLS) and confocal laser scanning microscopy (CLSM), which provides information about the shape of the cells at the surface [226]. This allows for the correlation between the cell-shape information from CLSM and the cell-surface interaction measurements from OWLS.

The evanescent field in planar waveguide structures reaches into the cell usually not more than 300 nm. Using a reverse symmetry waveguide presented in [95–97], where an evanescent field of about 1 μm is obtained, is sufficiently enough to penetrate into the cell body. The reverse symmetry waveguide uses a cladding material with a higher refractive index than the substrate (nanoporous silicon), thus enabling the optical mode field in the waveguide to turn around and penetrate the cladding compared to a conventional waveguide design, where the optical mode field decays gradually into the direction of the substrate.

Integrated ring resonator biosensors have attracted attention recently. Their compact size and the resonance effect enable the detection of very small quantities of substances [21, 40, 133, 134, 243], the combination of these integrated devices with living cells in microfluidic devices is of focus to the optofluidics community.

Optofluidic system technology is more than a LoC device. It is the integrating technology, which eventually can lead to a product. Optofluidic system technology integrates the LoC platform and combines it with fluid control systems. Therefore an optofluidic system, per definition, always consists of a photonic LoC device, a light source, a detector (e.g., photodiode, spectrometer) and fluid control elements. In order to pursue the path toward an integrated system for a specific application, certain peripheral devices such as physical and optical measurement principles are needed to automate the application and to provide a true system.

A definition of an optofluidic system is shown in the following Fig. 1.1, where the optical sensor element in combination with an appropriate fluidic interface forms the optofluidic system. The mechatronic layout of the optofluidic system needs to be developed according to the specific application, providing, of course, numerous possibilities from macro-to-nano optofluidic systems.

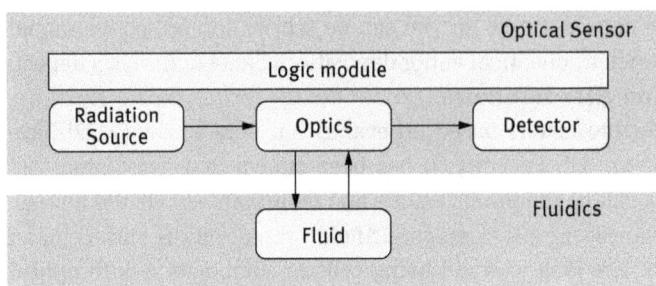

**Figure 1.1:** Definition of an optofluidic system.

An optofluidic system can be divided into five basic modules:
- A radiation source module, which provides the necessary wavelength or wave-
  length range for the measurement.
- An optics interaction module or optics measurement principle module, which
  defines how the radiation is used and transferred to the analyte. The main optical
  measurement and detection principles are as follows:
  - Absorption
  - Reflection
  - Scattering
  - Fluorescence
- Examples of a fluidic interaction module, which provides the necessary condi-
  tioning and preparation of the analyte, are as follows:
  - Dosing of a required reagent
  - Mixing of reagents or analytes
- A detection module. Examples are as follows:
  - Photodiode
  - Spectrometer
- A so-called logic module, which provides the controlling of the measurement,
  enough calculation power for the analysis of the measurement and for commu-
  nicating the result to the outside world.

It is important to note that each module is independent and fully operational.
Because of the increasing amount of available micro electro mechanical switches
(MEMS) and micro opto electro mechanical switches (MOEMS) sensors, these
described modules can be made very complex and highly functional, leveraging
the application of optofluidic systems in real-world surroundings.

The book is organized as follows: Chapter 2 provides an introduction to materials, especially polymers, and compares different technologies for modifying the
refractive index of suitable polymers by X-Ray, ion beam, photolocking, and DUV
radiation to create photonic waveguides in polymers, necessary for an optofluidic
system platform; Chapter 2 also provides an introduction to smart materials and their

potential applications in optofluidic devices; Chapter 3 provides an introduction to photonics, including waveguide fabrication, integration of active devices and hybrid silicon-polymer waveguides; Chapter 4 provides an overview on fluidics, including valves, and fluidic channel fabrication and provides a basic understanding of fluid control theory, necessary for optofluidic systems; Chapter 5 provides details on biology, including cell "integration"; Chapter 6 provides an overview of sensors in optofluidic systems ranging from standard physical measurement principles to analytical sensors to spectroscopy; Chapter 7 provides the necessary system aspect in order to take, for example, the polymer Bio-Fluidic-Photonic Platform toward an automated system and describes examples of optofluidic systems.

The book concludes with an outlook chapter providing visions for future research areas of this exciting and challenging topic. In order to provide a real system handbook, the appendix contains a wealth of gathered information from various sources providing a glossary on selected optofluidic terms and definitions, a chemical resistance chart of selected materials and a chapter on international standards and regulations important for optofluidic systems when bridging the gap from research to application.

# 2 Materials

The choice of the appropriate material for a specific type of application is the first important step toward a functional optofluidic system in the end. Several key factors need to be taken into account, which are generally application-driven, such as the property of the fluids to be handled, especially chemicals in contact with the fluidic parts of the system: biocompatibility, and compatibility with other used materials, just to name a few. When opto-microfluidics is concerned, further requirements are needed due to the micron-scale system of reference.

Materials for opto-microfluidic applications should, in fact, combine excellent optical properties together with high handability and machining compatibility.

For microfluidic applications, it is required to be compatible with molding or engraving processes to host microfluidic patterning as well as to show compatibility with the fluids it is in contact with. Therefore, depending on the final application, the microfluidic material should present chemical resistance or time durability and, when needed, even biocompatibility. Finally, they should be easily connected with external environment, allowing stable flux fluids injection by way of output holes and, eventually, suitable valves for an active flow control.

In the frame of reference of applied optics and integrated optics, instead, a material should be obviously optically transparent in the wavelength spectrum of interest, minimizing optical losses and optical-induced modifications promoted by light interaction as well. Basically, this is achieved by materials with a refractive index that can be both tailored and controlled upon request, showing optical-finished interfaces with low roughness and suitable compositional and morphological homogeneities.

It could be consequently believed that only the subset of materials with suitable optical properties among those used in microfluidics (and vice versa) can be of interest in opto-microfluidic applications but this is not the case. Although it would lead to a high level of integration, thanks to the monolithical combination of both optical and microfluidic features on the same substrate, this is not the only way to get an opto-microfluidic platform. Several alternatives have been currently developed by bonding optical materials to those hosting microfluidic circuitries, leading to hybrid solutions. Moreover, many applications exist using microfluidic circuitries and lab-on-chip platforms where optical investigations are performed by coupling external optical devices such as microscopes, laser sources in macro-sized optical benches as well. In the last case, we are still dealing with opto-microfluidics, while the term integrated opto-microfluidics should be addressed to the monolithical combinations or hybrids ones [113].

Generally, optofluidic materials are divided into categories depending on their mechanical response (rigid or soft materials), nature (inorganic, polymeric, etc.), final configuration (single materials, integrated or hybrids materials) and batch-processing

https://doi.org/10.1515/9783110546156-002

compatibility. Costs of production strongly limit the final commercialization of a product; therefore, it is mandatory to keep in mind whether a material is still at a level of interest for research purposes, prototypes or commercialized devises, respectively. In the first case, the priority relies on the material's overall performances (including its versatility), while ready-to-market solutions are upon consideration, instead, the fabrication costs, the elative easy-to-use and reliability are of relevant importance.

In the following chapter, the main materials exploited in optofluidics will be briefly described. For discussions regarding materials of interest in optics only, the reader is invited to specialized books on this topic. It is beyond the scope of this book that aims, instead, to help a newer in understanding the interplay between optics and fluidics/microfluidics requests.

In this frame of reference, silicon and glass are the first that were being used for microfluidic applications and in parallel in optics and integrated optics as well. The advances in the technology and material developing and processing, however, allowed new materials including polymer substrates, composites or paper to enter in the microfluidic scenario and later in the opto-microfluidic one. As a consequence, the chapter will prioritize on the material composition to later open a deeper understanding of its application in the field of photonics, fluid control system to sensors, respectively.

## 2.1 Inorganic materials

### 2.1.1 Silicon

Widely used in microelectronics with suitable doping to tailor its electrical conductivity to reach micro-scale down to nano-devices, silicon has played a key role in optoelectronic integration for computing, communications, sensing, and solar harvesting as well. The inventions of the bipolar transistor and the integrated circuit in fact have deeply revolutionized the world bringing technology advances in many different fields, that have in common the concept of a "planar process"-based approaches [53], where many functionalities can be integrated on the same slide and require the local material modifications with techniques that can be downscaled in the micron/nano-scale. They natively include the crystal technology, crystal dopant techniques and surface patterning (lithography), materials deposition processes, pattern transfer mechanisms, metal interconnect technology, and material passivation technology, respectively.

Due to the combination of its excellent material properties and the compatibility with metal–oxide semiconductor (CMOS) fabrication processing technology, silicon has played a great role in photonic and optoelectronic circuits with low cost, ultra-compact device footprint, and high-density integration [281]. An overview of its potentialities was recently reviewed in [251]. In the following, only a brief summary of its properties will be outlined, those of which are interested in the opto-fluidics field.

### 2.1.1.1 Physical and chemical properties

Silicon is a solid semiconductor material that crystallizes into a diamond crystal-line structure (density close to 2.3 g/cm$^3$, melting point close to 1,414 °C) with an energy gap at room temperature close to 1.12 eV. It is grown as single crystals in the form of ingots by the Czochralski techniques and then sliced in wafer commercially available up to 5" diameter. Silicon can be obtained even in the forms of crystallites, normally grown on pure silicon pre-existing seeds. One of the most known techniques is the Siemens process: it grows high-purity silicon crystallites directly on the surface of pure silicon seed rods by a chemical vapor deposition (CVD) that exploits the chemical decomposition of Si containing gases. More recently, the fluidized bed reactor (FBR) technique has also been introduced to reduce costs despite of the achievable lower polysilicon quality, in any case not so much to be a limiting factor in the photovoltaic applications. Its physical properties, temperature dependent, can be tailored by suitable doping that brings it to $n$-type or $p$-type: the description of their relevance is beyond the scope of this book.

In general, the silicon surface chemistry is based on the silanol group (-Si-OH) having reached a high level of standardization and mass production: large panels of surface accomplished via silanes chemistry are in fact available. Chemical modification of the silicon surface has been already widely exploited to adapt it to different applications, even to reduce non-specific adsorption of species on its surface. Silicon easily oxides and consequently at the surface it presents SiO$_2$ amorphous layer that can be removed by chemical etching. When combined to lithography masking procedures and chemical etching (refer Chapter 3), Si can be, therefore, easily patterned at a micro-scale down to nano-scale with great performances and resolutions.

From an optical point of view, Si is transparent to the infrared (IR) spectral region but absorbing in the visible (Figure 2.1). Its transmission strongly depends on its doping level and its thickness. The absorption in the visible range has been strongly exploited for photovoltaic applications as well as light detection sensors development (refer Chapter 3) while its transparency in IR opened the way to its use in the telecommunication field.

### 2.1.1.2 Silicon in microfluidics

Thanks to its popularity in microelectronics, it was the first natural choice of substrate even for microfluidic applications because of its high thermo-conductivity, significant resistance to organic solvents and is compatible with metal coasting. Microfluidic channels or even more complicated geometries were achieved by both additive methods and subtractive ones. In additive methods, metal/dielectric/

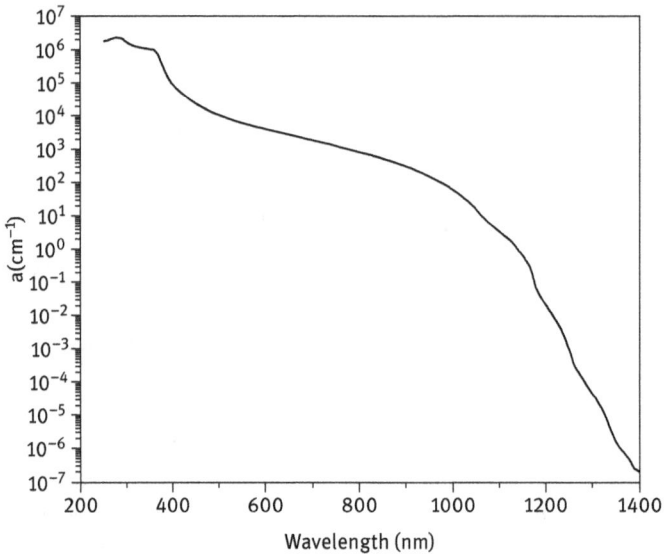

**Figure 2.1:** Absorption coefficient of Si at 300 K (data available in Ref. [88]).

insulate deposition(s) were performed on Si substrates to grow what needed. Subtractive methods, instead, were carried out to remove control-sized patterns by way of wet and/or dry etching. In this case, engraved channels were demonstrated on patterned configuration previously designed by lithography approaches (refer Chapter 3 for some details on fabrication processes). The Si microfluidic systems, however, suffered from complications connected to the sealing processes. As a matter of fact, direct Si-to-Si bonding requires a multi-step process: after a pre-processing of the Si wafers to remove any impurities at the surface by various cleaning procedures, a second step of pre-bonding at room temperature is performed. The wafers are aligned and put into contact to get a pre-bonding. Its strength is, however, not enough to resist to fluids pressures as needed in microfluidics applications and therefore a post-annealing at higher temperatures is necessary to strengthen the bond resistance (in the range between 150 and 800 °C to promote Si-Si covalent bonding). Current approaches, such as Reactive Ion Etching (RIE) for plasma exposure to activate surface for sealing purposes or high temperature, high pressure parameters for anodic bonding methods, are not always suitable but depends on the final application (e.g., they cannot be performed in biological ones). Silicon displays also a high elastic modulus (130–180 GPa) that limits its applications due to its overall fragility. When vibrations are concerned, in fact its durability is severely compromised. In particular, Silicon cannot host active microfluidic components such as pumps and valves and being quite expensive respect other materials. Finally due to these optical properties, fluorescence detection or imaging cannot be directly performed across this material in the visible range. For these reasons, silicon was considered less attractive in microfluidics than other solid materials such as glasses and polymers. However a "renaissance" for Si-based detectors for microfluidic systems has been back when successful studies on sealing process of Si channels with transparent materials, such as glass or polymers as well, were published giving birth to hydride configurations. Hybrid approaches in most cases currently consist of realizing a pressure-free soft contact with the rigid silicon part that needs either oxygen plasma exposure or thermal treatment for sealing and device completion. Despite of the multi-step procedures that are needed to get a final-sealed device, the high resolution of Si nanofabrication capabilities realized by either electronic beams or nano-imprint lithography technologies and Si-enriched surface chemistry has opened the way to complicated but high performance devices toward a high integration and miniaturization level.

### 2.1.2 Glass in opto-microfluidics

As stated in [64, 65], the name glass normally refers to a solid material that undergoes the solidification transition phase in a smooth way: the liquid viscosity increases progressively to reach the solid phase, meanwhile its constituents arrange in a random way, producing a disordered aggregation state in a network of cross-linked chemical bonds. This transition is named glassy or vitreous transition and is normally achieved by way of a rapid cooling or quenching the melt to the temperatures below the glass transition to prevent crystallization [134].

A glass differs from a crystal because of its structural arrangements of constituents: in crystals, they are ordered in space, being possible to identify basic units that repeat in space in given patterns and a transition melting point is present. A glass, instead, is an amorphous solid without long-range order on a scale greater than a few nm. The glasses are normally prepared by mixing the components in form of raw materials and with their subsequent melting. After melting, the homogenization and refining (removal of bubbles) procedures are carried out. Different techniques were developed depending on the final application to get flat glasses or shaped ones, normally achieved by blowing or pressing procedures.

#### 2.1.2.1 Physical and chemical properties
A glass does not have a fixed melting temperature but rather a range in which its viscosity decreases significantly. As an order of magnitude, already reviewed in [178], the viscosity $\eta$ is of the order of $10^{-4}$–$10^{-2}$ PI below the melting temperature upon further cooling the viscosity increases dramatically above $\eta = 10^{12}$ PI. The liquid–solid transition is, therefore, represented formally by a glass transition temperature, named $T_g$ (glass transition temperature), that reasonably represents this transition talking place [241]. Normally, it is taken as the midpoint of the temperature range in which the transition occurs: depending on

the glass composition, it can range in wide interval, 150 °C in chalcogenide glasses (GeSbTe) to 600 °C in sodalime glasses (containing mainly $Na_2O$-$SiO_2$-CaO oxides) just to cite a few. In general, below $T_g$, the liquid would be too viscous to flow on a year-based timescale. Very often glassy and vitreous are considered synonymous referring to the glass transition, amorphous instead should indicate the random disorder of constituents (a glass is amorphous but there exist amorphous materials that are not glass). It is worth mentioning that since the glass transition can undergo at different speed rates, the final glass state differs: each glass, therefore, has its own history depending on its preparation process (Figure 2.2).

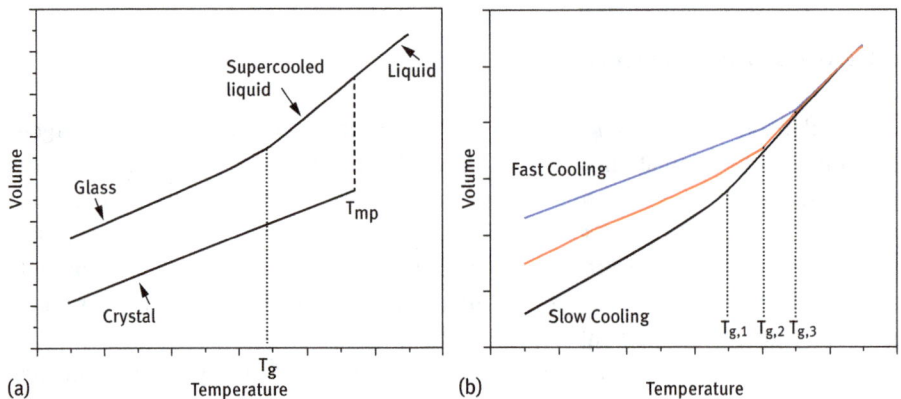

**Figure 2.2:** Specific volume variation with temperature: (a) glass in comparison with crystals, (b) glass under different cooling rates [241].

The overall properties of glasses are strongly dependent on their structure at a microscale and on their composition as well. In general, a glass is made of network formers, intermediates and modifiers: formers are those components that can constitute a glass as single component. Common formers are oxides of silicon, boron, and germanium, respectively. Among the others, silica is the most popular glass: made of $SiO_2$, tetrahedral oxygen atoms surrounded by Si atoms without long-range order (silica is, therefore, the amorphous state of quartz). Silicon oxide is not a mandatory ingredient in glasses: others exist made of polymers as it will briefly outlined in the following section. Modifiers, instead, are components entering in the glass network and modifying it without being able to form a glass by themselves (Figure 2.3). In common, oxide-based glasses calcium, lead, lithium, sodium, and potassium behave as modifiers, entering in the network as ions.

In oxide glass networks, there are bridging oxygen those are completely bonded to the glass network and non-bridging oxygen are those bound by one covalent bond

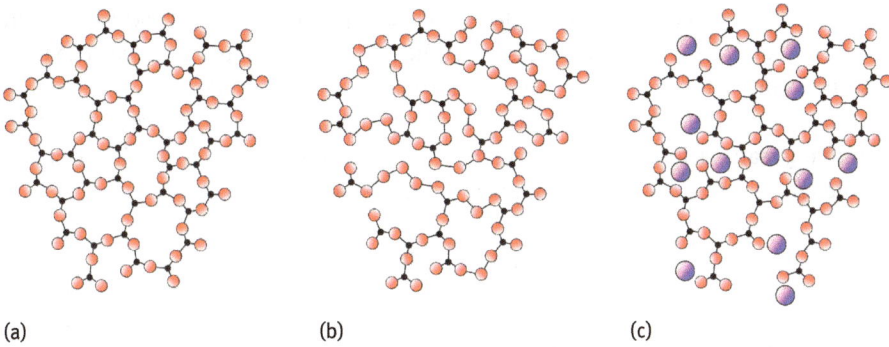

**Figure 2.3:** Typical network-forming glasses: (a) a stoichiometric glass former ($SiO_2$, $B_2S_3$) whose structure and network connectivity can be altered by the addition (b) of two-fold coordinated atoms (usually chalcogens, S, Se) that lead to cross-linked chains. The structure can also include modifiers (c) by the addition of a network modifier (alkali oxides or chalcogenides, $Na_2O$, $Li_2S$, etc.). [178].

to the glass network and the other holding negative charge that compensate for modifiers' one (Figure 2.4). Non-bridging oxygens, therefore, somehow weaken the overall glass network, decreasing the viscosity of the melt and lowering the melting temperature, favoring their moldability.

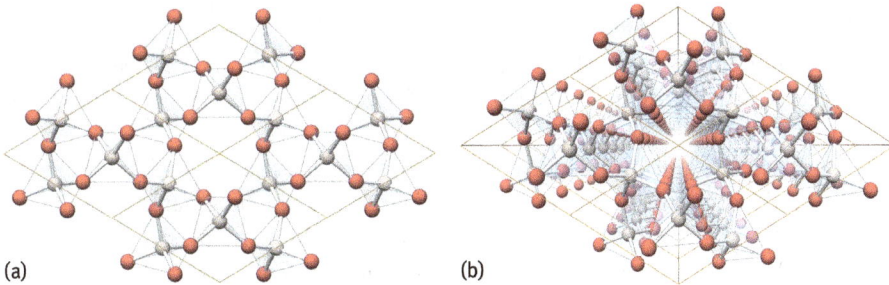

**Figure 2.4:** $SiO_2$ network: tetrahedral configuration looked by different perspectives (a) and (b) (kind courtesy of Prof. Amir C. Akhavanas, http://www.quartzpage.de/gen_struct.html)

Intermediates are those constituents that can play the role of formers or modifiers depending on the glass composition: typical intermediates are $Pb^{4+}$ replacing $Si^{4+}$, titanium, aluminum, zirconium, beryllium, magnesium, and zinc just to cite the most known. In general, the glass properties depend strongly on its network topology, that is, the way bonds and angles arrange to lead to a connected atomic network.

Glasses have been playing a great role in the production of optical materials, because most of the glasses are natively transparent in the visible without appreciable scattering. They can be colored adding dopants (such as FeO and $Cr_2O_3$ to get green, ruby using gold chloride just to cite some examples) or additives or dispersed particles in the nano-scale region to color the glass (such as in photo-chromic glasses). Color can also be achieved by way of light scattering (Mie-scattering) or metal nanoparticle light absorption mediated by Surface Plasmon resonance (such as in gold case), [3]. In the majority of the cases, glasses present a refractive index around 1.5. The most common identifying features for characterizing an optical glass are the refractive index $n_d$ in the middle range of the visible spectrum and the Abbe value $A_d$

$$A_d = (n_d - 1)/(n_F - n_C)$$

as a measure for dispersion, where blue F and red C are hydrogen lines of light emission, respectively ($\lambda_F$ = 486 nm, $\lambda_C$ = 656.3 nm). The difference of $n_F - n_C$ is called the principal dispersion. They can be coated with thin films to get tailored optical response (antireflective) or mechanical response (hardening) or chemical resistance to corrosion: thanks to their limited costs and high handability, they have gained a great market as insulator substrates. Finally, oxides glasses are normally compatible with biological samples. From the chemical point of view, glasses are quite resistant to attack and corrosion apart from contact with fluoridric acid. In general, glasses containing alkalines tend to be less resistant to corrosion because these elements present a higher diffusivity in the material due to the weaker bond to the glass network. They are also not permeable to gases and have relatively low non-specific adsorption.

### 2.1.2.2 Glasses in microfluidics
Thanks to the previously quoted properties, glass or quartz capillaries were largely employed in fluidics applications especially for gas chromatography and Capillary Electrophoresis (CE) [220].

They have also ben exploited to engrave or pattern glass substrates by way of wet/ dry chemical etching made on selected areas defined by photolithography techniques, as described in Chapter 3. In general, independent of etching techniques used, etched glass channels generally have rounded sidewalls and present a not very sharp definition [246], reason why in this case silicon presented a higher performance. Similarly to silicon, fluidic valves cannot be easily achieved in glass/silicon chips due to their hardness. Consequently, their applications were quite limited also because of the relative costs connected to the glass production respect other materials. These limitations are the origin to the development of alternative low cost chip materials that can be easily fabricated and are compatible for broader biological applications.

## 2.2 Polymers

Polymers are in general macro-molecules made of many repeating units that are molecules/macro-molecules linked by way of chemical bonds to form a network. It can be either 3D or 2D or 1D depending on the fact that they can occupy any direction in the space (3D) or on a plane/sheet (2D) or along a line (1D), respectively. Most commonly, the repeating basic units are made of carbon and hydrogen (in this case hydrocarbons), without excluding, however, the possibility of containing oxygen, nitrogen, sulfur, chlorine, fluorine, phosphorous and silicon as well. They are linked in "chains," where they are hooked up. In nature, several polymers exist and depending on their chemical composition they can form acid (such as deoxyribonucleic acid (DNA) and ribonucleic acid (RNA)) or protein-based polymers such as those that spider silk, hair and horn are basically made of. Similarly, saccharide-based polymers such as cellulose (($C_6H_{10}O_5)_n$) are very popular in nature.

Polymers can also be manufactured for specific needs and have gained a great role in the nowadays products since they are the constituents of plastics and rubber. They are produced by a polymerization process, that is basically chemical reaction of reagents (monomers): the final polymer is therefore made of repeating units, ending with final groups each bonded to only one repeating unit, that define the chain extremes.

The polymerization can occur by a step growth, a chain growth eventually initiated by absorption of visible or ultraviolet light, in which case is referred as photo-polymerization. In a step-growth process, polymers basically form by a stepwise reaction between functional groups of monomers. Chain-growth polymerization (or addition polymerization) instead involves the process of linking together the molecules/units.

In the case of photo-polymerization, the light may be absorbed either directly by the reactant monomer (direct photo-polymerization) or otherwise by a photosensitizer. In the last case, it absorbs the light and then transfers energy to the monomer. In general, photo-polymerization differs from the ordinary thermal polymerization only at the beginning; subsequent propagation, termination and chain transfer steps are normally remaining unchanged. While in step-growth photo-polymerization, absorption of light triggers an addition (or condensation) reaction between two co-monomers that do not react without light assistance, a propagation cycle is not initiated because each growth step requires the assistance of light.

Polymer can be classified basically depending on:
- composition (organic or inorganic);
- order configuration (amorphous, crystalline);
- response deformation (thermoset, thermoplastic and elastomers); and
- polymerization process (by addiction or condensation).

Most polymers also exhibit rather similar properties with respect to thermal (approximately $0.1-0.2$ WK$^{-1}$ m$^{-1}$) and electrical conductivity (resistivity on the order of $10^{12}-10^{18}$ $\Omega$ cm$^{-1}$). Consequently, chemical stability and surface chemistry are probably the most important material properties as they greatly define the applications in which the particular material can, or cannot, be used.

### 2.2.1 Role of polymer composition

Polymers can be viewed as long chains made by atoms linked one to another on a backbone where other atoms can attach in a sort of branched configuration. When the backbone contains carbon, they are usually named organic polymers; on the contrary they are normally referred to as inorganic polymers. Many polymers contain only carbon and hydrogen atoms, such as polyethylene, polypropylene, poly-butylene, polystyrene, and polymethylpentene just to cite the most known. Others, such as polyvinyl chloride (PVC), have chlorine attached to the all-carbon backbone, Teflon presents fluorine attached to the all-carbon backbone instead. Polymer backbones can include elements other than carbon: a typical example is nylon that contains nitrogen atoms in the repeat unit backbone, while polyesters and polycarbonates include oxygen in the backbone. Inorganic polymers, instead of having a carbon backbone, have silicon or phosphorous backbones: one of the more famous silicon-based polymers is Silly Putty®. In general by tailoring the molecular structure, different final properties can be achieved. For these reasons, various fillers, reinforcements and additives are often added into the base polymers, to expand the product performances.

### 2.2.2 Role of thermal response in polymers

As already mentioned, depending on their response to thermal treatment, polymers are usually divided in thermoset polymers (i.e., polymer networks that do not melt once formed, i.e., they cannot recover their original properties after re-melting due to their irreversible degrade) or thermoplastic polymers (i.e., chains that can be melted once formed, as in linear polymers). Epoxy resins used in two-part adhesives are known to be thermoset plastics, while plastic bottles, films, cups, and fibers are thermoplastic plastics, respectively.

### 2.2.3 Role of structure order – molecular arrangement

Polymers that arrange without a long-range order or form a configuration lacking in order are said to be amorphous: the typical picture used to describe it is a plate of "spaghetti" noodles. Very often it is achieved by a suitable control in the polymerization process and in the way of quenching the molten polymers so that no *configurational* ordering can be established, preventing that a repeating pattern can develop. Many amorphous polymers are also optically transparent, such as food wrap, plastic windows, headlight lenses, and contact lenses. Since glass is the typical amorphous material, these polymers are often referred to as glassy or glass as well. A crystalline configuration can also be achieved in polymers but in this case the repeating basic units are molecules, while in other materials (such as diamond, sapphire, quartz) they are atoms. Some materials can exist in both amorphous and crystalline states (silica and quartz, $SiO_2$ in amorphous or crystalline form respectively), while others cannot be crystalline or can be built to prevent it. In polymers, the crystallinity increases the polymer strength, stiffness, chemical resistance, and its stability despite of the degradation in the transparency to light.

### 2.2.4 Elastomers: PDMS

Polydimethylsiloxane (PDMS) is the most known elastomer in optofluidics, thanks to its cheapness and easiness to micro-fabrication. Belonging to a group of polymeric organo-silicon compounds, commonly referred to as silicones, PDMS's basic formula is $(C_2H_6OSi)_n$. It is optically transparent, non-toxic, and non-flammable and is viscoelastic. The PDMS has been widely exploited thanks to its elasticity and deformability under the application of forces or air pressure that allow molding and shaping by imprinting and soft-lithography. In particular, liquid PDMS pre-polymer is normally and thermally cured at temperatures in the range of 40–70 °C and casts with nanometer resolution from photoresist templates. Due to its low surface tension, its peeling from templates after being cured is relatively easy. The PDMS chip can be sealed by soft contact to another material such as PDMS or glass in both reversible and irreversible ways, respectively, in the last case using surface treatments such as plasma oxidizing. As reported in multilayer channel, structures were already fabricated also in 3D. It is worth mentioning that after polymerization and cross-linking, solid PDMS exposes a hydrophobic surface that can be turned to hydrophilic after suitable oxidation treatments (generally performed by Plasma Oxidation treatment to add silanol groups Si-OH). Since at a molecular level, PDMS behaves as a porous material, it presents a high degree of permeation of gases that facilitates its exploitation in biological application and cell-based microfluidics devices. Unfortunately, organic solvents can diffuse within, causing swelling of this material. Moreover, some drawbacks were reported due to absorption of small hydrophobic molecules into channel walls of PDMS-based microfluidic channels as well as the adsorption of biomolecules onto channel walls [220]. On the contrary, aqueous solutions are not able to infiltrate within the material and, for these reasons, its applications are restricted to this field. Problems, such as channel deformations, low solvent and acid/base resistivity, evaporation, sample absorption, leaching, and hydrophobic recovery, are the fundamental challenges associated with this material in microfluidic devices. Thanks to its elastomer nature, however, PDMS valve was invented to control micro-channel fluidic transportation, which also enables very large-scale integration in high-throughput applications. From an optical point of view, it presents a good optical transmission, thanks to its low absorption coefficient (Figure 2.5). It has recently applied also in the realization

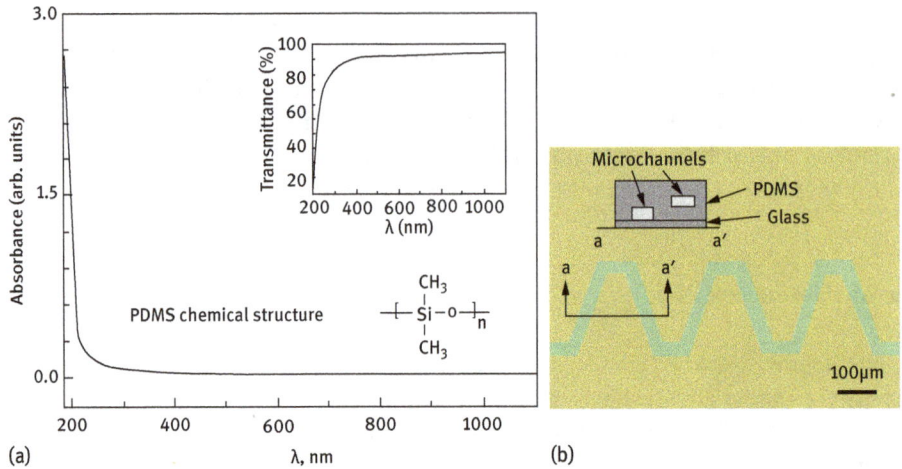

**Figure 2.5:** (a) PDMS Absorption coefficient in the visible range [245]. (b) PDMS micro-channel with green ink filled for visualization [282].

of opto-microfluidics devices such as PDMS microfluidic chip with integrated wave-guides for optical detection [80].

### 2.2.5 Thermoplastics: PMMA, PC, PVC CBC

Thermoplastics are synthetic polymers that have various surface properties for microfluidic applications, being essentially characterized by the native capability can be re-shaped multiple times by re-heating. This peculiarity opened them the way to a wide use in industry because they can be both molded and bonded, respectively. Typical thermoplastics are widely used in microfluidics for poly (methyl methacrylate) (PMMA), polycarbonate (PC), polystyrene (PS), poly-vinyl chloride (PVC), polyethylene terephthalate (PET), polyimide (PI), and the family of cyclic olefin polymers (i.e., cyclic olefin copolymer (COC), cyclic ole-fin polymer (COP), and cyclic block copolymer (CBC)) as well as perfluorinated polymers such as perfluoroalkoxy (Teflon PFA) and fluorinated ethylenepropylene (Teflon FEP), respectively. Thermoplastics are rigid polymer materials with good mechanical stability, a low water-absorption percentage, and organic-solvent, and acid/base resistivity, which is the critical factors in many bio-analytical microfluidic applications, such as high-pressure liquid chromatography (HPLC) microfluidic applications [17], that involve a high-pressure solvent injection procedure. From a structural point of view, they are glassy, softening at a glassy transition temperature ($T_g$) that allows them to be easily processable. Since at room temperature they are quite rigid, thermoplastics are normally purchased as solid and cannot be easily exploited to get fluidics diaphragm valves. In thermo-molding, thousands of replicas at high rate and low cost are produced using templates in metal or silicon; thermoplastics are surely excellent for commercial production but not economical for prototypic use. To overcome this limitation, Whitesides and Xia [274] proposed a rapid prototyping using transfer molding, where PDMS is a replication of intermediate and enables the transfer of micro-patterns to thermoplastics from easy-to-prepare photoresists. It is worth to mention that thermoplastics cannot form conformal contact with other surfaces consequently typical strategies for sealing their channels include thermo-bonding and glue-assisted bonding techniques. Respect to glass, plastic thermo-bonding generally requires much milder conditions, that is, lower temperature and pressure without cleanroom environments. From a chemical point of view, instead, they present a slightly better solvent compatibility than PDMS but fair resistance to alcohols as well as incompatibility with most other organic solvents, such as ketones and hydrocarbons. Most of them present lack in gas permeability, and in this case they cannot be exploited in microfluidic channels chambers if long-term cell studies are under concern but they can be used in combination with other materials as sealers.

Depending on their applications, thermoplastics surface can be modified by dynamic coating or surface grafting: the surface treatments of these materials appear quite stable in time. Hydrophilicity can, for example, be retained for years after a surface oxygen plasma treatment as well as they can be deposited by metal films. Consequently, they can be easily integrated with electrodes for flexible circuit, such as already demonstrated in is digital microfluidics applications to manipulate droplets by electro-wetting. As far as Teflons are concerned, they gained great interest in microfluidics because they can used to host microstructures with nanometer resolution. Apart being extremely inert to chemicals and solvents, nonsticky, and antifouling, they are optically transparent, soft enough to make diaphragm valves, and moderately permeable to gases [220]. Although their melting temperatures are high (over 280 °C), the whole Teflon chips show excellent compatibility with organic solvents, outstanding antifouling properties, and amenability for cell culture.

### 2.2.6 Thermoset: SU-8 photoresist, polyimide

Thermosets are polymer networks that cannot recover their original properties after re-melting due to their irreversible degrade. When heated or radiated, the thermosetting molecules cross-link to form a rigid network that cannot soften before decomposition and consequently they cannot be reshaped once cured. Thanks to their efficient photo-polymerization processing, some thermoset such as SU-8 and polyimide were first exploited as negative photoresists and later introduced into microfluidics. They are nowadays very common and can be used in 3D micro-fabrication technology allowing different shapes to be created that are stable in time. Since they show higher strength compared to other polymers, in fact, they allow the fabrication of high-aspect ratio and free-standing structures. These materials are normally stable even at high temperatures, resistant to most solvents, and optically transparent. On the contrary, however, due to their high stiffness, unfortunately thermosets are not the best choice for the fabrication of the diaphragm valve, also with their high cost, the applications in microfluidics are limited.

### 2.2.7 Smart polymers: Hydrogels

These are environmentally sensitive polymers; they can respond to different external stimuli such as temperature or pH change. They are widely applied in bio-technology, medicine, and engineering. An overview on smart polymers especially in application areas of bio-separation, protein folding, microfluidics and actuators, sensors, smart interfaces, and membranes is given in [155].

Smart polymers have three types of physical forms: (i) linear-free chains in solution: polymers undergo a reversible collapse, which is triggered by an external stress; (ii) covalently cross-linked reversible gels: these gels swell or shrink, depending on the environmental trigger, such as change in acidity, and (iii) chain-adsorbed or surface-grafted form (polymers are first grafted on a substrate and, subsequently, they swell or collapse reversibly on surface, once an external parameter is changed). Smart polymers are promising for future optofluidic systems and are thus the focus on current research activities. The following paragraph provides a closer look at a special group of smart polymers, hydrogels.

Hydrogels are cross-linked polymer network structures, which are capable of absorbing a significant amount of water. There is an extensive range of hydrogels, and they are classified based on different criteria of interest. The following section focuses on two types of classifications.

Depending on the bonding of the hydrogel polymer network, hydrogels can be classified as chemical or physical gel. When the hydrogel polymer network is linked by a stable chemical bond, such as a covalent bond, the gel network tends to be steady and permanent. Linking polymer chains with physical bonds, such as hydrophobic bonds, forms so-called physical hydrogels. Since these physical bonds are far less rigid than chemical bonds, the polymer chains have a certain degree of flexibility for mobility. In addition, this flexibility also gives the gel reversibility, which is often stimulated by external environment stress.

The most prominent usage of hydrogels is in the field of biomedical engineering. Since it was first invented for contact lenses [74] in 1999, doctors and researchers have been continuously exploring the potential of hydrogels in clinical treatment for tissue repairing. In the pharmaceutical industry, hydrogels are used as drug delivery systems to transport medicine to a targeted tissue. The third major group of hydrogels is environmentally sensitive hydrogels. Without additional actuation, they can

respond to environmental stimulus including temperature variations, pH fluctuations, electrical current changes, in various forms such as color switching and
dimension change. The hydrogels, which are being described further in the following
paragraph, are environmental-sensitive hydrogels (e.g., the hydrogel is sensitive to a
change in the pH value of the fluid surrounding it), which potentially have applications in optofluidic systems for controlling the flow [68], for example, as valves in
microfluidic channels (Figure 2.6).

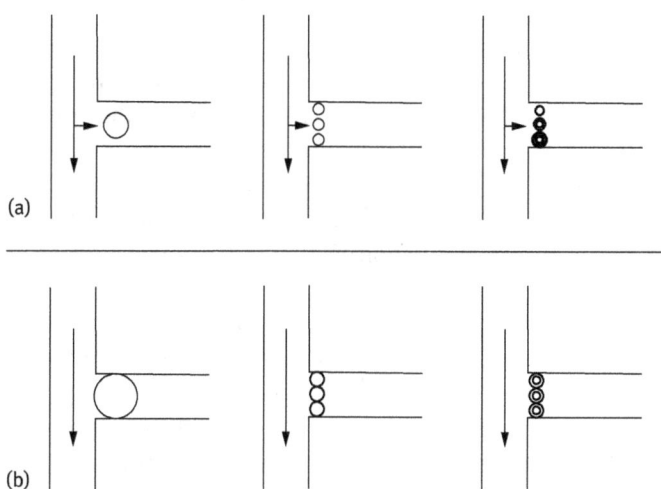

**Figure 2.6:** Illustration of (a) three hydrogel valve geometries including the single-post, multi-post
and multi-post jacket from left to right. The time necessary to close the valves as shown in panel
(b) is greatest with the single-post and least with the multi-post jackets. (Reprinted from Advanced
Drug Delivery Reviews 56 (2004) 199– 210, David T. Eddington, David J. Beebe, Flow control with
hydro-gels, with permission from Elsevier.)

An example of a pH-responsive hydrogel, which is being described further, is poly(2-
hydroxylmethacrylate) also referred to as HEMA [86]. Responsive to the pH value,
means that the gel responses to solutions with different pH by changing its mass.

Hydrogels based on 2-hydroxyethyl methacrylate (HEMA) polymers receive great
attention in biomedical applications. One of the major issues for this attention is the
fact that HEMA possesses excellent oxygen permeability properties. The most well-
known application of HEMA is the contact lens. The ability to supply oxygen has
always been a substantial criterion to select materials for contact lenses. The discovery of HEMA in 1999 provides a unique means to manufacture contact lenses with
superb oxygen permeability, such that sufficient oxygen can be supplied to the
eyeballs. Irritations due to the lack of oxygen are reduced.

The second major property of HEMA is its pH response capability. Once incorporated with other hydrophilic copolymers such as acrylic acid (AA), HEMA is able to swell and de-swell in aqueous solutions with distinguished pH values. For example, in the case of drug delivery, drugs and functional nanoparticles are often incorporated into the hydrogel. When a piece of hydrogel travels through an acid environment (low pH), such as the stomach, it is able to de-swell and keep the drugs inside the gel. When it reaches the target organ site such as the intestine, which has a higher pH value internally than the stomach, it will be able to swell and, consequently, release the encapsulated medication. Therefore, as an example, by the usage of this hydrogel, local drug delivery to a specific organ is achievable.

The HEMA is made by photo-polymerization. The precursor includes HEMA (monomer), acrylic acid (copolymer), Irguacure 651 (photoinitiator) and Eth-ylene-glycol dimethacrylate known as EDGMA (cross-linker). The precursor is the mixture of the chemical solutions to create the polymer before the photo-polymerization step. An important issue during the hydrogel formulation is the water concentration. It has been shown in literature that the synthesis in the presence of water may alter the degree of swelling. More specifically, the proportion of water has to be below 40–60% to obtain transparent hydrogels [279].

During the initiation step, UV light excites the photoinitiator Irguacure 651 to create free radicals. These free radicals are very reactive and will react with each monomer (HEMA) and copolymer, to create a long polymer chain.

Once individual polymer chains are formed, they are free-floating polymer chains, which do not give the hydrogel its glutinous properties. The next step is to introduce cross-linkers. These cross-linkers will then interconnect the linear polymer chains and form a polymer network. It is only with this network structure that the hydrogel is able to retain or expel water molecules.

The HEMA polymer network alone can swell, but acrylic acid (AA) is added to increase the swelling ratio, due to the fact that AA increases the hydrophilicity of the polymer.

The ability of the hydrogel to swell primarily depends on the chemical nature of the gel polymer. In other words, whether or not the used polymer and copolymer, which make up the hydrogel, have a strong affinity toward water molecules is the precondition that determines the gel's ability to attract water molecule and to cause the swelling behavior.

The HEMA hydrogel example, copolymerized with AA (HEMA-co-AA), is anionic, which means that it easily becomes negatively charged. The HEMA polymer possesses a hydroxyl group (-OH), which is polar and electronegative. Acrylic acid has a carboxylic group (-COOH) as a functional group and is, therefore, also electronegative.

In any acid solution, there is a significant amount of protons ($H^+$) and the majority of $H_2O$ molecules are hydronium ions ($H_3O^+$). When the gel is immersed in an acidic environment, the carboxylic group of AA and HEMA are both protonated by the $H^+$ from the surroundings. Therefore, as the functional groups of AA and HEMA are positively charged, they no longer have an affinity toward the positive hydronium molecules. And by osmosis pressure, the water molecules will migrate from the inside of the gel to the outside and thus cause the gel to shrink [161].

In any basic solutions, there is an extensive amount of $-OH$ ions. These negatively charged hydroxide anions have a high tendency to catch hydrogen groups on molecules, to deprotonate molecules and to undergo a neutralization reaction with the hydrogen group. In this case, in the basic environment, the hydroxyl group of HEMA and the carboxyl group are both deprotonated by $-OH$ ions. As a result, both HEMA and AA become negatively charged anionic molecules and possess a strong affinity toward the hydrogen on water molecules. This affinity drives a large amount of water molecules to flow into the gel and causes the gel to swell.

An example of a photonic crystal pH sensor based on template photo-polymerized hydrogel inverse opal is presented in [242].

### 2.2.8 General properties of polymers in microfluidics

Polymers present many inherent properties that can be further enhanced by a wide range of additives to broaden their uses and applications. In general, their physical properties depend on the size or length of the polymer chain (often quantified in the number of repeating units therein contained). Longer chains normally present higher melting and glass transition temperatures, higher viscosity but lower mobility. Polymer's great advantage relies on the light in weight and easiness to be processed in various shapes, spanning micro-sized details to macro-scale objects. By extrusion, in fact, fibers, films, or different shapes (such as pipes, bottles, etc.) can be easily achieved, while by injection molding very intricate parts or large car body panel are produced as well. In the micro-fabrication process of interest for microfluidics, the polymers should present high versatility to be molded in the micron-scale and be compatible with other materials the chip is made of. As a consequence, among the others, the polymer materials typically used are polydimethylsiloxane (PDMS) and thermoplastics, respectively. The choice of the right polymer depends on the precise application it is addressed to, by a proper balancing of benefits/drawbacks to get the best compromise. As recently reviewed in [47], several key aspects should be taken into account such as the mechanical and thermal properties as well as the chemical resistance and relative cost. As reported in Figure 2.7 of this review paper, it clearly emerges that polymers present in general good biocompatibility and the relative cost is below 150 \$/Kg, typically less than25\$/Kg in case of thermoplastics. In the following, the most popular polymers used in microfluidics will be summarized.

In microfluidics, a great role is played by the interaction of the fluids with the material surface. If it is hydrophobic, some complication can arise when bio-analysis are carried out on biological targets, such as proteins. High hydrophobicity can lead

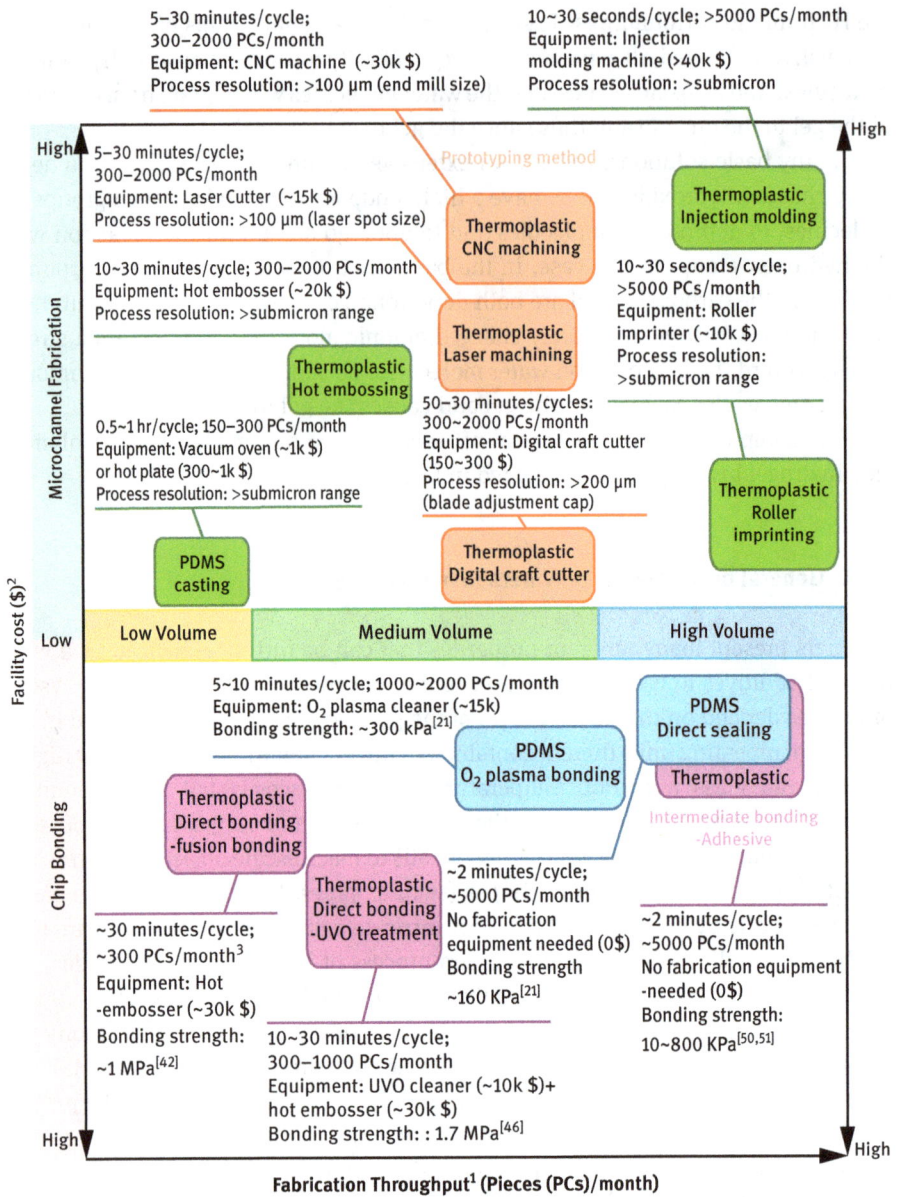

5–30 minutes/cycle;
300–2000 PCs/month
Equipment: CNC machine (~30k $)
Process resolution: >100 μm (end mill size)

10~30 seconds/cycle; >5000 PCs/month
Equipment: Injection
molding machine (>40k $)
Process resolution: >submicron

High

High

5–30 minutes/cycle;
300–2000 PCs/month
Equipment: Laser Cutter (~15k $)
Process resolution: >100 μm (laser spot size)

Prototyping method

Thermoplastic
Injection molding

Thermoplastic
CNC machining

10~30 minutes/cycle; 300–2000 PCs/month
Equipment: Hot embosser (~20k $)
Process resolution: >submicron range

Thermoplastic
Laser machining

10~30 seconds/cycle;
>5000 PCs/month
Equipment: Roller
imprinter (~10k $)
Process resolution:
>submicron range

Thermoplastic
Hot embossing

0.5~1 hr/cycle; 150~300 PCs/month
Equipment: Vacuum oven (~1k $)
or hot plate (300~1k $)
Process resolution: >submicron range

50~30 minutes/cycles:
300~2000 PCs/month
Equipment: Digital craft cutter
(150~300 $)
Process resolution: >200 μm
(blade adjustment cap)

Thermoplastic
Roller
imprinting

Microchannel Fabrication

PDMS
casting

Thermoplastic
Digital craft cutter

Facility cost ($)[2]

Low

| Low Volume | Medium Volume | High Volume |

5~10 minutes/cycle; 1000~2000 PCs/month
Equipment: $O_2$ plasma cleaner (~15k)
Bonding strength: ~300 kPa[21]

PDMS
Direct sealing

PDMS
$O_2$ plasma bonding

Thermoplastic

Thermoplastic
Direct bonding
-fusion bonding

Intermediate bonding
-Adhesive

Thermoplastic
Direct bonding
-UVO treatment

Chip Bonding

~2 minutes/cycle;
~5000 PCs/month
No fabrication
equipment needed (0$)
Bonding strength
~160 KPa[21]

~30 minutes/cycle;
~300 PCs/month[3]
Equipment: Hot
-embosser (~30k $)
Bonding strength:
~1 MPa[42]

~2 minutes/cycle;
~5000 PCs/month
No fabrication equipment
-needed (0$)
Bonding strength:
10~800 KPa[50,51]

10~30 minutes/cycle;
300~1000 PCs/month
Equipment: UVO cleaner (~10k $)+
hot embosser (~30k $)
Bonding strength: : 1.7 MPa[46]

High

High

**Fabrication Throughput[1] (Pieces (PCs)/month)**

[1] Estimate value based on process. And the fabrication throughput is evaluate based on 40 process hours in a week.

[2] Fabrication facilities cost is an estimate value, price may vary a lot depending on brand of tool, region, and fabrication resolution of the tool

[3] Process time directly depending on the complexity of the device design, Process temperature or treatment time

**Figure 2.7:** Summary of physical properties and suppliers for common polymer microfluidic materials [47].

to an undesired and non-specific adsorption of the proteins onto the micro-channel surface. In the case of polymers, various physical (prominently oxygen plasma treatment) or chemical (prominently coating by polyacrylamide) surface modifications were, therefore, proposed to overcome these issue, often by way of treatments that have a temporary effect or require post-processing, respectively. On the contrary, hydrophobicity can be used to favor water-based droplets generation in a dispersed fluid phase by mixing immiscible fluids by way microfluidics circuitries. In general, it is important to keep in mind that any change in hydrophobicity can affect the surface charge and vice versa. Greater hydrophilicity is often related also to the presence of more positive or negative charges on the surface: depending on the final application, therefore, coatings resulting in non-ionic surfaces also exist. Nevertheless, any post-processing treatment required to tailor the surface chemistry toward better biocompatibility increases manual work load and affects the analytical reproducibility. There is still a great demand on new, more economic polymer materials that hold inert and stable surface chemistry with inherent resistance to bio-fouling and that are, therefore, truly applicable to mass production.

## 2.3 Papers

Paper is a highly porous matrix made of cellulose, excellent in wicking liquids [171]. When certain areas of a paper are modified hydrophobically, aqueous solution applied to the paper was found to be guided through the hydrophilic region by the capillary effect. Paper-based microfluidic devices are, therefore, promising in portable and low-cost analysis, especially for bioassay-based personalized medical care [171]. The fabrication of paper-based microfluidic devices is quite simple because it is compatible with any method to deliver hydrophobic patterns. Patterning methods revised in Chapter 3 can be applied with relatively high-resolution but still remain expensive. In contrast, the printing (cutting) methods seem to be more competitive because they require simpler equipment to create hydrophobic barriers without pre-exposure of the channel area to reagents to get the final patterning. Among the advantages of exploiting papers, surely the low costs play a great role. In addition, papers can be easily stacked to form multilayer microfluidic channels. Although with these benefits, papers still have some drawbacks due to the presence of fabric matrix in the microfluidic channel area that can limit the detection methods quite severely: colorimetry is the most commonly used coupled with imaging of the detection zones. Indeed, the sensitivity of detection is still a limiting factor as the fabric matrix of the channel can somehow hinder the fluid flux and/or dilute the sample during transportation. Moreover, channel defined by hydrophobicity cannot be efficient in confining liquids with low surface. Few typical microfluidic applications (e.g., CE, droplets and laminar flow) were demonstrated on paper chips [170, 171] (Figure 2.8) but, as already reported, high-density integration is hard to realize and, up to now, it

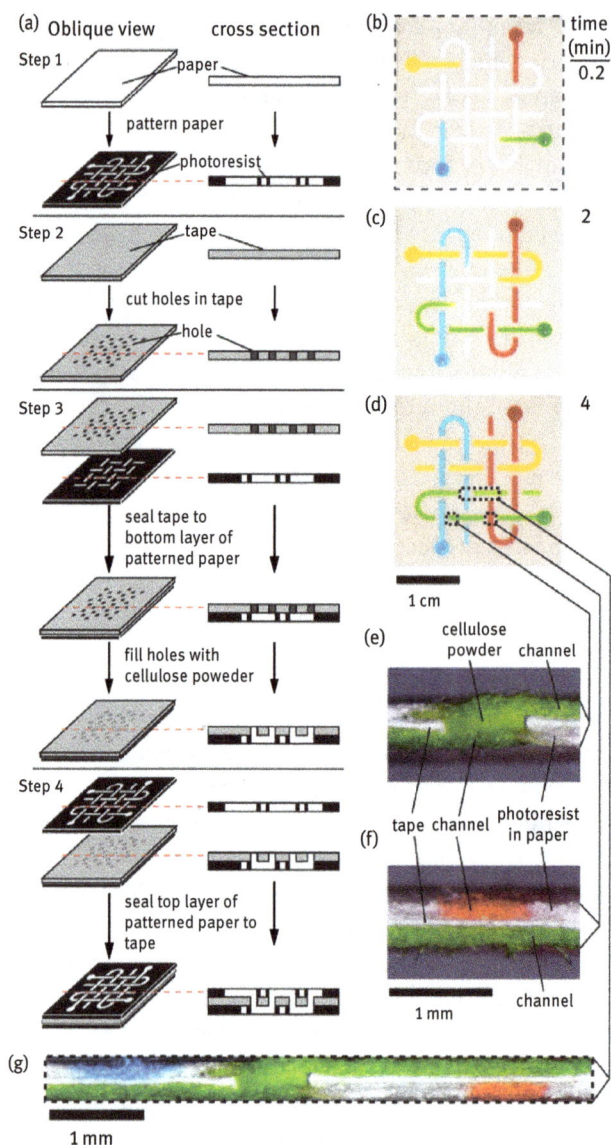

**Figure 2.8:** As reported in [171]: Preparation and demonstration of a 3D µPAD. (*A*) Fabrication. (*B*) Photograph of a basket-weave system 10 s after adding red, yellow, green, and blue aqueous solutions of dyes to the sample reservoirs. The dotted lines indicate the edge of the device. (*C* and *D*) Photographs taken 2 (*C*) and 4 (*D*) min after adding the dyes. The streams of fluids crossed each other multiple times in different planes without mixing. The dotted lines in *D* show the positions of the cross sections shown in *E*, *F*, and *G*. (*E*) Cross section of the device showing a channel connecting the top and bottom layers of paper. (*F*) Cross section of the device showing the three layers of the device with orthogonal channels in the top and bottom layers of paper. (*G*) Cross section of the device showing the layers and the distribution of fluid (and colors) in each layer of the device shown in *D*. The dotted lines indicate the edges of the cross section.

seems that the minimum channel width cannot be easily decreased below 200 µm. Finally, the integration of small-sized valves is still a challenge to face as well as the evaporation of liquid from open channels.

## 2.4 Comparisons

In summary, some comparisons of the material properties in use in the optofluidic field are reported in Figure 2.9. The proper choice of a material depends on its final application: in the opto-microfluidics field both the optical properties and microfluidics one must be taken into account as well the material durability and relative cost.

## 2.5 Smart materials

This section provides a brief introduction into the area of smart materials and their potential application in optofluidic systems. Examples of smart materials are provided which have a link to optofluidic system technology. All examples of smart materials have in common, that one or more properties can be significantly altered under controlled condition. Smart materials can be defined as [132] materials that receive, transmit, or process a stimulus and respond by producing a useful effect that may include a signal that the materials are acting upon.

### 2.5.1 Shape memory materials

Shape memory materials react to an external stimulus, which can be temperature for example. The effect by which the shape of the material is changed by temperature is called the thermally induced shape memory effect. Some examples of materials are described in the following section.

#### 2.5.1.1 Shape memory alloys (SMA)

This is a group of metals or metal alloys, which are capable of "remembering" its shape at a certain temperature. The shape changing of the material is due to the switching mechanism between its two crystalline structures: martensite and austenite. At low temperatures, the alloy has a crystalline structure called martensite. This structure can easily be deformed. Once the material is deformed by load, and consequently subject to heating, its crystalline structure is switched to austenite, and at macroscopic level, it can be seen that the material changes back to its original shape. For a standard material, 10 degrees of difference around the switching in temperature is needed to initialize the change. Examples of SMA materials are NiTi

| property | silicon/glass[a] | elastomers | thermoset | thermoplastics | hydrogel | paper |
|---|---|---|---|---|---|---|
| Young's (tensile) modulus (GPa) | 130–180/50–90 | ~0.0005 | 2.0–2.7 | 1.4–4.1 | low | 0.0003–0.0025 |
| common technique for microfabrication[b] | photolithography | casting | casting, photopolymerization | thermomolding | casting, photopolymerization | photolithography, printing |
| smallest channel dimension | <100nm | <1μm | <100nm | ~100nm | ~10μm | ~200μm |
| channel profile | limited 3D | 3D | arbitrary 3D | 3D | 3D | 2D |
| multilayer channels | hard | easy | easy | easy | medium | easy |
| thermostability | very high | medium | high | medium to high | low | medium |
| resistance to oxidizer | excellent | moderate | good | medium to good[c] | low | low |
| solvent compatibility | very high | low | high | medium to high | low | medium |
| hydrophobicity | hydrophilic | hydrophobic | hydrophobic | hydrophobic | hydrophobic | amphiphilic |
| surface charge | very stable | not stable | stable | stable | N/A | N/A |
| permeability to oxygen (Barrer[d]) | <0.01 | ~500 | 0.03–1 | 0.05–5 | >1 | >1 |
| optical transparency | no/high | high | high | medium to high | low to medium | low |

**Figure 2.9:** Summary of physical properties and suppliers for common polymer microfluidic materials (as presented by K. Ren et al in Accounts of Chemical Research Vol. 46, No. 11, pp. 2396–2406 (2013) [220].

[a]Photosensitive glass can be considered as thermoset. [b]Most of the materials can be fabricated by laser ablation, but compared with those obtained with lithographic or molding methods the ablated features usually have a rougher surface and are often misshaped. [c]Excellent for Teflon. [d]1 Barrer = $10^{-10}$[cm$^3$O$_2$ (STD)] cm cm$^{-2}$s$^{-1}$ cmHg$^{-1}$ .

(Nickel–Titanium), CuZnAl, and CuAlNi. Applications of shape memory alloys are actuators (e.g., for micro valves – see the following Chapter), sensors, and dampers (e.g., Miga Motor Company, www.migamotors.com). The commercial availability of the standard material can be regarded as high.

A review on shape memory alloy mini actuators is given in [187].

### 2.5.1.2  Shape memory polymers (SMP)

This is a group of polymers that are capable of "remembering" its specific shape at a certain temperature. The characteristics of the polymers are that they are physically and chemically cross-linked.

Physically cross-linked means the two molecules are mechanically linked, but the chemically linked polymer use chemicals as their links. Examples of materials are PEO-PET block copolymers, (PET: polyethylene terephthalate, PEO: polyethyleneoxide).

Several applications exist already, some of them are foam seal window, TEMBO® EMC, and TEMBO® Foams are being qualified for a variety of products for use in space, on aircraft and unmanned air vehicles (UAVs), for marine and maritime structures, in oil and gas equipment, automotive components, medical devices, composite tooling, and other industrial applications. The commercial availability can thus be regarded as high. Companies manufacturing or using this type of materials are, for example, Composite Technology Development Inc. (www.ctd-materials.com), Cornerstone Research Group (www.crgrp.com), Endoshape Inc. (www.endoshape.com).

A comparison between SMA and SMP is given in Table 2.1.

**Table 2.1:** Comparison between SMA and SMP.

|  | SMPs | SMAs |
|---|---|---|
| Density (g/cm³) | 0.9–1.1 | 6–8 |
| Extent of deformation | up to 800% | <8% |
| Required stress for deformation (MPa) | 1–3 | 50–200 |
| Stress generated upon recovery (MPa) | 1–3 | 150–300 |
| Transition temperatures (°C) | −10..100 | −10..100 |
| Recovery speed | 1s –minutes | <1s |
| Processing conditions | <200 °C Low pressure | >1,000 °C High pressure |
| Costs | <$10/lb | ~$250/lb |

A sub-category of shape memory polymers is so-called shape memory composites, which are shape memory polymers with reinforcement such as fibers. An overview of shape memory polymers is given in [142].

### 2.5.2 Smart fluids

In analogy to shape memory materials, smart fluids are fluids whose flow properties can be influenced by an electric or magnetic field. Therefore also two classes of smart fluids are described in the following paragraphs, electro-rheological and magneto-rheological fluids.

### 2.5.2.1 Magneto–rheological Fluids

The viscosity of this type of fluid can be changed instantly when it is subject to a magnetic field (Figure 2.10). The viscosity is a function of the strength of the magnetic field.

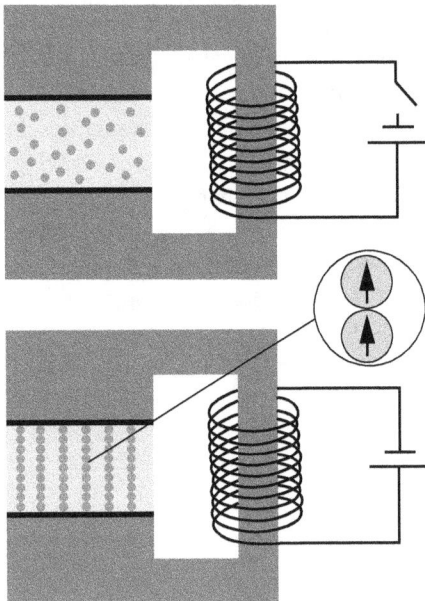

**Figure 2.10:** Schematic of a magneto-rheological material (Courtesy Holger Boese, www.cesma.de).

Magneto–rheological fluids consist of magnetically polarizable particles in soft elastomer (a type of polymer that has a high elasticity) matrix. When a magnetic field is applied, these particles align themselves along the magnetic flux. As a result, the movement of the liquid at the microscopic level is restricted, and thus the fluid is stiffened at macroscopic level.

Smart fluids are mainly fluids with ferromagnetic particles (iron, nickel, cobalt and their alloys). Applications are, for example, seismic damper, shock absorber, or optical surface finishing (e.g., www.lord.com). The commercial availability can be regarded as high.

### 2.5.2.2 Electro–rheological fluids

Viscosity of this type of fluid is changed instantly when it is subject to electric field (Figure 2.11). Viscosity is a function of the strength of the electric field. Suspension of electrically polarizable particles is dispersed in a non-conductive fluid.

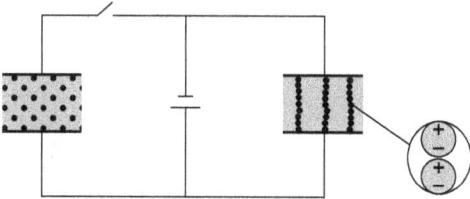

**Figure 2.11:** Schematic of an electrorheological material (Courtesy Holger Boese, www.cesma.de).

When there is the presence of electric field, particles align in an electric field, which rapidly changed the viscosity characteristics of the fluid.

The materials, which are used, need to be electrically active, for example, ferroelectric or conducting materials. Applications of such type of fluids are, for example, fast acting clutches, brakes, shock absorbers, and controllable features. The commercial availability is regarded as medium to high.

### 2.5.3 Magnetostrictive materials

Magnetostriction is a property of ferromagnetic materials such as iron, nickel, and cobalt. When placed in a magnetic field, these materials change size and/or shape, due to alignment of the magnetic domain. The operation of magnetostrictive materials, such as Terfenol-D, can be understood using a simple ellipse model. The material may be thought of as to contain many ellipses. When an external magnetic field is applied to the ellipses, the magnetization of the particles rotates to align with the direction of the applied field. Then, we observe the overall shape change of the material as a result. Applications of this type of material are, for example, magneto-mechanical sensors, actuators, and acoustic and ultra-sonic transducers. The commercial availability is regarded as high. Companies fabricating materials and/or using them in applications are, for example, Metglas (part of Hitachi Metals Ltd), www.metglas.com, Etrema products Inc., and www.etrema.com.

Design and applications of magnetostrictive materials are described and reviewed in great detail in [190].

### 2.5.4 Electrostrictive

When placed in an electric field, these materials change size and/or shape. Electrostriction is a property of all dielectric materials. Originally, there are randomly arranged domains in the dielectric material. When they are aligned within an external electric field, domain particles that have opposite poles attract each other and align themselves. As a result, there will be a net shape change for the material at the macroscopic level. An example of a material is lead magnesium niobate also referred to as PMN. An application is an actuator with small displacement. The commercial availability is regarded as high. An example for a company applying smart materials is Mide, www.mide.com.

### 2.5.4.1 Piezoelectric vs. electrostrictive materials

The piezoelectric phenomenon only applies to some types of electrostrictive materials, which have a reversible mechanical and electrical deformation. Electrostriction is a quadratic effect, whereas the piezoelectric is a linear effect. Also, the electrostrictive effect cannot be reversed, whereas the piezoelectric can be reversed. That is to say, mechanical deformation cannot induce an electric field. The main advantage of electrostrictive materials over piezoelectric materials is their low hysteresis parameter. This could be especially beneficial in high-frequency dynamic applications, which could involve considerable hysteresis-associated heat dissipation. The main disadvantage of electrostrictive materials is the temperature dependence of their properties.

### 2.5.5 Chromogenic materials

These types of materials [156] change their color when subject to, for example, electrical, optical or thermal changes. Many different sub-categories exist for these types of materials and some of the most common ones are explained in the following paragraphs. An overview of chromogenic smart materials and their applications is described in an open access article in [157].

### 2.5.5.1 Thermochromic

A color change of this type of material is subject to a temperature change. Two compositions are distinguished, liquid crystals and leuco dyes. Liquid crystals can be regarded as tiny capsules (liquid crystal particles that make up the entire liquid crystal material). Liquid crystals exist in several so-called phases. One of these phases is known as chiral nematics. In this phase, the liquid crystals arrange themselves in a very specific helical structure. These structures reflect certain wavelength of light. As the liquid crystal is heated up, the orientation of the helices change and will reflect light with another wavelength. As the crystal cools down, the crystals re-orientate themselves into the initial structural arrangement, responding to temperature with higher accuracy.

Leuco dyes are more robust and less expensive than liquid crystals and can be used in a wider range of inks, plastics, etc. Because leuco dyes are not very sensitive they do not make accurate thermometers, but are used in applications where larger temperature differences are present or when temperature accuracy is less important. Sunlight, extreme temperatures, and contact with aggressive solvents adversely affect a leuco dyes life span. Applications of leuco dyes can mainly be found in the food and beverage industry, for example when a beer is cooled to a certain temperature enough for drinking, the label will present a different temperature than at room temperature. The commercial availability is high which is resembled by a larger number of manufacturers for example for leuco dyes: Fraunhofer IAP, liquid crystals: Merck KGaA, B&H liquid crystal.

### 2.5.5.2 Photochromic

This type of chromogenic material changes color when exposed to electromagnetic radiation. At the presence of UV radiation, molecules inside the photo-chromic materials will undergo chemical reactions through which their chemical structure is altered. Originally, they are transparent, as they allow most of the light to be transmitted. However, when they have undergone structural changes, they will only be able to absorb partial of the light and reflected the rest. Thus, the material is observed to have a different color. Examples of materials are Agcl, Spiropyrans, spirooxazines, Diarylethenes, Azobenzene, and Photochromic quinines. Examples of applications are sunglasses and photochromic dyes. The commercial availability is regarded as high.

### 2.5.5.3 Electrochromic

The material changes color on the detection of ions or electrons. When the voltage is off, particles randomly distribute inside the material. They absorb light, so no light can pass through the film. It appears to be opaque. When the voltage is on, particles align under the electric field. Then light can pass through, so the material tends to be transparent. Examples of materials are suspended particle devices, liquid crystal, and polymer-dispersed liquid crystal (PDLC). Typical compounds are tungsten oxide ($WO_3$), viologens, polyoxotungstates, and polyaniline. Application can be in building windows or automobile glazing.

### 2.5.5.4 Piezochromic (piezochromic luminescence)

Piezochromic luminescence describes the tendency of certain materials to change color when a pressure is applied. Examples of materials are polymeric materials, mostly polymers with metal-complexes incorporated, poly(thiophenes) and poly(alkyl silanes). Examples of chemicals are derivatives of imidazole pyrrole, bianthrone, and xanthylidene anthrone. An article on piezochromic polymer materials displaying pressure changes in bar ranges is presented in [210]

### 2.5.5.5 Magnetochromic

The molecules of this type of material reorient themselves, when subject to a magnetic field. As a result, the molecules will refract light in a different way so that the external color displayed is changed. Magnetochromic microspheres (Figure 2.12) are described in an article in [71].

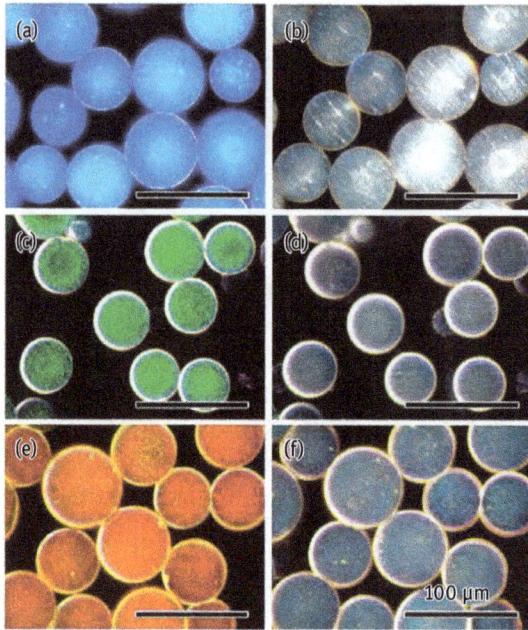

**Figure 2.12:** Optical microscopy images (500×) of magnetochromatic micro-spheres with diffractions switched between "on" (a, c, e) and "off" (b, d, f) states using external magnetic fields. These microspheres are prepared using (a, b) 127, (c, d) 154, and (e, f) 197 nm $Fe_3O_4@SiO_2$ colloids. (Reprinted with per-mission from Magnetochromatic Microspheres: Rotating Photonic Crystals, Jianping Ge, Howon Lee, Le He, Junhoi Kim, Zhenda Lu, Hyoki, Kim, James Goebl, Sunghoon, Kwon, and Yadong Yin. Copyright 2009 American Chemical Society).

## 2.6 Quantum dots (QD)

The QDs are nano-aggregates of the typical dimension in the nano-scale (typically 2–10 nm, i.e., 10–100 atoms) where the physical properties are described in the field of quantum physics. Quantum wells, quantum wires, and QDs are confined in one, two, and three dimensions, respectively [1, 2]. The confinement can be created due to electrostatic potentials, the presence, for example, of an interface between different semiconductor materials, and the presence of a semiconductor surface. When the electrons are confined in all dimensions, and the physics that governs the behavior of the QDs, it is described by quantum mechanics. Material examples are cadmium selenide, cadmium sulfide, indium arsenide, and indium phosphate to name a few. The QD have various applications. In the following paragraph, only the sensing aspect is highlighted.

The measurement principle is based on Fluorescence Resonance Energy Transfer (FRET). The FRET is a technique through which energy is transferred by dipole oscillation of molecules. The QDs appear to behave like any other type of fluorophore with regard to energy transfer. The FRET will occur between QDs. For instance, a CdSe QD can be a donor, and a rhodamine molecule can be an acceptor. The FRET donor is at the excited state, and the acceptor is at the ground state. If the vibration wavelengths of the donor overlap the vibration wavelengths of the acceptor, then energy will be transferred.

Application of QD FRET is, for example, to measure the distance of two sites on DNA, protein. If we label a DNA pair on a gene, the distance can be calculated by the energy transfer efficiency. The FRET does NOT involve the transfer of photons. So fluorophores can exchange energy with another fluorophore, if their frequencies are the same. Advantages of QDs over Organic Dye: Organic fluorophores often experience photobleaching, especially those with high illumination intensity. The QDs are mostly inorganic and have a long lifetime. In addition, the absorption spectrum is long. For instance, certain fluorophores such as Cy3 and Cy5 do not absorb at 400 nm, but QDs do. Common types of QDs are semiconductor nanoparticles and lanthanides.

## 2.7 Nanowires

A nanowire is a nanostructure, with a diameter in the order of a nanometer ($10^{-9}$ m). At these scales, quantum mechanical effects occur and need to be taken into account, leading to the term "quantum wires". Many different types of nanowires exist, including metallic (e.g., Ni, Pt, Au), semiconducting (e.g., Si, InP, GaN, etc.), and insulating (e.g., $SiO_2$, $TiO_2$). Molecular nanowires are composed of repeating molecular units either organic (e.g., DNA) or inorganic. A potential application could be, for example, in a gas sensor. The drawback of traditional gas sensors is the sensitivity; it takes a longer period of time to get ready for sensing. For example, nitrogen dioxide gas on a thin film might diminish the current by only 2%, which is much harder to measure than a 50% decrease observed in a zinc oxide nanowire [163]. Another advantage is that it takes only minutes to restore the original shape. In transistors made of zinc oxide nanowires, the presence of foreign substances alters the wire's ability to conduct the current. For example, nitrogen dioxide gas will reduce the conductivity of the wire, whereas carbon monoxide will increase it. Different substance increase or decrease the current by different degrees, so specific chemicals can be identified by how much they affect the flow of current. For a detailed insight into this exciting topic refer, for example, Lu, J.G.

## 2.8 Conclusion

Inorganic materials, polymers and smart materials are the important class of materials especially suited for solving applications in optofluidic systems. The focus on a polymer platform for creating waveguides and potentially being able to serve as a biocompatible fluidic platform at the same time is essential to reduce complexity in future optofluidic systems, as one can focus on the application and does not need to solve fundamental waveguide or fluidic channel fabrication issues. However, other materials have been widely exploited and represent a valid alternative in those fields in which polymers still have some drawbacks. The best material is, therefore, the one satisfying the compromise between usability, cost, and best performances (Figure 2.13).

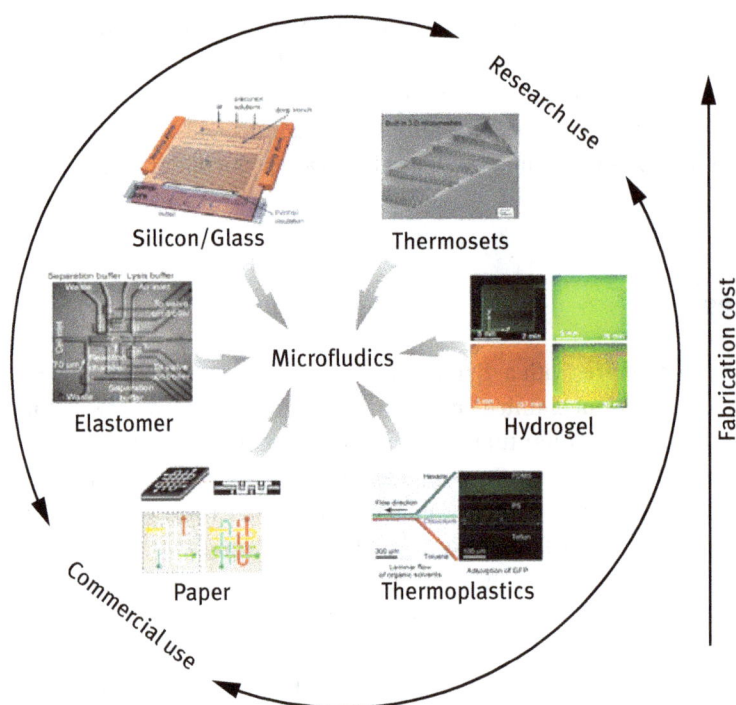

**Figure 2.13:** Microfluidics and materials, presented by K. Ren et al in Accounts Of Chemical Research Vol. 46, No. 11, pp. 2396–2406 (2013) [220]

This chapter only provided a small glimpse into this vast field, which will definitely play an important role in the future when it comes to real-world devices.

## Further reading

Fridman ML. Polymer Processing. Springer, 1990
Lee KS. Polymers for Photonics Applications I. Springer, 2002
Lee KS. Polymers for Photonics Applications II. Springer, 2003
Michler P. Single Quantum Dots. Springer, 2003
Schwartz M. Encyclopedia of Smart Materials. Wiley, 2014
Seeboth A, Lötzsch D. Thermochromic and Thermotropic Materials. Pan Stanford Publishing, 2013
Utracki LA. Polymer Blends Handbook. Kluwer Academic Publishers, 2002
www.thermochromic-polymers.com
Introduction to Glass Science and Technology, J.E. Shelby Royal Society of chemistry, 2005, ISBN
    0854046399, 9780854046393
"Advances in Microfluidics – New Applications in Biology, Energy, and Materials Sciences", book
    edited by Xiao-Ying Yu, ISBN 978-953-51-2786-4, Print ISBN 978-953-51-2785-7, Published:
    November 23, 2016
Advances in Microfluidic Materials, Functions, Integration and Applications by Pamela N. Nge
    et al., Chem Rev vol.113(4) pp. 2550–2583, 2013

## Self-assessment questions

1.  Which are the basic requirements for a material to be considered eligible for opto-microfluidics applications?
2.  How is it possible to classify the materials exploited in microfluidics?
3.  List the different types of material that are used in microfluidics.
4.  List the main advantages and drawbacks of silicon when exploited in microfluidics.
5.  What is the definition of a glass?
6.  How is a typical glass obtained and formed by?
7.  Describe the main differences of a glass respect a crystal.
8.  List the main characteristics of glass in opto-microfluidics, with special focus on advantages and drawbacks.
9.  Which are the main optical properties of a glass and which are the physical quantities used to describe them?
10. What is a polymer? Draw possible configuration of a polymer.
11. How is a polymer obtained? Describe the polymerization process and how it can be activated.
12. How can a polymer be classified?
13. Highlight the characteristics of polymers.
14. List the polymers that are commonly used in microfluidics, citing advantages and drawbacks.
15. What is PDMS and why is it widely used in microfluidics?
16. List the physical and chemical properties needed for microfluidics applications and those that are an advantage.

17. Sketch a table to compare the thermoplastic properties putting: column 1 = name of the material; column 2 = physical properties; column 3 = applications; column 4: strengths and weaknesses of the material
18. What are the differences between elastomers, thermosets and thermoplastics?
19. Sketch a table to compare the thermosets properties putting: column 1 = name of the material; column 2 = physical properties; column 3 = applications; column 4: strengths and weaknesses of the material
20. What is a hydrogel? List its characteristics.

# 3 Materials fabrication and patterning techniques

In this section, the main common fabrication techniques used to integrate an optical device in a material will be presented, focusing on those able to maintain the compatibility with microfluidics requirements. Successful fabrication technologies are needed to guarantee high reproducibility, efficient batch, and low operation costs. Basically, two main approaches can be exploited:
- deposition of the optical stage on the surface of a substrate (eventually hosting the microfluidics stage to get an optofluidic platform);
- modification of the substrate properties at a local level to host the optical stage. As it will be revised in Chapter 4, it means basically to generate a local change of refractive index following the design of the optical stage requirements.

In the following, a brief review on the most used technique to deposit optical layers or multilayers will be presented limiting the attention on those that are technologically implemented and are of interest for opto-microfluidics. Some technologies can be applied to both inorganic and organic materials or hybrids, others instead are limited to some materials only. In the case of inorganic materials such as glass, silica and crystals, the integrated optics stage can be achieved either by:
- depositing a layer (or multilayers) on the surface substrate, eventually with a pattern design to get the desired optical properties. In this case, so-called physical techniques or chemical-based ones have been widely exploited and optimized;
- modifying at a local level, the material properties as required by the optical application in suitable defined configurations.

The local modification can be realised in many different ways but in optics and integrated optics it has mainly been achieved by:
- thermal diffusion of dopants, from solid or a vapor phases. In the first case, the deposition techniques are usually those exploited to get a film deposition with the film playing the role of dopant source. Its thickness is, therefore, suitable chosen to be exhausted by the thermal process, so that to let the surface be "recovered back", that is, with the original properties as much as possible;
- ion exchange (IE) of chemical elements from the materials with those having the suitable required properties. This is normally achieved by thermal diffusion from a liquid source;
- ion implantation of the desired chemical element into the material matrix itself.

https://doi.org/10.1515/9783110546156-003

All these local doping techniques can be coupled with patterning methodologies to achieve a doping limited to given areas depending on the desired final design. In the following, a sort review on the fabrication and patterning technologies will be, therefore, provided to complete the brief overview on the state of the art and perspectives in the opto-microfluidics field.

## 3.1 Fabrication techniques in opto-microfluidics: Film deposition

Deposition of films is a widely exploited approach to grow a film or coating over the top of a given material maintaining its adhesion. It includes processes such as evaporation, sputtering, chemical vapor deposition, sol–gel coating, and epitaxial growth and hetero-epitaxial layer growth just to cite the most common at an industrial level. In the following, only those actually used in opto-microfluidics field will be shortly revised, suggesting the reader to get more details in specialized books as outlined in the References section.

### 3.1.1 Evaporation

Evaporation is a common method exploited to deposit thin solid films: it consists basically in evaporating a material and let the evaporated species be deposited on another substrate. This process can be achieved only under vacuum conditions to allow the evaporated particles to move from the source to the target object (substrate), where they can condense into a solid state. The deposition by evaporation is not straightforward to get a good uniform distribution, especially if the substrate has a rough surface and blocking parts that can lead to a sort of shadowing effect known as "step coverage". Finally, the possible aggregation of the evaporated species can compromise the overall quality of the deposited film, including the control on its thickness and its uniform composition. Special geometries of the evaporators have been developed to limit these drawbacks, enhancing the symmetry of the deposition path and the evaporation conditions. Great advances were successfully demonstrated by choosing the best heating system that allows for the thermal treatment of the source and the evaporation process. Among the others, thermal heating is used as electron beam bombardment (E-beam): in Table 3.1, a comparison of these two systems is reported. Evaporation is commonly used to deposit metal films but can be exploited to get also insulating films, such as $SiO_2$ and $TiO_2$. In general, it is also used to deposit thin solid film that are later diffused into the material by thermal diffusion and locally modify the substrate material (refer following sections). The deposition rate is not so high (ranging between 1 and 100 A/s) with pureness of the deposited film that strongly depends on the heating system used and the vacuum conditions established during the process.

**Table 3.1:** Evaporation: comparison on the performances.

| Deposition | Material | Common evaporated specie | Deposition rate | Temperature range | Cost |
|---|---|---|---|---|---|
| Thermal | Metals and low melting points materials | Au, Ag, Al, Cr, Sn, Sb, Ge, In, Mg, Ga, CdS, CdSe, NaCl, AgCl, CaF$_2$ | 1–20 A/s | Around 1,500 °C | Low |
| E-beam | Metals and dielectrics | Above and Ni, Pt, Ir, Th, Ti, V, Zr, W, Ta, Mo, SiO$_2$, SnO$_2$, TiO$_2$, ZrO$_2$ | 10–100 A/s | Around 3,000 °C | High |

### 3.1.2 Sputtering techniques

Among the techniques exploiting physical vapor deposition (PVD) methods, the sputtering is the most common to get thin films and coatings. It relies on the bombardment of a target plate (or cathode) by energetic ions to promote the removal of target species leading to the so-called "sputtering" process. Atoms, ions, and electrons are sputtered from target surface that is, therefore, eroded: the sputtered atoms present a wide energy distribution, typically up to tens of eV, and only a small fraction is ionized (on the order of 1%). The removed species can deposit, creating a thin film, on another substrate placed close to the eroded target provided that the experimental conditions are suitable controlled [140]. Although the erosion can be obtained also by ion beam bombardment, the film deposition by the sputtering process is mainly achieved by a plasma bombardment. Plasma is basically a "region containing balanced charges of ions and electrons" as described by the chemist Langmiur, who first reported on it. Being an electrically neutral medium of unbound positive and negative particles whose overall charge is almost zero, the moving charged particles inside the plasma can generate an electric current within a magnetic field. The plasma condition is not, therefore, straightforward to be obtained since it requires the coexistence of positive and negative charges compensating in the overall ensemble and need vacuum condition to be sustained. In the sputtering depositions, gas plasma is often used (frequently Ar, as a non-reactive specie) and secondary electrons emitted from the target surface as a result of the ion bombardment play an important role in maintaining the plasma conditions as well.

In general, the sputtering process is normally characterized by low deposition rates (of the order of 1–100 A/s depending on the material to be deposited), low ionization efficiencies in the plasma, and high substrate heating effects. These limitations have been overcome by the development of magnetron sputtering technique. Magnetrons exploit suitable magnetic fields configured parallel to the target surface to constrain the secondary electrons motion close to the target. This configuration is achieved by arranging the magnets in such a way that one pole is positioned at the central axis of the target and the second pole is formed by a ring of magnets around the outer edge of the target [140] or even in more complicated configurations (Figure 3.1).

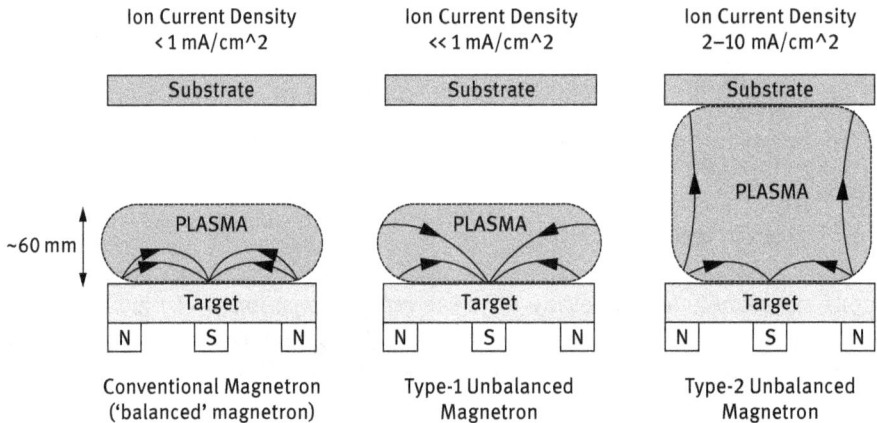

**Figure 3.1:** Schematic representation of the plasma confinement observed in conventional and unbalanced magnetrons [140].

The presence of the trapped electrons, therefore, constrained increases the probability of electrons – atom collisions and, consequently, the ionization efficiency significantly improves resulting in a dense plasma impinging on the target surface. The denser it is the higher the sputtering rate and, consequently, the higher deposition rates of the sputtered species at the substrate surface. In magnetrons, the increased ionization efficiency allows the discharge to be maintained at lower operating pressures (typically, $10^{-3}$ mbar, compared to $10^{-2}$ mbar) and lower operating voltages (typically for the order of 500 V, compared to 2–3 kV) common in other sputtering mode configurations [140].

In the case of insulating targets, where the charge build-up can prevent the target erosion due to discharges, radio frequency (RF) sputtering is used where the sign of the anode–cathode bias is varied at a high rate (generally 13.5 MHz). The sputtering principle is quite similar apart the exploitation of the RF power is suitable to produce insulating oxide films: although any type of film can be RF sputtered, deposition rates are still low and RF power is not so straightforward, power supplies being quite expensive and requiring additional circuitry is needed.

For the shake of completeness, the sputtering process is promoted as reactive gas plasma (such as oxygen), the sputtered particles undergo a chemical reaction before depositing on the substrate and, therefore, it is called a reactive sputtering. Finally, it is worth mentioning that the sputtering process can be achieved also by bombarding the surface with an ion beam; in this case, it is referred to Ion-beam sputtering (IBS) method. Pulsed laser deposition (PLD) is a variant of the sputtering deposition technique as well, in which a laser beam is focused inside a vacuum chamber to bombard a target of the material to be deposited. This interaction leads to the material vaporization in a plasma plume, which deposits as a thin film on a substrate. This process can be carried out in ultra-high vacuum (UHV) or in the presence of a controlled gas atmosphere, such as oxygen which is commonly used when depositing oxides. Lasers in PLD range from the mid infrared (e.g., a $CO_2$ laser, 10.6 µm), through the near infrared and visible (e.g., the Nd-YAG laser, with fundamental and second harmonic outputs at 1,064 nm and 532 nm, respectively) and down into the ultraviolet (UV). Much current PLD works use excimer lasers, which operate at a number of different UV wavelengths (e.g., 308 nm (XeCl), 248 nm (KrF), 193 nm (ArF) and 157 nm ($F_2$)). The growth and quality of the deposited film eroded by PLD generally depend on a number of fundamental parameters, including the choice of substrate, the substrate temperature, and the absolute and relative kinetic energies and/or arrival rates of the various constituents within the plume. The PLD is now established as one of the simpler, cheaper, and more versatile methods of depositing thin films, of a very wide range of materials (e.g., metals, carbon, and numerous more complex systems including oxides and other ceramics, ferroelectrics, high-$T_c$ super-conductors and materials exhibiting giant magneto-resistance), on a wide variety of substrates, at room temperature. In general, films deposited on room temperature substrates are usually amorphous, while their crystallinity can often be improved by deposition at higher $T_{sub}$. Almost all materials can be ablated, leading this technique to be quite flexible.

### 3.1.3 Chemical vapor deposition and plasma-enhanced chemical vapor deposition – PECVD

Among the chemical-based deposition techniques, the CVD is one of the most known. It involves the chemical reactions of gaseous reactants on or nearby a (heated) substrate surface to promote films and coatings deposition (Figure 3.2). This method can provide pure materials deposition with structural control at atomic or nanometer scale levels [51], in form of single layer, multilayers as well as composite, nanostructured, and functionally graded coating materials at a relative low processing temperatures.

In general, the CVD process involves the following some key steps (Figure 3.3):
- generation of active gaseous reactant species and their transport into the reaction chamber;
- gaseous reactants undergo gas phase reactions forming intermediate species;

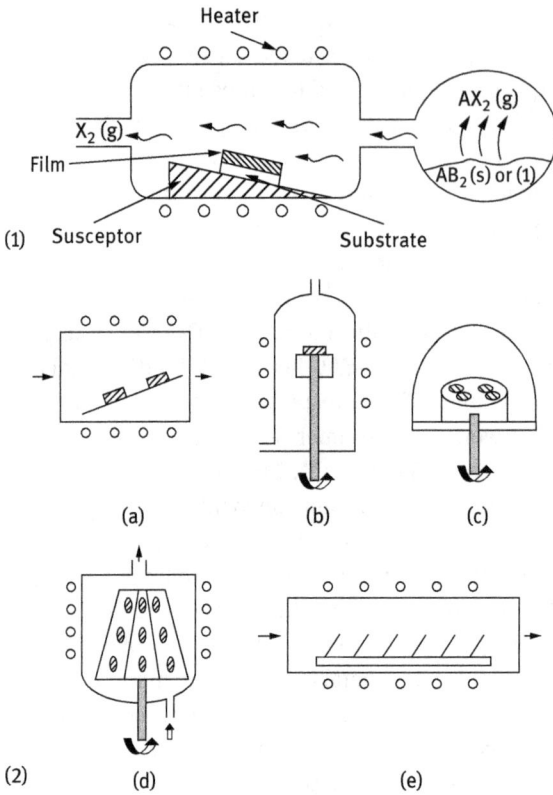

**Figure 3.2:** (1) Schematic diagram of the CVD process. (2) Various CVD reactor configurations: (a) horizontal; (b) vertical; (c) semi-pancake; (d) barrel; (e) multiple wafer [51].

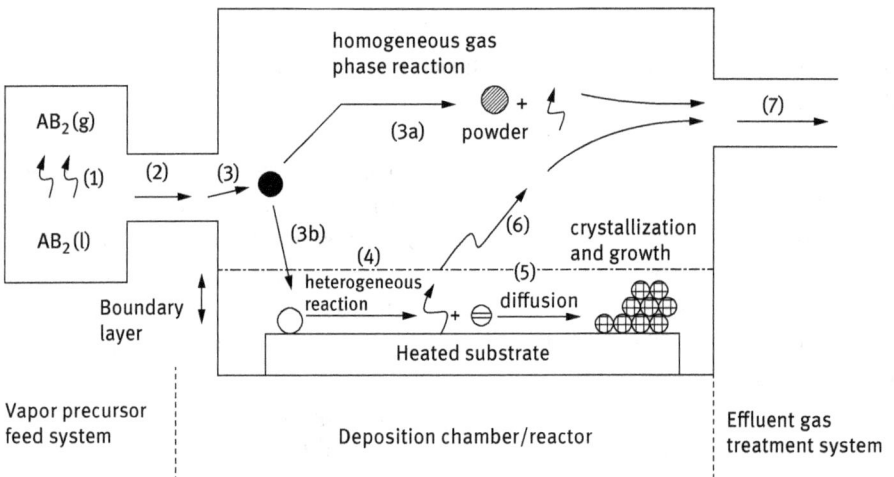

**Figure 3.3:** A schematic illustration of the key CVD steps during deposition [51].

- absorption of gaseous reactants onto the heated substrate, and the heterogeneous reaction occurs at the gas–solid interface (i.e., heated substrate), which produces the deposit and by-product species;
- the deposits diffuse along the heated substrate surface forming the crystallization center and growth of the film;
- gaseous by-products are removed from the boundary layer through diffusion or convection;
- the unreacted gaseous precursors and by-products will be transported away from the deposition chamber.

The precursors that are used in the CVD process are commonly metals and metal hydrides, halides, and halohydrides, and metalorganic compound as well. Generally, metal halides and halohyrides are more stable than the corresponding hydrides, while the metalorganic precursors offer the advantage of lower reaction and deposition temperatures than halides and hydrides and are less toxic and pyrophoric.

It is worth mentioning that the main CVD process parameters, such as the temperature, pressure, reactant gas concentration, and total gas flow, require accurate control and monitoring. The CVD occurs through chemical reactions that include pyrolysis, oxidation, reduction, hydrolysis, or a combination of these, and may be catalyzed by the substrate and the defined chemical reaction determines the operating temperature range required. Therefore, the thermodynamics and kinetics need to be defined.

Many are the advantage of CVD, such as:

- deposit uniform films with good reproducibility and adhesion at reasonably high deposition rates;
- high versatility allowing the deposition of thin films and coatings for a wide range of applications, including semiconductors (e.g., Si, Ge, $Si_{1-x}Ge_x$, III–V, II–VI) used for microelectronics, optoelectronics, energy conversion devices; dielectrics (e.g., $SiO_2$, AlN, $Si_3N_4$) used for microelectronics; refractory ceramic materials (e.g., SiC, TiN, $TiB_2$, $Al_2O_3$, BN, $MoSi_2$, $ZrO_2$) used for hard coatings, protection against corrosion, oxidation, or as diffusion barriers; metallic films (e.g., W, Mo, Al, Au, Cu, Pt) used for microelectronics and protective coatings; and fiber production (e.g., B and SiC monofilament fibers) and fiber coating;
- reasonable processing cost for the conventional CVD technique;
- deposition rate can be adjusted depending on the desired film thickness: low deposition rate is the best for the growth of epitaxial thin films for microelectronic applications, while for thicker protective coatings, a higher deposition rate is preferred (greater than tens of mm per hour);
- ability to control crystal structure, surface morphology, and orientation of the CVD products by controlling the CVD process parameters;

The disadvantages of the technique are mainly the following:
- chemical and safety hazards caused by the use of toxic, corrosive, flammable, and/or explosive precursor gases;
- difficult to deposit multicomponent materials with well-controlled stoichiometry.

When a CVD process exploits a plasma discharge rather than thermal energy supplied by high process temperature to activate the gas phase chemistry, it is denoted as plasma-enhanced CVD (PECVD) or alternatively plasma-activated CVD (PACVD) [165]. In a PECVD process, the gas phase chemistry, that is, the decomposition of the precursor gases and subsequent gas phase reactions are mediated by energetic species in the plasma, predominantly by electron impact collisions. As a consequence, precursors with low reactivity can be used together with low substrate temperatures during deposition, enabling deposition on temperature sensitive substrates. A PECVD system is classified as a direct plasma system, when the substrate is placed between the discharge electrodes (anode/cathode) and thus directly immersed in the plasma. A remote system is when the substrate is placed at some distance away from the zone of plasma generation, that is, not directly subjected to the plasma discharge generated between the anode and cathode [165].

The most common means of initiating plasma in PECVD are by discharges between two capacitively coupled parallel plates or by inductive coupling between the plasma gas and an applied RF field as already seen in the sputtering techniques. The PECVD has received particular attention in microelectronics for several decades, and it has now penetrated into a large number of other sectors. In certain areas (e.g., in optics), its industrial acceptance was originally slow, mainly owing to the complexity of the plasma–chemical reactions, plasma–surface interactions, and process control. However, thanks to fundamental and applied research and the development of new instrumentation tools, recent advances in plasma processing, and in PECVD in particular, have greatly increased the interest in PECVD for the fabrication of different coating systems.

The main advantages of the PECVD techniques are:
- it is compatible with different types of film fabrication equipment;
- the deposition process occurs at low temperatures (typically ranging from near room temperature (RT) with no intentional heating, to about 350 °C, when additional heating is applied);
- provide materials with functional; and
- characteristics similar to those obtained by their PVD and non-vacuum counterparts.

Unfortunately, it is a quite expensive techniques respect to the sputtering and evaporation, supporting its applications to the production of high-quality films especially dielectrics ones.

## 3.1.4 Comparisons

The choice of the suitable deposition technique depends on several parameters such as the composition, the quality, the thickness, deposition time, and the cost of the deposit as well. In general, the sputtering-based deposition is more flexible because that materials with very high melting points, and consequently quite difficult to be evaporated, are instead easily sputtered. The sputtered films typically have also a better adhesion on the substrate than evaporated films. Instead, it is more difficult to be combined with a lift-off for structuring the film to get patterned deposition. Other techniques such as CVD and PECVD guarantee a better control of the composition and a lower degree of contamination from undesired species in the deposited film, with a good control also in the film density. In Table 3.2, a comparison on the performances of some of previously described techniques is reported.

**Table 3.2:** Comparison on the performances of some of deposition techniques.

| Process | Material | Homogeneity | Film density | Substrate temperature | Impurity | Deposition rate |
|---------|----------|-------------|--------------|-----------------------|----------|-----------------|
| Thermal evaporation | Metals and low melting points materials | Poor | Low | 50–100 °C | High | 10–200 A/s |
| Sputtering | Metals and dielectrics | Very good | Good | Typically close to 200 °C | Very good | Metal close to 100 A/s; Dielectrics close to 1–10 A/s |
| PECVD | Mainly dielectrics | Good | Good | 200–300 °C | Good | 10–100 A/s |

## 3.1.5 Sol–gel coating

In the sol–gel process, a gel is formed from a sol and this gel is suitably deposited on a substrate to solidify into a film. A sol is a colloidal suspension, that is, a dispersion of one phase in another where IUPAC underlines that the molecules or poly-molecular particles dispersed in a medium have at least in one direction a dimension roughly between 1 nm and 1 μm. A gel state is generally defined as a non-fluid 3-dimensional (3D) network that extends through a fluid phase.

Sol–gel chemistry is the preparation of inorganic polymers or ceramics from solutions through a transformation from liquid precursors to a sol and finally to the gel network structure [60]. Sol–gels can be classified in different types. Although the sol formation has been mostly achieved through hydrolysis (i.e., chemical breakdown of a compound due to reaction with water) and condensation reaction of metal alkoxide precursors, in principle it should not be considered the only way to prepare it. It is beyond the scope of this book to treat in detail the chemistry of sol–gel, we, therefore, address the reader to more specialized books as cited in the reference section. However, aside from precursor preparation, the sol–gel process can be summarized in the following key steps:

- synthesis of the 'sol' (normally achieved by hydrolysis and partial condensation of alkoxides);
- formation of the gel via poly-condensation (normally to form metal–oxo–metal or metal–hydroxy–metal bonds);
- synthesis or 'aging' where condensation continues within the gel network, often shrinking it and resulting in expulsion of solvent;
- drying the gel either to form a dense 'xerogel' via collapse of the porous network or an aerogel, for example, through supercritical drying;
- removal of surface of M–OH groups eventually present through calcination at high temperature (normally even up to 800 °C, if required).

In general, the gel can be deposited by Spin Coating or dipping on a surface. Spin coating consists, in the application, of a thin film (a few nm to a few microns) on the surface of a substrate by coating (casting) a solution of the desired material in a solvent while it is rotating (Figure 3.4).

The rotation of the substrate at high speed (usually >10 rotations per second = 600 rpm) allows the centripetal force combined with the surface tension of the solution to pull the liquid coating into an even covering. During this time, the solvent evaporates, drying and removal steps are undertaken after the deposition to guarantee a final good-quality film. Therefore, in addition to the chemistry to form a gel, many other aspects must be taken into account to exploit this technique to have high-quality film deposition. Evaporation during gelation can have a substantial impact on the gel structure and heat treatment is also important for drying gels as well as removing surface chemical groups (such as hydroxyl), densifying the material to produce a ceramic monolith or converting to a crystalline material, if crystallinity is needed. In general, both single layers and multilayers can be achieved even in multistacks configuration, the constraints coming eventually only in the chemical affinity of the deposited layers and on the maximum thickness that can be deposited without any stress occurring into the single layer that can lead to cracks and adhesion problems.

Sol–gel chemistry offers some particular advantages, centered on the ability to produce a solid-state material from a chemically homogeneous precursor. It can host nanoparticles suspended into the gel still keeping high compatibility with flexible composition and mixturing. The sol–gel chemistry should enable greater control over particle morphology and size but in reality, producing a homogeneous precursor at room temperature does not ensure homogeneity throughout a reaction and many

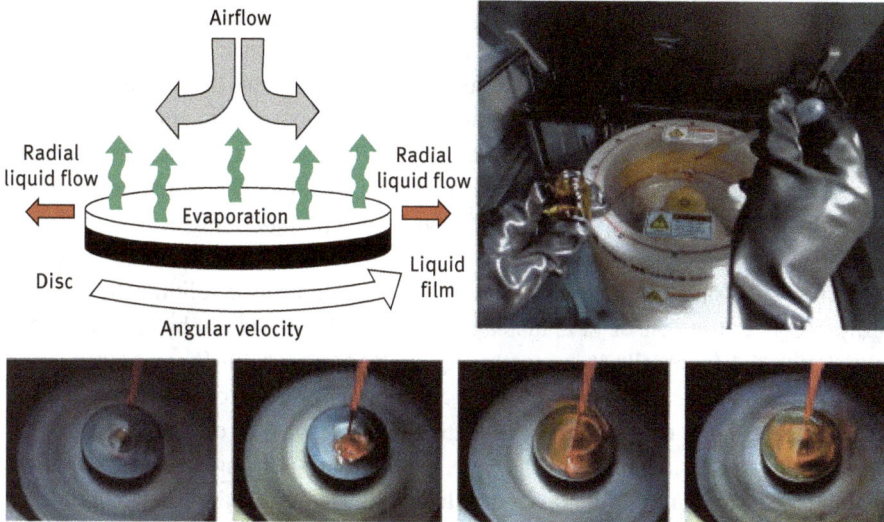

**Figure 3.4:** Schematic illustration of the spin-coating (top left) along with a photograph of a typical spin-coating operation in a glovebox environment (top right) and high-speed images showing application of a solution of MEHPPV to a rotating substrate and film formation. The images were recorded at 300 images s–1 (below). The timing of the images (from left to right) after impact of the first drop is: t = 17, 100, 137 and 180 ms [courtesy of F.C. Krebs in Solar Energy Materials and solar cells, vol.93 (4) pages 394–412 (2009)].

sol–gel routes have, therefore, been designed to combat or control phase segregation during synthesis. In fact, some of the most interesting advances in the sol–gel field in recent years have come from gels with some degree of ordering and structure.

### 3.1.6 Epitaxial growth and heteroepitaxial growth

Epitaxy growth refers to the growth of layers with an ordered structural arrangement respect to the substrate: an oriented growth of epilayers on a single crystalline substrate is achieved when the atomic planes in contact have matched lattice cells. Epitaxy is, therefore, a term applied to the process of growing a monocrystalline film on a substrate: homoepitaxy regards to cases when a crystal is grown epitaxially on a substrate of the same material, as in silicon film grown on silicon substrate. It is termed as heteroepitaxy when a crystal is grown on a foreign substrate, as in a silicon film grown on sapphire or gallium arsenide grown on a silicon substrate. The typical film thickness can range to single monolayers (ML) up to microns, depending on the applied technology. Many different techniques have been developed to achieve an epitaxial growth. The most prominent among these are liquid phase epitaxy (LPE), liquid phase electro-epitaxy (LPEE), vapor phase epitaxy (VPE), organometallic vapor phase epitaxy (OMVPE), molecular beam epitaxy (MBE), chemical beam epitaxy (CBE), atomic layer epitaxy (ALE), respectively. The use of epitaxial growth in general reduces the growth time, wafering cost, and eliminates the wastages respect to the bulk growth. The major advantage of the epitaxy is the uniformity in the composition, controlled growth parameters, and better control of the growth and the relative properties. The main difference among the previously quoted techniques is the physical principle exploited for achieving the epitaxial growth.

The LPE is based on the growth of thin films from metallic solution on an oriented crystalline substrate, the growth occurring from the controlled solidification from a liquid phase. In the case of VPE, instead, the growth is carried out by vaporizing the source material (generally it is mostly exploited to growth III–V compounds) and promotes its solidification in an ordered arrangement on the crystalline substrate. Growth is, therefore, controlled by the partial pressures of each of the components of the source materials. Finally, the MBE is a process of depositing epitaxial thin films from molecular or atomic beams on a heated substrate under UHV conditions. Although vacuum evaporation was already used as early as in the 1950s for preparing semiconductors, epitaxial growth conditions were not achieved until improvements occurred in UHV technology that are nowadays well established. In MBE, a typical growth rate might be only 1 ML per second, and near-equilibrium growth often occurs and guarantees very high purity in the film growth despite its low deposition rates. It is beyond the scope of this book to treat the physics beyond these epitaxial techniques: the reader is addressed to specialized books for further details, few being cited in the reference section.

## 3.2 Fabrication techniques: Local modification of bulk material properties

In integrated optics, other techniques are also widely used to provide for light confinement instead of depositing a film with the required optical properties. They rely on the principle to modify the refractive index in given regions to confine the light. This local refractive index change can be achieved by either introducing dopants (eventually replacing chemical species within the material) or by modifying the chemical bonding inside it in a local way. In the first case, thermal diffusion from thin films, IE, and ion implantation are the most widely exploited techniques. In the second case, UV irradiation and femtosecond laser writing are upon consideration. Since the optical confinement needs a channel configuration, that is, a refractive index jump into a 2D geometrical path to get the optical waveguide, these fabrication techniques must be compatible with pre-processes or post-processing to define a fixed geometry, such as those given by photolithography, chemical etching, and eventually thermal treatments (i.e., the so-called annealing).

### 3.2.1 Thermal diffusion

In general, diffusion refers to the process of movement of particles induced by a gradient in their chemical potential (or decrease in the Gibbs energy), that is, by gradient of concentration of the same particle specie, gradient of temperature or by applied electric fields active as drifts in the case of ionized species. Diffusion leads to a net transfer of matter that normally is accompanied by a random motion that spreads the particles far away in the direction addressed by the gradient. Atoms, ions, and molecules can undergo diffusion: it results that the random thermal motion rate is species dependent. In the case of ion diffusion, their movement induces electric current and the consequent building up of space charge fields. The complex phenomena that, therefore, can establish are quite difficult to be described and strongly depend on the environment, the dynamics of the ion motion itself and the system constraints. Atomic diffusion is a process where the random thermally activated movement of atoms in solid results in the net transport of atoms.

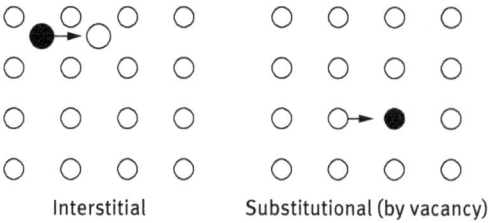

Interstitial       Substitutional (by vacancy)

In the crystal solid state, it occurs by either by interstitial or *substitutional* mechanisms and is referred to as lattice diffusion. In interstitial lattice diffusion, a "diffusant" diffuses in between the lattice structure of another crystalline element. In *substitutional* lattice diffusion (self-diffusion for example), instead, atom can only move by substituting another atom or taking the place of a vacancy throughout the crystal lattice. In this case, diffusing particles migrate from point vacancy to point vacancy by the rapid, essentially random jumping about (jump diffusion).

The diffusion process (in one direction) is normally well described by the Fick's law:

$$\frac{\partial C(x, t)}{\partial t} = \frac{\partial}{\partial x}\left(D\frac{\partial}{\partial x}C(x, t)\right)$$

where C(x, t) is the dopant concentration at the depth x and at the time t, D is the diffusion coefficient that can be dependent on both the temperature T and the concentration C(x,t) as well. The dependence of D on the temperature T is usually modeled by an exponential law according to Arrhenius law:

$$D = D_0 \exp(-E_a/K_B T)$$

where $E_a$ is the activation energy and $K_B$ the Boltzmann constant. The solution of the Fick's law is quite complicated when D depends on the concentration C. An analytical solution of the Fick law in the case of two inter-diffusing semi-infinite alloys was found by Matano in 1932. In case of dilute solutions, D can be considered independent of C, and the analytical solution of the Fick's law can be estimated. When diffusion originates a source of thickness $l_m$ at a constant temperature T, two extreme conditions can be achieved:

- $l_m$ is not completely exhausted; therefore, it can be approximated as an infinite source of dopant because at the surface of the crystal the concentration is constant and equal to the saturation concentration $C_s$. The solution can be described by an error function:

$$C(x, t) = C_s \mathrm{erfc}\left(\frac{x}{\sqrt{4Dt_d}}\right)$$

- $l_m$ is completely exhausted, that is, the diffusion time is long enough, the dopant source at $t = 0$ can be considered as a very thin layer with negligible width. The solution to the Fick's equation has a Gaussian shape:

$$C(x, t) = C_0 \exp\left(-\frac{x^2}{4Dt_d}\right)$$

where $C_0$ is the concentration at the surface of the substrate. In the case of an optical waveguides (in-diffused planar waveguide), the general condition to be achieved is that of a completely diffused dopant.

### 3.2.2 Ion exchange

The IE is a thermodynamic process consisting in a transfer of chemical elements between two adjacent phases, driven by the gradient of the chemical potential. Usually, the transfer takes place between a liquid phase (containing the dopant element, $A^{a+}$) and a solid one (the substrate-to-dope, containing the mobile ion $B^{b+}$). The IE depends on both the thermal agitation and the non-zero mobility of certain ions in the crystals at sufficiently high temperature and on the electric potential difference set up across the crystal causing a ion current to grow. Qualitatively, the IE can be obtained immersing the substrate in a molten salt containing the dopant ion (refer Figure 3.5).

At the substrate-melt interface, both ion concentrations initially drop suddenly from finite values to zero. This is clearly a non-equilibrium situation. The thermal agitation at the interface produces random collisions in which a dopant ion

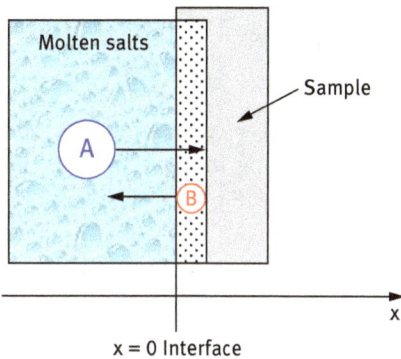

x = 0 Interface

**Figure 3.5:** Ion exchange principle [228].

replaces a matrix ion and then gradually diffuses inside the substrate. The matrix ion released to the molten salt can rapidly move away from the surface and is lost in what can be considered an infinite reservoir. The process accelerates at higher temperature because of stronger thermal agitation and also because the matrix is less rigid. When the substrate is lifted out of the melt but is kept at high temperature, the diffusion of the dopant continues tending to redistribute the dopant, which moves in deeper but with decreasing surface concentration. The process becomes infinitely slow only when the source of heat is removed and the substrate allowed to cool at room temperature. The rate of the IE may be influenced by mass transfer of exchanged species to and the removal of products from the interface between the solid and liquid phases, by the kinetics of the exchange process at the surface and by the transport of the dopant in the substrate. At the interface, the IE process can be thus described in terms of a chemical reaction, the process rate being limited by the reaction kinetics at the liquid–solid interface and depends on the species transport inside both the phases. In the solid solution, the mass transfer is mainly due to diffusion. In the liquid phase, the mass transfer is driven by diffusion and convection processes, the first being the most relevant in the interface nearby region. Convection can be enhanced by stirring the melt; however, even in forced convection case, a region may exist near the solid-melt interface where no convective mixing occurs because of fluid friction. To maintain the charge neutrality, in fact, the migration of ion $A^{a+}$ into the solid phase occurs together with action $B^{b+}$ out diffusion into the liquid phase. As the driving force of the ionic species exchange is the chemical potential, or, in other words, the concentration gradient and electric potential gradient induced by the local non-balance of charges ($E_{loc, charge}$), the dynamics of the process is quite complex. Some models of the process have been developed, the most known being the Nernst–Planck model. It describes the ion flux $j_A$ of the replacing ions $A$ with a charge $q_A$ and a concentration $C_A$ occurring at a temperature $T$ in terms of the driving forces (concentration gradient and $E_{loc, charge}$):

$$j_A = -D_A \frac{\partial C_A}{\partial x} + C_A \frac{D_A q_A}{K_B T} E_{\text{loc, charge}}$$

the IE condition requires the charge balance: $q_A C_A + q_B C_B$ = constant and a total electric current that must be zero due to balancing leading to $aj_A + bj_B = 0$. By exploiting the continuity equation and these constrains, the IE process is, therefore, well described by the following equations:

$$\frac{\partial C_A}{\partial t} = \frac{\partial}{\partial x} \left\{ \widetilde{D_A} \frac{\partial}{\partial x} C_A \right\}$$

where $\widetilde{D_A}$ is an effective diffusion coefficient depending mainly on $C_A$ and on the diffusion coefficients of the species A and B, respectively [228]. Most of the equipment needed for IE process are relatively simple and inexpensive. Basically, they consist of a furnace capable of reaching temperatures up to 800 °C and a suitable sample holder for immersing the crystal in the molten bath contained in a crucible. Although tube furnace may be used, we performed IE process in a more convenient set-up where a crucible furnace with a vertical chamber accessible from above is employed. It is widely used to locally modify the glass surface properties, especially to get optical waveguides.

## 3.3 Material implantation and irradiation

This section provides an introduction into the main types of radiation used to modify a material, focusing on those leading to a change of the refractive index as this plays an essential role throughout this book. The ability to control the refractive index is essential to be able to fabricate waveguides and waveguide-based components and thus lays the foundation stone for optofluidic devices and systems (refer Chapter 4). The different types of radiation methods will be described depending on the irradiation source: Ions (ion irradiation and implantation), X-ray, photolocking, and deep ultraviolet (DUV).

### 3.3.1 Ion implantation

Ion implantation is well-known technique to introduce different chemical elements (dopants) into matrix in a given region (limited in depth and in lateral space): for this reason, it is among the local doping techniques used in integrated photonics and modern communication systems. First proposed by William Shockley in semiconductor materials in his patent in 1954, ion implantation enables the direct control of essential process parameters, such as concentration and depth of the desired dopant. This technology is being used not only in research centers world wide but has also found entrance into numerous commercial applications. Therefore, only a basic description is being given here in this section and the reader is advised to look for application-specific publications in this exciting field.

It mainly consists of bombarding a surface at a given angle respect to the surface normal with ions at an impact energy $E_o$ so that they end their trajectory within the material itself. After releasing their energy, the impinging ions stop at a given depth from the surface, implanting at the so-called projected range, $R_p$ (Figure 3.6).

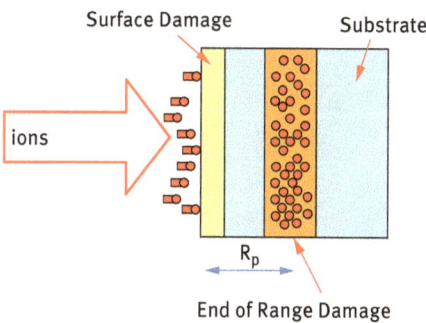

**Figure 3.6:** Ions beam implantation and projected range $R_p$.

During implantation, the energy of the bombarding ion beam is transferred to the implanted substrate by way of two main contributes depending on the impinging ions velocity: energy loss coming from the electronic excitation and the nuclear collision losses, respectively. The first contribute is quantified by the so-called electronic stopping power Se = $(dE/dx)_e$ and dominates at higher velocity. The bombarding ion's trajectory in the material is almost straight, being not affected by collisions with the material constituents. At lower velocity regime (energy typically lower than 8–10 KeV/amu), instead, the nuclear collisions are significant and the ion path is consequently characterized by abrupt changes in direction. In this case, the energy loss is quantified by the nuclear stopping power $S_n$ = $(dE/dx)$. Because of the energy and momentum transfer during the nuclear collisions, separated point defects and disordered regions within a collision cascade are generated, inducing displacement damage. In a non-relativist approach where the nuclear collisions are pictured as elastic interactions, the nuclear stopping power was predicted to increase linearly with energy up to a maximum value at an energy $E_{max}$ and then to decrease according to the law $S_n \propto lnE/E$. In the case that the transferred energy T overcomes a certain value $E_D$ (displacement energy), the knocked atom is moved from its lattice position, creating even damage in the material structure, defects, and amorphization. By considering the contribute of the secondary displacements caused by each primary displaced atom, in average it was found to depend linearly on the $\ln(T_{max}/E_D)$. Consequently, as far as the impacting ions slow down in the material, they experience a different regime of energy loss that depending on the energy: in Figure 3.7, the nuclear and electronic stopping powers are reported in $SiO_2$ in function of the energy of the implanted ions, in this case Au.

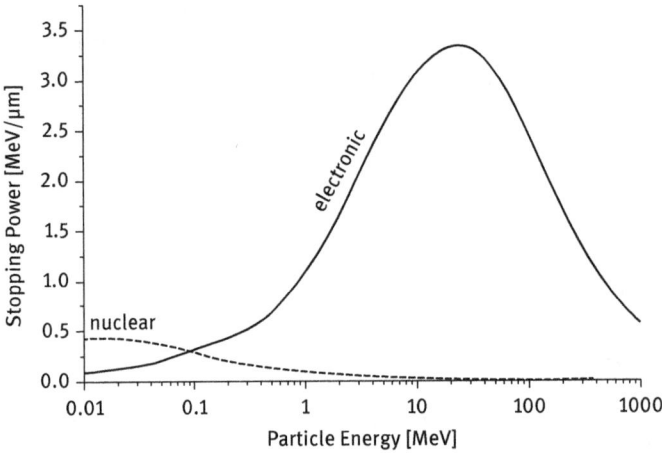

**Figure 3.7:** Nuclear and electronic stopping in function of the implanting ions [kind courtesy of H. Paul, 2007].

The energy release and the effect on the material were studied in detail, with accurate simulations to trace the ion track within the material.

Their physical principle is quite simple: after extracting the ions to be implanted from a source (typically solid or gaseous), they are accelerated by means of suitable electric potential difference (in the range of Volt to MV) to and undergo mass/energy selection before bombarding the material. The selection criteria are ion mass and charge, respectively. Further acceleration provides the ions with their final kinetic energy: to obtain a uniform dosage over the target, a final beam shaping takes place and the dose is monitored in situ with appropriate sensors. By exploiting standard beam scanning techniques, a homogenous irradiation is obtained all over the sample surface, with typical implanted beam diameter in the range of 20–100 μm. When combined with photolithographic patterning of the surface, this procedure allows for selective implantation in 2D patterns. Alternatively, when the ion beam is lowered down to about 100 nm, Focused Ion Beam (FIB) implantation can be successfully obtained allowing directly selective implantation to spatially defined regions. In the following, the effect on the optical properties induced by ion implantation will be reviewed with special care to works performed in the last five years.

### 3.3.1.1 Ion Implantation: effects on the optical properties

As a consequence of the energy loss in the material, the material undergoes to structural modifications as well as compositional changes due to the implantation of the bombarding ion in the end of range region. Both of these phenomena can modify the optical response of the material and, in particular, were widely exploited for modifying the refractive index that, as it will be described in Section 4, it allows to obtain optical waveguides.

Depending on the interactions between the bombarding ions and the material, that is, on the impact of the electronic energy loss and the nuclear one and their eventual interplay, different optical response can be obtained:

–   at the implantation region where the ions stop their trajectories, a variation in the refractive index was reported in many materials. In this case, the role of the nuclear stopping power must be taken into account since it is the displacement energy to play a key role. Many efforts to model this phenomenon were proposed investigating any cause for the refractive index to change the effects induced by implantation: (i) the change in the molar polarization $a_M$ of the implanted specie; (ii) the strain developed in the implanted region that can be related to a change of the volume V of the material. In crystals, the volume increases in the implanted region so at the end of the ion track, the displacement damage is characterized by a decrease of material density; (iii) the modification in the material-spontaneous polarization Ps is present in the material (that decreases toward zero in the amorphous region). Depending on the material and on the implanted specie, each contribute has a different weight and the refractive index consequently changes more often decreasing its value especially in crystals where the induced strain plays a great role. Since ion implantation is often followed by thermal treatments (annealing) to recover the crystallinity, the overall refractive index change is strongly dependent on both the implanted species and the process and post-processes. When ion implantation is performed in amorphous material such as glasses and polymeric materials, the net effect is mainly due to the molar polarization of the implanted species and the effects in the chemical bonding, mostly leading to an increase in the refractive index of the implanted region;

–   along the implantation track, while entering in the material, the refractive index can undergo to modifications correlated to the electronic energy loss. The effect of ion-beam damage in the electronic stopping regime for many oxide, semiconductor, and even metallic materials, is the formation of heavily defective and even amorphous linear tracks around the trajectory of single ion impacts.

It is worth mentioning that, depending on the implant process conditions, defects are normally produced along the ion trajectory and color centers (optical absorption sites) by electronic energy deposition can, therefore, form. Consequently, an annealing treatment is commonly necessary for all the ion-implanted waveguides in crystals to recover the optical properties of the material but preserving the others properties if present (such as electro-optic or non-linear optic properties just to cite the most common). Commonly, conventional annealing furnace, rapid thermal anneals, and laser beam processing have been exploited to reach this goal. In the case of thermal treatments, they are performed in the range of temperature far below the melting of the material but enough to remove color center and/or reduce the structural strains therein induced by the implantation process. The rapid thermal annealing was instead preferred in the case that undesired thermal diffusion of the chemical species is needed and recently, laser beam processing was presented as a valid alternative to get a local thermal treatment by exploiting the absorbed radiation at the surface and, therefore, preventing re-growth or solid phase epitaxy.

### 3.3.1.2 Ion implantation: 1D and 2D geometries

The planar implanted region may be fabricated by directly ion implantation into the sample, in which case no specific masks are required because 1D configuration only needs entire implantation of the target surface. However, for 2D local doping formation, several other techniques are used to properly mask the implanted region. In the earlier works, masks were made by first depositing thin photoresists or metals (e.g., Au), and using standard photolithography technique. In this case, the thickness of the mask was chosen so that the incoming ion underwent to energy loss while crossing the mask deposit and could stop at the correct depth, generating a decrease of the refractive index that can act as side-wall optical barrier. Meanwhile, the incoming ions passing in the unmasked region stop at a depth suitable to realize an optical barrier at the bottom of the waveguide. Alternatively, the common way to proceed and get a better lateral confinement starts from an implantation in a planar waveguide configuration, followed by a photolithographic step to cover those region where the channel waveguide should be obtained. After masking, a series of successive implantation can be performed with tailored energy to build up index-decreased sidewalls: in this way, the waveguides are formed in the mask-protected regions, being confined the bottom barrier (of the planar wave-guide implantation) and sidewalls, together. Alternatively, other solutions were proposed to mount the sample of a rotatable holder and implant it through a movable metal slit.

An overview of the implantation in different material is available in many specialized books (please refer the reference section). In the following, a case study on the application of the ion implantation technique on polymers will be presented to show the potentiality of the process.

### 3.3.1.3 A case study: Ion beam implantation in polymers

Among the first publications, describing the use of ion implantation to fabricate polymer waveguides is shown in [153, 154]. Li+ and N+ ions are implanted into aliphatic PMMA, polyvenylalcohol (PVA), aromatic polyimide (PI), and polycarbonate (PC) polymers. The waveguides exhibit losses less than 2 dB/cm at a wavelength of 633 nm. The change of the refractive index is obtained in the energy range between 100 and 130 keV. Waveguides with acceptable parameters are fabricated for doses below $10^{14}$ cm$^{-2}$. A summary of the refractive index changes is shown in Table 3.3.

**Table 3.3:** Refractive index changes, obtained by ion implantation in different polymers. For PVA and PI the TE values are given. (Reprinted with permission from Ion implantation, a method for fabricating light guides in polymers, J. R. Kulish, H. Franke, Amarjit Singh, Roger A. Lessard and Emile J. Knystautas J. Appl. Phys. 63, 2517. Copyright 1988, http://dx.doi.org/10.1063/1.341032, AIP Publishing LLC.).

| Polymer | Li+: PMMA | N+: PMMA | Li+:PVA | N+:PC | Li+:PI |
|---------|-----------|----------|---------|-------|--------|
| $N_0$   | 1.548     | 1.550    | 1.563   | 1.586 | 1.584  |
| $n_s$   | 1.490     | 1.490    | 1.528   | 1.584 | 1.582  |
| $\Delta n$ | 0.058  | 0.060    | 0.035   | 0.002 | −0.002 |

The refractive index in glassy isotropic polymers which is related to the physical and chemical properties can be described by the Lorentz–Lorenz relation [18]:

$$\frac{n^2 - 1}{n^2 + 2} = \frac{\rho R}{M}$$

where $n$ = refractive index, $\rho$ = density, $M$ = molecular weight, and $R$ = molar refractivity. An increase in the molecular refraction by, for example, the replacement of an aliphatic C-C bond by a double bond or by an aromatic C–C bond would eventually lead to an increase in the refractive index.

The implantation of high energy protons in PMMA to fabricate optical waveguides is described in detail by [108]. The samples were irradiated by 350 keV protons with ion fluences of $2 \times 10^{14}$, $6 \times 10^{14}$ and $1 \times 10^{15}$ ions per cm$^2$, respectively. The relation between the number of modes and the ion implantation fluence is shown in Figure 3.8 for a wavelength of 633 nm.

In [107], it is stated that the molecular weight of the polymer molecules is decreased by crosslinking and increased by scission. The scission effect is dominant in this energy range of protons. The irradiated region is compacted which is observed as well leading to the measured refractive index increase. The question which arises when dealing with ion implantation is the question to which depth protons are able to penetrate into the material. In [107], the depth profile of the refractive index of PMMA

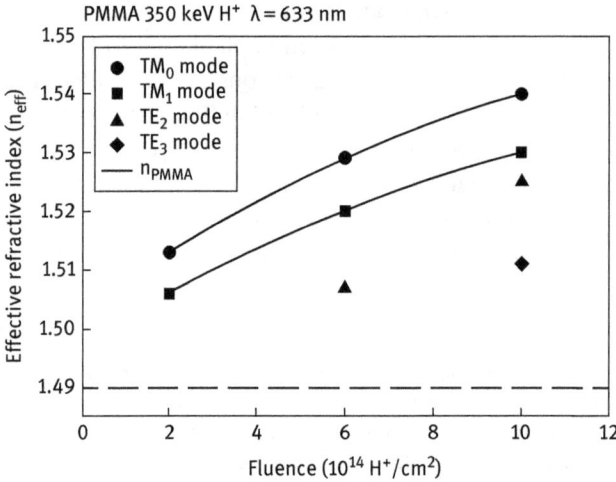

**Figure 3.8:** Variation of the effective index of PMMA for 633 nm along with implantation fluences of 350 keV protons. (Reprinted from Applied Surface Science 169–170 (2001) 428–432, Optical property modification of PMMA by ion-beam implantation, Wan Hong, Hyung-Joo Woo, Han-Woo Choi, Young-Suk Kim, Gi-dong Kim with permission from Elsevier).

irradiated by 350 keV protons is simulated with a fluence of $6 \times 10^{14}$ ions per cm$^2$ and the stopping power curve of 350 keV protons. The profile is shown in Figure 3.9.

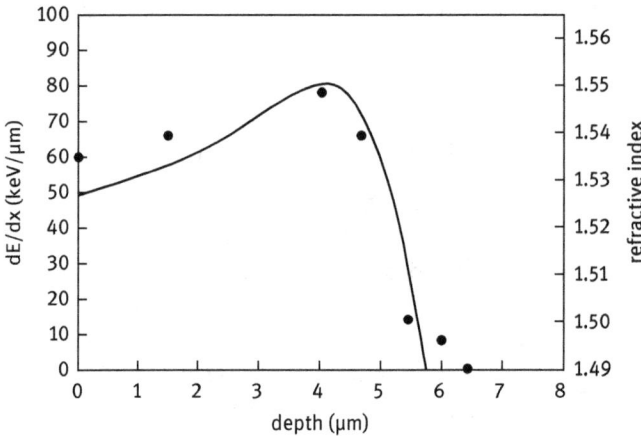

**Figure 3.9:** Comparison of the refractive index depth profile and the simulated stopping power curve of 350 keV protons in PMMA. The solid line represents stopping power curve. Fluence was $6 \times 10^{14}$ ions per cm$^2$ (Reprinted from Applied Surface Science 169–170 (2001) 428–432, Optical property modification of PMMA by ion-beam implantation, Wan Hong, Hyung-Joo Woo, Han-Woo Choi, Young-Suk Kim, Gi-dong Kim with permission from Elsevier).

The obtained refractive indices for different implantation fluence are shown in Table 3.4.

**Table 3.4:** Light attenuation (in ions per cm²) of PMMA waveguides for 635 nm along with implantation fluence.

|  | $2 \times 10^{14}$ | $4 \times 10^{14}$ | $6 \times 10^{14}$ | $8 \times 10^{14}$ |
|---|---|---|---|---|
| Loss (dB/cm) | 1.2 | 1.7 | 2.0 | 2.3 |

The surface of polymers has been modified in several publications with all types of ions. Another example, which has a potential for application, is described in [118]. In this work, Ar+ ion beam irradiation is used to reduce the polymer chain mobility of PET from the surface to the bulk. The irradiation is followed by an $O_2$ plasma treatment. The samples were ultrasonically washed in ethanol and deionized water for 5 min before ion beam radiation. The plasma treatments were carried out under 100 mTorr oxygen pressure for 1 min and at a RF power of 180 W. In [118], the authors also came to the conclusion that the polymer surface undergoes crosslinking when irradiated at energies above 250 eV. This is due to free radicals, created in the absence of oxygen or other free radical processes. The ions break up the C–C and C–H bonds, which lead to the observed free radicals. The increase in the refractive index is shown in Figure 3.10.

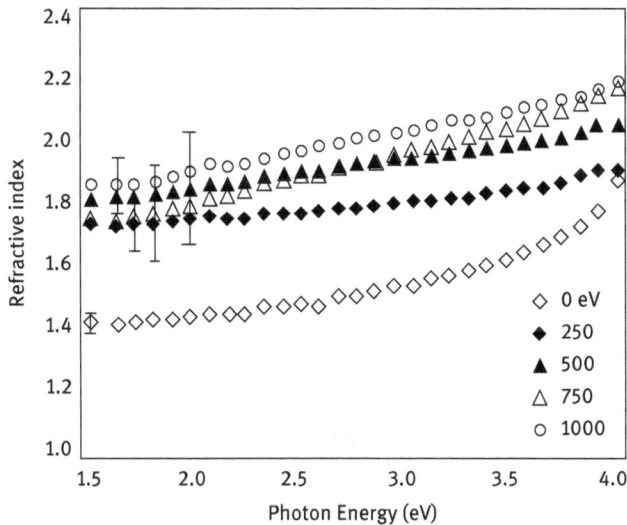

**Figure 3.10:** Refractive index data for PET films irradiated with Ar1 ion at different energies as indicated. The base pressure was 1026 Torr, the running pressure was 1024 Torr, the flow rate was 25 sccm, and the treatment time was 1 min (Reprinted from Journal of Applied Polymer Science 2000, 1679–1683, Jinho Hyun, Philip Barletta, Kwangok Koh, Sangduk Yoo, Jaehwan Oh, David E. Aspnes, Jerome J. Cuomo, Effect of Ar ion beam in the process of plasma surface modification of PET film with permission from John Wiley and Sons).

The increase in the refractive index is attributed to a densification of the surface of the polymers, which is common to the polymers dealt with in this chapter irradiated by the described methods. The main effect is the crosslinking of the polymer chains thus producing a denser surface in the irradiated regions. Crosslinking occurs, when two free dangling ion or radical pairs on neighboring chains unite.

Several ion implantation examples have so far been dealt with. To give a general overview of ions used, hydrogen ($H^+$ and $H_2^+$) should not be left out as well as B+ ions. The authors [145] performed experiments with films of PMMA, polystyrene, polyimides, and positive photoresists spun coated on silicon substrates leading to a thickness between 0.5 and 1 μm of polymer material. The values obtained for changes of the refractive index (n) and extinction coefficient (k) for PMMA and PS films irradiated with B and N ions are listed in Tables 3.5 and 3.6. The extinction coefficient k is the imaginary part of the complex index of refraction. The coefficient k is also an indicator for the amount of light absorption in the material.

**Table 3.5:** The values of refractive index (n) and extinction coefficient (k) of PMMA and PS films irradiated by light ions. (Reprinted from Nuclear Instruments and Methods in Physics Research B 191 (2002) 728–732, Ion implantation for local change of the optical constants of polymer films, F.F. Komarov, A.V. Leontyev, V.V. Grigoryev, M.A. Kamishan with permission from Elsevier).

| $H_2^+$ (250 keV) PMMA ® | | | $B^+$ (100 keV) PMMA ® | |
| --- | --- | --- | --- | --- |
| D (cm$^{-2}$) | n | K | D (cm$^{-2}$) | N |
| $1 \times 10^{12}$ | 1.49 | – | $1 \times 10^{12}$ | 1.48 |
| $1 \times 10^{13}$ | 1.51 | – | $5 \times 10^{12}$ | 1.50 |
| $1 \times 10^{14}$ | 1.54 | – | $1 \times 10^{13}$ | 1.51 |
| $5 \times 10^{12}$ | 1.57 | – | $5 \times 10^{13}$ | 1.53 |
| $7 \times 10^{14}$ | 1.60 | – | $7 \times 10^{13}$ | 1.54 |
| $1 \times 10^{15}$ | 1.66 | – | $1 \times 10^{14}$ | 1.55 |
| $5 \times 10^{15}$ | 1.85 | – | $3 \times 10^{14}$ | 1.58 |
| $7 \times 10^{15}$ | 1.86 | 0.002 | $5 \times 10^{14}$ | 1.64 |
| $1 \times 10^{15}$ | 1.91 | 0.007 | $7 \times 10^{14}$ | 1.68 |
| | | | $1 \times 10^{15}$ | 1.76 |

As can be seen from the described examples of ion implantation, it is possible to change the optical properties of polymers by irradiation with ions, leading to a change in the refractive index of the respective polymer. An advantage is the use of a small fluence, changing the refractive index only by a small amount. The application is straight forward as the described method can be used with masking and lithography technologies to create integrated optical waveguide circuits.

Knowing how to manufacture a waveguide is as equally important as what the material properties are in terms of material behavior, in our case due to radiation. An important aspect in the choice of suitable polymer materials is the determination of the increase of the refractive index. The theoretical background

**Table 3.6:** The values of refractive index (n) and extinction coefficient (k) of PMMA and PS films irradiated by light ions (Reprinted from Nuclear Instruments and Methods in Physics Research B 191 (2002) 728–732, Ion implantation for local change of the optical constants of polymer films, F.F. Komarov, A.V. Leontyev, V.V. Grigoryev, M.A. Kamishan with permission from Elsevier).

| $N^+$ (300 keV) PMMA ® | | | $N^+$ (380 keV) PS ® | | |
|---|---|---|---|---|---|
| D (cm$^{-2}$) | n | K | D (cm$^{-2}$) | n | k |
| $1 \times 10^{12}$ | 1.49 | 0 | $1 \times 10^{12}$ | 1.59 | – |
| $6.7 \times 10^{12}$ | 1.50 | 0 | $3 \times 10^{13}$ | 1.61 | – |
| $1 \times 10^{13}$ | 1.53 | 0 | $5 \times 10^{13}$ | 1.64 | – |
| $6.7 \times 10^{13}$ | 1.54 | 0 | $1 \times 10^{14}$ | 1.66 | – |
| $1 \times 10^{14}$ | 1.55 | 0.005 | $2 \times 10^{14}$ | – | 0.01 |
| $2 \times 10^{14}$ | 1.60 | 0.006 | $5 \times 10^{14}$ | 1.71 | 0.03 |
| $6.7 \times 10^{14}$ | 1.70 | 0.08 | $1 \times 10^{15}$ | 1.96 | 0.118 |
| $1 \times 10^{15}$ | 1.71 | 0.08 | | | |

in doing so is derived from the Lorentz–Lorenz equation, which provides a connection between the real part of the refractive index and the material properties of the used polymer.

$$R_m = \frac{n^2 - 1}{n^2 + 2} \cdot \frac{M}{\rho} = \frac{N_A}{3\varepsilon_0} \cdot \alpha_e$$

where $R_m$ is the molar refraction, $\alpha_e$ is the polarizability of the electrons, $M$ is the molar mass of the monomer, $\rho$ is the density of the material, $N_A$ is Avogadros number and $\varepsilon_0$ is the dielectric constant. The refractive index is only dependent on the relationship between the molar refraction and the molar volume given by:

$$V = \frac{M}{\rho}$$

leading to the refractive index n:

$$n = \sqrt{\frac{1 + 2\frac{R_m}{V}}{1 - \frac{R_m}{V}}}$$

This equation demonstrates the feasibility of the aforementioned waveguide fabrication techniques as either the molar refraction is changed and/or the molar volume of the material.

Another important method of increasing the refractive index is by a method called photolocking, which is one of the most widely used techniques.

### 3.3.2 Modification of polymers by radiation

Practical applications for ionizing radiation, which induce the modification of polymer materials, have been evolving for many decades. Photoreactions are successfully applied to control, for example, the solubility, the polarity, the surface energy, and mechanical strength of polymers (e.g., [45]). The use of these and other processes has found entrance into many applications and they are widely applied today, not only in photonics to control the refractive index, but also in microfluidic applications where the surface properties of fluidic channels plays an important role so as to provide a hydrophobic or hydrophilic surface.

To date only few publications are available about the intrinsic radiation-induced modification of the optical properties of polymers for the fabrication of planar waveguide circuits. The authors in [23] showed that polyvinylcinnamate (PVCi) undergoes a (2+2) cycloaddition under UV irradiation, which leads to a reduction of the refractive index $n_D$ by approximately 0.02, which is sufficient to produce waveguides [23]. Furthermore, it is known that the refractive index of poly (diethylene glycol bis(ally carbonate)), common name CR 39, increases on irradiation with γ–rays or ion beams [7]. For ion beams, the increase can be sufficient to produce optical waveguides. Furthermore, silicone polymers with dithienylethylene moieties along the main chain were reported to give large refractive index changes ($n_D = 0.04$) upon UV irradiation [109]. Poly(4-vinylbenzyl thiocyanate) (PVBT) and a copolymer of styrene and 4-vinylbenzyl thiocyanate (PST-co-VBT) also show large refractive index changes in the range of 0.03, which is applied to obtain volume gratings in polymer films [157].

One has to note that optical waveguides, which are formed by the irradiation of inorganic materials, are known to have low losses (0.1–1.0 dB/cm) and good thermal stability; however on the other hand, the irradiation doses needed to change the value of the refractive index for about 1% can be very high in the order of $0.5$–$6.0 \times 10^{16}$ cm$^2$ and Δn can be either positive or negative.

The following sections will highlight different radiation methods and provide details on their ability to induce useful changes to the refractive index of the bulk material.

### 3.3.2.1 X-Ray in polymers

X-rays, which have been first discovered in 1895 by Conrad Röntgen, have found their way into modern medicine and have numerous applications. The wavelength of X-rays can be described to be between UV light and gamma rays ($10^{-8}$ und $10^{-12}$ m) with photon energy between 100 eV und 250 keV. It is thus obvious that X-rays have been used to change the properties of materials, especially polymers, one of the first reports on using X-rays to modify the refractive index of polymers, in this example

PMMA [266]. The authors not only changed the refractive index of PMMA but also fabricated graded index waveguides.

Refractive index changes and optical waveguide propagation losses were measured in spin-coated PMMA as a function of X-ray exposure. The thickness of the PMMA film is 6 µm. The irradiation dose is 12 J/cm² at 65° to the surface normal with an exposure time of approximately 10 min. The PMMA was deposited on 3 inch in oxidized (100) silicon wafers that were then coated with ~ 0.5 µm of electroplated nickel. The wafers were then heated to 160 °C at 1 °C/min and annealed for one hour. Wafers were placed under 20 torr He pressure and exposed to a broadband X-ray source (~900–1,300 eV at the Alladin Storage Ring, University of Wisconsin). The waveguide loss was determined to be around 0.5 dB/cm using a HeNe laser source at a wavelength of 632.8 nm. The change of the refractive index is only measurable a few micrometers below the surface. This means that the irradiation penetrates only a certain distance into the material enabling the creation of surface waveguides in PMMA.

Another detailed experimental investigation providing a quantitative explanation of the radiation-induced refractive index changes in PMMA by X-rays is given by [4]. The irradiated samples used in the experiment were PMMA from Roehm, PLEXIGLAS; GS 233 and PLEXIGLAS SUNACTIVE; GS. The samples were cut from large sheets with a buzz saw. The dimensions of the samples were $20 \times 20 \times 10$ mm³ for GS 233 and $20 \times 20 \times 8$ mm³ for SUNACTIVEGS. The advantage of this fabrication method is the availability of material in optical grade quality and with basically no restriction in size. Another advantage is the fact that it only takes a buzz saw to cut the material to the desired size.

The authors in [4] used the electron stretcher accelerator (ELSA) of the Physical Institute of the University of Bonn, operated in the 2.3 GeV storage-ring mode as a synchrotron light source. The radiation intensity was distributed over a range of 1–20 keV with a maximum at 3 keV. To reduce damage at the exposed sample surface, a 12.5-mm-thick KAPTON™ foil was used as a preliminary filter. For high-dose exposures, a 50-mm-thick KAPTON foil plus a 1-mm-thick PMMA sheet were used instead for pre-filtering. The refractive index modulation $\Delta n(x,y)$ of a sample was determined by interferometric measurements utilizing the dependence given in the following equation:

$$\Delta n(x,y) = \frac{\lambda_i}{2\pi d_0}\Delta\phi(x,y) - \frac{n_0 - n_{air}}{d_0}\Delta T(x,y)$$

where $\lambda_i$ is the vacuum wavelength of the measurement light, $d_0$ the sample thickness, $n_0$ the refractive index of the material, and $n_{air}$ is approximately 1, the refractive index of air. The coordinate system is used, where $x$ and $z$ are coordinates perpendicular to the X-ray beam and $y$ is parallel to it. The radiation induced refractive index change for several grades of PMMA are described in [4]

In [4], it is stated that the X-ray-induced refractive index changes in PMMA result predominantly from the change of the material density, and only at higher doses additionally from a chemical modification. Diffusion processes are not significant in the bulk material, but are presumably the reason for the profile changes due to the presence of surfaces. A scheme of all major radiation-induced reactions is presented in this publication.

It is also noted that outgassing of volatile reaction products leads to a densification that increases the refractive index. This fact is also observed in other irradiation procedures, for example, the DUV modification. Radicals, which are formed during irradiation, remain in the material and cause further reactions if not speeded up by heating after the irradiation is performed. This annealing step is necessary to obtain a stable refractive index profile.

### 3.3.2.2 Photolocking

Photolocking is a method by which a polymer substrate is coated with another polymer, which in turn contains a high amount of monomer and what is most important, contains a photoinitiator, which serves as the dopant. By selective photolithography, the dopant initiates a polymer chain reaction leading to a local increase of the refractive index. The word photolocking was first proposed in 1974 [42]. As can be seen from this publication, the technique is quite mature and widely used in the fabrication of optical waveguides.

The sequence for fabricating waveguides by photolocking [194] is as follows:
- Starting Structure (from bottom to top):
  - Substrate
  - Cladding layer: Polymer
  - Core layer: Polymer with dopant
- Dopant "locked" by UV-light (structured by appropriate mask) to polymer chain realizing a local increase of the refractive index and thus realizing waveguides.
- Annealing: diffusion of unreacted dopant
- Add cladding

Suitable materials are, for example, Benzildimethylketal/PMMA. When using this waveguide fabrication method, two negative aspects should be known and considered:
- Refractive index inhomogeneity can cause segregation and stress, leading to a high optical waveguide loss.
- Dimensional stability of the waveguides is affected by diffusion, which can lead to a broadening of the optical field. This is especially important when designing single- or multimode photonic waveguides.

In summary, this is another well-known method, which enables the fabrication of waveguides in an easy way, if the materials are well chosen.

The following paragraph will provide an insight into the waveguide fabrication method by DUV radiation. This method will be used and referenced throughout the book, as it is one method of creating a true optofluidic system platform.

### 3.3.2.3 DUV radiation

This chapter is reprinted with permission from "Modification of polymethylmetha-crylate by DUV radiation and bromination for photonic applications, P. Henzi, K. Bade, D. G. Rabus, J. Mohr, J. Vac. Sci. Technol. B 24 (4), 1755–1761, Copyright 2006, American Vacuum Society".

An extremely useful class of low-cost thermoplastic polymers, which exhibit a significant increase in refraction index through application of ionizing radiation, such as ion and deep UV (DUV) radiation, are methylmethacrylate polymers (PMMA). It was shown that optical waveguides can be created in polymethylmethacrylate homopolymers by ion [107, 227] and DUV radiation [82, 233–235].

In this section, progress for realizing integrated optical circuits, which uses the DUV-induced refractive index modification of methylmethacrylate polymers, is given. This technology has several advantages with respect to other methods because only a single polymer layer is required, which serves as the substrate and waveguide as well and no further etching, development step or filling with higher index polymer is needed. To control and optimize the properties of the UV-induced waveguides, a thorough understanding of radiation-induced processes and resulting physical and chemical properties of the modified material is essential. Thus, besides investigation of the waveguides by various optical characterization methods such as m-line spectro-scopy, near-field measurements, or cut-back method, a fundamental study using spectroscopic methods (IR- and UV-VIS-spectroscopy) and differential scanning calori-metry (DSC) has also been carried out, with the aim of gaining a better understanding of the structural modifications caused by irradiation of the used materials. The para-meters for waveguide fabrication are optimized by changing the primary methylmetha-crylate polymer properties, such as molecular weight, copolymer composition, and environmental process condition. A postexposure reaction of the irradiated zones of the material with bromine is investigated to achieve an enhancement and a stabiliza-tion of the refractive index changes.

Although extensive research is reported on photochemistry on methylmethacry-late polymers [49, 129, 181, 233–235, 260], it appears rather difficult to extract reliable information from literature for a fundamental understanding of photochemical pro-cesses, which influence the refractive index modification. This may, in part, be caused by the fact that the materials may contain impurities or co-monomers that intervened in photochemical reactions. If absorption by impurities or co-monomers contributes essentially to the initiation of chemical changes, the photochemistry changes remark-ably compared to photoinduced reactions of a pure polymer. To understand and to optimize the process of structural modification with respect to waveguide generation, the investigation of photoinduced chemical reactions of different homo- and copoly-mers of methylmethacrylate in air and vacuum (or inert gas) are given.

For DUV-modification, a commercial UV-exposure equipment (UVAPRINT CM, Dr. Höhnle GmbH) is used. A 100 W/cm mercury xenon arc lamp combined with a cold mirror with reflectance in the range of 220 nm–420 nm is used in the exposure system. The resulting output is 0.8 mW/cm² at 240 nm. The exposure system is equipped with a vacuum chamber. The investigated PMMA types are Hesa@Glas, a homopolymer from Notz-Plastic, Lucryl G77 Q11 a copolymer of 95 wt.% methylmetha-crylate and 5 wt.% methylacrylate from BASF and PMMA 950 K, a homopolymer from MircoChem Corporation. The UV spectra of films on a quartz disk, before and after irradiations, were obtained with a Perkin-Elmer Lambda 2 UV-VIS-spectrometer. The FTIR spectra were recorded with a Brucker IFS 48 V Fourier-transform infrared spectro-meter. Refractive index measurements were carried out by m-line spectroscopy, which allows the calculation of refractive index depth profiles. M-line spectroscopy is a method, which relies on the waveguiding of modes in a material with the help of a prism coupling arrangement. Details of the measurement are described in Chapter 3.2.

In Figure 3.11, typical variations of the effective refractive index at 633 nm as a function of irradiation dose for the copolymer Lucryl G77 are shown.
With increasing dose, more and more modes are able to propagate. For a dose higher than 15 J/cm² only the effective index of the lowest order mode is increasing, the effective index of higher order modes stays constant. No birefringence can be found in the range of the measurement resolution of $10^{-4}$. After annealing the samples at 70 °C for 15 h, the mode spectra remains unchanged. Independent of environmental condition (air/vacuum) during modification, the investigated PMMA types show comparable behavior, but slightly different dose dependencies. High humidity levels and water immersion is a critical test to ensure refractive index stability under highly humid environment. By a simple water immersion of the samples for 24 h at room temperature, only the samples exposed under vacuum show good resistance and unchanged mode spectra. In contrast, the samples irradiated under ambient air show a corrosion of the waveguide surface and no turning back to the previous state can be observed by air-drying or baking. From the measured mode spectra, the refraction index profile was reconstructed using an inverse WKB-method [266]. It can be seen in Figure 3.12 that for irradiation doses higher than 6.48 J/cm² a nearly exponential decreasing profile, with a decay con-stant of approximately 5 μm, can be found.

Higher index contrast and higher thermal stability of the waveguides can be achieved by introducing bromine substituents into the polymer material. It is well known that the high molar refraction of a C-Br bond results in a higher refractive index compared to C-C bonds [150]. Since during UV-induced modification of PMMA, the generation of C=C bonds takes place and since unsaturated C=C are suited for electrophilic addition of halogens, a post-exposure reaction of bromine with the irradiated zones of the materials was investigated to achieve an enhancement and

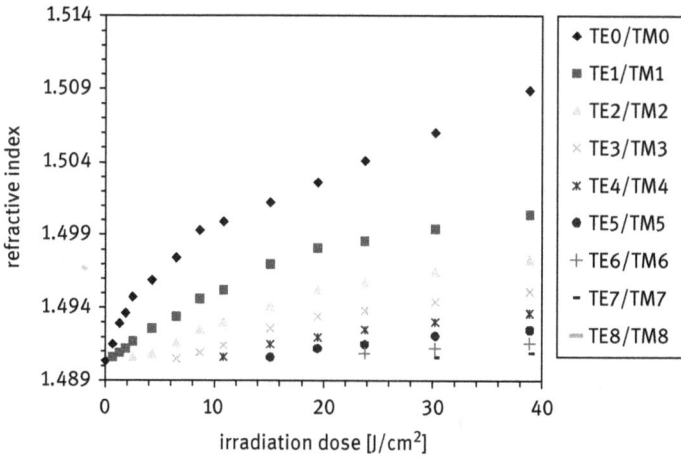

**Figure 3.11:** Effective refractive index as a function of irradiation dose. The indices TE or TM indicate transversal electrical or transversal magnetic polarization.

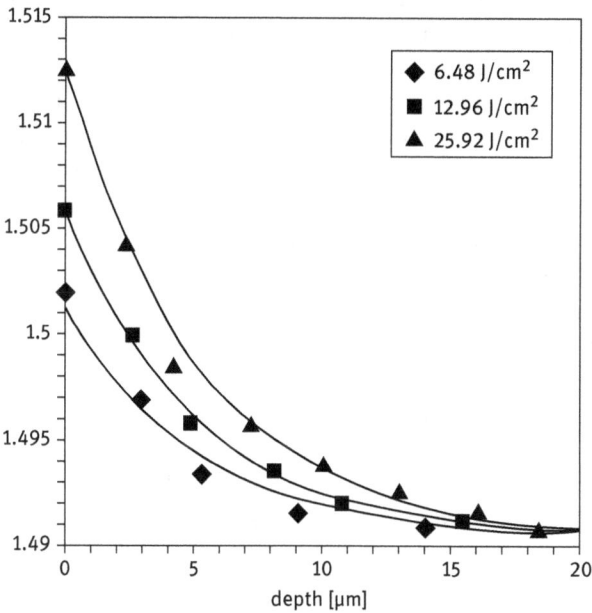

**Figure 3.12:** Refractive index profile as a function of depth calculated with the inverse WKB method of [266].

a stabilization of the refractive index changes. These reactions were carried out in a 0.1 vol.% aqueous solution of bromine, which leads to an addition of Br and OH. In Figure 3.13, a comparison between the effective refractive index of the lowest order mode for under vacuum-modified and afterwards brominated material is shown.

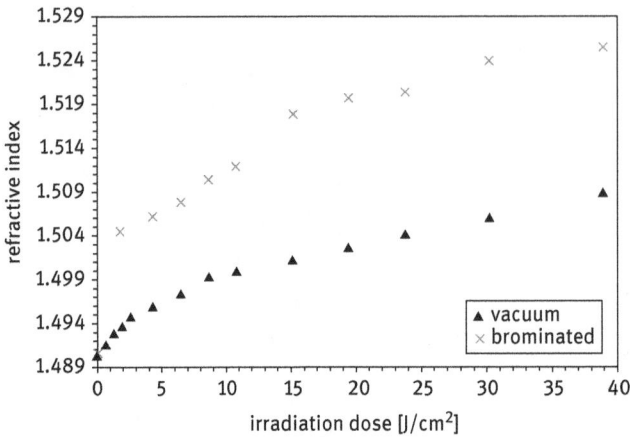

**Figure 3.13:** Effective refractive index of the lowest order mode in dependence of irradiation dose for samples modified under vacuum and after 5 h in a 0.1 vol.% aqueous solution of bromine.

The brominated material shows a similar function of the UV dose, but approximately up to 2 times higher index contrast is shown. Therefore, to obtain a sufficient high index, contrast for wave guiding a smaller irradiation dose is needed. Long-term stability tests have not been carried out specifically, but what is observed is that samples which have been prepared nearly one year ago still show the same optical properties. The γ-lactone is expected to be more stable compared to the reactive carbon-carbon double bond. But attack of bases will lead to opening of the lactone ring and therefore to an additional hydroxyl-group, which can reduce the performance of the waveguide. The introduced bromine, which is yielding the increased refraction index, is expected to be stable.

The UV absorbance spectra were measured in transmission using a 3-μm-thick PMMA 950 K film on a quartz disk after different irradiation doses (Figure 3.14). Irradiation was performed in ambient air and vacuum.

The increasing absorption peak below 200 nm is consistent with the generation and reaction of unsaturated bonds, which are also determined by Raman spectroscopy. This peak is due to a $\pi$–$\pi^*$ transition of C=C bonds [102, 103]. In air, this absorption peak decreases for an irradiation dose higher than 2.16 J/cm². This indicates the oxidative decomposition of the double bonds. At incident wavelengths above 220 nm, the wavelength region where photo degradation takes place, the optical penetration depth rapidly decreases with increasing irradiation dose to some few micrometers. This means that a few micrometer thin surface layer can be modified in a thicker PMMA bulk layer with a higher refractive index. The increasing branch of the absorption peak to longer wavelengths is essential for single mode waveguide fabrication, since the penetration depth of the UV-light must fall under a certain value to guarantee single mode propagation. Therefore, not only the refractive index increase determines the necessary irradiation dose but also the existence

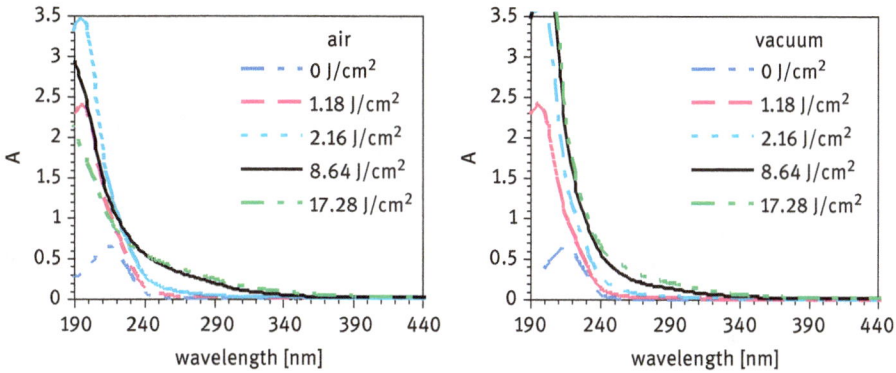

**Figure 3.14:** UV absorption spectra of a 3-μm-thick PMMA 950 K film after different irradiation doses (measured at 240 nm) in ambient air and vacuum [103].

of a certain threshold dose, which must be exceeded to guarantee a thin enough modified layer for single mode propagation. The UV spectra of the copolymer show qualitatively similar features to those of pure PMMA.

The corresponding FTIR spectra of the PMMA 950 K films are shown in Figure 3.15.

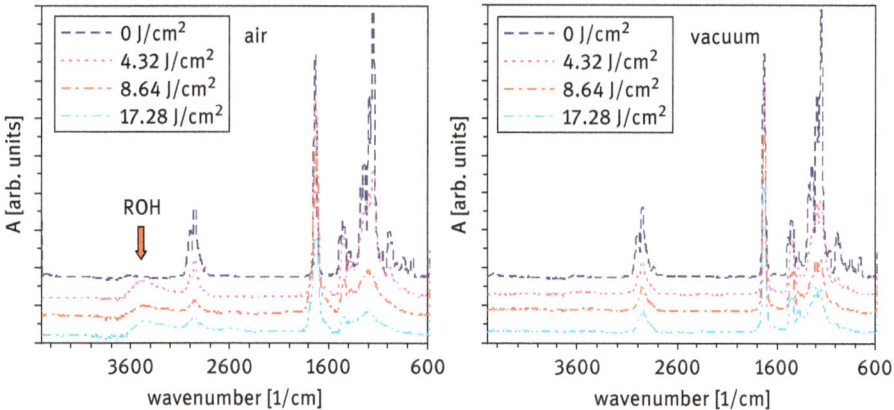

**Figure 3.15:** IR absorption spectra of a 3.2 μm thick PMMA 950 K film after different irradiation doses in ambient air (left) and vacuum (right).

With an increasing dose, both in air and in vacuum, the carbonyl band in the pendant group (-COOCH$_3$) of PMMA at 1,735–1,750 cm$^{-1}$ decreases and shifts toward higher absorption frequencies, indicating a change of the chemical environment. Furthermore the C–H stretching and bending absorption bands at 2,847–2,931 cm$^{-1}$ and 1,500–1,350 cm$^{-1}$ and the C–O stretching bands at 1,270–1,150 cm$^{-1}$ decrease substantially. This behavior is consistent with literature and was explained to a removal of the ester side group followed by a main chain scission or the generation of unsaturated bonds in the polymer chain ([49, 129, 181, 233–235, 260]).

The yield of double bond formation in the main chain is larger than the main chain scission with UV irradiation. The copolymers of methylmethacrylate and methylacrylate show similar characteristics, indicating similar dominant photochemical processes. In contrast, it has been reported in literature that the mechanism of photolysis of the MMA/MA copolymer is different from that of pure polymethylmethacrylate [87]. Whereas pure polymethylacrylate (PMA) undergoes permanent crosslinking due to hydrogen abstraction, a copolymer of MMA/MA will show a decreased crosslinking rate with decreasing content of methylacrylate. In our case, the content of 5 wt.% methylacrylate is too low for efficient crosslinking. The dominant process is furthermore the cleavage of pendant methylester groups (side chains) followed by the generation of C=C double bonds in the polymer chain. However, during photolysis in ambient air, photo-oxidation products can be detected in form of hydroxyl groups at 3,500 cm$^{-1}$. The absorption peaks do not vanish even when annealing the samples at 100 °C. For waveguide fabrication, it is very important to avoid these oxidation products, because higher order modes of OH-vibration increase the absorption of the waveguides in the optical window for waveguide application at 1,550 nm. The oxidation products can be avoided by exposure in an inert gas atmosphere or vacuum as shown in the right part of Figure 3.10. The characteristics described here are independent of the material used. Evidence for the electrophilic addition of bromine is given by an increasing peak in the UV spectra due to a shift of the n–π* transition of the C=O group to longer wavelengths before annealing (Figure 3.16). After annealing the samples at 70 °C, a shift in the IR spectra of the carbonyl band toward higher wavenumbers (Figure 3.17) and the detection of evolution of gaseous products, such as methanol, indicates a γ-lactone formation.

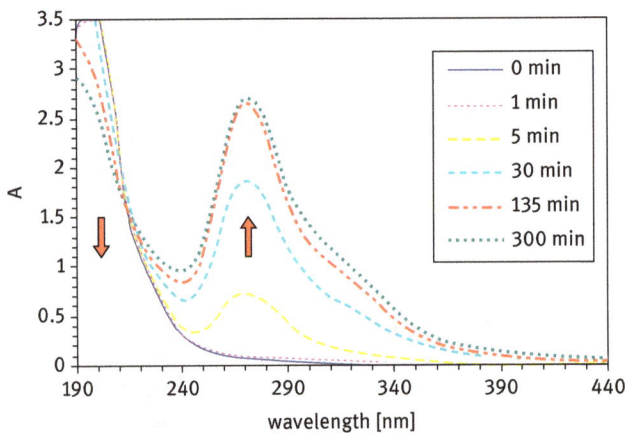

**Figure 3.16:** UV absorption spectra after different bromination times in a 0.1 vol.% aqueous solution of bromine.

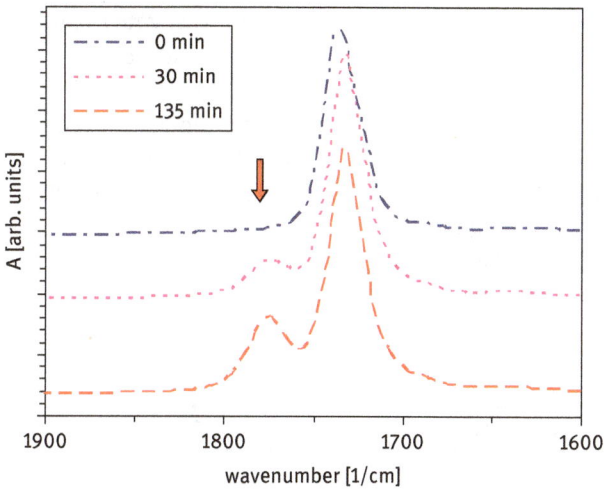

**Figure 3.17:** IR absorption spectra after different bromination times in a 0.1 vol.% aqueous solution of bromine.

Another type of methacrylate polymer has been used in [119] to fabricate waveguides with the same procedure as described in this Chapter. The materials used here were alicyclic methacrylate copolymers, which are obtained from Hitachi Chemical Co., Ltd as OPTOREZ-Series (OZ-1000, OZ-1100, OZ-1310 and OZ-1330). Similar results have been obtained as for PMMA.

The absorption of the shoulder peak of OZ-1000 becomes larger as expected with an increase of the exposure dose. The increasing absorption peak below 200 nm represents the generation and reaction of unsaturated bonds. This peak is due to a π–π* transition of C=C bonds. To achieve a single-mode waveguide by a given refractive index profile, the geometry of the waveguide structure must be adjusted. The width of the waveguide is determined by the structural design of the photo mask used in the standard UV lithography process. If the process tolerances are known, then the structural dimensions on the photo mask can be adjusted to produce the exact width required for single-mode waveguide propagation. The depth of the waveguide is controlled by the penetration depth of the DUV light into the polymer. To achieve a single-mode waveguide propagation behavior in this direction, it is needed that the penetration depth can be accurately adjusted. As is the case with PMMA, the absorption peak becomes larger with increasing UV exposure dose. That means, it is possible to control the penetration depth by the exposure dose. Thus, after being exposed with a certain UV exposure dose, UV light is absorbed in a very thin layer at the polymer surface and the UV light does not reach the deeper area of the polymer. The depth of the waveguide is now controlled by adjusting the exposure dose. By increasing the exposure dose, the effective refractive indexes become larger in all samples. No birefringence can be found in the range of the measurement resolution of $\pm 10^{-4}$. Independent of the OZ grade used, the investigated material types show comparable behavior, but slightly different dose dependences.

Structural changes of the polymer also mean changes in its thermo-mechanical properties. To characterize the thermo-mechanical properties, the glass transition temperature (Tg) was determined with DSC. Table 3.7 shows the change of Tg of OZ-Series before and after exposure. The exposure dose was 4.32 J/cm$^2$ by the UVAPRINT setup, and the samples' thickness was 20 µm. Except for OZ-1310, $T_g$ decreases after exposure, but the decrease of $T_g$ is relatively small.

**Table 3.7:** Glass transition temperature of OZ-Series before and after exposure.

| Sample name | Tg before exposure (°C) | Tg after exposure (°C) |
|---|---|---|
| OZ-1000 | 125.3 | 110.2 |
| OZ-1100 | 130.5 | 123.7 |
| OZ-1310 | 129.9 | 130.8 |
| OZ-1330 | 120.1 | 107.0 |

The waveguide and waveguide/fiber coupling loss were determined using the cut-back method. The attenuation of the waveguides at 1,550 nm is measured to be 1.5 dB/cm. The polarization dependent loss is less than 0.15 dB. The coupling loss to a single-mode fiber was 0.5 dB/facet.

Waveguide patterns may be inscribed into polymer films by lithographic techniques using standard photomasks (Cr on glass) or using pre-structured polymer plates, as will be described in Section 3.5 [99–103].

The UV-irradiation results in a local and controllable increase of the refractive index in the illuminated areas of the polymer surface generating the integrated-optical waveguide structures in a polymer plate. For pre-structured polymer plates, embossed ridge or groove structure serves as masking structures for the ensuing flood exposure. If the penetration depth of the evanescent field of these waveguides is smaller than the distance between them and the planar waveguides at the surface, no light can couple from one waveguide into the other. Either ridge or groove waveguides can be used as stripe waveguides. For strong guiding application, the ridge structure is preferred, for weak guiding application the groove structure is used. Details on the types of waveguides will be given in Chapter 4.3.

Additional to the waveguides, fiber alignment grooves can be produced in one fabrication step. By this approach, a time-consuming active pigtailing can be avoided.

In summary, waveguiding enabling fabrication methods have been introduced focusing on polymer materials.

## 3.4 Patterning techniques

The term patterning normally refers to all those techniques able to introduce a controlled modification of the morphology of a surface with a desired design. It contains, therefore, featuring with lateral dimensions spanning over the nano- to millimeter range. The microelectronics industry and need for smaller and faster computing systems have pushed this development during the last two decades, mainly focused on obtaining patterns with the smallest possible lateral dimensions via optical lithography in its multiple variants [62]. Recently, increased effort has been devoted to fabrication technologies allowing the production of structured surfaces with greater geometrical complexity at reduced operation time and cost. These include patterns made of materials possessing elongated features in the vertical dimension (aspect ratio > 3), exhibiting several hierarchy levels, or in intricate tilted, suspended, or curved 3D arrangements [62]. Such surface structures find applications in emerging fields like biosensors with increased sensitivity and throughput due to higher effective surface area. In the following, a brief summary of the patterning techniques exploited in opto-microfluidics will be presented.

### 3.4.1 Photolithography

Lithography has its origin in the two Greek words, "lithos" meaning "stone" and "graphein" meaning "write", leading to "pattern writing in stone".

Photolithography is one of the most widely known multi-step processes to transfer a given pattern to a material by way of light interaction with a photosensitive medium. It is used in integrated optics as well as in micro-fabrication processes to mask regions respect to others where to deposit films or coating. Dated back to 1959, photolithography is basically a process that exploits a high-intensity light to induce photo-polymerization process of photo-sensible monomers/polymers masked to be exposed to light only in given area. The photomask is, therefore, a device that allows the light to "pass through" given areas to project an image on a surface. Photolithography is consequently a process leading to a patterned polymer network deposited on a substrate. The polymer network is formed by chancing the chemical–physical properties of a photosensitive material (normally referred as photoresist), by chancing its solubility: it comes soluble or insoluble when exposed to UV light. Photoresist can be:
- positive photoresists when made of insoluble polymers that modify to a soluble ones when exposed to UV light;
- negative photoresists that change to insoluble polymers upon UV exposure.

The main steps of photolithography involving a negative photoresist can be summarized as follows:

- a thin layer of photoresist is spin-coated onto the surface with a typical thickness < 1 μm;
- a photomask with the desired structural feature is positioned above the photoresist;
- UV light is focused onto the photoresist through the photomask. The light passing through the photomask (clear area) causes the photoresist to polymerize into insoluble polymers. The covered and protected areas (dark areas) remain instead soluble to solvents because they are not interacting with the light;
- the sample is treated in solvent: the dark areas are dissolved away, leaving those exposed to light making the resulting pattern;
- photoresist strip (removal of the photoresist using for example plasma etching);
- final inspection.

A typical example of photolithography with negative photoresist is shown in Figure 3.18 to create a pattern [199].

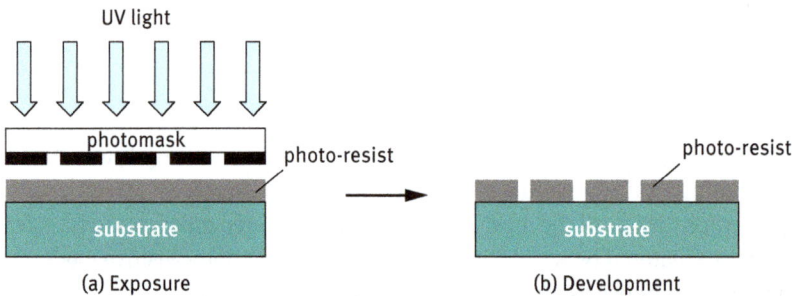

**Figure 3.18:** Schematic illustration of the main steps in photolithography. (a) exposure step: photoresist coated on the substrate is exposed to UV light, (b) development step: the exposed photoresist is removed by immerging into a developer [199].

In general, the photomask can be put into contact with the substrate or suspended above it. Several techniques have been developed, each with given advantages and drawbacks. In general, three main approaches are used: contact printing, proximity printing, and projection printing as schematically illustrated in Figure 3.19.

Contact and proximity printings put the photomask in contact with or in a close proximity to the photoresist [199]. Generally, contact and proximity printings are capable of making patterns as small as a few micrometers and consequently they are typically used in the fabrication of moderate resolution patterns. A projection printing system (often called "stepper") uses an optical lens system to project a DUV pattern from an excimer laser (typically with wavelength of 193 or 248 nm) on the photoresist with a pattern size reduction by 2–10 times. It is capable of fabricating high-resolution

UV light

UV light

UV light

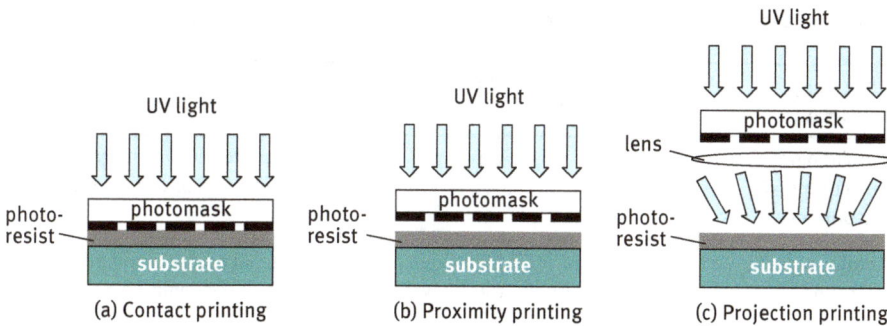

**Figure 3.19:** Schematic illustration of three forms of photolithography: (a) contact printing; (b) proximity printing; and (c) projection printing [199].

patterns as small as a few tens of nanometers (around 30–40 nm) at a high throughput (even better than 60 wafers/hr). However, since it requires a sophisticated optical-lens system and precise control systems of temperature and position resulting in a very expensive setup, it is employed in manufacturing of advanced products. In recent years, immersion lithography resolution enhancement technology and extreme UV have been developed to improve the lithography resolution of projection printing. A comparison on the overall performances is reported in Table 3.8.

Most semiconductor processes nowadays use a positive photoresist, leaving a "positive" image of the mask pattern on the surface of the substrate. The photomask is usually made out of soda lime, borosilicate glass or fused quartz. Advantage of using quartz is the fact that it is transparent to the DUV radiation and has a low thermal coefficient of expansion. This is why for the DUV process described earlier the photomask is made out of quartz.

The goal of the lithography community is to improve the resolution of the lithography system. Two approaches are being pursued: either use optical or processing techniques such as increase the optical aperture of the used optics in lithography systems or use double-patterning techniques; the other approach is decrease the wavelength of the light used in patterning (mercury lamp i-line: 365 nm; ArF laser-DUV 193 nm; EUV 13.5 nm). Further details can be achieved in more specialized books.

## 3.5 Replication

The emergence of optofluidic devices is largely enabled by recent advances in microfabrication, microfluidics, and polymer-processing technologies. The methods of choice are micromachining, soft lithography, and embossing techniques, which enable the fabrication of micron-scale fluidic channels in silicon, glass, polymer, and elastomer materials.

**Table 3.8:** Specifications and applications of the major lithography techniques [199].

| Lithography Technique | Minimum Feature Size | Throughput | Applications |
|---|---|---|---|
| Photolithography (conact & proximity printings) | 2-3µm [22] | very high | typical patterning in laboratory level and production of various MEMS devices |
| Photolithography (projection printing) | a few tens of nanometers (37 nm) [2] | high - very high (60-80 wafers/ hr)[1] | commercial products and advanced electronics including advanced ICs [1], CPU chips |
| Electron beam lithography | < 5 nm [23] | very low [1, 3] (8 hrs to write a chip pattern) [1] | masks [3] and ICs production, patterning in R&D including photonic crystals, channels for nanofluidics [23] |
| Focused ion beam lithography | ~20 nm with a minimal lateral dimension of 5 nm [2] | very low [3] | patterning in R&D including hole arrays [125, 134], bull's-eye structure [132], plasmonic lens [137] |
| Soft lithography | a few tens of nanometers to micrometers [2, 13] (30 nm) [2] | high | LOCs for various applications [13, 96] |
| Nanoimprint lithography | 6-40 nm [14, 15, 18] | high (> 5 wafers/ hr) [1] | bio-sensors [17], bio-electronics [18], LOCs: nano channels, nano wires [97, 102, 104] |
| Dip-pen lithography | a few tens of nanometers [39, 40, 43] | very low – low possibly medium [39] | bio-electronics [43], bio-sensors [40], gas sensors [42] |

Polymers have been accepted as the material of choice for the integration of photonic integrated circuits and fluidic devices, mainly due to their increasing performance, rapid processibility, capability for precise tailoring of their optical properties, and their comparatively low cost. Another important aspect is the biocompatibility of polymer materials and the fact that these materials are already in use in many bio- and non-bio laboratories, which increases the acceptance of polymer-based optofluidic devices. This advantage requires the improvement of fabrication technologies as well as the development of application-specific tailored materials. As stated before, polymer optical waveguides have been fabricated by various techniques, such as dry etching, UV curing and soft lithography replica molding [94], and embossing.

## 3.5.1 Replica molding

Replica molding creates stamps with shapes complementary to patterns etched in silicon chips by photolithography. Elastomeric stamps with a surface topography

complementary to the etched surface are often created by a replica-molding technique. It consists in the deposition of a liquid pre-polymer, normally PDMS, that is, cast
on top of the etched photoresist pattern, polymerized, and peeled off [117]. The PDMS
stamp can be used for micro-contact printing in order to achieve even microfluidic
organs-on-chips as proposed by Bathia and Ingber in 2014 [17]. For example, even

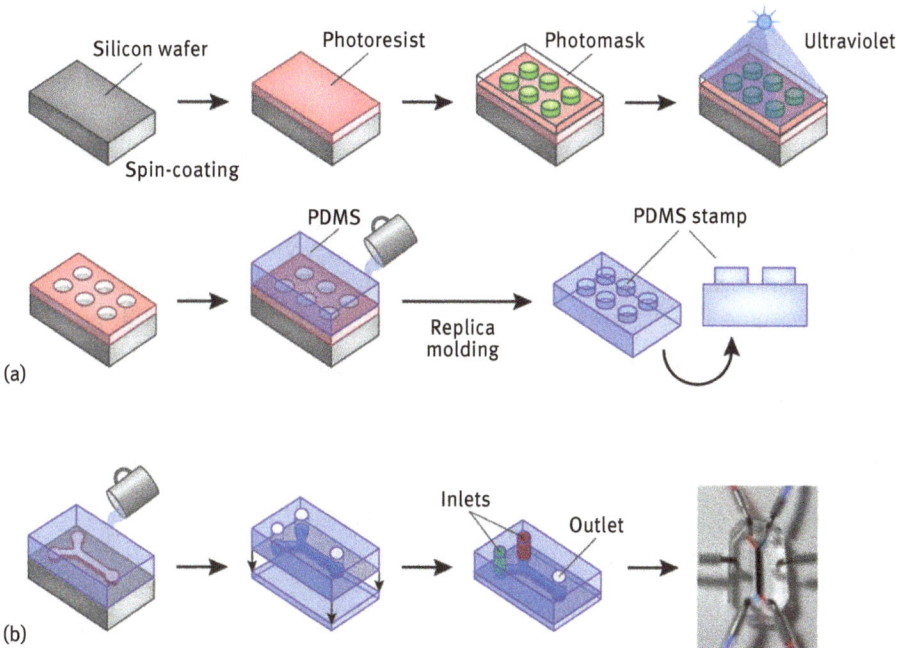

**Figure 3.20:** Replica molding: and example applied to PDMS. Replica molding creates stamps with
shapes complementary to patterns etched in silicon chips by photolithography. A thin uniform film of
a photosensitive material (photoresist) is spin-coated on a silicon chip, which is then overlaid with a
photomask (e.g., a transparent glass plate patterned with opaque chrome layers) bearing a
microscale pattern generated with computer-assisted design software. The photomask protects
some regions of the photoresist and exposes others during exposure to high-intensity UV light. The
UV-exposed material dissolves in a developer solution, leaving the microscale pattern etched into
the photoresist. Elastomeric stamps with a surface topography complementary to the etched surface
are created by a replica-molding technique in which liquid pre-polymer of PDMS is cast on top of
the etched photoresist pattern, polymerized and peeled off. The PDMS stamp can be used for
micro-contact printing of ECM molecules on any substrate, including those within microfluidic
devices (not shown). (b) A single-channel microfluidic device is fabricated by making a PDMS stamp
with two inlets, a single main channel and one outlet and conformally sealing it to a flat glass
substrate. A photograph of a two-chamber microfluidic culture device, with red and blue dye are
perfused through upper and lower channels, is shown at the right. The clear side channels are used
to apply cyclic suction to rhythmically distort the flexible central membrane and adherent cells Fig. 1
in [17], Microfluidic organs-on-chips, ny S.N. Bhatia and D.E. Ingber in Nature biotechnology, Vol.32 n.8
2014 (DOI:10.1038/ntb.2989).

Extracellular matrix (ECM) can undergo to micro-printing by PDMS stamp achieved by replica molding. In this case a single-channel microfluidic device can be fabricated by making a PDMS stamp with two inlets, a single main channel and one outlet and conformally sealing it to a flat glass substrate. A scheme on the replica molding process is reported in Figure 3.20.

### 3.5.1.1 Conclusions
The main common fabrication techniques used to integrate an optical device in a material were presented to allow the reader to understand the wide-spread opportunities that are available to get new materials properties. Some of these techniques, especially when combined to patterning techniques, can be exploited in microfluidics and integrated optics. The recent advances in these fields are amazing and open the way of sustainable miniaturization in compact opto-microfluidics platforms. In this case, it is fundamental to couple the light in an integrated optical circuit embedded and coupled with the microfluidic one. In the next chapter, therefore, some insight into photonics and optical waveguiding will be briefly revised.

## Further reading

Ion Implantation Science and Technology, J.F. Ziegler, 1988, ISBN: 9780323161657

Handbook Of Thin-Film Deposition Processes And Techniques Principles, Methods, Equipment and Applications, Krishna Seshan Ed. ISBN: 0-8155-1442-5, 2002

Sol-Gel Technologies for Glass Producers and Users. Springer, Boston, MA, Aegerter M.A., Mennig M. (eds), ISBN 978-0-387-88953-5 (2004)

http://diffusion.uni-leipzig.de/books.html

Photolithography, Rainer Leuschner and Georg Pawlowski, 2013, DOI: 10.1002/9783527603978. mst0258

## Self Assessment questions

1. What are the main techniques used to deposit a film? Make a list and describe the main features.
2. What is the main difference between physical deposition techniques and chemical ones?
3. Describe the basic principles of sol gel and its preparation steps.
4. What are the main characteristics in the thermal evaporation process?
5. How is the sputtering process achieved and what are the advantages of this technique

6.   What is ion implantation and how is it achieved?
7.   Find the diffusion coefficient of Boron in Silicon at a given temperature and calculate the diffusion time to get 100 nm diffusion profile at that temperature. Determine the decrease of the diffusion coefficient if the temperature is decreased of 50 °C respect the previous value.
8.   What are the main parameters that the thermal diffusion and the ion exchange depend on?
9.   How is it possible to achieve a patterned deposited region?
10.  What is the main difference between replica molding and photolithography?

# 4 Photonics and biophotonics

This chapter provides a basic introduction into optical waveguide theory and discusses relevant fabrication methods and waveguide-based devices so-called photonic integrated circuits (PICs). The PICs are divided into passive and active devices. Passive devices are, for example, co-directional couplers, multi-mode interference (MMI) couplers, arrayed waveguide gratings. Active devices are for example lasers and photodiodes. This chapter provides not only a basic introduction into photonics but also the essentials of the design of different types of waveguides and the way they are fabricated.

## 4.1 Wave guiding basics

An optical waveguide is a material or a portion/region of it where light can be confined and propagated within even for long distances with minimized losses. This can be achieved when this region presents a modification in the refractive index respect the surrounding so that light confinement can be achieved by total internal refraction of the light therein propagating. Typical optical waveguides are optical fibers, that is, wires made of silica-based materials or polymers. Other typology of optical waveguides can be achieved by locally modifying the refractive index of a material in a given region

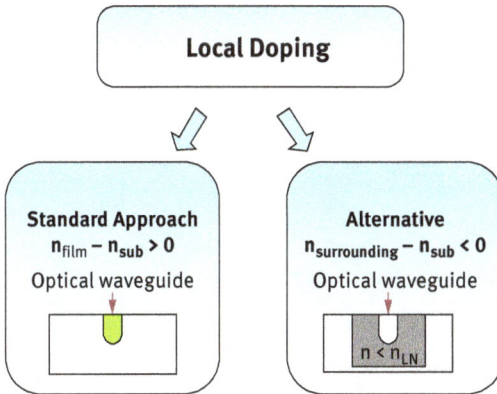

The local doping allows to modify the refractive index in a given region: in the Standard Approach (left) the refractive index is increased in the region where the optical confinement takes place. See green area: section of an optical waveguide, where the beam is propagating along a direction normal to the page sheet plane. As Alternative (right), the optical confinement can be achieved by lowering the refractive index of the surrounding. See white area: same section of the green one, same configuration of beam propagation. The refractive index jump needed to confine the beam is here achieved by lowering the refractive index as depicted by the grey area.

https://doi.org/10.1515/9783110546156-004

(channel or stripes) so that to allow the light confinement by total internal refraction generated at their boundaries. The local doping can increase the refractive index in the waveguiding region, therefore, miming what produced in optical fibers but in a bulky material or it can decrease the refractive index outside the waveguiding region in which the confinement is needed: independently of the approach, the optical waveguide must have a higher refractive index than the surrounding and the geometrical configuration fulfills the conditions to allow total internal refraction.

Optical waveguides are extremely useful devices used for sending optical signals over very long distances with very little attenuation [254]. Although developed and used primarily for use in the telecommunications industry, waveguides, especially fiber optics, have proven to be useful in other areas including biological and chemical sensing [76, 139]. The basic principle of light guiding in optical waveguides is derived from the reflection and refraction principle which is commonly known as Snells law which he discovered in 1621. An incident beam of light will be (in general) partially reflected and partially transmitted as a refracted ray (Figure 4.1).

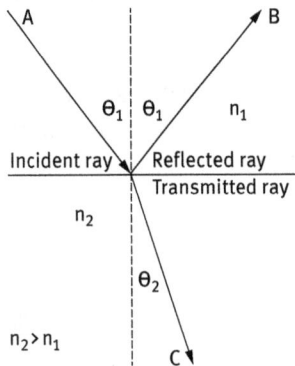

Figure 4.1: The basic principle of reflection.

Snells law can thus be written as:

$$\frac{sin(\theta_1)}{sin(\theta_2)} = \frac{n_2}{n_1}$$

where $n_{1,2}$ are the refractive indices of the media where the light gets reflected and refracted.

The fact that the angle of incidence is equal to the angle of reflection is also known as "the law of reflection". In 1662, Pierre de Fermat added his "principle of least time" theory, which proved that a ray of light passing from a rarer to a denser medium follows the path, which takes the least time. For example, the theoretical direct optical path between points A and C (Figure 4.1) will not take the least time, because the ray AC, will be spending a larger part of its path moving slower in the denser medium.

Fermats theory is calculated as follows:

$$L = \sqrt{a^2 + x^2} + \sqrt{b^2 + (d-x)^2}$$

Taking into account, that the speed of light is constant, results in the minimum time path being the minimum distance path. Setting the derivative of $L$ with respect to $x$ equal to zero leads to:

$$\frac{dL}{dx} = \frac{1}{2}\frac{2x}{\sqrt{a^2 + x^2}} + \frac{1}{2}\frac{-2(d-x)}{\sqrt{b^2 + (d-x)^2}} = 0$$

resulting in:

$$\frac{x}{\sqrt{a^2 + x^2}} = \frac{d-x}{\sqrt{b^2 + (d-x)^2}}$$

which is the "law of reflection":

$$sin(\theta_i) = sin(\theta_r) \rightarrow \theta_i = \theta_r$$

Fermats principle can further be used to derive Snells law. The time required for the light to travel the distance from A to C is calculated as follows:

$$t = \frac{\sqrt{a^2 + x^2}}{v} + \frac{\sqrt{b^2 + (d-x)^2}}{v'}$$

The derivative with respect to the variable $x$, setting $x$ equal to zero and using the previous equations leads to:

$$\frac{dt}{dx} = \frac{x}{v\sqrt{a^2 + x^2}} - \frac{d-x}{v'\sqrt{b^2 + (d-x)^2}}$$

resulting in:

$$\frac{sin(\theta_1)}{v} - \frac{sin(\theta_2)}{v'} = 0$$

which is Snells law.

Assuming that when light is incident upon a medium of lesser index of refraction, the ray is bent away from the normal (opposite to the previous example), so the exit angle is greater than the incident angle. Such reflection is commonly called "internal reflection". The exit angle will then approach 90° for some critical incident angle $\theta_c$. The critical angle is calculated from Snells law by setting the refraction angle equal to 90°. Total internal reflection occurs for incident angles greater than the critical angle.

Optical waveguides operate on the principle of total internal reflection, whereby light in a core material with refractive index $n_2$ is totally reflected at the interface with a substrate (refractive index $n_1$) and cover material (refractive index $n_3$) resulting in light being guided within only the core material.

Snell's Law can be used to show that light within the core material with refractive index $n_2$ can be completely contained, as long as the substrate material (refractive index $n_3$) has a refractive index $n_3 < n_2$, the cover material has a refractive index $n_1 \leq n_3 < n_2$, and the angle $\theta$ of the light is less than a critical angle $\theta_c$ defined by:

$$\theta_c = arcsin\left(\frac{n_3}{n_2}\right)$$

At the interface of the core material and cover material an evanescent field is generated that extends some hundreds of nanometers into the cover medium where light is totally internally reflected. This field is caused by the fact that electromagnetic fields cannot be discontinuous at a boundary, as would happen if there were no evanescent field. Evanescent fields are at the same wavelength as the light used to generate them. The intensity of evanescent fields decays exponentially as the distance from the interface is increased according to:

$$I(z) = I_0 e^{-\frac{z}{d}}$$

where $I_0$ is the intensity of light at the interface, $z$ is the distance from the interface and $d$ is the penetration depth of the evanescent field, defined as the distance from the interface where the amplitude of the electric field has decayed to $1/e$ of its original intensity, and given by:

$$d = \frac{\lambda}{4\pi\sqrt{(n_1^2 sin^2\theta - n_2^2)}}$$

where $n_1 > n_2$, $\lambda$ is the wavelength of the incident light in vacuum.

The electromagnetic waveguide fields, which superimpose in the core of the waveguide, can be written as:

$$exp[-ikn_2(\pm xcos(\theta) + zsin(\theta))]$$

where $k$ is the wavenumber, given by:

$$k = \frac{2\pi}{\lambda}$$

The wave vector component of the propagating wave is given by:

$$\beta = \frac{\omega}{v_{ph}} = kn_2 sin(\theta)$$

where $v_{ph}$ is the phase velocity in propagation direction. Due to the fact that the angle $\theta$ is larger than the critical angle in the case of total internal reflection in waveguides, the condition for the wave vector component can be written as:

$$kn_3 \le \beta \le kn_2$$

Using the previous equations, a so-called "effective refractive index" is defined as:

$$n_{eff} = \frac{\beta}{k} = n_2 sin(\theta)$$

leading to the condition:

$$n_3 \le n_{eff} \le n_2$$

The phase condition at the interfaces of the waveguide of core to upper and lower claddings must be steady resulting in the waveguide dispersion equation:

$$2 \cdot k \cdot n_2 \cdot h \cdot cos(\theta) - 2\phi_2 - 2\phi_3 = 2 \cdot \pi \cdot m$$

where $h$ is the thickness of the core and $m$ is a positive number, $\Phi_{2/3}$ are the phase shifts from core to upper cladding and core to lower cladding, respectively.

To solve the waveguide dispersion equation for the general case, a few more parameters are introduced. The normalized frequency $V$ is given by:

$$V = k \cdot h \cdot \sqrt{n_2^2 - n_3^2}$$

The phase parameter $B$ is defined as:

$$B = \frac{n_{eff}^2 - n_3^2}{n_2^2 - n_3^2}$$

The asymmetry parameter for transverse electric (TE) – and transverse magnetic (TM) – waves is given by:

$$\alpha_{TE} = \frac{n_3^2 - n_1^2}{n_2^2 - n_3^2}$$

and

$$\alpha_{TM} = \frac{n_2^4}{n_1^4} \cdot \frac{n_3^2 - n_1^2}{n_2^2 - n_3^2}$$

For symmetrical waveguides $n_3 = n_1$ and thus $\alpha = 0$. The normalized waveguide dispersion equation can then be rewritten to:

$$V \cdot \sqrt{1-B} = m \cdot \pi + arctan\sqrt{\frac{B}{1-B}} + arctan\sqrt{\frac{B+\alpha}{1-B}}$$

The equation can be solved graphically and is also known as the normalized $\omega$-$\beta$ diagram. For every $V$- and $\alpha$- values (geometry and structure of the waveguide), a corresponding $B$ value can be found. The diagram in Figure 4.2 is only valid for symmetrical waveguides where $n_1 = n_3$ and assuming that the refractive index of the film is approximately the same as the refractive index of the cladding ($n_2 \approx n_1$).

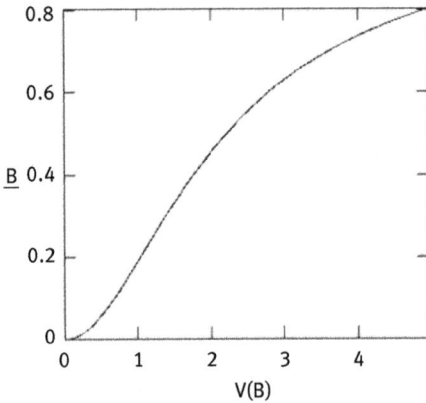

**Figure 4.2:** V(B) diagram for a symmetrical waveguide where $m = 0$ and $\alpha = 0$.

The so-called cutoff frequency $V_0$ for the zero-order mode, where no waveguide mode is able to propagate in the film and leaks into the substrate, is calculated by setting $B = m = 0$:

$$V_0 = tan^{-1}\sqrt{\alpha}$$

Using equations, the previous equations leads to:

$$\left(\frac{h}{\lambda}\right)_0 = \frac{1}{2 \cdot \pi}(n_2^2 - n_3^2)^{-\frac{1}{2}} \cdot tan^{-1}\sqrt{\alpha}$$

The cutoff frequency $V_m$ of the $m$ th order mode is given by:

$$V_m = V_0 + m \cdot \pi$$

To design and fabricate single-mode waveguides, it is essential that the parameter $V/\pi$ does not get greater than 1. The requirement for a single-mode waveguide can then be defined as being:

$$V < \pi$$

This condition can be fulfilled taking three parameters independent of each other into consideration when designing and fabricating waveguides:
- the operating wavelength of the waveguide-based device is chosen large enough
- the dimensions of the waveguide are chosen adequately small
- the refractive index difference between the core and the cladding of the waveguide is chosen sufficiently small

To fabricate a functional monomode waveguide, all these parameters have to be taken into account. The wavelength is in most cases already defined; therefore, it is even more critical to pay attention to the other parameters so as to calculate the condition for the next higher mode. The design parameters for the next higher mode are necessary to account for fabrication tolerances.

The following example demonstrates the straightforward approach when designing single-mode optical waveguides using the effective index method. The effective index method converts the original three-dimensional (3D) channel waveguides into effective two-dimensional (2D) planar waveguides. One 2D waveguide $I$ with a boundary for the light wave in x-direction and a height T; and a 2D waveguide $II$ with a boundary for the light wave in y-direction and a width W.

The normalized frequency in waveguide $I$ is calculated as follows:

$$V_I = k \cdot T \cdot \sqrt{n_2^2 - n_3^2}$$

where $0 < V_I \leq \pi$ (for a symmetrical waveguide $n_1 = n_3$). The normalized frequency in waveguide $II$ is derived as follows:

$$0 < V_{II} \leq \pi$$

$$V_{II} = k \cdot W \cdot \sqrt{N_1^2 - n_3^2}$$

where

$$N_I = \sqrt{n_3^2 + B \cdot \left(n_2^2 - n_3^2\right)}$$

inserting $N_I$ into $V_{II}$ and using the requirement, that $0 < V_{II} \leq \pi$, leads to:

$$0 < \frac{W}{T} \leq \frac{\pi}{\sqrt{B \cdot V_I}}$$

Using the following values obtained from literature as an example, $n_2 = 1.4819$ (modified polymer); $n_3 = 1.4770$ (polymer); $\lambda = 1.55$ µm; $T = 6.45$ µm, results in:

$$V_I = k \cdot T \cdot \sqrt{n_2^2 - n_3^2} = 2.92 \le \pi$$

which is smaller than $\pi$ and thus accomplishes one condition required for single-mode waveguides. From the $B(V)$ diagram (Figure 4.4) $B = 0.617$.

Inserting both values $B$ and $V_I$ leads to:

$$0 < \frac{W}{T} \le 1.37$$

and as $T$ is known, provides the second condition for single-mode waveguides in this example to:

$$0 < W \le 8.84\,\mu m$$

The dimensions of the fabricated waveguide have been determined to be:
–   Height: 6.45 μm
–   Width: 8.05 μm

These dimensions fulfill the criteria of single-mode waveguides as calculated.

Table 4.1 provides some examples of refractive indices of common substances for comparison. The refractive index is wavelength dependent; this is why the associated wavelength is provided next to the refractive index.

**Table 4.1:** Refractive indices of common materials at specific wavelengths.

| Substance | Refractive index (wavelength in [nm]) |
| --- | --- |
| Water | 1.3330 (589) |
| Benzol | 1.5013 (589) |
| Ethanol | 1.3617 (589) |
| Air at a pressure of 1013 mbar | 1.000272 (589) |
| BK1 glass | 1.51 (589) |
| PMMA | 1.4931 (589) |
| Diamond | 2.4170 (589) |
| Fused silica | 1.46 (633) |
| Microscope slide | 1.51 (633) |
| Gallium Arsenide (GaAs) | 3.6 (900) |

Another important aspect, which needs consideration, is the losses in waveguides. When designing a waveguide for a certain application, it is necessary to know the requirements. One of the requirements is the wavelength, which defines the waveguide material. The material already specifies the loss range of the waveguide, as one cannot get lower than this value. Additional waveguide losses can be categorized into three mechanisms:

- Waveguide loss through scattering at waveguide boundaries;
- Waveguide loss by absorption;
- Waveguide loss due to waveguide bends (e.g., in y-couplers or ring resonators).

Scattering is usually the dominant effect in glass or dielectric waveguides; whereas, absorption is predominantly present in semiconductors and crystalline materials. The quantum mechanical model is used to describe the loss mechanisms in waveguides, that is, a ray of light is not seen as a set of electromagnetic waves, but photons. Photons can be scattered and absorbed while passing through the optical waveguide and thus "contribute" to the waveguide loss. If photons get scattered, they only change their direction and do not loose their identity. The scattering at surfaces or interfaces can be categorized into two mechanisms, volume scattering and surface scattering. Volume scattering is mainly caused by impurities in the waveguide material itself. Volume loss can be derived in a straightforward manner, as the loss per unit length is proportional to the number of impurities per unit length. An important aspect to consider when designing waveguides is the fact that the volume loss is strongly dependent on the wavelength used. Impurities are usually considered as extremely small compared to the wavelength used and, therefore, can be neglected compared to surface scattering.

Surface scattering can also occur at very smooth surfaces or interfaces, especially in the case of higher order modes where the surface/interface and the propagating wave are in interaction with each other. The so-called film wave or core wave undergoes several reflections at the interface to the cladding and the substrate. The number of reflections $N_R$ taking place during a certain waveguide length $L$ can be calculated by:

$$N_R = \frac{L}{2 \cdot h \cdot \cot \kappa}$$

where $\kappa = 90° - \theta$

Loss can occur at every reflection at the interface. Higher order modes have a larger angle $\kappa$ and are, therefore, subject to higher losses. To quantify the waveguide loss, another parameter is defined, the waveguide loss coefficient $\alpha_{loss}$, which is directly proportional to the number of reflections at the surface/interface. The light wave intensity (energy per unit area) at every position of the waveguide is given by:

$$I(z) = I_0 \cdot e^{-\alpha z}$$

where $I_0$ is the light wave intensity in mW coupled into the waveguide at $z = 0$, $\alpha_{loss}$ is the waveguide loss coefficient [1/cm] and $z$ the length of the waveguide in cm.

The unit of the waveguide loss $D$ is usually decibel [dB]. $D$ is given by:

$$D = -10 \cdot log\left(\frac{I}{I_0}\right)$$

Absorption losses in polymer waveguides [285] occur mainly due to:
- Harmonic oscillations of polymer molecules, which consist of OH and CH bonds
- Impurities of atomic size, for example, such as metal ions whose absorption spectrum can effect the entire wavelength spectrum used in the waveguide

Waveguide losses also occur due to bends or curvatures in waveguide-based devices. These losses are due to optical distortion of the wave and result in scattering. These bend losses are waveguide-specific and define the smallest radius of curvature possible with a given waveguide structure. Which type of waveguides is to use for small bends is discussed later in this chapter. Bend losses depend exponentially on the bend radius as well as the number of reflections on the surface/interface of the wave guiding layer.

Extracting the waveguide loss and other relevant parameters from optical measurements is dealt with in the following paragraph.

## 4.2 Characterization of optical waveguide based-devices

A high end setup for the measurement of optical waveguide-based devices usually consists of the following devices.

An external cavity laser (ECL) is used as the light source, which is connected via a polarization controller (PLC) and a tapered or lensed fiber to the input waveguide. The xyz-axis is controlled with the help of a piezo-controller, which has sub-µm accuracy. The device under test (DUT) is monitored by a microscope with CCD camera to achieve optimum fiber-waveguide coupling. The waveguide-based device is placed on a Peltier element, which controls the temperature making sure, that the DUT is characterized at the same temperature throughout the measurement.

The transmitted signal is detected using, for example, the lock-in technique employing a photo diode and a lock-in amplifier. The advantage of employing the lock-in-technique is the suppression of unwanted noise. The optical signal, which is coupled into the waveguide is made periodic, for example, with the help of a chopper. If a laser is used, it can alternatively be modulated at a certain frequency, which the lock-in amplifier is able to detect. In this way, only the light, which is coupled into the waveguide, is detected and no interference with other effects or light sources takes place.

The light is coupled into the input waveguide using, for example, a tapered fiber, which can be adjusted by a three-axis piezo-drive. Tapered fibers are usually used to couple into strong guiding waveguides and reduce the fiber-to-chip coupling loss considerably compared to a standard butt fiber due to the reduction of the mode field diameter, which suits the high index contrast waveguides better than the large field diameter of the butt fiber. The near field of the output waveguide can then be focused on the photo diode or detector using a microscope lens, which has a sufficient aperture to guarantee correct power measurement. Instead of using a microscope lens on the output side, a tapered fiber can also be used to direct the light wherever needed, for example, to an optical power meter or an optical spectrum analyzer.

This setup can now be used to extract various device parameters. The coupling losses can, for example, be calculated from measuring the loss of different waveguide lengths of the same device. This internationally recognized reference test method is known as the cut-back technique. The measured insertion loss is plotted against different lengths of the measured devices. The slope of the obtained curve is determined by the propagation losses of the waveguide. The coupling losses are obtained from the point, where the extrapolated measurement curve crosses the y-axis at a device length of zero (unit length). The cut-back method is a destructive method to obtain the propagation and coupling losses of straight waveguides.

A non-destructive method for measuring the waveguide loss is the prism-sliding method. The prism-coupling assembly consists of input and output prisms. The waveguide loss can be determined with this setup for every mode, which is able to propagate in the waveguide. The modes are selected by the coupling angle chosen to couple the light into the prism and the waveguide.

The position of the input prism is kept the same throughout the measurement so that the input-coupling coefficient is constant. The output prism is moved toward the input prism. This is because while sliding the output prism a possible degradation of the waveguide might happen by slightly damaging the surface. Therefore, moving the output prism toward the input prism will not affect the loss measurement. It is recommended to make as much measurements as possible as this leads to more accurate results. Immersion oil is used between the prisms and the wave guiding layer to enhance the light coupling from the prisms into the waveguide. The prism-coupling angle determines the propagating mode in the waveguide. The zero-order mode (m = 0) or fundamental mode is usually chosen as this is the propagating mode having the highest intensity. The fundamental mode is found by varying the coupling angle. If more than one maximum is found, the fundamental mode is the one having the smallest coupling angle. Once the fundamental mode is found, the prism arrangement is adjusted and the measurement is started. Depending on the length of the waveguide, the data are collected and the waveguide loss is calculated by plotting the length of the waveguide on the x-axis and the measured loss on the y-axis. The slope is the waveguide loss.

Another nondestructive method, which is commonly used, is the Fabry-Perot method. Kaminow and Stulz [133] have first introduced this method in 1978. The chip is regarded as a Fabry–Perot resonator for waveguide losses < 1 dB/cm, where the facets of the chip serve as the mirrors of the resonator. The optical wave is reflected back and forth within the waveguide. The resulting "filter" spectrum depends on the intrinsic losses and the reflection factor of the facets. Varying the temperature of the whole chip or the wavelength exploiting the group velocity dispersion changes the optical length. The following calculations closely follow [77, 219]. The transmission spectrum of a Fabry–Perot resonator is given by:

$$\frac{I_T}{I_0} = \frac{\eta T^2 \exp(-\alpha L)}{\left(1 - \tilde{R}\right)^2 + 4\tilde{R}\sin^2\left(\frac{\phi}{2}\right)}$$

$$\tilde{R} = R\exp(-\alpha L)$$

Here, $I_T$ is the transmitted intensity, $I_0$ is the intensity of the input laser signal, $\eta$ is the coupling efficiency to the fundamental waveguide mode, $\phi = 2\beta L$ is the internal phase difference, $R$ is the end-face reflectivity, $L$ is the length of the resonator and $\alpha$ is the insertion losses of the waveguide. The contrast $K$ of the Fabry–Perot resonances, which is independent of $I_0$ and $\eta$ is given by:

$$K = \frac{I_{max} - I_{min}}{I_{max} + I_{min}}$$

The transmitted maximum and minimum intensities can be described by:

$$I_{min} = \frac{(1 - R)^2 \cdot \exp(-2\alpha L)}{(1 + R \cdot \exp(-2\alpha L))^2}$$

$$I_{max} = \frac{(1 - R)^2 \cdot \exp(-2\alpha L)}{(1 - R \cdot \exp(-2\alpha L))^2}$$

Using the previous equations, $K$ can be rewritten as:

$$K = \frac{2\tilde{R}}{1 + \tilde{R}^2}$$

leading to:

$$\tilde{R} = \frac{1}{K}\left(1 - \sqrt{1 - K^2}\right)$$

The value of $K$ is taken from the measurement of the Fabry–Perot resonances. The maximum value of $K$ is obtained for lossless waveguides. Increasing losses on the other hand reduces the contrast.

If the end-face reflectivity is known, the insertion loss of the device is given by:

$$\alpha = \frac{4.34}{L}\left(\ln R - \ln \tilde{R}\right)$$

This equation can also be written using $I_{min}$ and $I_{max}$ directly:

$$\alpha = \frac{1}{L}\left(\ln\frac{1+\sqrt{u}}{1-\sqrt{u}} + \ln R\right)\left[cm^{-1}\right]$$

$$u = \frac{I_{min}}{I_{max}}$$

When the reflection coefficient of the chip is unknown, it can be calculated from the insertion loss measurements plotted for different device lengths using the previous equation. The measurement is approximated by a straight line, from which the reflection factor can directly be taken. The slope of the curve is again the value for the intrinsic losses $\alpha$ of the waveguide, which is given by:

$$\ln\left(\frac{1+\sqrt{u}}{1-\sqrt{u}}\right)$$

in [dB/cm].

When the reflection coefficient of the end faces of the waveguide is known, the intrinsic losses of the chip are easily determined without the necessity to cut the waveguide several times for characterization as has to be done using the cut-back-method.

Another way of calculating the reflectivity of the end faces, which is valid for weak-guiding waveguide structures, is using:

$$R = \left(\frac{n_{eff} - 1}{n_{eff} + 1}\right)^2$$

In this case, the value of the effective refractive index needs to be known to determine the reflectivity. The effective refractive index of materials can be measured using a so-called Abbe refractometer (e.g., www.kruess.com for available instruments).

The refractive index of waveguides can be determined by so-called m-line spectroscopy. This method relies on the same principle as the prism sliding method as described before. The refractive index and/or the thickness of the waveguide are obtained by determining the effective refractive index $n_{eff}$. Laser light is coupled into the waveguiding layer with the help of a prism coupler. The coupling angle $\overline{\alpha_m}$ is measured. In the case of a single-mode waveguide, only the thickness or the refractive index can be derived. To calculate both values, several modes (at least two) need to be able to propagate the waveguiding layer. The accuracy of the obtained results increases with the number of modes able to propagate. The effective refractive index is calculated using:

$$n_{\text{eff}} = n_P \left[ \varepsilon + \sin^{-1} \left( \frac{\sin \overline{\alpha_m}}{n_P} \right) \right]$$

In addition to the input coupling angle $\alpha$, also the angle $\varepsilon$ and the refractive index of the coupling prism $n_p$ need to be known.

A simplified measurement arrangement is achieved, where laser light is directly coupled into the waveguiding layer and an output prism is used to guide the supported modes onto a screen.

The effective refractive index for each supported mode can be calculated using the previous equations.

Once $n_{\text{eff}}$ is known, the thickness of the waveguiding layer and/or the refractive index of the layer can be calculated. There exist three cases as follows:
– the waveguide only supports one mode and only the refractive index of the waveguide or the thickness is known.
– the waveguide supports two modes. In this case both the refractive index $n_f$ and the thickness $h$ can be determined.
– The waveguide supports more than two modes. Both the refractive index $n_f$ and the thickness $h$ can be determined with a high accuracy.

All three cases rely on the wave equation, which is given by:

$$2 \cdot \pi \cdot B \cdot h - m \cdot \pi = \phi_2' + \phi_3$$

and

$$n_f^2 = B^2 + n_{\text{eff}}$$

the angles $\phi_2'$ and $\phi_3$ are calculated as follows:

$$\phi_2' = \phi_2 + \sin(2\phi_c)\cos\left(2\phi_p\right)\exp(-4\pi P_2 s)$$

where

$$\phi_2 = \tan^{-1}\left(\frac{P_2}{B}\right)$$

$$\phi_3 = \tan^{-1}\left(\frac{P_3}{B}\right)$$

$$\phi_p = \tan^{-1}\left(\frac{P_2}{\sqrt{n_p^2 - n_{\text{eff}}^2}}\right)$$

$$P_2^2 = \left(n_{\text{eff}}^2 - n_2^2\right)$$

$$P_3^2 = \left(n_{\text{eff}}^2 - n_3^2\right)$$

$s$ is the distance of the prism from the waveguiding surface. The distance $s$ is usually neglected as, for example, in a practical measurement setup the material is pressed against the prism with the help of compressed air.

The solutions to the equations need to be found in an iterative manner if more than one mode is supported in the waveguide. A certain range for the refractive index $n_f$ of the guiding layer is assumed and inserted, giving:

$$2 \cdot \pi \cdot B \cdot h - m \cdot \pi - \phi_2' - \phi_3 = R$$

now $R \rightarrow 0$ by applying different refractive indices for $n_f$.

Both the prism sliding method and m-line spectroscopy can only be applied to surface waveguides without cladding. This, however, does not impose a difficulty when dealing with other types of waveguides as the basic material and waveguide properties can be derived from these fundamental measurements.

Other waveguide geometries will be dealt with in the following paragraph providing an insight into the numerous possibilities of designing waveguides.

## 4.3 Types of waveguides

Waveguides can be categorized in several different ways. In this context, the heading of this chapter: types of waveguides, refers to the 3D geometrical layout of the waveguide itself. From the previous chapter, the waveguiding principles, the essence of what is a waveguide and how wave guiding is realized are known. Waveguides can be fabricated in a vast number of materials and material combinations. The questions one has to solve as a waveguide designer and eventually as the one who needs to fabricate the devices are:

- What is the used wavelength? What are the used wavelengths?
- What is the targeted application?
  - This in combination with the known wavelength defines suitable waveguide materials or hybrid solutions
- Is it an active or passive device?
  - This defines if photodiodes, lasers or only trimming solutions need to be incorporated.
- What is the form factor of the device?
  - This defines the size of the device and eventually the type of waveguide used.

Once these basic questions have received answers, a suitable waveguide platform is chosen. A simple waveguide is a so-called slab waveguide, which consists of three layers. The waveguide is a 1D structure, which is referred to as planar structure. Light can propagate in the core in the entire area as the waveguide is only limited in z, but not in x and y. The slab waveguide serves as the basis for the channel waveguide, which is a 2D waveguide and referred to as wire or stripe structure. The light is confined to the core of the waveguide, although the evanescent field is present in both the lower and upper claddings. To concentrate the light further in the core of the waveguide and eliminating the evanescent field in the other layers, another type of waveguide is used, the so-called ridge waveguide. Here, the waveguiding layer is sandwiched between additional layers, which separate the core from the lower and upper claddings, thus minimizing the evanescent fields in these layers. Ridge waveguides are, for example, used for lasers.

Some of the most common channel waveguides are shown in Figure 4.3. The embedded waveguide is one of the simplest waveguide forms, which is, for example easy, to fabricate in polymers as only two types of materials or material modifications are needed. The raised strip, rib, ridge and strip loaded waveguides can be made with and without upper cladding and are used to concentrate the guided light in the core. With these types of waveguides, strong guiding of the light is obtained leading to smaller radius of curvature for waveguide-based devices and thus smaller form factors. The indiffused or graded index waveguide is obtained by modifying the surface of a material as described in Chapter 2 to locally increase the refractive index and create waveguides.

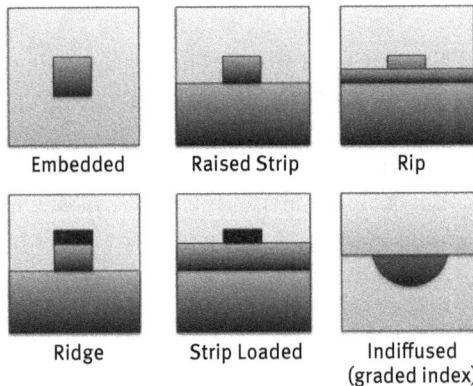

Embedded          Raised Strip          Rip

Ridge          Strip Loaded          Indiffused          **Figure 4.3:** Common types of channel
                                       (graded index)     waveguides.

The type of waveguide geometry chosen defines the way optical modes propagate along the waveguide. The numbers of modes able to propagate along a waveguide have been derived in the previous section of this chapter. Maxwells equations are used to calculate the mode patterns [117]. The transverse modes can be classified into the following types:

- Transverse electric (TE) modes do not have an electric field in the direction of propagation.

- Transverse magnetic (TM) modes do not have a magnetic field in the direction of propagation.
- Transverse electromagnetic (TEM) modes do neither have an electric nor a magnetic field in the direction of propagation.

As an example of modes in 2D, some of the common modes are shown in Figure 4.4. The modes are denoted $TE_{MN}$ where $M$ and $N$ are the radial and angular mode orders. $M$ and $N$ are integers.

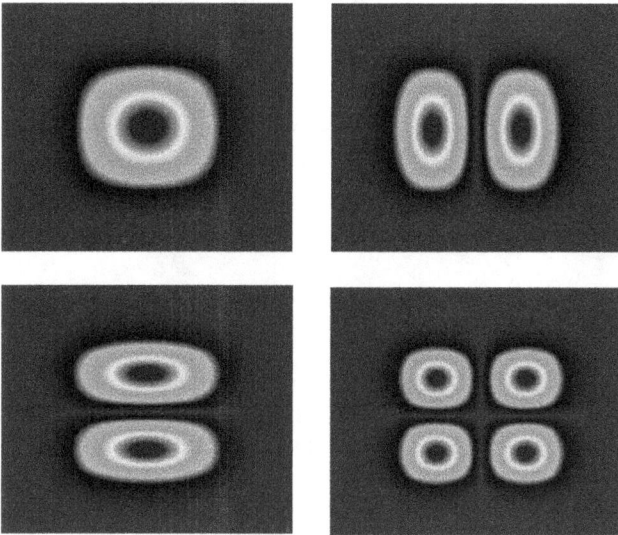

**Figure 4.4:** 2D plot of the TE00, TE10 mode, TE01 mode and TE11 mode (clockwise).

The TE and TM optical modes are both able to propagate in waveguides. To eliminate the so-called polarization effects, both modes need to be able to propagate equally well. This is realized in symmetrical waveguides. From a practical point of view, this needs to be taken into consideration. For example, rib or ridge waveguides as can be seen in Figure 4.3 are asymmetrical waveguides where either TE or TM modes are preferred. The knowledge of polarization sensitive and insensitive design can be used advantageous when choosing waveguide geometries for specific applications.

## 4.4 Types of optical fibers

Optical fibers are nowadays available in several materials and different geometries suitable for different types of applications ranging from telecommunication to spectroscopy. The main material of choice for fiber optics is of course glass. Main application for glass fibers is their use in fiber optic backbone networks for telecommunication. In the optical fiber attenuation curve of a glass fiber, there are minima in the transmission loss at wavelengths of 1,300 nm and 1,550 nm. This is one reason why these wavelengths were chosen as telecommunication wavelengths.

Attenuation of the light in an optical fiber can have several causes. One of the main contributions comes from linear scattering at inhomogeneities in the molecular structure of the fiber core, which is also known as Rayleigh scattering. It can be explained by local fluctuations of the density in an optical fiber. The attenuation is a ~ $1/\lambda^4$. The lowest value, as described before, is at a wavelength $\lambda \approx 1.5$ µm. Another scattering effect at optical inhomogeneities in the magnitude range of the wavelength is known as Mie scattering. Attenuation mechanisms in optical fibers can be reduced through technological measures, for example, the content of OH groups can be lowered resulting in low attenuation values in the near infrared (NIR) range. Fibers are selected according to their properties, which must suit the application. A wide variety of optical fibers exist and in the following paragraph a brief overview of the types of fibers available is given.

There are three major fiber types:
- Step index fibers
- Graded index fibers
- Single-mode fibers

In step index fibers, a significant number of modes are guided in the core due to the larger core diameter and/or a high refractive index difference $\Delta n$ between the core and the cladding compared to single-mode fibers. There exists nowadays a large variety of such fibers. A selection of material choices for realizing step index fibers is provided in Table 4.2.

**Table 4.2:** Selection of examples for step index fibers. (Source: Leoni Fiber Optics GmbH)

| Optical waveguide | Core material | Cladding material |
| --- | --- | --- |
| – POF | – PMMA | – Fluorinated PMMA |
| – PCF | – Fused silica glass | – Plastic (acrylate) |
| – Silica fibers (low OH,– high OH) | – Fused silica glass | – Fused silica glass |
| – Glass fibers | – Fused silica– or composite glass | – Doped silica or– composite glass |
| – MIR fibers | – Special glass– (fluoride glass,– chalcogenide glass) | – Special glass |

Step index fibers are robust and due to their relatively large cross-section of the core, it is easier to couple light into the fiber.

In graded index fibers, the graded refractive index profile minimizes the differences in delay times for all optical modes and provides an improvement in the bandwidth for certain wavelength ranges. This is one reason why they are chosen for telecommunication applications. The profile of the refractive index in the core can be regarded as parabolic. Graded index fibers are also used in sensor applications as the gradient of the refractive index can be used advantageously in creating a tailored evanescent field essential in these devices.

In a single-mode fiber, which is also known as a mono-mode fiber, optical signals are transferred through the fiber by transmitting the optical signals only in the fundamental mode of the fiber. The fundamental mode is the only mode capable of being propagated in the fiber. Single-mode fibers are preferred for applications involving long distances and wide bandwidths, as they provide the least amount of signal distortion.

A standard fiber used in telecommunication application is the monomode fiber E9/125. The fiber is used for wavelengths of 1,300 nm and 1,550 nm. The mode field diameter at a wavelength of 1,300 nm is 9.2 ± 0.4 µm and the cladding diameter is 125 ± 1 µm. The coating diameter is 245 ± 5 µm. The datasheet of this type of fiber is given in Table 4.3.

**Table 4.3:** Transmission properties of monomode fiber type E9/125 (Source: Leoni Fiber Optics GmbH)

| Transmission properties | Fiber type A for semi-tight and tight buffered fibers acc. to ITU-T G.652.D and ISO 11801 type OS 2 | |
| --- | --- | --- |
| Wavelength (nm) | 1,310 | 1,550 |
| Attenuation max. (dB/km) | 0.38 | 0.28 |
| Dispersion coefficient max. (ps/nm · km) | 3.5 | 18 |
| Zero dispersion wavelength (nm) | 1,300–1,322 | |
| Dispersion slope (ps/nm2 · km) | ≤ 0.092 | |
| Cutoff wavelength (cabled) (nm) | ≤1,250 | |
| Mode field diameter at 1,310 nm (nm) | 9.2 ± 0.4 | |
| Polarization mode dispersion (ps/sqrt(km)) | ≤ 0.1 | |
| Effective group of refraction | 1.4695 | 1.4701 |

Similar fibers are available for different applications and of course different wavelength ranges.

This is only a brief introduction into the vast area of fiber optics. Due to the tremendous applications and research pursued in this field not only in telecommunications but also in sensor technologies, further reading is recommended targeted at the specific needs of the reader. Fundamentals of optical fibers is described, for example, in [126] as well as in [39].

The following chapter deals with the fabrication methods and tools for the realization of waveguides and waveguide-based devices. The fabrication of optical fibers is beyond the scope of this book.

## 4.5 Fabrication methods for optical waveguides

There are basically two methods for creating waveguide layers. The first method is to produce the layers needed for fabricating waveguides beforehand, for example, by epitaxial steps or spin-coating several layers of polymers after each other. This means that the layers, which make up the waveguide, are basically present already, only the geometry of the waveguide (length and width) has to be defined with, for example, lithography and/or etching steps. The other method to create the layers for the waveguide is sequential by adding a layer as required and having other fabricating steps like lithography and/or etching steps in between.

Methods for fabricating waveguides for PICs, which can be applied for optofluidic systems have been described in great detail in the last years in literature. Some of the methods are [117]:

- Deposited thin films (glass, nitrides, oxides, polymers also organic polymers which are ideal for manufacturing and integrating organic lasers and organic photodiodes (OPDs) with optical waveguides)
- Photoresist films
- Ion bombardment in glass
- Diffused dopant atoms
- Heteroepitaxial layer growth
- Electro-optic effect
- Metal film stripline
- Ion migration
- Reduced carrier concentration in a semiconductor
    - Epitaxial layer growth
    - Diffusion counterdoping
    - Ion implantation counterdoping or compensation

This chapter describes the fabrication methods and tools for realizing waveguides and waveguide-based devices. The chapter starts with an introduction into basic micro-system technologies such as lithography, and continues with an introduction into replication technologies, ending with dry etching and similar fabricating steps.

### 4.5.1 Lithography

Lithography has its origin in the two Greek words, "lithos" meaning "stone" and "graphein" meaning "write", leading to "pattern writing in stone".

As already described in Chapter 3, a pattern on a mask is transferred in today's semiconductor or photonics industry manufacturing processes to a surface. Optical radiation is commonly used to image the mask on a semiconductor or polymer substrate either as shown also in the special case of DUV radiation, directly or with the help of a so-called photoresist. The photoresist is spun with the help of a spin-coater on to the substrate forming a thin layer on the surface. Most semiconductor processes nowadays use a positive photoresist, leaving a "positive" image of the mask pattern on the surface of the substrate. The photomask is usually made out of soda lime, borosilicate glass, or fused quartz. Advantage of using quartz is the fact that it is transparent to the DUV radiation and has a low thermal coefficient of expansion. This is why for the DUV process described earlier; the photomask is made out of quartz.

The goal of the lithography community is to improve the resolution of the lithography system. Two approaches are being pursued: either use optical or processing techniques such as increase the optical aperture of the used optics in lithography systems or use double-patterning techniques; the other approach is decrease the wavelength of the light used in patterning (mercury lamp i-line: 365 nm; ArF laser-DUV 193 nm; EUV 13.5 nm ... ) as already quoted.

### 4.5.2 Replication

The emergence of optofluidic devices is largely enabled by recent advances in microfabrication, microfluidics, and polymer-processing technologies. The methods of choice are micromachining, soft lithography, and embossing techniques, which enable the fabrication of micron-scale fluidic channels in silicon, glass, polymer, and elastomer materials.

Polymers have also been accepted as the material of choice for the integration of photonic-integrated circuits and fluidic devices, mainly due to their increasing performance, rapid processability, capability for precise tailoring of their optical properties, and their comparatively low cost. Another important aspect is the biocompatibility of polymer materials and the fact that these materials are already in use in many bio and non-bio laboratories, which increases the acceptance of polymer-based optofluidic devices. This advantage requires the improvement of fabrication technologies as well as the development of application-specific tailored materials. As stated before, polymer optical waveguides have been fabricated by various techniques, such as dry etching, UV curing, and soft lithography replica molding [94], and embossing.

To manufacture rib waveguide structures, several process steps are necessary [32–36]. Figure 4.5 summarizes the technology.

The template is fabricated by spin-coating photoresist onto a silicon substrate coated with a metal seed layer, and patterned using standard UV lithography (a). The molding tool can then be electroplated with nickel (b). It is necessary to deposit the metal seed layer for the electroplating before spin-coating the substrate to avoid metal deposition on the sidewalls. This can lead to voids in extremely narrow structures during the growth of the nickel. The electroplated template, or shim, is then fixed into a hot embossing machine (c).

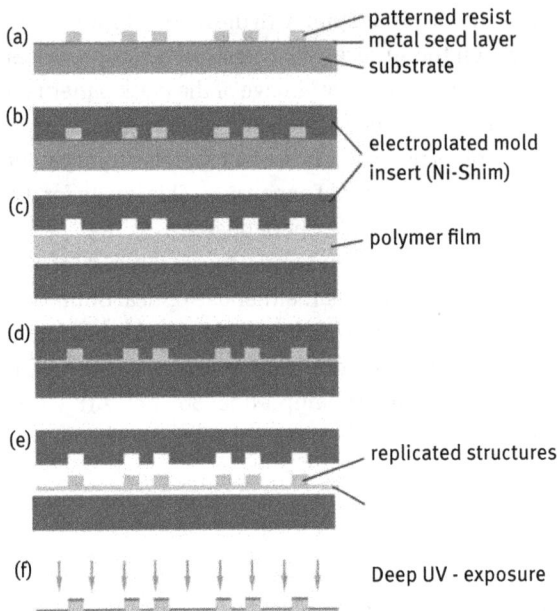

(a) — patterned resist — metal seed layer — substrate

(b) — electroplated mold insert (Ni-Shim)

(c) — polymer film

(d)

(e) — replicated structures

(f) Deep UV - exposure

**Figure 4.5:** Process steps for the fabrication of rib waveguides [32–36].

Shim fixation requires a special tooling to attach it to the upper plate of the hot embossing machine, as the shim is relatively thin (typically between 300 and 600 μm) and cannot be clamped from the side. Ni-shims used for component replication described in this work were adhesively joined to a counterplate. Special care has to be taken to maintain parallel surfaces between the tool and the polymer.

After inserting a thermoplastic polymer film into the setup, the molding tool is evacuated, heated above the softening temperature $T_g$ of the polymer, and pressed into the film. The microstructures are filled with polymer material and are replicated in detail (d). The setup is then cooled down and the tool is withdrawn from the polymer (e). The residual layer, which carries the structures and serves as a substrate for the following process steps is minimized during the process to reduce the effect of stress due to the different thermal expansion coefficients of the tool and the polymer. To induce waveguide structures, the replicated rib structures are modified by flood exposing them to deep UV (DUV) radiation (refer Section 2.1.5 for details), which increases the refractive index of a thin surface layer of the polymer.

For the fabrication of the mold insert, AR-N 4400 negative tone photoresist from Allresist was coated using a Hamatech spin coater on a 4" standard silicon wafer. The silicon wafer was pre-sputtered with an approximately 300-nm-thick titanium layer, which serves as the seed layer for the electroplating process. The resist was patterned using standard UV lithography on a commercial maskaligner (EVG620, EV Group). Electroplating was carried out in a nickel sulfamate electrolyte at a temperature of 52

°C and a current density of 1 A/dm². As waveguide material, poly(methyl methacry-late) (PMMA) from Notz Plastics (Brandname Hesaglas VOS) was hot embossed in a custom-made hot embossing machine (Jenoptik AG) at a temperature of 140 °C and a pressure of 90 kN. The replicated structures were then flood-exposed using the EVG620 maskaligner with a DUV configuration with a dose of 5 J/cm². Exposed light intensity was measured by an UV intensity meter (Model 1000, Karl Suss) at 240 nm. The structures were characterized using a Santec Laser TSL-210 at 1,550 nm in combination with an IR-Camera 13215 (Electrophysics Corporation) and a Hewlett-Packard 8163 lightwave multimeter. The photograph of replicated rib waveguides by hot embossing is shown in Figure 4.6.

**Figure 4.6:** Rib waveguides replicated by hot embossing [32–36].

The ideal geometry of the rib waveguide was determined by simulation. Waveguides and photonic components such as MMI couplers [141, 212] have been simulated (refer Section 4.6 for details). All numerical simulations of waveguides were performed at a wavelength of 1,550 nm using the beam propagation method (BPM) in the commercial software package BeamPROP™. Due to the absorption of the DUV radiation in the material, the refractive index cannot be assumed as a step function but must be described as an exponential decay of the refractive index from the maximum value at the top of the waveguide to the refractive index of the substrate. A corresponding model for a rib waveguide with a gradient index profile is chosen. The profile was designed utilizing BeamPROP's profile type "diffused" with a diffusion ratio set to the minimal value of 0.001 to describe a 1D exponential decay. The refractive index of the substrate material was taken to be 1.4797.

Critical geometry values for rib waveguides are width and height of the rib structure. As we aim to fabricate single-mode waveguides, the width has to be in a range that allows the propagation of only the TM0 mode. Rib waveguides, which are smaller than the width at the cutoff wavelength of this mode will not allow light to propagate, while ribs which are too wide will lead to a multi-mode behavior. A normalized B-V diagram was obtained from simulation. The cutoff wavelength for the TM0 mode was calculated to be between 10 and 11 µm. The upper limit for single-mode behavior is calculated to be a width of the rib of about 20 µm. A width of 17 µm was chosen for input/output waveguides and S-bends (S-bends are waveguides, which look like an S from the top). The rib height was determined by measuring the spread of the evanescent field in the intensity profile. The distance between the top of the rib and the surface of the substrate has to be larger than this specific calculated distance to avoid light coupling (refer Figure 4.5). The simulation shows that a minimum rib height of 17 µm is needed to fulfill this condition. To compensate computing errors and tolerances in the manufacturing process, a minimum structure height of 19 µm was chosen.

Based on the results for straight waveguides, a 1 × 2 MMI coupler was designed to serve as a 3 dB splitter. Figure 4.7 shows the intensity distribution of the field in the multi-mode waveguide.

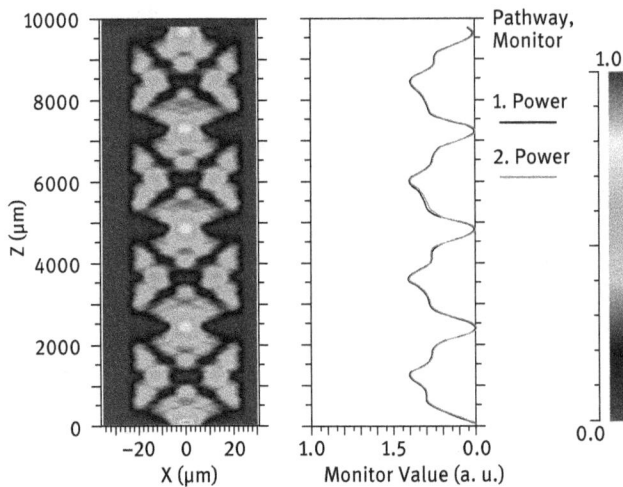

**Figure 4.7:** Simulation of intensity distribution in a MMI coupler at a wavelength of 1,550 nm [32–36].

It is evident that the pattern is repeated continuously along the z-axis (self-imaging). The optimal length of the MMI is reached when two maxima of the same intensity can be found along the x-axis. For our design, the length of the MMI was chosen to be 1,213 µm. At this point, both output waveguides reach a maximum in output power of 42% of the input power.

A single-mode profile of the embossed waveguides has been obtained for a width of 10 µm. The near-field photograph of the rib is shown in Figure 4.8. It can be noticed from the photograph that the intensity distribution along the x-axis is symmetrical, while the distribution along the y-axis shows a non-symmetrical spread of the evanescent field into the rib.

For comparison, waveguides with a width of 17 µm show a slight multi-mode profile. This does not match the simulation results, but can be explained by process deviations and imperfection of the model. However, further effort is necessary both on experimental and simulation side to reduce this divergence. No coupling of light into

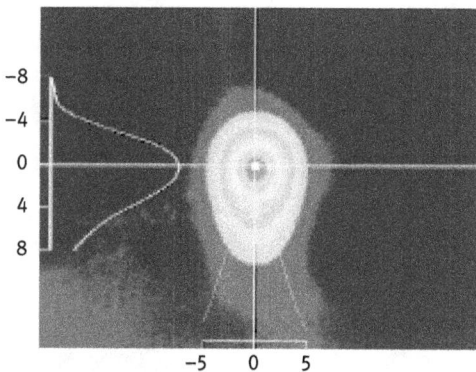

**Figure 4.8:** Near-field photograph pattern of rib waveguide of 10 µm width. All units are in µm [32–36].

the substrate surface was observed, which means that the rib height was chosen correctly.

The insertion loss of a waveguide of length 17.4 mm at a wavelength of 1,550 nm was determined using the cut-back method to be 4.9 dB including fiber-chip coupling loss. The fiber-to-chip coupling loss is calculated to be 3.0 dB for both facets, resulting in a waveguide loss of 1.1 dB/cm and a fiber-to-chip coupling loss of 1.5 dB per facet. The waveguide loss is comparable to planar waveguides fabricated with DUV technology, which demonstrates the high quality of the nickel shim and the potential for this technology. The insertion loss of the 3 dB 1 × 2 MMI devices measured at each of the two output ports is 9 dB. This means that MMI couplers have 1.1 dB additional losses compared to a rib waveguide of the same length. Expected additional loss derived from the simulation is 0.8 dB, the slightly higher measured value is a result of the different output geometry (S-bends were used to taper the output waveguides to a spacing of 250 µm) as well as imperfection of process and simulation model. The uniformity of the two output ports is approximately 0.5 dB.

It is believed that a further improvement of the simulation and the process will enable the reduction of the loss as well as the deviation. A near-field photograph of the coupler is presented in Figure 4.9. Figure 4.10 shows a photograph of the device.

In recent years, hot embossing of microcomponents has become a routinely used replication technology for thermoplastic polymers. Low flow rates and slow molding speeds ensure that even the smallest details in the nanometer range are replicated perfectly. Hot embossing is particularly suited for structuring planar plates and foils, as only a small amount of plastic has to be molded. In contrast to injection molding, the polymer flows a very short distance from the foil into the microstructure during hot embossing. As a result, very little stress is induced into the polymer and the molded parts are well-suited to optical applications, such as waveguides and lenses. The setup of the hot embossing machine is relatively simple. Setup times are short as the mold insert and the polymer are easily exchanged. Nickel shims of only a few hundred micrometers can be used for replication without major effort. The electroplating

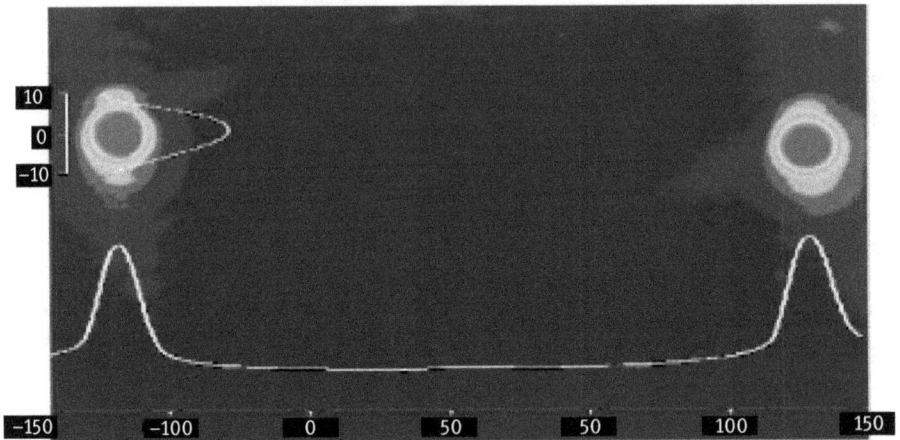

**Figure 4.9:** Near-field photograph pattern of a rib 1 × 2 MMI coupler. All units are in μm. [32–36].

**Figure 4.10:** Photograph of the output region of a 1 × 2 MMI taken by light microscope. The width of the MMI and of the output waveguides is 50 μm and 17 μm, respectively [32–36].

process for such shims takes much less time than for more compact tools, as the electroplating time increases linearly with shim thickness. Therefore, tools can be manufactured from an existing photomask design within several days.

Since fiber alignment and assembly are serious difficulties and cost factors in a mass production of integrated optical devices, an alternative process variation is described in the following paragraph allowing a passive alignment and assembly. The approach uses the LIGA-technique in a first step to pre-emboss the polymer substrate, similar to the process using nickel shims, only with a higher aspect ratio. At this stage, the embossed ridge or groove structure serves as the masking structure for the ensuing flood exposure. If the penetration depth of the evanescent field of these waveguides is smaller than the distance between them and the planar waveguides at the surface, no light can couple from one waveguide into the other waveguides. Either the ridge or groove waveguides can be used as stripe waveguides. Since the effect of UV-induced material modification typically produces refraction index differences that are small and causes these guides to be very lossy in bends, ridge waveguides are preferred for strong guiding application, for weak guiding application the groove structure is used. Additionally, waveguides and fiber alignment grooves can be produced in one fabrication step. Furthermore, it is possible to integrate functional sidewalls like mirrors or gratings. The fabrication of the micro-optical bench and the waveguides in the same process step assures very good alignment accuracy and guarantees a low-cost passive fiber-chip coupling [149] and assembly. The approach presented here provides the opportunity to have a mass fabrication process using replication technologies such as the previously described process of hot embossing or injection molding.

Figure 4.11 shows SEM photographs of pre-embossed substrates with fiber alignment structures and groove or ridge waveguides. Elastic ripples in the sidewalls of the fiber fixing structures, facilitate fiber insertion. They also make the alignment insensitive to variations in fiber diameter.

**Figure 4.11:** Pre-embossed substrate in PMMA containing waveguide and passive fiber alignment structures.

The proof of principle of the fabrication of waveguides structured by masking with pre-embossed polymer substrates and following flood exposure is demonstrated in the following figure (Figure 4.12). It shows a near-field photograph of the facet of a groove (top) and a ridge waveguide (bottom).

An intense single-mode profile is found. For this, the fibers are inserted into the alignment structures. Since the gap at the front face of the waveguide is 2 μm smaller

**Figure 4.12:** Near-field photograph of waveguides realized by masking by a pre-embossing substrate and ensuing flood exposure. Pre-embossed groove (a) and pre-embossed ridge waveguide (b).

than the fiber diameter, the fibers are clamped gently by the alignment structures, thus allowing easy handling during the ongoing characterization. The attenuation of the waveguides at 1,550 nm is found to be 4 dB/cm. This is mainly due because of a roughness of about 100 nm rms of the top surface of the waveguides due to end-milling imperfections of the two-level LIGA copper substrate, which leads to increased scattering losses compared to the waveguides fabricated by conventional lithography. This problem can be easily solved by polishing the substrate.

In summary, replication methods, especially hot embossing using polymers is an appropriate method for manufacturing basic waveguide components and fluidic channels at the same time for optofluidic systems. Hot embossing provides the required accuracy and has the potential for scaling up in an industrial environment. The same accounts for micro-injection molding and injection molding in general.

The following paragraph provides a brief introduction into dry etching methods also used for waveguide fabrication in semiconductor materials.

### 4.5.3 Dry & wet etching

Dry etching uses in contrast to so-called wet etching ions, transferring their physical momentum to the material being etched, resulting in an anisotropic, non-selective sputter etching. Wet etching uses "wet" chemicals, vapor, or plasma, resulting in usually isotropic and highly selective etching. A list of typical wet etchants is provided in Table 4.4.

**Table 4.4:** Typical wet etchants and their corresponding materials.

| Material | Wet Etchant |
|---|---|
| Si (a-Si) | $KOH$, $HNO_3$ + $H_2O$ + $HF$ |
| $SiO_2$ | $HF$, $BHF$ |
| $Si_3N_4$ | $HF$, $BHF$, $H_3PO_4$ |
| GaAs | $H_2SO_4$ + $H_2O_2$ + $H_2O$, $Br$ + $CH_3OH$ |
| Au | $HCL$ + $HNO_3$, $KI$ + $I_2$ + $H_2O$ |
| Al | $HCl$ + $H_2O$, $NaOH$ |

A combination of both methods is used in so-called ion-enhanced or reactive ion etching (RIE) combining the directionality and the selectivity of both. Isotropic etching is referred to the etching process if the etching rate is the same in both horizontal and vertical direction. Anisotropic etching is referred to the etching process if the etching rate is different in horizontal and vertical direction. The lateral etch ratio is defined as:

$$R_L = \frac{\text{Horizontal Etch Rate}(r_H)}{\text{Vertical Etch Rate}(r_V)}$$

Isotropic etching: $R_L = 1$
Anisotropic etching: $0 < R_L < 1$
Directional etching: $R_L = 0$

In dry etching terminology, the word bias is used to define the difference in the lateral dimensions between the patterns on the photomask and the actually achieved etched patterns on the substrate. With respect to the lateral etch ratio, a smaller $R_L$ leads to a smaller bias.

The work horse of semiconductor and polymer waveguide processes is RIE by which a gas is introduced into a chamber where plasma is created using radio-frequency power. Reactive species are created in the plasma resulting in radicals

(leading to chemical reactions with the used substrate) and ions, which are accelerated in the electric field of the plasma toward the substrate (bombardment) "drilling"
out substrate material. The gas is selected in such a way that the gas reacts with the
material to be etched. Table 4.5 lists some typical RIE gases and the material for
which they are used (refer also e.g., www.halbleiter.org).

Table 4.5: Typical RIE gases and their corresponding material.

| Material | Dry Etchant | Mask |
| --- | --- | --- |
| Si (a-Si) | $CF_4$, $SF_6$, $BCl_2 + CL_2$ | Resist, Metal (Cr, Ni, Al) |
| $SiO_2$ | $CHF_3 + O_2$, $CF_4 + H_2$ | Resist, Metal (Cr, Ni, Al) |
| $Si_3N_4$ | $CF_4 + O_2$ ($H_2$), $CHF_3$ | Resist, Metal (Cr, Ni, Al) |
| GaAs | $Cl_2$, $Cl_2 + BCl_3$ | $Si_3N_4$, Metal (Cr, Ni) |
| InP | $CH_4/H_2$ | $Si_3N_4$, Metal (Cr, Ni, Al) |
| Al | $Cl_2$, $BCl_3 + Cl_2$ | Resist, $Si_3N_4$ |
| Resist/Polymer | $O_2$ | $Si_3N_4$, Metal (Cr, Ni) |

There are of course more wet and dry etching processes such as, for example, ion
beam etching, plasma etching, chemically assisted ion beam etching, which are
readily described in literature and used in commercial semiconductor and polymer-
processing production lines.

The following chapter provides an insight into so-called planar waveguide-based
devices.

## 4.6 Planar waveguide-based devices

The characteristic of a planar waveguide is the fact that this type of waveguide has an
optical confinement only in one transverse direction, as it is the case in the DUV-
induced modificated polymer waveguides (Figure 4.13).

The following two chapters provide examples of planar and rib waveguide-based
devices.

### 4.6.1 All polymer waveguides

The expansion of high capacity optical transmission techniques into price-sensitive
areas such as telecom, datacom, access networks and of course optofluidic systems
requires a major reduction in the cost of optical components. Polymer waveguides are
attractive because they are very simple to process and are promising for low-cost or
even single-use devices [70, 119–123, 144]. The DUV-induced modification of the

**Figure 4.13:** Planar polymer waveguide based devices.

dielectric properties of polymers is not only a useful technique for low-cost realization of integrated optical circuits for telecommunication and sensor applications [82, 99–103, 213, 214] but has also been used to create gratings in polymer fibers [196].

To achieve single-mode waveguides, the width is determined to be 7.5 μm. The waveguide has a graded index profile with an exponential decay. The near-field image of a straight waveguide with a width of 7.5 μm is shown in Figure 4.14. All transmission experiments were carried out using randomly polarized light. A waveguide loss of 1 dB/cm at 1.55 μm has been measured using the cut-back method. The polarization-dependent loss is less than 0.15 dB. Excellent stability of the waveguides in the temperature range between 0 °C and 85 °C could be demonstrated under continuous operation.

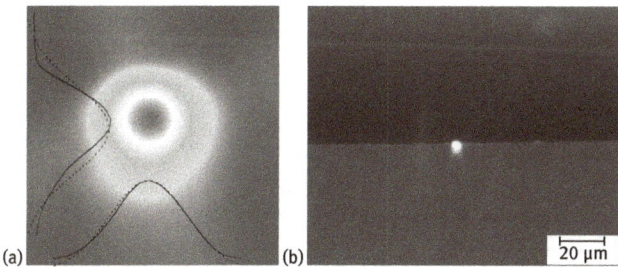

**Figure 4.14:** Near-field photographs of a straight waveguide, width 7.5 μm (line – measurement, dotted line – Gaussian fit; (a) image is close up of (b) image).

After waveguide fabrication the polymer substrates are separated by a wafer saw. No further polishing of the end faces is required. The fiber-to-chip coupling loss is 0.5 dB per facet.

A simple waveguide-based device is the Y coupler, consisting of a straight input waveguide branching out into two waveguide. This type of device is described in the following paragraph.

### 4.6.1.1 Y – splitter

Symmetrical 1 × 2 splitters (Figure 4.15) with different angles were fabricated by combining pairs of s-bend structures with the same radii, which compose of two identical arcs. For s-bends with a radius of curvature greater than 6 mm no additional propagation loss was observed.

**Figure 4.15:** Photograph of a y – splitter fabricated by DUV lithography.

Splitters were fabricated by varying the branching angle from 1° to 2.5°. For angles smaller than 1.5°, the measured losses are relatively high due to processing imperfections caused by a limited photolithographic resolution at the Y-junction. As an example, the spectral insertion loss (fiber-chip-fiber) characteristic of the two output ports of a 1 × 2 splitter with branching angle of 1.75° are depicted in Figure 4.16. For comparison, the spectral insertion loss of a straight waveguide is shown, too. The excess loss due to

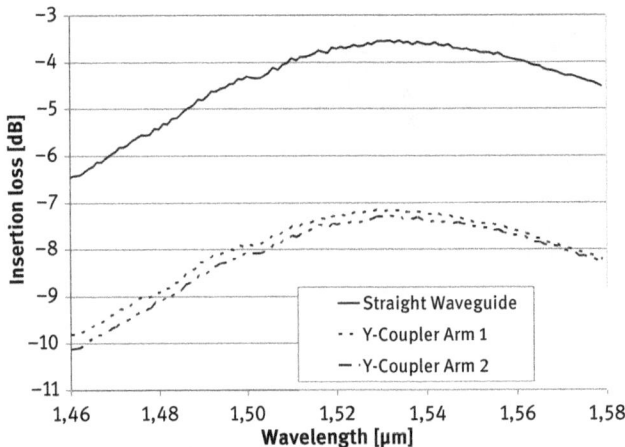

**Figure 4.16:** Measurement result of a y – splitter with a branching angle of 1.75°.

the Y-splitter is in the range of 0.5 dB. The uniformity of the output ports is within the measurement accuracy of 0.1 dB. The device length is 2.5 cm. The wavelength dependency of the insertion loss is due to the material loss in this range.

### 4.6.1.2 Co-directional coupler

Another important device is a 2 × 2 co-directional coupler (Figure 4.17). This type of coupler can be used for equal splitting of the power independent from the input channel in the whole wavelength region as well as to realize splitting ratios other than 3 dB. The separation between the waveguides in the coupling region is the critical parameter regarding the fabrication. The resolution of the photolithography defines the minimum coupling gap, which is 1 μm in our case.

**Figure 4.17:** Photograph of a directional coupler with a gap of 5 μm.

Since the structures are weak guiding, the coupling gap can be chosen in the range of 1 to 5 μm.

The measured result of a co-directional coupler at 1.55 μm with a coupling gap of 2 μm and 3 and different coupling lengths is shown in Figure 4.18. A PLC was used to measure the maximum and minimum polarization dependent values of the insertion loss. The polarization dependency is negligible, as can be seen from the measurement. The intensity at both output ports is normalized to the sum of the output intensities. The excess loss of the coupler is less than 0.5 dB. The measured values confirm the simulations quiet well.

### 4.6.1.3 Multi-mode interference coupler

Another basic component, which is implemented for a variety of optical signal processing and routing functions, is a MMI. The coupler consists of a broad center waveguide, which supports several modes depending on the width of the waveguide. The photograph of the input region of a MMI coupler is shown in Figure 4.19.

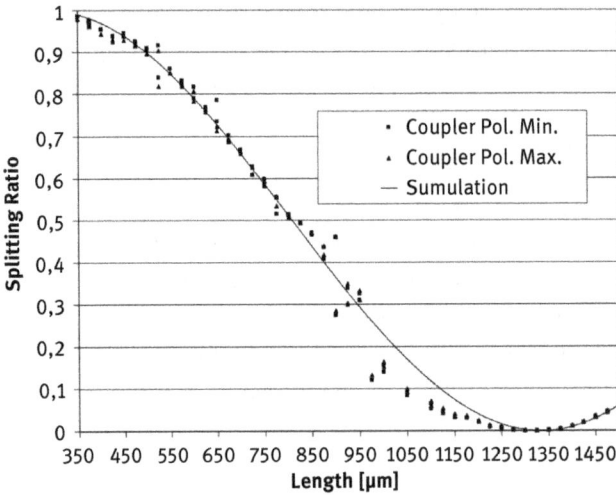

Figure 4.18: Measurement result of a co-directional coupler with a gap of 2 µm.

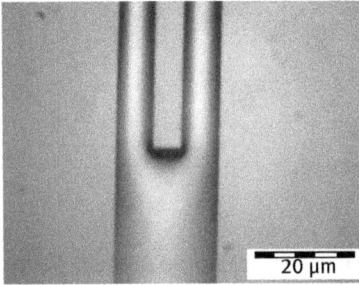

Figure 4.19: Photograph of the input region of a MMI coupler.

These MMI devices [116] offer important advantages such as compact size, low imbalance, and crosstalk. In comparison with co-directional couplers, the main advantage is the fabrication tolerance. Most of these devices have so far been fabricated in high-index contrast materials such as InP. For high-index contrast materials, this type of waveguide has nearly an ideal mode spectrum, which ensures a proper self-imagine. To date there has only been limited application of polymer materials for MMI devices. An important limitation on broader use is that typical polymer waveguides have low index contrast; therefore, the modes are so weakly confined that the eigenmode spectrum deviates greatly from the ideally required mode spectrum for self-imaging. Thus, the imaging quality is significantly degraded, exhibiting an increased loss and imbalance. However, our simulation shows at the best image point tolerable loss and imbalance, which justify to fabricate a weak-guiding MMI coupler. The designed MMI coupler has a width of 27 µm. As can be seen in Figure 4.20, there are four modes, which are supported in the broader multi-mode section. The mode solving is carried out via BPM using the correlation method. The multi-mode section should support more than three modes to assure an appropriate interference signal. The interference pattern of the four modes in transmission direction is shown in Figure 4.20 at the bottom. In the range between 1,650 and 1,950 µm, the intensity of the MMI is divided equally at both of the output ports.

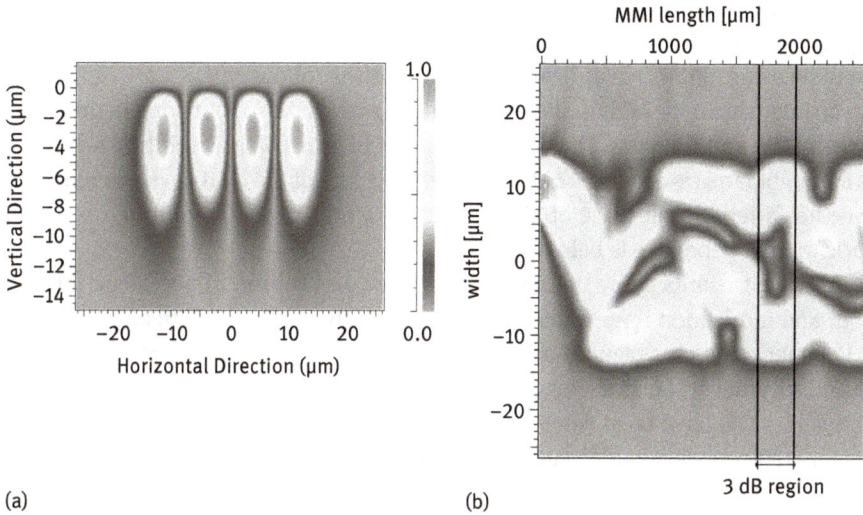

(a)

(b)

3 dB region

**Figure 4.20:** Simulated modeprofile of a MMI coupler, TE polarization (a). Simulation of the interference pattern in transmission direction, TE polarization (b).

Figure 4.21 illustrates the calculated output coupling intensities as a function of the length of the MMI region. The best image point in terms of 3 dB splitting ratio, low loss, and imbalance is around 1,800 µm. The device has a length tolerance of 150 µm

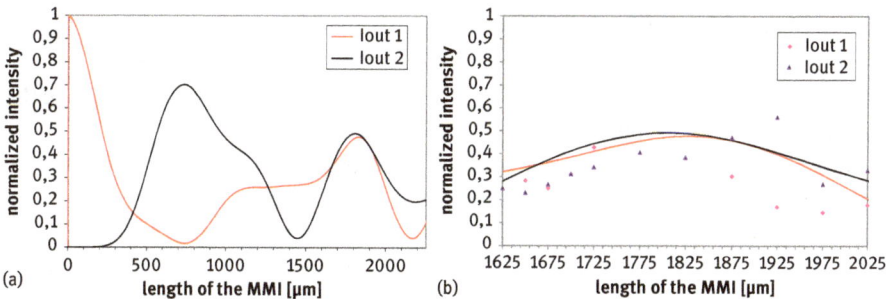

(a)

(b)

**Figure 4.21:** Simulation of the intensities at the output ports of a MMI with width of 27 µm in the multimode region (a). Measured intensities at the output ports of the MMI coupler at different lengths (b).

for a 5% imbalance in splitting ratio. The MMI is nearly polarization independent with respect to the 3 dB point. This is an advantage compared to co-directional couplers. The measured results of the MMI coupler as a function of the length of the MMI region around the 3 dB imaging point are shown on the bottom. The intensities of both outputs of the MMI coupler have been added and normalized.

Another example of a designed MMI coupler with a width of 21 µm has been calculated and fabricated.

There are 3 modes which are supported in the broader multi-mode section. The measurement result of the fabricated MMI reveals the following details:

The intensity of the MMI is divided equally at both of the output ports in the range between 1,215 and 1,315 µm. The best image point in terms of 3 dB splitting ratio, low loss and imbalance is around 1,250 µm. The excess loss of the MMI device is 1 dB. The device has a length tolerance of 100 µm for a 5% imbalance in splitting ratio. The polarization dependence is below 0.2 dB.

These first results show that MMI devices can be realized in polymers. Further design and fabrication improvements will increase the device performance.

Polymer planar waveguide-based devices [136–138] are demonstrated in literature and several components, for example, like arrayed waveguide gratings or Bragg gratings [71, 225] have been demonstrated. The following chapter focuses on hybrid polymer – silicon waveguides and their fabrication.

### 4.6.2 Silicon–polymer hybrid waveguides

Polymer-integrated reverse symmetry waveguides on porous silicon substrate fabricated using DUV radiation in poly(methyl methacrylate) (PMMA) are the focus of this chapter which is based on [211]. Several other silicon–polymer hybrid waveguides, platforms, and devices can be found in literature (for example [10]).

The layer sequence and geometry of the presented waveguide enables an evanescent field extending into the upper waveguide or analyte layer, enabling various integrated optical devices where large evanescent fields are required. The presented fabrication technique enables the generation of defined regions where the evanescent field is larger than in the rest of the waveguide. This technology can improve the performance of evanescent-wave-based waveguide devices.

Considerable interest has been focused on optical devices, which are based on the interactions associated with evanescent waves such as directional couplers and waveguide sensors. These devices make use of the electromagnetic field, which propagates in the waveguide, creating an evanescent field that extends into the surrounding layers present at the surface of the waveguide. For example, commercial success has been realized employing surface plasmon resonance (SPR) spectroscopy for optical biosensing of substances extending ~ 200 nm from the transducer surface. Several techniques have been presented to enhance the evanescent field to improve the performance of the devices particularly for optical studies of cells. The authors in [186] use a multilayer dielectric structure to enhance the evanescent waves. The authors in [208] use thin high-index films for evanescent field enhancement. A novel use of metamaterials is presented in [207] for increasing the evanescent field. Recently, a so-called planar reverse symmetry waveguide has been demonstrated [110–112], where the substrate has a refractive index less than the refractive index of the cladding layer enabling the extension of the evanescent field (> 1 μm) into the analyte present on the surface of the waveguide. The following paragraph describes a fabrication process utilizing standard wafer fabrication processing for realizing an integrated reverse symmetry (IRS) waveguide.

A conventional integrated waveguide is made up of a substrate, a waveguiding layer, and a so-called cladding layer, which covers the waveguiding layer (refer Section 4.3). In this configuration, the substrate index is usually higher than the cladding layer. The exponentially decaying evanescent field in this configuration typically penetrates into the cladding layer 100–300 nm, which is sufficient for many integrated optical devices but not enough for devices requiring large evanescent fields. The IRS waveguides solve this problem by choosing a lower cladding layer having a lower refractive index (RI) than the upper cladding layer, in our case electrochemically formed mesoporous silicon (PSi) [63]. The PSi layer not only enables a lower cladding having a lower refractive index but also optically isolates the waveguide core from the Si-substrate. As Si has a refractive index (~3.5) much higher than that of typical polymers (~1.5–1.7), a thick lower cladding layer must be used to avoid light coupling with the substrate. The authors in [191] have found that a 3-μm-thick porous silica layer as a lower cladding is sufficient to fabricate polymer waveguides on a silicon wafer.

The fabrication of the IRS waveguide is based on the DUV modification of a polymer described in Section 2.1.5, DUV Radiation, in this book (in our case poly(methyl methacrylate) (PMMA)) coated on top of PSi for realizing the waveguiding layer. The layer sequence and the dimensions of the IRS waveguide are shown in Figure 4.22.

IRS waveguide fabrication begins with electrochemical formation of a mesoporous silicon substrate layer in a p+ silicon wafer (0.01 ohm-cm) using a hydrofluoric acid (15%) containing electrolyte in ethanol (70%). The barrier layer is a thin <12 nm, low porosity layer (60%) etched using a current density of 5 mA/cm², which produces a pore structure with diameters less than ~5 nm (Figure 4.22(a)). This is followed by a thick 4–8 micron high porosity (80–90%) optical isolation layer etched using a current density of 50–70 mA/cm², which produces more open pore diameters ranging between 20 and 30 nm (Figure 4.22(b)). The PMMA layer (Resist MicroChem 950k PMMA A11) is spin-coated (Spincoater Hamatech: 2000/15/2 [upm/ramp/min]) onto the mesoporous silicon substrate to a thickness of approximately 3 μm. The sample is annealed at 170 °C for 30 min. A DUV lithography system (EVG620) is used for fabricating the waveguides under vacuum with a quartz/chromium mask. The illumination dose used is 3 J/cm². The samples are then annealed at 70 °C for 4 h. The

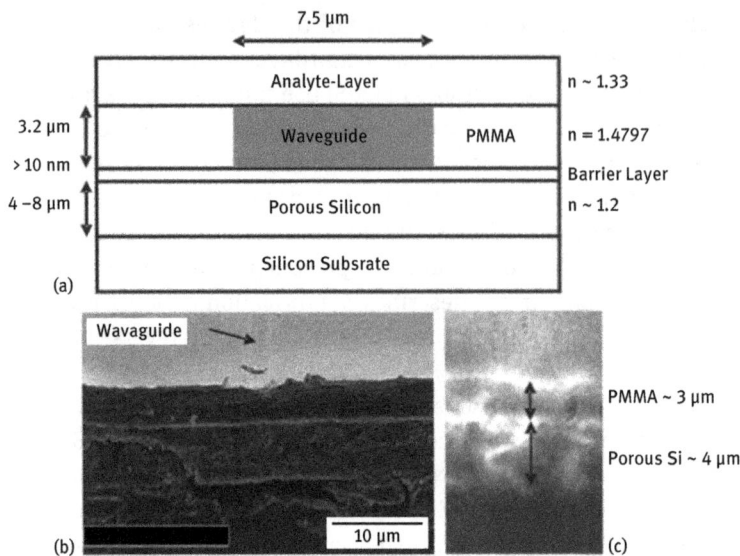

Figure 4.22: Integrated reverse symmetry waveguide. (a) Layer sequence and layout (b) Scanning electron microscope image of reverse symmetry waveguide (c) Near-field image of outcoupling waveguide facet [211].

final step is the dicing of the samples to provide waveguide end faces for optical measurements.

(c) Side view of bilayer structure illustrating anisotropic growth of pore channels and porosity contrast between barrier and optical isolation layers. (d) Magnified view of barrier layer grown intentionally thick (120 nm) to observe with SEM. (e) Magnified view of pore tip/silicon wafer interface.

Direct spin-coating of polymers is usually not feasible when using porous substrates, as the solvent and polymer can infiltrate the pores which will raise the refractive index. To prevent this, authors in [110–112] used a technique referred to as dip-floating, which requires several fabrication steps. The aforementioned process prevents polymer resist from infiltrating too deep into the optical isolation layer by monolithic integration of the PSi bilayer structure, described above, comprised of a low porosity barrier layer on top of the high porosity optical isolation layer (Figure 4.23). Bilayer formation leverages the ability to tune PSi morphology simply by altering current density during the electrochemical etch process. The PSi barrier layer is crucial for enabling waveguide performance. Several porous silicon configurations have been fabricated with and without the barrier layers. The fabrication parameters for the samples tested are summarized in Table 4.6. The refractive index of the optical isolation layer is estimated using the Bruggman effective medium approximation [37].

The IRS waveguides are characterized using an optical fiber mounted onto a piezo-electrically driven XYZ-stage for coupling into the waveguides. A microscope objective

**Figure 4.23:** SEM images of mesoporous bilayer structure. a) Pore structure of barrier layer (5 mA/cm²) (b) Pore structure of optical isolation layer (50 mA/cm²).

**Table 4.6:** Fabrication parameters of porous silicon layers [211].

| Sample | Layers | Etch Conditions Current Density (mA/cm²)/time (s) | Porosity (%) | Thickness (µm) | Refractive index |
|---|---|---|---|---|---|
| 2 | 2 | 5/1 & 50/138 | 60 & 77 | 0.010 & 4 | 1.83 & 1.38 |
| 3 | 1 | 70/108 | 84 | 4 | 1.21 |
| 4 | 1 | 70/216 | 84 | 8 | 1.21 |
| 5 | 2 | 5/1 & 70/216 | 60 & 84 | 0.010 & 8 | 1.83 & 1.21 |

(100X) is used on the outcoupling side, which focuses the nearfield image onto an IR camera (Hamamatsu Model C2400). All measurements are carried out at a wavelength of 1,550 nm. Shifting to other wavelengths is however possible, as only the design (width and DUV exposure dose) of the integrated reverse symmetry waveguides has to be adjusted to guarantee single-mode operation at the required wavelength. A near-field image of the facet of sample 3 is shown in Figure 4.23(c) without light from the laser source being coupled into the waveguide. The results from the near-field measurements on samples 2, 3, 4 and 5 are presented in Figure 4.24.

To demonstrate IRS waveguide performance, measurements are performed by taking the near-field image of samples with and without water on top of the waveguides to determine the expansion of the mode traveling in the waveguide. Water is chosen to be the cladding layer as the refractive index (n=1.333) is sufficient to demonstrate the performance and it is relevant to the future goals of utilizing this device for aqueous-based biosensing applications. Substituting the water by a suitable polymer having the appropriate refractive index is of course state-of-the-art. The water is placed directly on top of the waveguides and it is made sure that the water is also at the outcoupling facet, making it possible to see the portion of the mode penetrating into the water.

Left: without water                    Right: with water

Sample 5 (RI = 1.21; NP thickness = 8 µm; 12 nm layer)

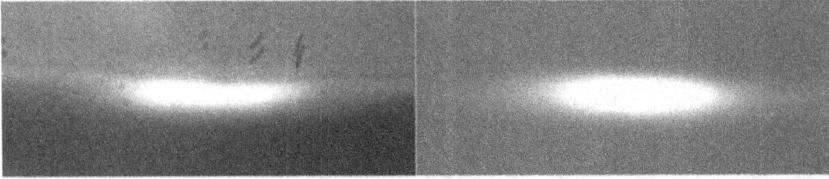

Sample 3 (RI = 1.21; NP thickness = 4 µm)

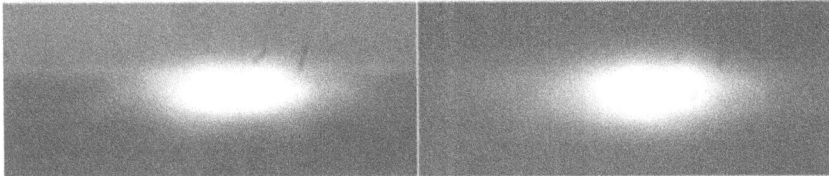

Sample 2 (RI = 1.38; NP thickness = 4 µm; 12 nm layer)

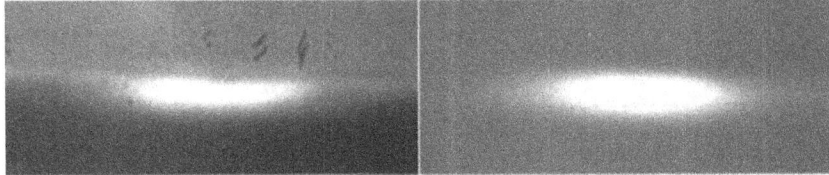

Sample 4 (RI = 1.21; NP thickness = 8 µm)

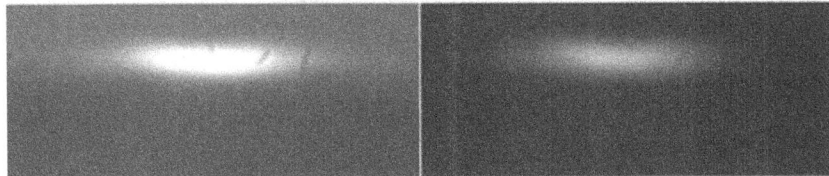

Modeprofile of bulk polymer DUV wavegise without water for comparison

**Figure 4.24:** Near-field analysis of integrated reverse symmetry waveguide samples. Sample 5: Optical effect observed, which is attributed to the 12 nm intermediate layer. Sample 3: Optical effect not observed, no barrier layer present! Field also decays more into the nanoporous layer when compared to sample 5. Sample 2: Optical effect observed despite the "lower" index of 1.38. Sample 4: Optical effect not observed, no barrier layer present. The illumination and camera parameters are unchanged throughout the entire near-field measurements to guarantee compatibility of the images.

Results find that only the integrated reverse symmetry waveguides with low porosity PSi barrier layer showed the desired optical effect of a deep penetration of the evanescent field into the upper analyte layer. The height of the mode field nearly doubles as can be seen in the near-field images of sample 2 and 5 in Figure 4.24 with the analyte layer present. Due to the fact that the mode is confined in the polymer layer which is ~3 µm in the absence of the analyte layer, the height of the evanescent field in the presence of the analyte layer can be approximated to be more then 3 µm. The evanescent field decays into the water even when the refractive index is estimated to be 1.38, which is slightly higher than that of water. The measurements also confirm that in absence of the barrier layer, the spin-coated polymer fills up the pores as visible, for example, in the near-field image of sample 3 in Figure 4.24 leading to an unwanted conventional waveguide having an evanescent field, which decays only a few 100 nm into the upper analyte layer.

The evanescent field has been simulated using Beamprop with and without water on top of the waveguide. The simulation results are shown in Figure 4.25 (Wavelength = 1.55 µm, width of the waveguide 7.5 µm, PMMA thickness 3.2 µm).

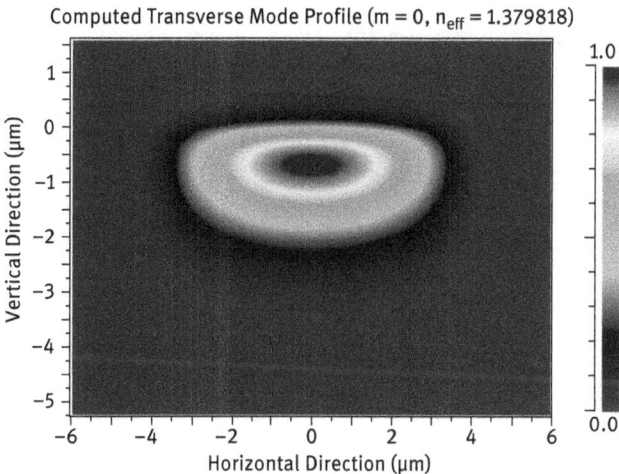

**Figure 4.25:** Evanescent field simulation of polymer silicon hybrid waveguide without water on top of the waveguide.

Efforts to characterize waveguide loss using the cut-back method proved difficult to determine accurately the losses as our samples are only 1 cm × 1 cm and dicing introduced more losses at the interface as can be seen in Figure 4.22(b). The losses are estimated by measuring 1-cm-long waveguides on several samples to be approximately 2 dB/cm and the fiber chip coupling is around 9–10 dB, comparable to the results in [191] (Figure 4.26).

Computed Transverse Mode Profile (m = 0, $n_{eff}$ = 1.350918)

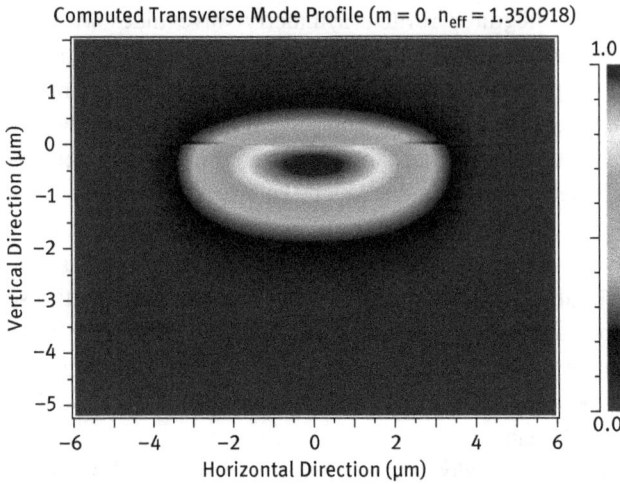

**Figure 4.26:** Evanescent field simulation of polymer silicon hybrid waveguide with water on top of the waveguide.

In conclusion, the DUV-integrated reverse symmetry waveguide is a novel addition to the DUV technology platform described in this book for realizing optical devices with a defined evanescent field area required in couplers or sensors. Moreover, defined patterning of the porous silicon layer, as, for example, described in [240], in combination with the described fabrication process enables the realization of defined waveguide regions having a larger evanescent field than in other parts of the integrated waveguide device, which can be used, for example, in cell-based biosensor studies.

The following chapters provide an introduction into the vast area of active photonic devices such as light emitting diodes, detectors, image sensors, and lasers, required for realizing optofluidic systems.

## 4.7 Active photonic devices

Active photonic devices are essential modules in optofluidic systems, as they provide the necessary illumination and detection for conducting optofluidic analysis. The following chapters provide a brief introduction into the most commonly used devices.

### 4.7.1 Light-emitting diodes (LEDs)

The following paragraphs are adapted from the opto-semiconductor handbook from Hamamatsu and the figures are reprinted with kind permission from Hamamatsu.

The LEDs are opto-semiconductors that convert electrical energy into light energy. The LEDs offer the advantages of low cost and a long service life compared to laser diodes. The LEDs are broadly grouped into visible LEDs and invisible LEDs. Visible LEDs are mainly used for display or illumination, where LEDs are used individually without combination with photosensors. Invisible LEDs, however, are mainly used in combination with photo sensors such as photodiodes or CCDs.

The LED chips are manufactured starting from an LED wafer containing internal PN junctions, which is then subjected to processes including diffusion and vapor deposition, and is finally diced into chips. In the LED wafer, PN junctions are first of all formed by vapor or liquid epitaxial growth. These PN junctions can be made from the same material, but using different materials makes it possible to create LEDs with high emission efficiency. For example, in structures where the GaAs active layer is sandwiched between clad layers of GaAlAs, both P-clad layer and N-clad layer are in a heterojunction, so this is called the double-heterostructure (Figure 4.27). In this type of structure, the injected electrons and holes are confined in a highly dense state by heterobarriers, where the electrons and holes have a high probability of recombining, so light emission efficiency will be high.

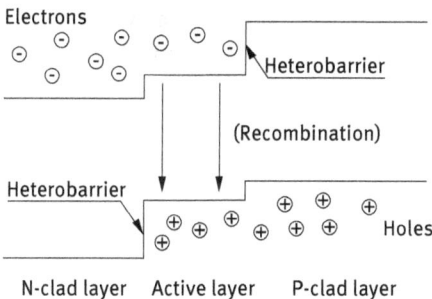

**Figure 4.27:** Double-heterostructure. (Courtesy Hamamatsu)

Next, gold electrodes are vapor-deposited on the upper and lower surfaces of the LED wafer and are subjected to a high temperature to form alloys that provide ohmic contacts between the gold and semiconductor. The electrodes on the chips' upper surface are then etched away, leaving only the minimum required sections for extracting light with high efficiency. The LED chips are usually die-bonded to a gold-plated metal base or a silver-plated lead frame and electrically connected by gold wire to wire leads. The gold wire is resin-coated or sealed with a cap for protection.

To enhance radiant power, some LEDs use a metal base with a concave area, which serves as a reflector, and the LED chip is mounted in that area.

When a forward voltage is applied to an LED (Figure 4.28), the potential barrier of the PN junction becomes smaller, causing movement of injected minority carriers (electrons in the N-layer, holes in the P-layer). This movement results in electron-hole recombination which emits light. However, not all carriers recombine to emit light

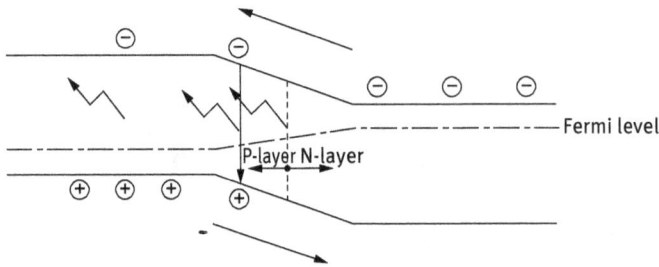

**Figure 4.28:** LED energy levels with forward voltage applied. (Courtesy Hamamatsu).

(emission recombination), and a type of recombination not emitting light (non-emission recombination) also occurs. The energy lost by recombination is converted into light during emission recombination but is converted into heat during non-emission recombination.

Light generated by electron-hole recombination travels in various directions. Light moving upward can be extracted from the upper surface of the chip with relatively high efficiency. Light moving sideways can also be converted to effective light with relatively high efficiency using a reflector to reflect the light forward. In light moving downward, if there is a GaAs substrate with a narrower band gap than the emission wavelengths, then light absorption will occur there. So, in liquid phase epitaxial growth, the GaAs substrate is sometimes etched away in the wafer process. In vapor phase epitaxial growth, however, the GaAs substrate cannot be removed because the epitaxial layer is too thin. To cope with this, a light-reflecting layer is formed beneath the emission layer to suppress light absorption in the GaAs substrate (Figure 4.29).

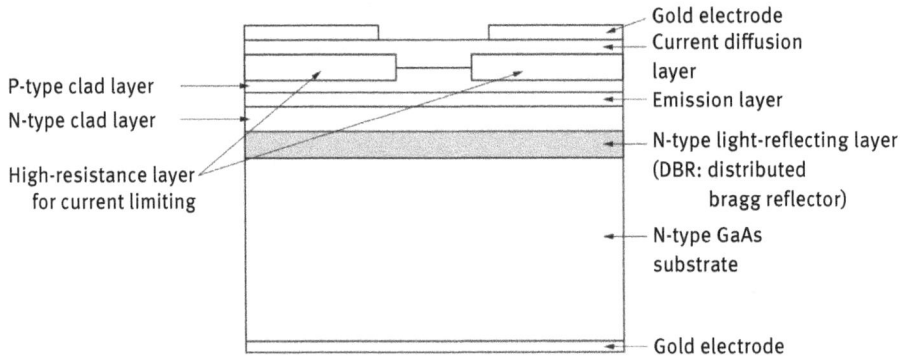

**Figure 4.29:** Cross-section of LED having light-reflecting layer. (Courtesy Hamamatsu).

When a light-reflecting layer is formed over and under the emission layer, the emitted light reflects repeatedly between the upper and lower light-reflecting layers, causing a weak resonance. This resonance light can be extracted from the upper side of the

LED chip by setting the reflectance of the upper light-reflecting layer lower than that of the lower light-reflecting layer. An LED with this structure is called the resonant cavity (RC) type LED.

## 4.7.2 Detectors

This chapter is adapted from and based on the opto-semiconductor handbook from Hamamatsu. Figures, which are taken from the opto-semiconductor handbook, are reproduced with kind permission from Hamamatsu.

Figure 4.30 shows a cross-section example of a Si photodiode. The P-type region (P-layer) at the photosensitive surface and the N-type region (N-layer) at the substrate form a PN junction, which operates as a photoelectric converter. The usual P-layer for a Si photodiode is formed by selective diffusion of boron, to a thickness of approximately 1 μm or less, and the neutral region at the junction between the P-layer and N-layer is known as the depletion layer. By controlling the thickness of the outer P-layer, N-layer, and bottom N+ -layer as well as the dopant concentration, the spectral response and frequency response can be controlled.

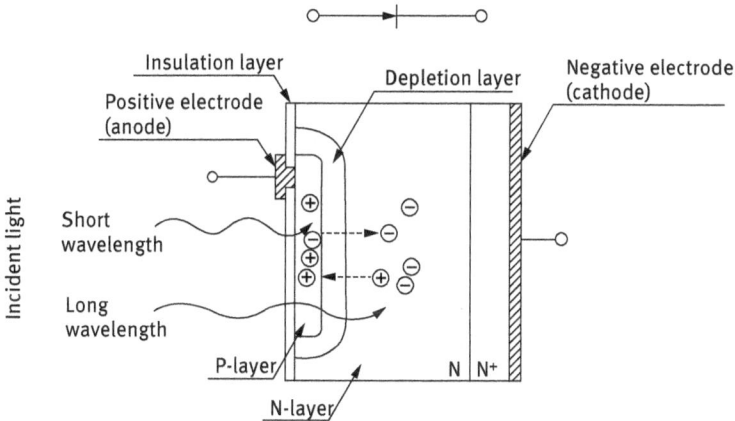

**Figure 4.30:** Schematic of a Si photodiode cross section (Courtesy Hamamatsu).

When a Si photodiode is illuminated by light and if the light energy is greater than the band gap energy, the valence band electrons are excited to the conduction band, leaving so-called holes in their place in the valence band. The so-called PN junction is shown in Figure 4.31.

These electron-hole pairs occur throughout the P-layer, depletion layer, and N-layer materials. In the depletion layer, the electric field accelerates these electrons toward the N-layer and the holes toward the P-layer. Of the electron-hole pairs

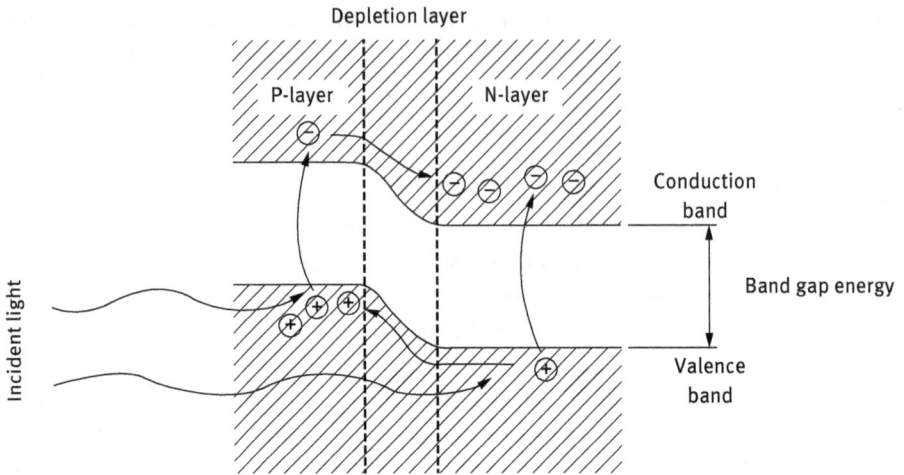

**Figure 4.31:** Si photodiode PN junction state (Courtesy Hamamatsu).

generated in the N-layer, the electrons, along with electrons that have arrived from the P-layer, are left in the N-layer conduction band. The holes at this time are being diffused through the N-layer up to the depletion layer while being accelerated, and collected in the P-layer valence band. In this manner, electron-hole pairs, which are generated in proportion to the amount of incident light, are collected in the N-layer and P-layer. This results in a positive charge in the P-layer and a negative charge in the N-layer. When an electrode is formed from each of the P-layer and N-layer and is connected to an external circuit, electrons will flow away from the N-layer, and holes will flow away from the P-layer toward the opposite respective electrodes, generating a current. These electrons and holes generating a current flow in a semiconductor are called the carriers.

From a system point of view, the optofluidic system designer is interested in the electrical and optical characteristics of the device. Focusing in a first step on the equivalent electrical circuit, for example, of a Si photodiode leads to the following (Figure 4.32):

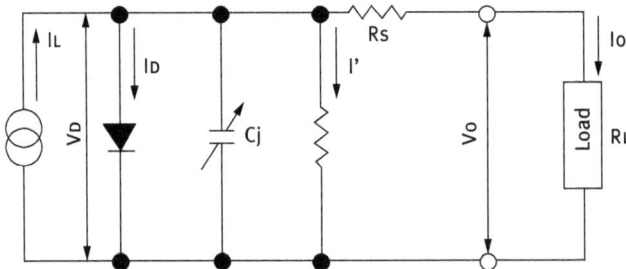

**Figure 4.32:** Si photodiode equivalent circuit (Courtesy Hamamatsu).

Where, $I_L$ is the current generated by incident light (proportional to the light level), $V_D$ is the voltage across the diode, $I_D$ is the diode current, $C_j$ is the junction capacitance, $R_{sh}$ is the shunt resistance, $I'$ is the shunt resistance current, $R_S$ is the series resistance, $V_O$ is the output voltage, and $I_O$ is the output current.

Using the above equivalent circuit, the output current ($I_O$) is given by the following equation:

$$I_O = I_L - I_D - I' = I_L - I_S \left( exp \left( \frac{q \cdot V_D}{k \cdot T} \right) - 1 \right) - I'$$

where $I_S$ is the photodiode reverse saturation current, $q$ is the electron charge, $k$ is the Boltzmann's constant, and $T$ is the absolute temperature of the photodiode.

The open circuit voltage ($V_{OC}$) is the output voltage when $I_O = 0$, and is expressed by the following equation:

$$V_{OC} = \frac{k \cdot T}{q} ln \left( \frac{I_L - I'}{I_S} + 1 \right)$$

If $I'$ is negligible and since $I_S$ increases exponentially with respect to ambient temperature, $V_{OC}$ is inversely proportional to the ambient temperature and proportional to the log of $I_L$. However, this relationship does not hold when detecting a low-level light. The short circuit current ($I_{SC}$) is the output current when load resistance ($R_L$) = 0 and $V_O = 0$, and is expressed by the following equation:

$$I_{SC} = I_L - I_S \left( exp \left( \frac{q.I_{SC}.R_S}{k \cdot T} \right) - 1 \right) - \frac{I_{SC}.R_S}{R_{sh}}$$

In the above equation, the 2nd and 3rd terms become the cause that determines the linearity limit of the short circuit current. However, since $R_S$ is several ohms and $R_{sh}$ is $10^7$ to $10^{11}$ ohms, these 2nd and 3rd terms become negligible over quite a wide range.

Focusing now on the optical characteristics of a photodiode, the principle of operation of a photodiode is to absorb light at specific wavelength and convert it to an electrical current. When the energy of absorbed light is lower than the band gap energy, for example, of Si photodiodes, the photovoltaic effect does not occur. The cutoff wavelength ($\lambda_C$) can be expressed by the following equation:

$$\lambda_C = \frac{1240}{E_g} [nm]$$

where $E_g$ is the band gap energy.

In the case of Si at room temperature, the band gap energy is 1.12 eV, so the cutoff wavelength is 1,100 nm. For short wavelengths, however, the degree of light absorption within the surface diffusion layer becomes very large (refer Figure 3.57 PN junction). Therefore, the thinner the diffusion layer is and the closer the PN junction is to the surface, the higher the sensitivity will be. For normal Si photodiodes, the cutoff wavelength on the short wavelength side is 320 nm, whereas it is 190 nm for UV-enhanced Si photodiodes. The cutoff wavelength is determined by the intrinsic material properties of the Si photodiode, but it is also affected by the spectral transmittance of the light input window material. For borosilicate glass and plastic resin coating, wavelengths below approximately 300 nm are absorbed. If these materials are used as the window, the short-wavelength sensitivity will be lost. When detecting wavelengths shorter than 300 nm, Si photodiodes with quartz windows are used. Measurements limited to the visible light region use a visual-sensitive compensation filter that allows only visible light to pass through it.

In summary, the desirable characteristics of a photodiode:
–   High sensitivity at the operating wavelength range
–   Short response time
–   Linearity
–   Stability (in time and with temperature changes)
–   Low cost and high reliability

Another form of a photodiode is the so-called PIN, which is a diode with a wide, lightly doped "near"-intrinsic semiconductor region between a P-type semiconductor and an N-type semiconductor region. The PIN diode is suitable for application as attenuator, fast switch, photodetector, and high voltage power electronics applications.

Table 4.7 lists important parameters of silicon (Si), Germanium (Ge), and Indium Gallium Arsenide (InGaAs), which are commonly used for manufacturing photodiodes.

**Table 4.7:** Important parameters of silicon (Si), Germanium (Ge), and Indium Gallium Arsenide (InGaAs), which are commonly used for manufacturing photodiodes.

| Parameter | Si | Ge | InGaAs |
|---|---|---|---|
| Wavelength (nm) | 300–1,100 | 500–1,800 | 1,000–1,700 |
| Peak response (nm) | 800 | 1,550 | 1,700 |
| Peak responsivity (A/W) | 0.5 | 0.7 | 0.9 |
| Dark current (nA) | 1 | 200 | 10 |
| Typical risetime (ps) | 500 | 100 | 300 |

Another type of commonly used photodiode in several applications is the so-called avalanche photodiode (APD). The APD is a high-speed, high-sensitivity photodiode that internally multiplies photocurrent by applying a reverse voltage. Compared to PIN photodiodes, the APD provides a higher signal to noise ration (S/N) and is used in a wide variety of applications such as optical rangefinders, free space optics (FSO), and scintillation detection. The APD can multiply a low-level light signal into a large electrical signal. However, it is not always simple to use because a high reverse voltage is needed and the multiplication ratio (gain) is temperature dependent.

A special form of APDs is so-called multi-element type APDs (Figure 4.33). Multi-element Si APDs have an array of active areas. The avalanche layer formed just below each active area on the APD array multiplies the light incident on the active areas. However, carriers generated outside these active areas cannot pass through the avalanche layer so their signal is small. This means that APD arrays have lower crosstalk than photodiode arrays because of their gain. The internal structure of a multi-element Si APD is shown in the figure below.

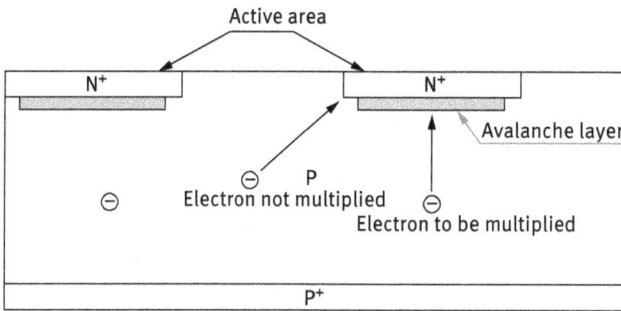

**Figure 4.33:** Internal structure of a multi-element Si APD (Courtesy Hamamatsu).

When light enters a photodiode, electron-hole pairs are generated if the light energy is higher than the band gap energy. The ratio of the number of generated electron-hole pairs to the number of incident photons is called the quantum efficiency (QE), commonly expressed in percent (%). The mechanism by which carriers are generated inside an APD is the same as in a photodiode, but the APD has a function to multiply the generated carriers. When electron-hole pairs are generated in the depletion layer of an APD with a reverse voltage applied to the PN junction, the electric field created across the PN junction causes the electrons to drift toward the N+ side and the holes to drift toward the P+ side. The drift speed of these electron-hole pairs or carriers depends on the electric field strength. However, when the electric field is increased, the carriers are more likely to collide with the crystal lattice so that the drift speed of each carrier becomes saturated at a certain speed. If the reverse voltage is increased even further, some carriers that escaped collision with the crystal lattice will have a great deal of energy. When these carriers collide with the crystal lattice, ionization takes place in which electron-hole pairs are newly generated. These electron-hole pairs then create additional electron-hole pairs in a process just like a chain reaction. This is a phenomenon known as avalanche multiplication. The number of electron-hole pairs generated during the time that a carrier moves a unit distance is referred to as the ionization rate. Usually, the ionization rate of electrons is defined as "$\alpha$" and that of holes as "$\beta$". These ionization rates are important factors in determining the multiplication mechanism. In the case of silicon, the ionization rate of electrons is larger than that of holes ($\alpha > \beta$), so the electrons contribute more to the multiplication. The ratio of $\beta$ to $\alpha$ is called the ionization rate ratio ($k$) and is used as a parameter to indicate APD noise:

$$k = \frac{\beta}{\alpha}$$

Photodiodes and APDs are widely used in ever increasing applications, designs, and material combinations which are documented in literature and a more detailed description would be beyond the scope of this book. However in the following paragraph, a practical example of the implementation of a special form of a photodiode is given, which is known as a position-sensitive detector (PSD). Various methods are available for detecting the position of incident light, including methods using an array of many small detectors and a multi-element detector (such as CCD which will be described in Section 4.7.3). In contrast to these, the PSD is a monolithic device designed to detect the position of incident light. Since the PSD is a non-segmented photosensor that makes use of the surface resistance of the photodiode, it provides continuous electrical signals and offers excellent position resolution, fast response, and high reliability. The PSD is used in a wide range of fields such as measurements of position, angles, distortion, vibration, and lens reflection/refraction. Applications also include precision measurement such as laser displacement meters, as well as optical remote control devices, distance sensors, and optical switches.

A PSD basically consists of a uniform resistive layer formed on one or both surfaces of a high-resistivity semiconductor substrate and a pair of electrodes formed on both ends of the resistive layer for extracting position signals. The active area, which is also a resistive layer, has a PN junction that generates a photocurrent by means of the photovoltaic effect.

Above an N-type high-resistivity silicon substrate, a P-type resistive layer is formed that serves as an active area for photoelectric conversion and a resistive layer. A pair of output electrodes is formed on both ends of the P-type resistive layer. The backside of the silicon substrate is an N-layer to which a common electrode is connected. Basically, this is the same structure as that of PIN photodiodes except for the P-type resistive layer on the surface. When a light spot strikes the PSD, an electric charge proportional to the light level is generated at the light incident position. This electric charge flows as photocurrents through the resistive layer and is extracted from the output electrodes $X_1$ and $X_2$, while being divided in inverse proportion to the distance between the light incident position and each electrode.

The relation between the incident light position and the photocurrents from the output electrodes $X_1$ and $X_2$ is as follows:

When the center point of the PSD is set as the origin

$$I_{X1} = \frac{\frac{L_X}{2} - X_A}{L_X} \cdot I_O$$

$$I_{X2} = \frac{\frac{L_X}{2} + X_A}{L_X} \cdot I_O$$

$$\frac{I_{X2} - I_{X1}}{I_{X1} + I_{X2}} = \frac{2X_A}{L_X}$$

$$\frac{I_{X1}}{I_{X2}} = \frac{L_X - 2X_A}{L_X + 2X_A}$$

When the end of the PSD is set as the origin

$$I_{X1} = \frac{L_X - X_B}{L_X} \cdot I_O$$

$$I_{X2} = \frac{X_B}{L_X} \cdot I_O$$

$$\frac{I_{X2} - I_{X1}}{I_{X1} + I_{X2}} = \frac{2X_B - L_X}{L_X}$$

$$\frac{I_{X1}}{I_{X2}} = \frac{L_X - X_B}{X_B}$$

where $I_{X1}$ is the output current from electrode $X_1$; $I_{X2}$ is the output current from electrode $X_2$; $I_O$ is the total photocurrent ($I_{X1} + I_{X2}$); $L_X$ is the resistance length (length of active area); $X_A$ is the distance from the electrical center of the PSD to the light incident position; $X_B$ is the distance from the electrode $X_1$ to the light incident position. By finding the values of $I_{X1}$ and $I_{X2}$ from the equations above, the light incident position can be obtained irrespective of the incident light level and its changes. The light incident position obtained here corresponds to the center-of-gravity of the light spot.

Another example of an implementation of photodiodes with respect to the focused polymer platform in this book is the integration of silicon photodiodes with polymer waveguides manufactured by DUV lithography [119] another example in polymers is presented in [48]. The integrated structure of a PMMA waveguide and a p+n photodiode is shown schematically in Figure 4.34.

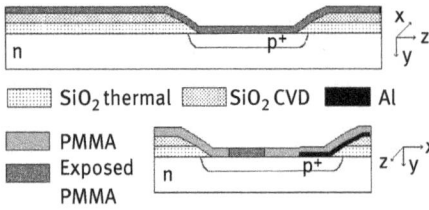

Figure 4.34: Integrated PMMA waveguide with silicon p+n photodiode.

The optical isolation between the PMMA waveguide and the silicon substrate is provided by a SiO$_2$ layer with a thickness of 1.5–2 $\mu$m. On top of the active area of the photodiode, this layer is removed to allow a gradual transfer of the optical power from the waveguide into the photodiode by leaky wave coupling [56]. A controlled degree of coupling within a reasonable coupling length can be achieved with a tapered geometry, and, if necessary, with a thin remaining isolation layer. The leakage parameter, $\alpha_g$, is defined as the amount of the power that is lost per unit length divided by the power carried in the guide. The lost power is transferred in the silicon photodetector, as the silicon refractive index is greater than that of the waveguide. In the coupling region, the power carried in the waveguide is:

$$P_g(z) = P_0 \cdot \exp\left(-\alpha_g \cdot z\right)$$

and the power density transferred in the photodetector is:

$$\frac{\partial P}{\partial z} = P_0 \cdot \alpha_g \cdot \exp\left(-\alpha_g \cdot z\right)$$

For multi-mode waveguides, there is a leakage parameter for each mode. The required parameters were determined using OptiFDTD software (Optiwave). It was found that the coupling efficiency increases when the core thickness decreases. The structures were fabricated on n-type Si wafers 3–5 $\Omega$cm using p-well CMOS compatible processes.

A microscope image of waveguides on a silicon photodiode is shown in Figure 4.35.

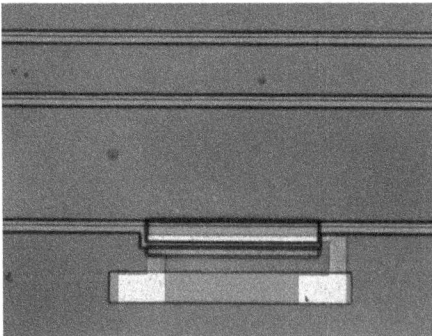

Figure 4.35: Microscope image of a waveguide on a silicon photodiode.

The atomic force microscopy (AFM) measurements proved that the PMMA wave-guide follows the topography of the photodiode smoothly: While the uncoated photo-diode shows a local step of 1.2 μm from the top of the $SiO_2$ layer to the photodiode surface, after coating with PMMA and subsequent generating of the waveguide, this step showed a height of still more than 900 nm. Thus, the light guiding layer is still thin enough to allow an efficient leaky wave coupling. For near-field pattern measurement, PMMA waveguides were fabricated directly on the SiO2 layer without any photodiode devices. The length of the straight waveguide was 4 mm and the width was 7.5 μm. Preliminary tests with a none-optimized waveguide showed a detectable photocurrent at the photodiode and could eventually pave the way for implementation in optofluidic system when combined with the previously described integrated reverse symmetry waveguides.

The following chapter provides an introduction into image sensors that are important devices for imaging and spectroscopy applications and eventually opto-fluidic system applications.

### 4.7.3 Image sensors

This chapter is adapted from and based on the opto-semiconductor handbook from Hamamatsu. Figures, which are taken from the opto-semiconductor handbook are reproduced with kind permission from Hamamatsu.

The CCD area image sensors are semiconductor devices invented by Willard Boyle and George Smith at the AT&T Bell Laboratories in 1970. The CCDs are image sensors grouped within a family of charge transfer devices (CTD) that transfer charges through the semiconductor using potential wells. Most current CCDs have a buried channel CCD (BCCD) structure in which the charge transfer channels are embedded inside the substrate. As shown in Figure 4.36, a CCD potential well is made by supplying one of multiple MOS (metal oxide semiconductor) structure electrodes with a voltage which is different from that supplied to the other electrodes. The signal

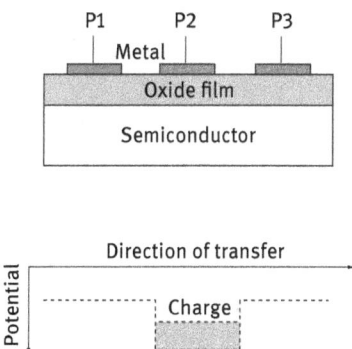

**Figure 4.36:** CCD basic structure and potential well (Courtesy Hamamatsu).

charge packed in this potential well is sequentially transferred through the semiconductor toward the output section. Because of this, the CCD is also called an analog shift register. The CCDs are essentially semiconductor devices through which a signal charge is transferred. Currently, however, the term "CCD" has come to signify image sensors and video cameras since CCDs are widely used as image sensors.

In general, CCDs are designed to receive light from the front side where circuit patterns are formed. This type of CCD is called the front-illuminated CCD. The light input surface of front-illuminated CCDs is formed on the front surface of the silicon substrate where a Borophosphosilicate glass (BPSG) film, poly-silicon electrodes, and gate oxide film are deposited. Light entering the front surface is largely reflected away and absorbed by those components (Figure 4.37). The quantum efficiency is, therefore, limited to approx. In total, 40% at the highest in the visible region, and there is no sensitivity in the UV region.

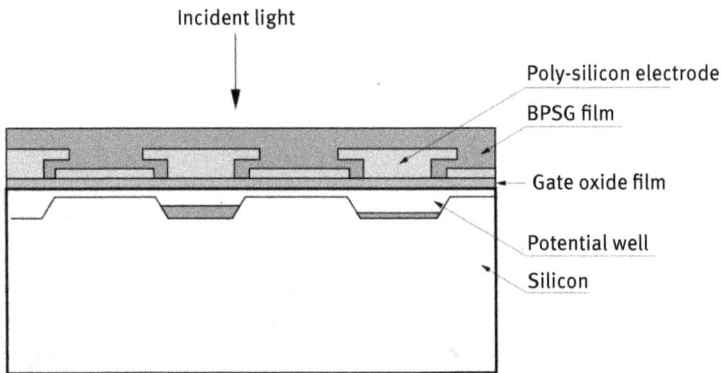

**Figure 4.37:** Front-illuminated CCD (Courtesy Hamamatsu).

Back-thinned CCDs were developed to solve such problems. Back-thinned CCDs also have a BPSG film, poly-silicon electrodes, and gate oxide film on the surface of the silicon substrate, but they receive light from the backside of the silicon substrate.

Because of this structure, back-thinned CCDs deliver high quantum efficiency over a wide spectral range. Besides having high sensitivity and low noise which are the intrinsic features of CCDs, back-thinned CCDs are also sensitive to electron beams, soft X-rays, UV, visible, and NIR light.

In order for back-thinned CCDs to achieve high sensitivity, it is essential to thin the silicon substrate and to activate the active area. The active area is activated by forming an internal potential (accumulation) so that signal charges generated near the backside light input surface are smoothly carried to the CCD potential wells without recombining.

Back-thinned CCDs (Figure 4.38) have high quantum efficiency in the UV to visible region. However, since the silicon substrate is about 15–30-μm-thick, the

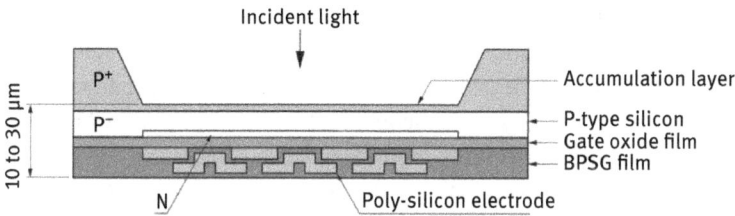

**Figure 4.38:** Internal structure of back-illuminated CCD – back thinned type (Courtesy Hamamatsu).

quantum efficiency in the NIR region is low. For example, the quantum efficiency at a wavelength of 1 μm is approximately 20%. Thickening the silicon substrate improves sensitivity in the NIR region, but the resolution decreases due to diffusion of charges that are generated when the input photons are absorbed in the neutral region (undepleted region) near the backside light input surface.

Fully depleted back-illuminated CCDs (Figure 4.39) were developed to solve such problems. These fully depleted back-illuminated CCDs use an ultra-high resistance N-type wafer to form a thick depletion layer. Using a thick silicon substrate also causes the dark current to increase, so the CCD must be cooled to −70 to −100 °C. Since the silicon substrate thickness for back-thinned CCDs is about 10 to 30 μm, it is difficult to fabricate large-area devices. However, fully depleted back-illuminated CCDs use a silicon substrate whose thickness is 100 to 300 μm over the entire area, so large-area devices can be easily fabricated.

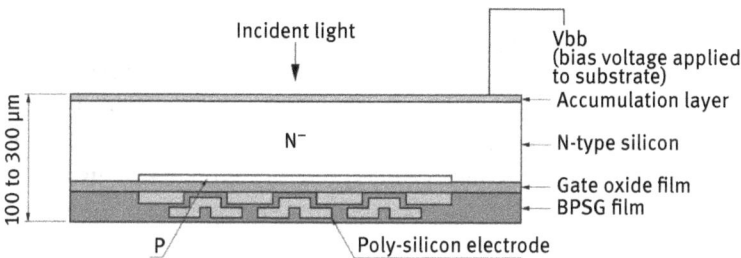

**Figure 4.39:** Internal structure of back-illuminated CCD – fully depleted type (Courtesy Hamamatsu).

In summary, front-illuminated CCDs have no sensitivity in the UV region, and the maximum quantum efficiency in the visible region is approximately 40%. In contrast, back-thinned CCDs deliver very high quantum efficiency, which is approximately 40% in the UV region and approximately 90% at a peak wavelength in the visible region. The fully depleted back-illuminated CCDs use a thick silicon substrate which allows higher sensitivity in the wavelength range from 800 to 1,200 nm than back-thinned CCDs. The fully depleted back-illuminated CCDs also have high sensitivity in the visible region from 400 to 700 nm due to use of a special AR (anti-reflection) coating process; however, the UV sensitivity is low. The spectral response range is determined on long wavelengths by the silicon substrate thickness and on short wavelengths by the sensor structure on the light input surface side.

The following chapter provides an insight into several common laser types.

### 4.7.4 Semiconductor lasers

Lasers (light amplification by stimulated emission of radiation) play an important part in our daily lives and will play in the future.

In this section, the focus is purely on providing an introduction into semiconductor lasers suitable for optofluidic systems. Numerous of books can be found in literature on principles and backgrounds of several different types.

This paragraph deals with Fabry–Perot, distributed feedback and distributed Bragg reflector semiconductor lasers, ring resonator lasers, and vertical cavity surface emitting lasers (VCSEL), which can be used for optofluidic sensing applications. Recently, the research focus is also on so-called optofluidic biolasers and a review on the current progress is provided in [75].

Semiconductor lasers can be generally described by certain typical characteristics. These characteristics have to be kept in mind when choosing a semiconductor laser for optofluidic system applications.

The first selection criteria for Optofluidic system applications are of course the operating wavelength.

The second selection criteria for choosing a laser are the number of longitudal modes or number of optical frequencies as laser emits. This issue will be reflected in the price of the laser, as single-mode lasers are more expensive than multi-mode lasers.

To determine the quality of a single-mode laser, the side-mode suppression ratio (SMSR) is being taken into account. A high SMSR is preferred as this value (usually in dB) reflects the amount by which side modes of the laser are not present in the wanted output laser wavelength.

The minimum current required to switch on the laser is known as threshold current, which should be as low as possible to decrease transmitter power dissipation and thus reduce the performance of the laser.

A measure of how random the optical laser output power is over time is described as laser noise. This parameter is especially important in optical communication networks where a stable laser source is a key to an optimum data transfer rate.

A measure of how "noisy" a laser is can be taken from the laser linewidth, which can provide a good indicator for the overall performance regarding the laser line stability of the laser.

The modulation bandwidth of a laser is of importance in optical data communication systems, as this parameter determines the bit rate that can be achieved by modulating the current of the laser. In this respect, the chirp is mentioned, which is a measure of how the optical frequency changes with current modulation and thus has an impact on the transmission rate.

Other important issues to think of when evaluating laser sources for optofluidic systems are the fiber output power, if the measurement to be performed can be done using optical fibers; the wavelength tenability, if several wavelengths are needed for the application and finally the long-term stability of the laser for certain parameters of interest defined by the application, for example, the wavelength and the output power.

The photograph of a so-called ridge waveguide laser is shown in Figure 4.40.

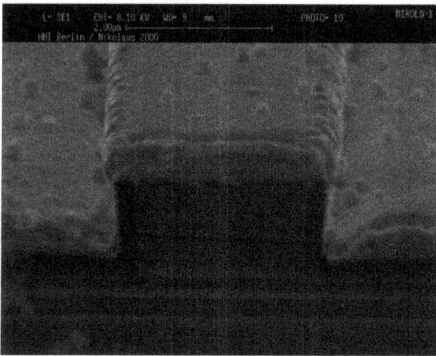

**Figure 4.40:** SEM photograph of a semiconductor ridge waveguide laser.

Ridge waveguide lasers are the basis for Fabry–Perot lasers, DFB and DBR lasers. A Fabry–Perot laser can be easily fabricated out of a ridge waveguide laser by cleaving the laser bars. The facets of the semiconductor chip serve as the mirrors and form a resonator, the Fabry–Perot resonator.

In the case of a distributed feedback laser, a Bragg grating is implemented on the ridge waveguide by semiconductor manufacturing processes, which acts as a longitudal mode filter as the output laser wavelengths has to be a multiple number n of modes which can resonate in the Fabry–Perot resonator from the overall structure and which can resonate in the Bragg grating on top. Typical values for a high power semiconductor ridge waveguide DFB laser are: output power (400–800 mW), line width (< 1 MHz), and side mode suppression ratio (> 60 dB).

Distributed Bragg reflector lasers are also manufactured on the basis of ridge waveguides, but the DBR laser consists of several sections, for example, the gain section and the reflector section. These sections are combined to form the laser. State of the art DBR lasers consists of up to five sections, which includes, for example, the gain section, the phase section, and the absorbing section as well.

Another type of laser, which has also gained attention in the optofluidics community is the ring resonator laser. In contrast to semiconductor-based ring resonator lasers, optofluidic ring resonator lasers are made in polymers and use an organic dye as the gain medium.

Another important type of laser is the vertical surface emitting laser (VCSEL). As the name suggests, this laser emits in the vertical direction and not from the cleaved facets as it is the case for Fabry–Perot or DFB lasers. This is a key advantage of VCSELs, as they can be tested on the chip before separation. The laser resonator is realized using layers of different refractive indices, thus manufacturing, for example, a Bragg resonator in the vertical direction.

Compared with other laser diodes, VCSELs have the following additional features. They can be used to fabricate monolithic optical cavities, for which cleaving is not needed. The laser light is coupled easier to optical components, because circular beams with small radiation angles are obtained.

The following paragraph deals with organic active components.

### 4.7.5 Organic LEDs, organic PDs and organic lasers

The use of organic optoelectronic devices such as organic light-emitting diodes and OPDs in applications for optofluidic systems is discussed in this chapter, which is based on [202–205].

Device characteristics including emission spectra, I–V curves and the dynamic behavior are presented. In the combination with a polymeric optical fiber (POF), a transmission line comprising of organic LED and OPDs is demonstrated. An important step toward integration is realized by coupling the amplified spontaneous emission of an organic semiconductor material into a single-mode polymethylmethacrylate (PMMA) waveguide, which is fabricated using the aforementioned DUV lithography method.

Optoelectronic components based on organic semiconductors are opening new possibilities for a great diversity of systems and applications. There is a wide range of organic semiconductor devices. Light emitters, photodiodes, solar cells, transistors as well as organic lasers have been demonstrated. The versatility of organic components and their application in integrated systems attract much attention for this class of materials.

Organic LEDs (OLEDs) are already having a significant impact on the display market. Their features, such as low energy consumption, full colors, and a wide viewing angle, render them ideal for small consumer electronics devices, for example, MP3 players. Electronic circuits based on organic semiconductors will have a big impact. Radio frequency identification (RFID) tags will be used for low-cost electronic tagging. Photodetectors based on organic materials are heavily investigated in the field of photovoltaics. Organic solar cells prepared by low-cost processes are expected to bring down the costs for solar electricity. Efficiencies of more than 5% have recently been demonstrated. This progress closes the efficiency gap to amorphous silicon solar cells being on the market.

The integration of organic semiconductors into optofluidic systems is of interest lately, which has led to several optofluidic lasers and devices (Handbook of Optofluidics). The combination of different organic devices for integrated optofluidic systems, promises completely new products and an improvement of existing systems. The ease of processing and the possibility to deposit these devices basically on any substrate render organic semiconductors attractive for integrated systems in data communication and sensing application.

Two basic applications for integrated organic semiconductor devices, namely, optical interconnects [22, 183] and lab-on-a-chip devices, are discussed in the following paragraph.

The rapid development of computer and telecommunication systems has led to dramatic improvements in computing power and data transfer rates. One possible solution for forthcoming challenges including higher data transfer rate, transmission distance, reduced cross talk, and heat-dissipation is using optical interconnects (OI). These systems can be used at several levels of a computer or a communication system like cabinet-to-cabinet, board-to-board, and chip-to-chip. Optical interconnects are basically the combination of a modulated light source and a photo detector using some kind of optical waveguide in between.

The light source can either be modulated actively or can be used in continuous wave (cw) mode using an additional modulator. Waveguiding can be realized in planar form as well as in fiber systems. The optical receiver decodes the transmitted signal utilizing a photodetector and subsequent electronics. Several components of OI based on organic semiconductors have been investigated recently. The authors in [189, 270] and [104] demonstrated both OLEDs and organic photodiodes (OPDs) on polymeric waveguides for data transfer purposes. Polymeric waveguides and polymer optical fibers (POF) are already used in commercial systems [93]. Electrooptic modulators based on organic materials have reached modulation frequencies in the 100 Gigahertz regime [46, 59, 72]. A complete system made from organic materials could provide a highly integrated and very cost-effective solution for optical interconnects. Organic devices such as OLEDs and OPDs can be fabricated on all kind of substrates including silicon, glass, and plastics. Fabricating organic devices on plastics enables flexible devices. Direct patterning of organic devices on waveguides would significantly simplify the assembly process, avoiding the most expensive fabrication steps.

Sensor systems on the microchip level have gained great interest in recent years. Especially, the term "lab-on-a-chip" is commonly used in this context. Devices with "lab-on-a-chip" characteristics are small sensor systems with a high level of integration ideally suited for application in optofluidic systems. Possible applications for such optofluidic systems are in the areas of medicine, drug discovery, and environmental monitoring. There is a clear trend toward decentralization of analytical tasks, that is, directly at the point-of-care in medical analysis instead of clinical laboratories [1, 83]. Optical analysis, such as absorption measurements and laser-induced fluorescence (LIF), are widely used in life science.

A very important part is the excitation source for the analysis. Here, incandescent lamps and various lasers (diode and non-diode) are often used. Recently, both inorganic and OLEDs are employed to minimize both the system size as well as the costs [41, 52, 106, 280]. The light is guided to the analyte by conventional free-space optics or integrated waveguides. The analyte is either placed in a special detection chamber or in a more complex microfluidic system that provides the transport into the detection chamber. The resulting optical signal is then monitored by the detector. Photomultiplier tubes and semiconductor photodiodes are the most common choice on the detector side. The use of organic devices in "lab-on-a-chip" platforms promises completely integrated systems containing all necessary components for an analysis. A monolithical integration of OLEDs and OPDs with polymeric waveguides and microfluidic systems on a single substrate would open the way to smaller and cheaper optofluidic systems. Organic devices can be fabricated on a broad range of substrates including materials often used in microsystems, for example, glass, poly-dimethylesiloxane (PDMS), polymethylmethacrylate (PMMA), and other plastic materials.

In the following paragraph, the fabrication and characterization of different OLEDs as well as organic photodiodes is described.

Organic semiconductors for OLEDs can be separated into two classes: small molecules and polymer materials. The first class refers to the low molecular weight of the materials. High vacuum thermal evaporation is used for the deposition of these molecules. Polymer materials on the other hand consist of long chains of conjugated monomers. These polymers are dissolved in an organic solvent and processed from solution by wet processing, for example, spin-coating, doctor-blading, or inkjet-printing. Different materials for both classes of applications are investigated. Processing from solution promises a more cost effective fabrication, whereas vacuum deposition is providing a very controlled deposition of ultrathin organic layers without the use of solvents. One of the goals is to cover a wide range of emission wavelengths using different emitters for the fabrication of OLEDs.

The OLEDs have been fabricated on indium tin oxide (ITO) glass substrates. The ITO substrates are cleaned by ultra sonication in acetone and isopropyl alcohol and cleaned subsequently in an oxygen plasma chamber. Hole transporting layers are triphenyl diamine (TPD) derivatives for the small molecule devices and poly(3,4-ethylenedioxythiophene)-poly (stryrenesulfonate) (PEDOT:PSS) for the polymer OLEDs, respectively. High vacuum thermal evaporation was used for deposition of the small molecule materials tris-(8-hydroxyquinoline)-aluminum (Alq3) (undoped and doped with the laser dye (4-(dicyanomethylene)-2-methyl-6-(p-dimethylaminos-tyryl)-4H-pyran) (DCM)). The active layers of the polymer OLEDs are a polyfluorene-derivative (poly(9,9-dioctylfluorenyl-2,7-diyl) – end capped with N,N-Bis(4-methyl-phenyl)-aniline) (Sigma-Aldrich PF6/2am4)) as well as a PPV-derivative (poly[2-methoxy-5(3,7dimethyloctyloxy)-1,4-phenylenvinylen] (Covion PDO 121)). These materials are deposited by spin-coating in an inert gas atmosphere. In both cases, the cathode materials are thermally evaporated onto the organic layers. In Figure 4.41, the layer configuration and thicknesses of an example device is shown.

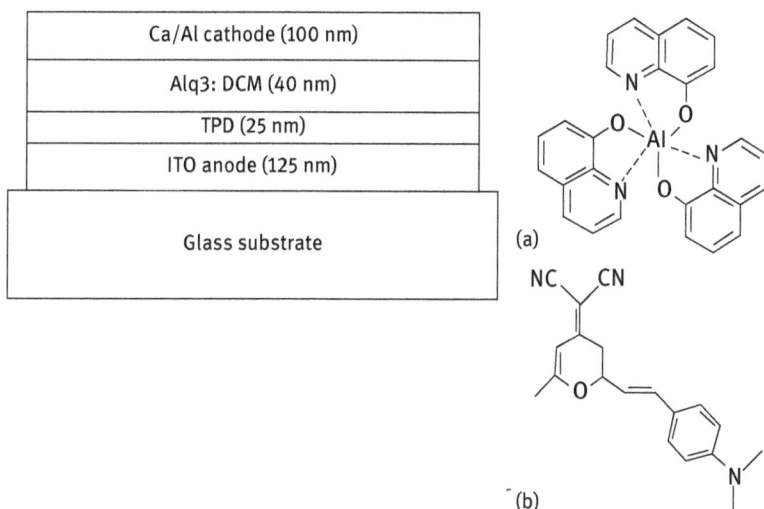

**Figure 4.41:** Alq3:DCM-OLED scheme and chemical structure of the molecules (top:Alq3; bottom: DCM).

The devices are tested using a computer-controlled characterization system. This system consists of an automated rotation stage with sample holder, a fiber optic spectrometer (StellarNet EPP2000), and a source measurement unit (Keithley SMU 238). The setup is calibrated using a standard of spectral irradiance (1000 W FEL type) and provides accurate spectroradiometric and photometric quantities. For sensor applications, the total radiant flux (Watts), the radiant existence (Watts/m²), and the emission spectra of the OLEDs are the most relevant quantities. The spectral characteristics of the different OLEDs are shown in Figure 4.42.

All OLEDs show the characteristic broad spectra of organic emitter materials. Similarly to the case of dye lasers, the whole visible range can be covered with different materials. As a blue emitter polyfluorene is chosen, with an emission maximum at approximately 420 nm. The green emitter is undoped Alq3 that exhibits an emission maximum at a wavelength of 520 nm. For the orange part of the spectrum, OLEDs based on the polymer PDO 121 can be used while a deeper red is generated in devices using the small molecule Alq3 doped with the laser dye DCM as the emissive guest material.

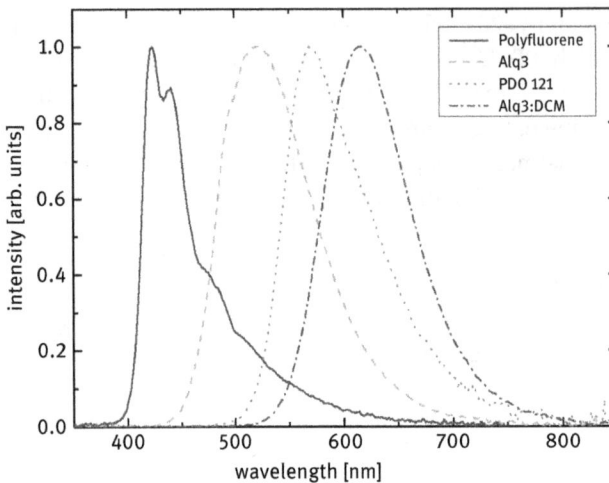

**Figure 4.42:** Electroluminescence spectra of the different OLEDs.

For the application in optical interconnects as well as in sensor systems the dynamic response of an OLED is of crucial importance. A dynamic characterization of a polyfluorene OLED is carried out. The OLED is fabricated as described in the previous paragraph with a smaller active size (approximately 1 mm²) to reduce the capacitance of the device. Patterning of the OLED is realized by depositing an insulating $SiO_2$ layer through a shadow mask between the ITO anode and the hole transport layer. The OLED is driven by a signal generator (Wavetek Model 801) with square pulses at different modulation frequencies. The resulting optical signal is detected using a Si-PIN photodiode (Thorlabs DET 210) and monitored by a digital sampling oscilloscope (Welleman Instruments PC scope). Figure 4.43 depicts the modulation characteristics of the OLED when a 100 kHz square pulse train (Input voltage: 15 V) is applied.

By applying a forward bias voltage in the order of the turn-on voltage of the OLED, the dynamic response can be improved.

The OLED follows the modulation frequency up to 300 kHz. The dynamic behavior of an Alq3:DCM-OLED is also analyzed. A modulation frequency of up to 1 MHz is obtained.

The thinner active layers in small-molecule OLEDs are resulting in smaller carrier transit times thus allowing a faster modulation.

Another important part in optical interconnects and sensor systems is the photo detection. The composite system made of the conjugated poly-(3-hexylthiophene) (P3HT) and the fullerene derivative [6,6]-phenyl-C61 butyric acid methyl ester (PCBM) for the use as organic photo detectors is investigated and described.

A schematic of the organic photodiodes used is shown in Figure 4.44.

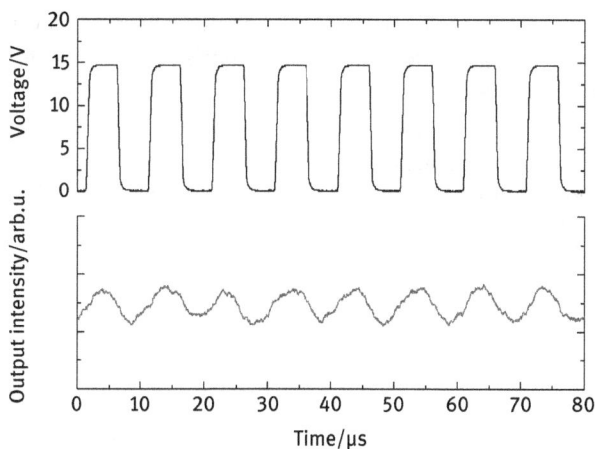

**Figure 4.43:** Modulation of a polyfluorene-OLED with 100 kHz pulses from a signal generator.

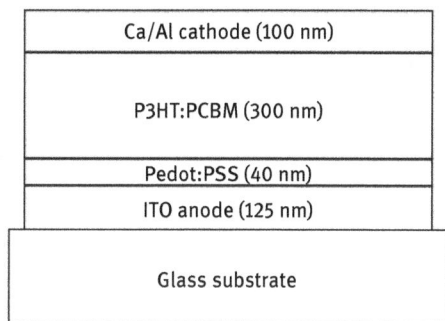

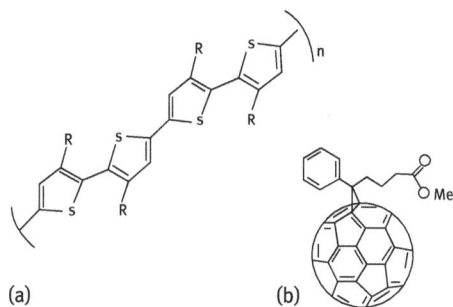

| Ca/Al cathode (100 nm) |
| P3HT:PCBM (300 nm) |
| Pedot:PSS (40 nm) |
| ITO anode (125 nm) |
| Glass substrate |

(a)          (b)

**Figure 4.44:** Device structure of the OPD and chemical structure of the active materials (top: P3HT; bottom: PCBM)

On top of cleaned glass substrates, a sputtering process deposits an ITO anode in a commercial deposition system (Leybold Univex 350). The ITO films show a good transmission rate (more then 85%) and a sheet resistance in the order of 200 Ohms/ square without an annealing process. The active size (approx. 1 mm²) of the OPD is defined by an intermediate layer made from insulating $SiO_2$ between ITO and the spin-coated hole transport layer (HC Starck PEDOT:PSS). The photon-absorbing layer is a blend of P3HT and PCBM with a weight ratio of 1:0.9. The polymer P3HT acts as an electron donor upon incident light, while the fullerene derivative PCBM is the electron acceptor. This enables an effective charge carrier separation. Finally, an Al electrode (thickness 100 nm) is thermally evaporated onto the organic layer. The absorption spectrum of the active materials is shown in Figure 4.45. In visible, the absorption is dominated by P3HT.

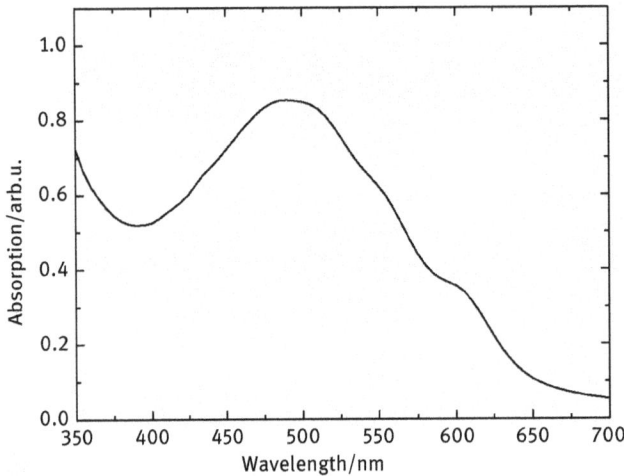

**Figure 4.45:** Absorption spectrum of P3HT:PCBM.

For the dynamic characterization of the OPDs, pulsed inorganic LEDs with a peak emission wavelength of 466 nm are used as illumination sources. The LEDs are modulated by a frequency generator (Wavetek Model 801). The dependence of the photoresponse on applied bias voltages is also investigated. In Figure 4.46, the I–V characteristics of the OPD in the dark state and under illumination with a simulated solar spectrum are demonstrated.

The device shows a very low dark current in the reverse bias region and a good photo-response under illumination. The photoresponse of the OPD under pulsed optical illumination is shown in Figure 4.47.

The device shows a good response to the 20 kHz input modulation. The dependence of the photoresponse with different applied bias voltages is shown in Figure 4.48.

The device reaches a 3dB cutoff frequency of 17 kHz without an applied bias. A reverse voltage of -0.6 V is improving the response thus enabling the detection of modulated signals of up to 50 kHz. A further improvement of the dynamic response is possible with even smaller active areas and hence reduced capacitances.

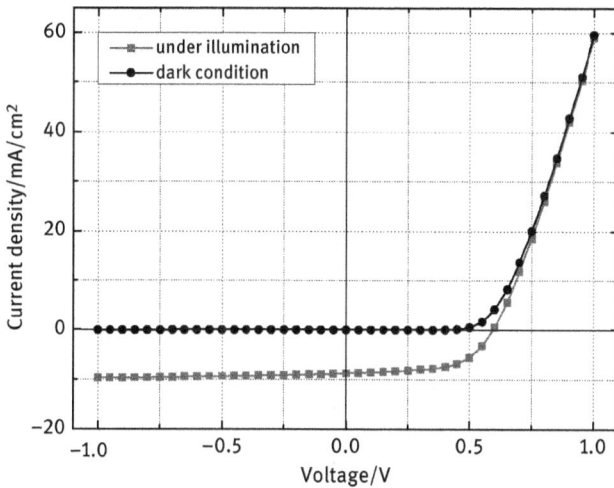

**Figure 4.46:** I–V characteristics of the OPD in dark condition and under illumination.

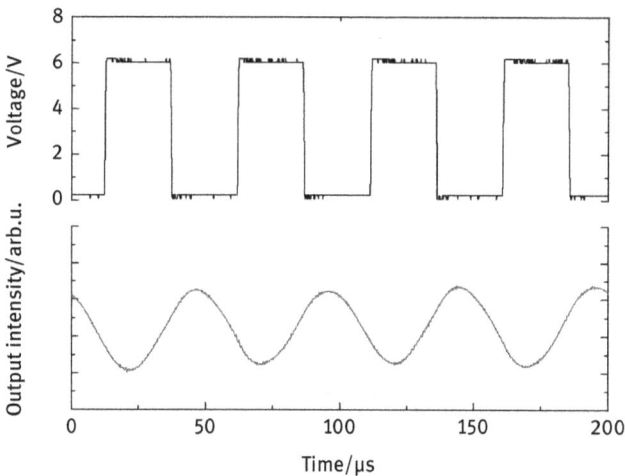

**Figure 4.47:** Photoresponse of the OPD under illumination with a blue LED.

Alternatively, lateral organic photodiodes with extremely short electrode spacing might lead to shorter response times [54].

The following paragraph will provide an overview of realized example systems comprising organic optoelectronic devices. At first part, an optical interconnect consisting of an OLED, a polymeric optical fiber (POF), and an OPD is demonstrated. In the second part, an organic semiconductor lasers coupled to waveguides is described.

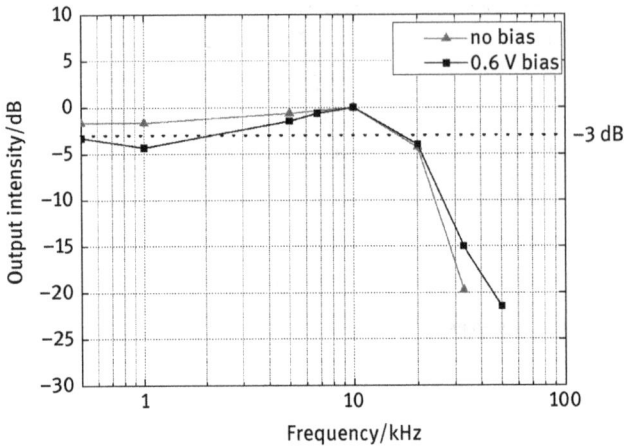

**Figure 4.48:** Dependence of the output intensity vs. the illumination frequency and bias voltage.

An optical interconnect using a polyfluorene OLED and a P3HT:PCBM OPD as described in the previous paragraph is fabricated. Both organic devices are linked via a polymeric optical fiber (core size 0.5 mm). The fiber ends are polished and attached to the glass substrate of the organic devices with a special glue.

The OLED was driven by a signal generator with an applied voltage of 17.5 V. The resulting photoresponse of the OPD was fed into a transimpedance amplifier and recorded using a digital sampling oscilloscope. Using this simple passive alignment, we were able to transmit a pulse train with a modulation frequency of up to several kHz.

Another class of organic devices has gained much interest in recent years: organic semiconductor lasers [95, 96, 230, 231, 243, 275]. The unique properties of such lasers render these devices very interesting for applications such as optical sensing in optofluidic systems. Like OLEDs, organic lasers can cover a wide range of light emission wavelengths spanning from the UV [135, 222] throughout the whole visible range. Even using only one active laser material, the tunability can be as high as 50 nm [222]. Up to now optically pumped organic lasers are realized, and research toward an electrically pumped organic laser is ongoing [12, 148, 185, 198, 221, 278]. Organic lasers are ideally suited, for example, for LIF applications. The possibility of the integration of such lasers in "lab-on-a-chip" devices would allow to build small, all-integrated devices at very low-cost for optofluidic systems. A step on the way to such devices is the coupling of mirrorless lasing (amplified spontaneous emission, ASE) of an organic material into a polymeric waveguide.

For the coupling experiment, a patch of the organic laser material Alq3:DCM (thickness 350 nm) is fabricated onto a PMMA substrate with integrated waveguides (width 10 μm, DUV lithography) in a thermal evaporation process under high vacuum conditions. The organic material has a refractive index of approximately 1.76 thus building a slab waveguide on the PMMA substrate (n = 1.5). The material is then optically pumped with a short-pulse UV-laser (Crystal laser FTSS 355-Q; wavelength 355 nm) under inert gas atmosphere. At high enough pump intensities, the gain in the active material exceeds the losses and spontaneously emitted photons are amplified in the waveguide. A collapse of the emission spectrum is observed, since only those photons are amplified whose energy coincides with the spectral position of the maximum material gain. Figure 4.49 shows the resulting gain narrowing.

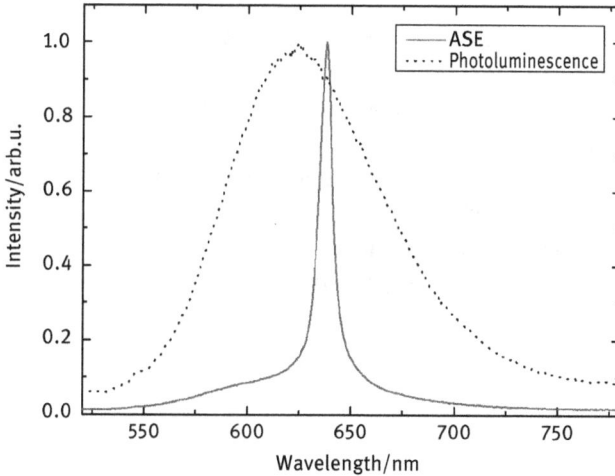

**Figure 4.49:** ASE in Alq3:DCM layers (ca.100 μJ/cm²).

Even without a resonator structure, the resulting emission spectra of the Alq3:DCM material is spectrally narrowed. This might be an interesting alternative for semi-conductor laser and LED excitation of fluorescence marker molecules. The amplified spontaneous emission in the Alq3:DCM patch on the PMMA substrate is coupled into the waveguide via the evanescent field.

The basis for small and portable optical sensor systems is inexpensiveight sources. Light emission is utilized in optical detection schemes such as laser-induced fluorescence, absorption, or evanescent field sensing. While organic and inorganic LEDs are comparatively cheap and small, certain applications require a laser light source [202–205]. Here, organic solid-state lasers are a promising choice as their fabrication involves only few steps and they cover a wide spectral range from UV to NIR.

The fabrication of optically pumped organic solid-state lasers consisting of a nanostructured substrate coated with an active laser material is demonstrated. The periodically nanostructured substrate forms a 1D- or 2D-photonic crystal resonator providing the necessary optical feedback for lasing. The feature size of the resonator structure is as small as 200 nm. Aluminum tris(8-hydroxyquinoline) (Alq$_3$) doped with the laser dye 4dicyanomethylene2-methyl-6-(p-dimethylaminostyryl)-4H-pyran (DCM) is used as active material. As the refractive index of the active layer is chosen to exceed that of the substrate, a planar waveguide is formed supporting the laser mode.

The fabrication of the laser resonators by UV and thermal nanoimprint technologies is described. One goal of the work is the development of a process for the cost-efficient fabrication of organic laser in a wafer-scale production. Furthermore, the combination of the organic lasers with polymeric waveguides is demonstrated. This opens the way to a further integration into chip-based microfluidic sensors for optofluidic systems.

The master structures for both imprint technologies are fabricated by e-beam lithography on an oxidized silicon wafer. The 70-nm-deep resonator structures are transferred onto a silicon wafer via RIE. The wafer is then coated with a metal seed layer and electroplating forms a nickel shim.

For UV nanoimprinting, the inorganic–organic hybrid material Ormocer® is used as the resonator material. Ormocer® is a very stable material with low absorption in the visible wavelength region. To protect the nickel shim, a secondary master fabricated by hot embossing in the Cyclo-Olefin-Copolymer Topas® is employed. The picture of the imprinted structures can be seen in the inset of Figure 4.50.

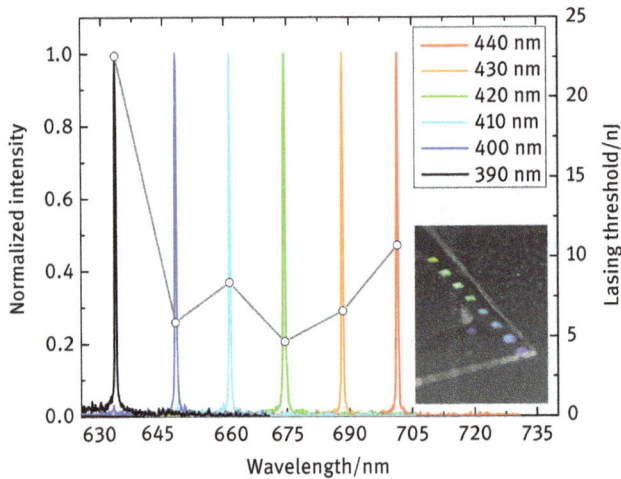

**Figure 4.50:** Emission spectra and lasing thresholds of the Alq3:DCM lasers with different resonator periodicities. Inset: Photograph of the imprinted structures.

After high-vacuum deposition of the active organic material Alq$_3$:DCM on top of the Ormocer® resonator, the material is optically excited by a UV-laser. Figure 4.50 shows the dependency of the laser wavelength and threshold on the resonator periodicity.

This demonstrates the feasibility of the UV nanoimprint technology for the fabrication of high-quality laser resonators. Fabrication of multi-mode waveguides with

Ormocer® using UV lithography [40] has been demonstrated, thus the combination of laser resonator structures and waveguides for integrated systems is possible.

In a second step, organic lasers on polished poly(methyl methacrylate) (PMMA) wafers using thermal nanoimprint lithography (also referred to as hot embossing: refer Section 3.5) are realized.

The imprint technology and the DUV-modification of PMMA are combined to fabricate waveguide-coupled organic lasers. After imprinting the resonator structures directly with the nickel shim, waveguides are defined by a subsequent lithographic step using a DUV light source (200–240 nm) and a chromium mask. The waveguides are patterned either in the immediate vicinity or across the resonator structure as shown in Figure 4.51.

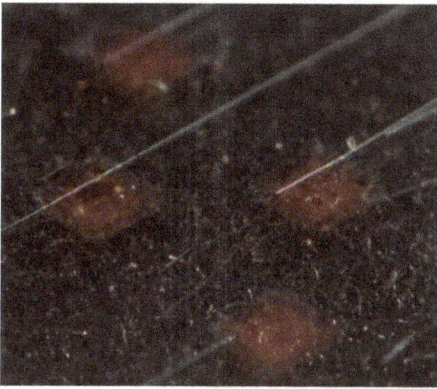

**Figure 4.51:** Photograph of the waveguide-coupled organic lasers. The patches are the laser resonators with the active material on top, the thin lines are the waveguides defined by DUV-modification in PMMA.

For the characterization, the active material is pumped optically using a short pulse UV-laser. The resulting laser emission is coupled into the waveguides by evanescent field coupling. The guided laser light is detected at the end facet of the waveguide using a fiber-coupled spectrometer. The laser emission is centered at the Bragg wavelength of 643 nm with a full width-half maximum of 0.4 nm (refer Figure 4.52).

To fabricate a complete optofluidic system, microfluidic channels can be combined with the organic laser sources by a second imprinting step or by using a tool, which combines channel and resonator structures. This is a step toward the wafer-scale production of an integrated optofluidic system.

## 4.8 Biology

A chapter with the heading biology in a predominantly engineering book can already be regarded as not something special as in recent years, the merger of disciplines has led to a number of novel applications not only in the biosensor area. Cells are grown

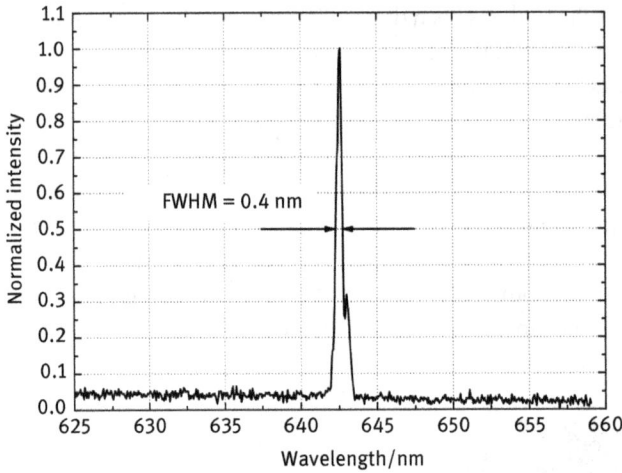

**Figure 4.52:** Spectra of the emission of an Alq3:DCM laser measured at the end facet of the waveguide.

on different substrates to eventually serve as spare parts in medicine. This trend will definitely continue in the future and, therefore, this chapter is devoted to cell-patterning focusing on two types of cells, fibroplasts, and neural cells. This chapter shall be regarded as a basic introduction as further literature is readily available.

The focus of cell patterning is due to the fact that cells need to be in contact with the optical waveguide to be analyzed. The principle of evanescent field type wave-guide sensors is shown in Figure 4.53

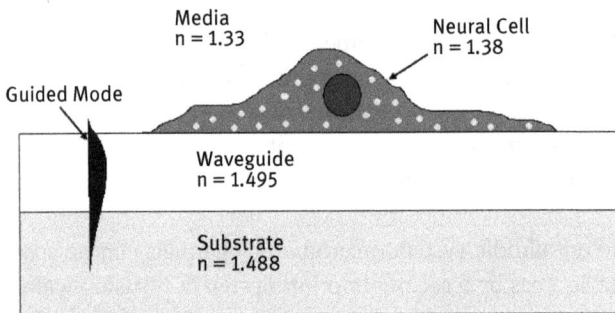

**Figure 4.53:** Evanescent wave excitation of fluorescent molecules in living cells patterned on optical waveguides fabricated by DUV lithography (Courtesy R. A. Seger).

The example in Figure 4.53 shows a waveguide fabricated with DUV lithography and the guided optical mode. As can be seen from the schematic, the evanescent field only touches the bottom of the cell. Depending on the type of application, this might already be sufficient. If a larger penetration is required, integrated reverse symmetry waveguides can be used as described in a previous chapter.

The following chapter provides a short overview on different cell types which can be used for experiments. The information is based and adapted with kind permission from R.A. Seger.

## 4.8.1 Cell types

The focus in this chapter is on cell types, which can be used for experiments in laboratories throughout the world, especially for interfacing with the artificial world.

### 4.8.1.1 Helix aspersa cultures

It is useful to have a supply of readily available, robust cells that can be easily obtained for use in experiments. The ganglion ring from Helix aspersa fits these requirements quite well. Helix aspersa, also known as the common brown garden snail, is one of the best known terrestrial mollusks. Originating in Europe, Helix aspersa was imported to the USA in the early 1850s to be used in escargot as they were used in France. They were not well received by the public and some were released. After proving to be well adapted to life on the pacific coast Helix aspersa became a common pest in Californian gardens [16]. Specimens are easily obtainable and simple to care for. Neurons obtained from the animal can be maintained at room temperature and tend to be relatively large (~30–100 µm). Because the cells are large they tend to have large action potentials, and so can be used to help troubleshoot electrophysiology setups [128].

Central nervous system (CNS) neurons from Helix aspersa are a useful cell for use in experiments for several reasons. However, there are drawbacks as well. The CNS neurons from Helix aspersa have many useful qualities for use in experimentation. They are large, robust, easily obtainable, and inexpensive. Due to their large size (~30–100 µm), the cells are more easily manipulated and can more easily interface with micropatterned devices, such as microelectrodes, than smaller cells might be able to. Larger cells additionally indicate more ion pores in the cell membrane, leading to more current flowing across the cell membrane, which can lead to the detection of larger signals. Their large size contributes to their robust nature. Aspersa neurons can be handled relatively aggressively when not attached to a rigid substrate. The cells can also be grown in ambient conditions, at room temperature and without a modified atmosphere. This directly leads to the added benefit of being able to run an entire culture within a fume hood without ever having to remove the sample, minimizing the risk of contamination.

Due to the fact that Helix aspersa is essentially a pest in California, they are easy and free to obtain. Likewise, CNS cells from aspersa can be obtained through a relatively simple cell culture procedure outlined below.

Although there are many positive attributes associated with CNS cells from Helix aspersa, there are several downfalls which keep these cells from being an entirely ideal cell for the use when developing devices for use with mammalian tissue. Because the work presented here is meant to lead to the future development of in vivo tools for neurological applications, neurons that are very similar to human cells are the most ideal. In this regard, Helix aspersa cells are ill suited as they are very poor models of mammalian cellular systems for multiple reasons. The size and speed of cell growth between aspersa cells and mammalian cells tend to be different and can be significantly so [14, 250]. This can lead to unreasonable expectations in mammalian cell studies. Aspersa cells are monopolar, meaning that they have only one axon and no dendrites extending from the cell body. Additionally, because the cells obtained are mature at the time of culture, they generally do not form chemical synapses, but rather electrical synapses. As most synapses tend to be chemical in nature it would be more useful to have a cell model that formed chemical synapses between cells.

The CNS of Helix aspersa consists primarily of a ganglion ring encircling the gut. Nerves radiate outward from this ring to innervate the rest of animal body.

Cell culture of Helix aspersa CNS cells is relatively straightforward and only a brief overview is presented here.

Helix aspersa is nocturnal and prefer warm, wet environments. Optimal times to look for them are at night directly after a rain, or after sprinklers have been running in the spring or summer. During the day, snails tend to be dormant and can often be found on the undersides of leaves in gardens. After capture aspersa can be fed carrots. After several days, this will lend an orange color to the gastrointestinal tract allowing distinction between the gut and the ganglion ring surrounding the gut. As it is ultimately the ganglion ring that is of interest, this can be a useful technique to aid in dissection. It should be noted that all dissections should be performed in a laminar flow hood to minimize the chance for infection, as not fungicides or antibiotics are used in the tissue culture. Generally, cultures are well attached and can be used roughly after five days in dissociated cultures.

### 4.8.1.2 Primary rat central nervous system cultures

Due to the limitations discussed above, it is useful to have a model cell more closely related to those found in humans. Neural cells obtained from laboratory rats are sufficiently similar to human cells to be useful in making limited predictions about interactions with human tissues. However, rats, with a mature nervous system make fewer synapses, are harder to culture and exhibit less plasticity than cells obtained from immature specimens. Therefore, cells are obtained from embryonic rats from either the hippocampus or the cortex, both of which are obtained in areas of the brain, which have direct analogs in the human brain. Sprague-Dawley rats are used for the experiments presented in [11]. The neurons are used to make a large number of connections and will spontaneously fire given the right conditions.

While the rat brain is inherently simpler than a human brain, the individual cells function basically identically to those found in humans, and the basic structure of the rat brain is very similar to that of a human's. Two different primary cells are described in the following paragraph. Both hippocampal and cortical neurons are investigated. Both types of cells can be grown in cultures without a co-culture and with other supporting cells. They reach maturity in roughly 10 days in culture and can begin spontaneously firing within this timeframe. These cells will make many connections with neighboring cells and will generally form chemical synapses between one another, as seen in Figure 4.54.

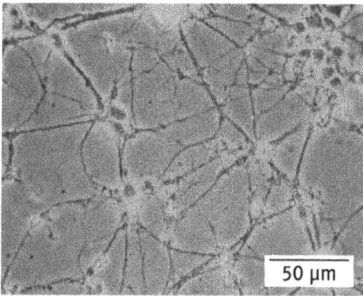

**Figure 4.54:** Phase contrast micrograph showing multiple rat hippocampal neurons grown on a glass coverslip. Notice many connections are made between the cells in the image, which is typical of this type of culture (Courtesy R. A. Seger).

50 μm

These traits make them ideal for in vitro experimentation. Although these cells are useful as human cell models, they are not without their drawbacks. The long incubation period before the cells are electrically active requires a substantial investment in materials (e.g., microelectrode arrays or optofluidic devices) that cannot be used while the cells mature (assuming the cells are grown directly on the devices).

The long incubation time presents a problem of infection as well. Because cell characteristics can be altered by the presence of antibiotics or fungicides, both are excluded from cell cultures performed here, resulting in cultures, which are extraordinarily susceptible to culture destroying infection [237]. Consequently, all culture practices must be performed under sterile conditions. The cells are smaller than their Helix aspersa counterparts, with a soma of roughly 10–20 μm in diameter. This has the potential effect of action potentials being smaller in magnitude, due to fewer total ion channels combined with a smaller likelihood that an extracellular microelectrode will be completely covered.

All experiments were performed with cells obtained from BrainBits, LLC (www. brainbitsllc.com). Cells are dissected from E18 Sprague-Dawley rats on the morning of every Tuesday and Thursday, and shipped overnight for delivery on Wednesday or Friday, ensuring maximum cell viability. Cells are shipped in Hibernate®, a proprietary media created expressly to keep neural cells alive and viable under atmospheric conditions and under refrigeration [31]. Viable cultured neurons in ambient carbon dioxide and hibernation storage for a month. Cells can be stored in such conditions for up to a week. Tissue is shipped as a whole portion of the brain (e.g., entire and intact hippocampus or cortex) and needs to be dissociated before cells can be plated.

Application of adherent biomolecules, such as polylysine or laminin, is used to aid in cell attachment and growth [257]. Cells are typically plated at densities ranging from 160 to 300 cells/mm$^2$. By maintaining sterile conditions when culturing cells, antibiotics and fungicides are unnecessary leading to a closer approximation of in vivo conditions for cells. Additionally, by culturing in the suggested media (B27/ Neurobasal +0.5 mM glutamine +25 um glutamate), the need for co-culturing with astrocytes is obviated, as all essential nutrients are included. If cell cultures are too dense it becomes difficult to differentiate individual cells, too sparse and cells may not spontaneously fire. An order of cells from BrainBits will have at least $1 \times 106$ viable cells, and cortices may have up to double that number. One half of the media should be changed once every 3 to 4 days.

The following chapter describes the patterning of different cell types suing different methods.

## 4.8.2 Patterning of living cells

Cell patterning is of great interest not only in the biology field but also in neuroscience. Being able to control cell growth, predefined networks can be grown and studied in vitro, which enables the study of neural cells and neural cell networks process information.

Patterned arrays of living animal cells are essential for studies of a variety of defined cell (co-)cultures, cell differentiation, intercellular communication, and for engineered tissues as highly functional bioartificial organs for therapy and drug research [2, 21, 200, 226, 244].

Several methods exist for culturing cells and specifically neuronal cells in predefined patterns. The focus in the following chapters is DUV lithography, microcontact printing (µCP), and microelectromechanical systems (MEMS) printing. These methods show good resolution, and the ability to grow neural cells in predefined patterns on various substrates. Especially, polymer substrates are the focus of this chapter as they are essential if a optofluidic system platform is envisaged. The method of choice for cell patterning depends as usual on the application.

The following chapters provide an insight into the patterning of fibroblasts and neural cells using DUV lithography, followed by a basic introduction into microcontact printing and MEMS printing.

### 4.8.2.1 Fibroblast patterning by DUV lithography

The surface modification was performed by an UV irradiation of polymer samples in air using a low pressure mercury lamp (Heraeus Noblelight, NNQ lamp, $\lambda = 185$ nm, 15 W) at 10 cm distance for 30 min. For patterned exposure, a chromium mask on quartz was placed in contact to the polymer surface [262, 263].

The UV irradiation [57, 124, 174, 255] induces: 1. The *formation of peroxides* at the polymer surface allows graft coupling. Peroxides decompose within a couple of days during storage. 2. *Formation of carboxylic acids* affecting cationic dye binding, wettability and surface energy contributions. 3. *Surface ablation.* Due to corrosion at longer irradiation times, a constant COOH surface density is obtained at t > 60 min. Carboxylic acids groups were found to be stable for shelf lives of several months.

Due to these changes in surface chemistry, the adsorption of albumin (BSA) is hindered on irradiated as compared to unmodified regions. In addition, the viscosity of the surface-bound albumin layer on PS ($\rho = 7,6 \pm 1,3$ [$10^{-3}$ Ns/m$^2$]), being a cell-repellent material in its native state, is lowered on irradiated surfaces ($4,5 \pm 0,4$) [108, 172]. Laminin or other proteins from fetal calf serum (FCS) are preferentially adsorbed on photo modified areas and serve as adhesion stimulus for different cells (hepatoma cell line HepG2, L929 fibroblasts and others).

Masked DUV irradiations of polymers (PS, PMMA, PC) opened a simple, fast, and economical route to obtain chemically patterned substrates for structured cell adhesion. UV patterning is advantageous as compared to alternative techniques [43], Detrait et al. 1998, [28, 175] due to the elimination of any wet chemical treatment, the clean room compatibility, the small size of achieved structures, and possibility to produce gradients.

The described cell patterning allows in combination with planar optical waveguides produced with the DUV method described in Section 2.1.5 the non-invasive online monitoring of living cells in culture [259]. Examples of successful patterning on waveguides and perpendicular to waveguides are shown in Figures 4.55 and 4.56.

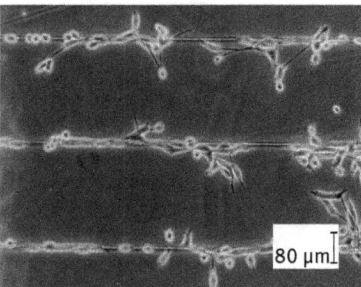

**Figure 4.55:** Fibroblasts on polymer waveguides.

The following chapter provides details on the cell culturing of neural cells with the help of the DUV lithography method.

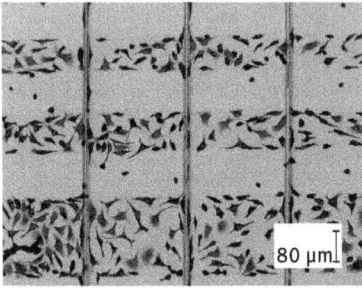

**Figure 4.56:** Fibroblasts patterned perpendicular on polymer waveguides using a PMMA cladding layer.

### 4.8.2.2 Neural cell patterning by DUV lithography

In a first step, Polystyrene (PS) Petri dishes were used as the substrate for the following experiments to validate the patterning process. The base of the Petri dish is exposed to UV light in air through a quartz/chromium mask with a specific pattern as described in the previous section. The UV light source used was a low pressure mercury lamp with $\lambda$=185 and 245 nm and 15 W output at a distance of 10 cm with the chromium mask in contact with the substrate [210]. By exposing PS to UV light, unstable peroxides are formed on the surface along with more stable carboxylic groups. These patterns are stable for several months after fabrication. Additionally, this is a self-sterilizing procedure, whereby the light used for surface modification will sterilize the substrate of any unwanted biological contaminants.

To achieve controlled cellular attachment and growth, the surface of the substrate needs to be functionalized. A method of competitive binding is used to define patterns such that cell growth is controlled. First, the substrate was incubated for 1 h with 1 mg/ml bovine serum albumin (BSA) and 1 mg/ml Pluronic F-68 in phosphate-buffered saline (PBS) at 37 °C and 5% $CO_2$. The substrate was then washed with PBS for 15 min and then incubated again for 1 h with 5µg/ml laminin in 1mg/ml Pluronic F-68 in PBS at 37 °C and 5% $CO_2$. The substrate was washed one final time with PBS for 15 min. The laminin adsorbs more strongly to the polar, irradiated areas of the substrate than does the BSA. Likewise, the BSA adsorbs more strongly to the non-polar, unirradiated areas than does the laminin.

Good cellular attachment and growth of rat hippocampal neurons was seen in irradiated areas of PS substrates. Cells were grown as described in Section 6.2.3.

Cells were seen to remain in isolated predefined patches as seen in Figure 4.57 without branching out onto the surrounding substrate. Additionally, cells were observed to follow predescribed lines as seen in Figure 4.58.

Islands for cell bodies are defined with thinner lines running out, connecting the islands. It was observed that cell bodies attached to the islands and sent processes out along the lines connecting the islands. It was also observed that some cell processes extended over unirradiated areas, thus not following the predescribed pattern. This is due to lines being placed too close to each other, allowing cell processes to bridge unirradiated areas. It was observed that cell bodies and processes

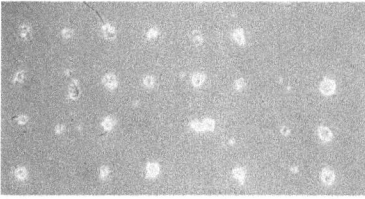

**Figure 4.57:** Phase micrographs of hippocampal neurons grown directly of DUV-modified PS substrates. Good attachment was seen in irradiated areas (Courtesy R. A. Seger).

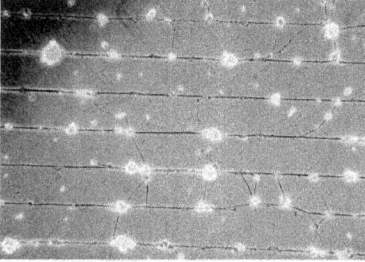

**Figure 4.58:** Phase micrographs of hippocampal neurons grown directly of DUV-modified PS substrates. It can be seen that processes were able to extend over non-irradiated areas, but exhibited poor adhesion to the substrate. Greater line spacing alleviates this problem (Courtesy R. A. Seger).

that were above irradiated areas were well attached to the substrate, whereas processes above unirradiated areas were not attached to the substrate, but were in fact floating above it.

This method provides a high resolution method for growing neuronal cells in predefined patterns on PS with single-cell precision. This method can be potentially integrated with other measurement methods described subsequently, for a hybrid device capable of monitoring cells in specific patterns and configurations.

To transfer the previous method to polymer waveguides fabricated using the DUV lithography method (refer Chapter 3), cell culture chambers are added to the PMMA substrates to maintain cell growth for several weeks. A photograph of the setup is shown in Figure 4.11.

Chambers are fabricated out of polytetrafluoroethylene (PTFE) to an area of ~500 mm$^2$ and a height of 10 mm. Chambers are then attached to the substrate using polydimethylsiloxane (PDMS) as seen in Figure 4.59. The PDMS is degassed under vacuum and then cured at 58 C overnight.

**Figure 4.59:** Image of a completed device. Two culture chambers are fabricated out of PTFE and attached to the PMMA substrate using PDMS. Waveguides run perpendicular to the long edge of the device (Courtesy R. A. Seger).

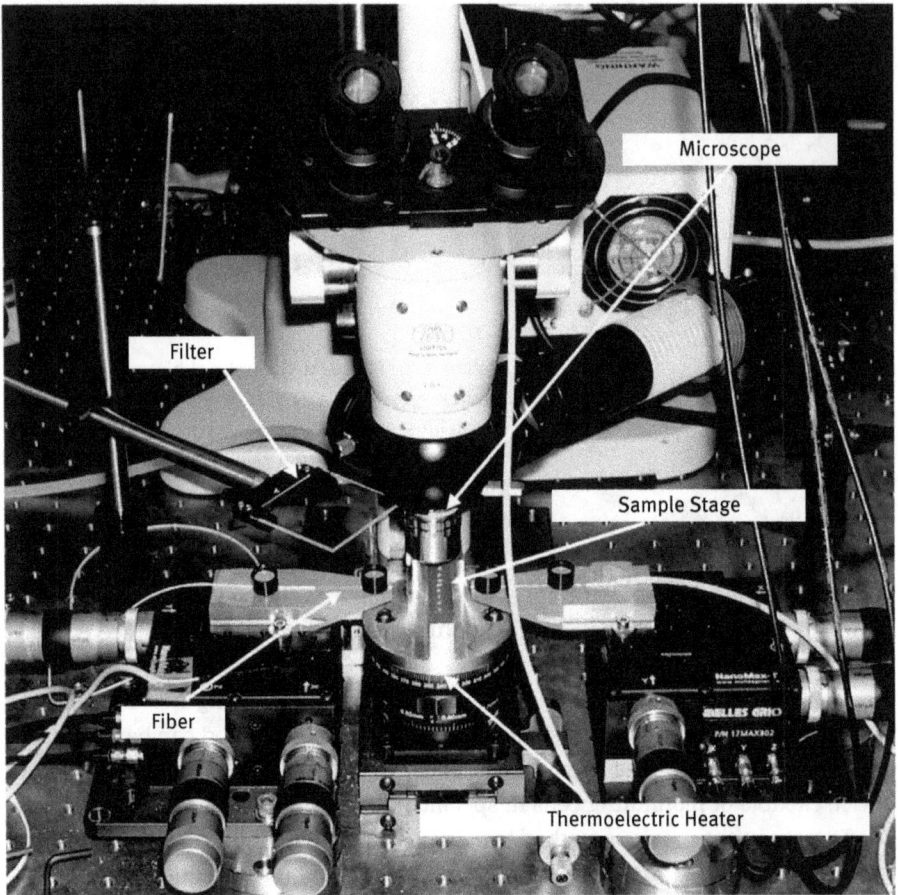

**Figure 4.60:** Image of the experimental setup used to monitor evanescent field excitation of cell-bound fluophores. An upright, long working distance microscope is used to image cultures from above. A heated vacuum chuck directly below the objective is used to secure and maintain cell cultures. Not shown is a DPSS laser operating at 473 nm to the left of the microscope. Light is free-space coupled into an optical fiber, used to insert light into any specified waveguide (Courtesy R. A. Seger).

Cell cultures are mounted to a custom designed monitoring system as seen in Figure 4.60.

Culture chambers are held down using a custom-fabricated vacuum chuck. Attached to the chuck is a Peltier heater used to maintain the cell culture at 37 C. Fiber optics can then be brought in from either side using a conventional XYZ stage, enabling excitation light to be inserted into the waveguides as well as be extracted. A Zeiss upright microscope with long working distance objectives is used to align optical fibers to the waveguide facets as well as monitor fluorescent activity of cells on the waveguides. Excitation light is provided by a diode-pumped solid-state (DPSS) laser operating at 473 nm, 100 mW output, and coupled using a graded-index (GRIN) lens (CFC-11-A-APC, Thorlabs, Inc., Newton, NJ). Fluo-4 [85] has an excitation peak at 493 nm and an emission peak at 516 nm. At 473 nm, Fluo-4 is excited at 42% quantum efficiency when compared to its maximum excitation wavelength. A low-pass optical filter with cutoff frequency of 495 nm can be moved in and out the line of sight of the microscope to remove excitation light and transmit only emitted light.

The evanescent field is predicted to be ~200 nm above the surface of the wave-guide. As the cell membrane is 5–10-nm thick, this is deep enough to penetrate into the cell body, while still only interacting with molecules near the cell surface [137]. Waveguides are tested using PS fluorescent microspheres of various diameters. Microspheres with diameters of 10 μm, 4 μm, and 0.1 μm are used. Microspheres are maintained in a solution of deionized water with 0.15 M NaCl, 0.05% Tween® 20 and 0.02% thimersal with $3.6 \times 10^6$ particles/mL for 10 μm spheres, and deionized water with 2-mM sodium azide and $1.4 \times 10^7$ particles/mL for 4 μm spheres and $1.8 \times 10^{11}$ particles/mL for 0.1 μm spheres. All sizes of microspheres have an absorption peak at 505 nm and an emission peak at 515 nm, which matches quite closely to the fluorescent dye used in the cells as described below. A 0.1-mL droplet of micro-spheres is pipetted onto the surface of the waveguides and visualized using a tradi-tional epifluorescence microscope (Olympus IMT-2). After 30 sec, microspheres have settled out of the solution and onto the surface of the PMMA. Fluorescent activity is verified first with the fluorescent microscope. Upon verification of fluorescent excit-ability, the sample is moved to the optical setup and excitation light is coupled into a single waveguide. All sizes of microsphere show good fluorescent excitation directly above optical waveguides, with significantly reduced excitation from beads not directly above waveguides as seen in Figure 4.61.

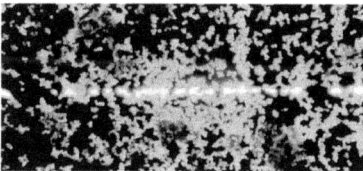

**Figure 4.61:** Fluorescent micrograph of PS microspheres on planar waveguide substrate illustrating fluorescent excitement using a waveguides intrinsic evanescent field (Courtesy R. A. Seger).

The 10-μm spheres are particularly bright and selective excitation can be seen with-out filtering the excitation light out, due to their high efficiency and the fact that the excitation light is mostly confined to the waveguide itself. An increase of up to 4 dB is seen in the fluorescent intensity of spheres on waveguides compared to spheres off waveguides. After successfully testing fluorescent microspheres, cell cultures were prepared with the methods presented previously.

At 4 to 14 days, cells are loaded with Fluo-4 (F-14217, Invitrogen Corp., Carlsbad, CA), a calcium-sensitive dye. Briefly, cells are incubated in pre-warmed Ringer's solution (in sterile HBSS 2.5 mM $CaCl_2$, 1.3 mM $MgCl_2$, 10 mM HEPES, pH adjusted to 7.4) with 10 mM Fluo-4 for 30 min in the dark at room temperature. Cells are then incubated for another 30 min in indicator-free Ringer's solution at 37 °C and 5% $CO_2$. Ringer's solution is designed to maintain healthy cell cultures outside of an incubator for a short period of time. Experiments are carried out within 1 h of the cells being loaded with Fluo-4.

Figure 4.62 shows cells grown directly on planar waveguides. The cells appear to be well attached and have good morphology. Cells directly on top of waveguides are illuminated directly by the waveguide, demonstrating the feasibility of this method.

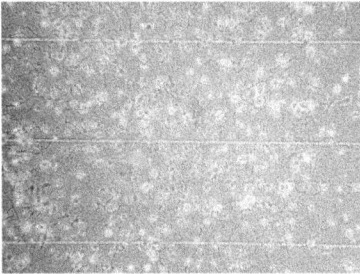

Figure 4.62: Micrograph of hippocampal neurons growing directly on planar waveguides. Cells are seen to be healthy and well spread, indicating compatibility between waveguide substrate and cell tissue (Courtesy R. A. Seger).

Figure 4.63 shows cells illuminated directly above and to the sides of ridge waveguides. It is noted that cells tended to grow on the sides of the ridges, as opposed to the tops. However, this does not affect the mechanism of fluophore excitation by evanescent excitation. An up to 4 dB increase in fluorescent intensity was observed for both planar and ridge type waveguides.

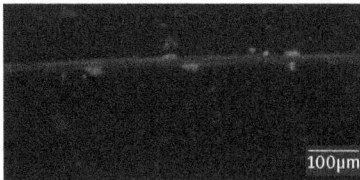

Figure 4.63: Fluorescent micrograph of fluorescent excitation of cells on ridge waveguides. Cells can clearly be seen attached to the sides of the waveguides as opposed to the tops, indicating a preference to grow along the edges of the waveguides and verified under phase contrast microscopy (Courtesy R. A. Seger).

Based on a refractive index of the PMMA waveguides of 1.495 and a refractive index for cells of roughly 1.38, the evanescent field is calculated to be between 125 and 175 nm deep into the cell. This is deep enough to penetrate the cell membrane to excite fluophores within the cell, while simultaneously being shallow enough to offer an image of the basal membrane of the cell without much interference from fluophores within the rest of the cell. Cells that grow in direct contact with the integrated waveguides are clearly seen to be fluorescently excited when compared to cells not in contact with the integrated waveguides. This indicates that cells are being fluorescently excited using the evanescent field associated with the waveguides.

Although there is some small amount of fluorescent excitation observed in cells not located directly in contact with waveguides, it is up to 4 dB lower than cells in contact with waveguides. This parasitic fluorescence can be explained by stray excitation light that is scattered at the interface where light is injected into the waveguide from the fiber optic. For this reason, there is a need to improve the waveguide-fiber coupling. In this way, cells can be selectively visualized by choosing the specific waveguide where the cell of interest resides. In addition, using cell-patterning techniques, cells can provide even more sensitive results by placing cells in an exact relation to one another and to the waveguides themselves. It was noticed that cells grown on ridge waveguides tended to grow on the large areas between waveguides and along the sides of the waveguides, but not on the tops of the waveguides. It was unclear if this was due to preferential adsorption of proteins onto etched portions of the substrate, or due to topographical cues from the ridge waveguides being raised several microns above the surrounding substrate. However, ridge waveguides have strong guiding and have the possibility of directing growth solely on the tops of the waveguides through the use of microcontact printing, using a completely featureless stamp, thus negating the need to fabricate a stamp master. In either case, extra care may need to be taken to ensure good contact of cells to the ridge waveguides.

The following section will focus on the topic of microcontact printing.

### 4.8.2.3 Microcontact printing (μCP)

The most common-place method is microcontact printing (μCP), which is described in this paragraph. Briefly, an elastomeric stamp is used to lay cell adhesion molecules in a predefined pattern on a substrate, resulting in controlled cellular growth [55, 127]. A multitude of molecules have been used with μCP, including laminin, polylysine, and extracellular matrix (ECM) [192]. Controlled differentiation of neural cells has even been demonstrated through both chemical means as well as through the actual geometric design of the pattern laid out. Submicron feature sizes have been reported using this method [247].

Microcontact printing is a relatively inexpensive, fast, and effective method for controlling cellular attachment and outgrowth. An elastomeric stamp made using polydimethylsiloxane (PDMS), and used to stamp a pre-described chemical pattern onto a substrate is the basis for this method. Whitesides et al. first described this method in 1994 as a method to easily pattern gold on silicon substrates [268]. This method has since gained popularity in the biological sciences as a method to control cellular attachment and outgrowth.

The elastomeric stamps used for μCP are made by use of a master mold, typically made from etched silicon. The silicon master is the negative of the final stamp desired. Once the master is made, PDMS is poured onto the mold and degassed in vacuum. Sylgard 184 (Dow Corning, Midland, MI) is the type of PDMS used and comes as a two-part system: a base elastomer and a hardener. They are mixed in a ratio of 10:1. The

PDMS is used for its elastomeric properties, optical clarity, being non-toxic, and the ability to cure at room temperature. After degassing, the stamp is cured in either at room temperature for 48 h or for 45 min at 100 °C. The stamp is then peeled off of the master mold. The PDMS bonds strongly to bare silicon and requires a method of release. One method is to fluorinate the surface, which acts as a release agent for the PDMS [236]. The method used in these examples is to have a master etched into a layer of SU-8 on top of a silicon wafer [261]. There is very little bonding between SU-8 and PDMS, negating the need for a release agent. The stamp is then mounted onto a glass backing. Attachment is achieved by exposing the mounting surfaces of both the PDMS and an appropriately sized piece of a glass slide to oxygen plasma for 1 min. This results in a strong hydrophilic bond developing between the glass and the PDMS, and adding a rigid backing to the stamp. The addition of a rigid backing has been shown to improve the resolution of µCP to 500 nm [192]. Similarly, by increasing the rigidity of stamp sidewalls with polymer reinforcements, the resolution limit can be decreased to 100 nm [247]. Since these resolutions are not required for the described examples presented here, only a rigid backing is employed for resolution enhancement.

In order for µCP to be successful at printing biomolecules capable of controlling cell growth, particular attention needs to be paid to the chemistry of the protein being stamped. The largest influence determining successful pattern transfer is the hydrophylicity of the stamp, the substrate, and the particular molecule being printed. As PDMS is natively hydrophobic, it is most useful for printing non-polar molecules [105]. This can achieve cellular outgrowth by printing hydrophobic proteins, and performing a subsequent wash of the substrate with a polar molecule, in a self-masking procedure where only areas without the non-polar molecules experience attachment of polar molecules. However, the hydrophilicity of PDMS can be controlled by exposing it to oxygen plasma. The oxygen plasma forms silanol groups to the surface of the PDMS, rendering it hydrophilic. The stamp can then be used directly to print polar molecules, which allow direct attachment of cells to the surface of a substrate. The PDMS surfaces are also cleaned using a rigorous cleaning technique to remove contaminants. First stamps are sonicated for 10 min in 100% acetone, followed by sonication for 10 min in 100% propanol. Finally, the stamp is sonicated for 10 min in ultra-pure water for 10 min. After these surface treatments, the stamp is ready to be used for protein pattern transfer.

Stamps were soaked for 15 min in 10 µg/ml polylysine ECM diluted 1:100 in DMEM with 10 µg/ml rhodamine. The stamps were then dried in a clean stream of air until no liquid remained. Stamps were then pressed to the surface of a substrate with a 50 g weight holding them in contact. After 15 sec, the stamp was removed and the substrate could be used in cell-culturing experiments. By viewing under a fluorescent microscope, the quality of the pattern transfer could be determined as seen in Figure 4.64.

Controlled cellular growth with rat hippocampal neurons was obtained using µCP on several different patterns. Cells were grown using the following protocol.

**Figure 4.64:** (a) Fluorescent micrograph of polylysine patterned on a glass substrate. (b) Phase contrast micrograph of hippocampal neurons grown on the pattern indicated in (a) (Courtesy R. A. Seger).

1. One order of hippocampal or cortical tissue from day 18 embryonic rat (BrainBits # hp2 or cx), from Brain Bits (www.brainbitsllc.com) arrives with one 2 ml tube containing one hippocampal pair in 2 ml B27/Hibernate (Hibernate is a proprietary media used to keep neuronal cells viable for up to a week under refrigeration) containing at least one million cells, and one 15-ml centrifuge tube with 12-ml B27/Neurobasal +0.5 mM glutamine +25 uM glutamate [14, 31]. Tissue can be stored at 4–8 C for up to one week.
2. Remove 1 ml media from tube containing cells and place into a fresh centrifuge tube until step 3.
3. Using a 9" Pasteur pipette with the tip fire polished to roughly 50% its original diameter, suck the tissue and media into the pipette and immediately dispense the contents back into the container. Repeat this step 10 times or until most of the tissue is dispersed.
4. Add 1 ml of tissue and media into centrifuge tube used in step 1.
5. Centrifuge tube at 1,100 rpm for 1 min.
6. Aspirate off supernatant.
7. Flick the tube to break apart the pellet of cells.
8. Add 1 ml of B27/Neurobasal +0.5 mM glutamine +25 uM glutamate and resuspend cells.
9. Count cells in a hemacytometer to determine how many cells per milliliter.
10. Cover substrates with 0.2 ml/cm$^2$ B27/Neurobasal +0.5 mM glutamine +25 uM glutamate.
11. Add cells to substrate to achieve density desired, typically 160 cells/cm$^2$.
12. Incubate at 37 °C and 5% $CO_2$. Change one half of the media every three to four days.

However, non-uniformity in pattern transfer resulted in a corresponding non-uniformity of cell growth. As seen in Figure 4.64, biomolecules were occasionally more densely deposited around the edges of pattern features as opposed to uniformly across the features. As expected, neurons tended to attach and grow along regions with the highest concentrations of polylysine, as seen in Figure 4.64 [206]. This can be alleviated by pre-stamping, whereby a stamp is used to stamp on a blank substrate that will not be used. This effectively removes the excess biomolecules, resulting in much more even pattern transfer onto the actual substrate to be used. Overall, this method of cell patterning is one of the best methods available. However, if a new pattern is needed it requires a new mask and new master mold to be made, which can be expensive and is not useful for trying to prototype designs.

### 4.8.2.4 Microelectromechanical systems (MEMS) printing

Inkjet printing has gained popularity recently as a means to quickly and inexpensively prototype MEMS devices [253]. Using metallic inks, patterns can be laid down with resolutions of only 25 μm [193]. Inkjet printers work by expelling a single, small drop of ink onto a substrate in a controlled manner. By placing these droplets close enough together, arbitrary patterns can be laid down on a substrate.

Patterns for inkjet printing were designed using L-Edit software, and loaded into a PC computer used to control a Dimatix DMP-2800 MEMS inkjet printer. The DMP-2800 uses a piezoelectric, all-silicon print head, and acoustic membrane to form ink droplets. A mixture of ECM diluted 1:100 in DMEM, 10 μg/ml polylysine and 10 μg/ml rhodamine was loaded into print cartridges capable of delivering 10 pL drop sizes. Prior to loading, the media was sterile filtered to remove both biological contaminants as well as any air that may have been present in the system. The DMP-2800 then printed the pattern onto a prepared substrate.

Special coverslips are used for printing onto 25-mm round coverslips shown to support neuronal growth are used as the substrate (Fisher Scientific #1943- 10025). The coverslips are cleaned by sonicating for 10 min in ultra-pure water, followed by a 90 min, dry-heat sterilization at 350 °F.

Good printing results were obtained, although with lower resolution than other previously described methods. Good uniformity was observed with little protein transfer to areas outside of the prescribed pattern as seen in Figure 4.65. Notice that the drops visible in Figure 4.65 that appear smaller than 25 μm are satellite droplets and cannot be printed reliably.

Because the print heads are optimized for liquids with significantly higher viscosities, a higher resolution was difficult to achieve. However, it has been suggested that by printing with a hydrophobic material of higher viscosity, followed by a subsequent wash of the substrate with adhesion proteins, could possibly lead to higher resolutions, potentially down to 5 μm.

This solution could potentially solve another problem with the technique. Sterility is an issue affecting the usefulness of this technique. As no antibiotics are used in the cell cultures performed, the substrate has to be sterilized if it is exposed to an environment that is not sterile, such as the print chamber of the DMP-2800. As most

**Figure 4.65:** Fluorescent micrograph illustrating typical pattern formation from an inkjet printer. Notice the minimum line width is ~25 µm (Courtesy R. A. Seger).

sterilization techniques available to us will inherently denature the proteins laid down, and culturing without sterilization led to infection, no cell growth was ever observed on inkjet-printed substrate. Sterilization could also be maintained by enclosing the entire DMP-2800 inside a laminar flow hood. The DMP-2800 could potentially be sterilized using several methods including, wiping it down with 70% ethanol and exposing the interior to UV light. It would be necessary to make sure that all of the components used would hold up well to large doses of UV irradiation with this method. Additionally, substrates may be potentially sterilized by soaking in antibiotics and fungicides after printing and before plating of cells. However, based on the good uniformity of the pattern transferred, the definition of the pattern transferred and successful experiments using previously described methods, controlled cell growth is a reasonable expectation, provided sterility can be re-established prior to cell culture, and with no denaturation of the relevant biomolecules. Inkjet printing shows promise as technique for prototyping patterns without the expense associated with the development of a new photolithographic mask, as long as a lower resolution limit is acceptable.

## 4.9 Conclusion

Photonics is a driving force for future sensors and the availability of technologies is continuing this trend toward miniaturized optofluidic systems. This chapter has barley touched the surface of this exciting field and has provided the necessary background for understanding the principle components and manufacturing processes useful for implementation in optofluidic systems and some basics on biology and examples in biophotonics. In fact, cell patterning can be used to describe specific patterns for cell growth. Three distinct methods for cell growth were presented each with its own advantages and disadvantages. Additionally, two of the three methods were shown to be compatible with neural cell growth for long-term studies. These methods may be combined with measurement methods described subsequently, to create a hybrid device where cells can be arbitrarily placed to affect the greatest signal from the measurement devices present.

The following chapter provides an introduction into fluidics and fluid control systems equally necessary for realizing true optofluidic systems.

## Further reading

Al-Azzawi. Photonics. CRC Press, 2006

Aziz-Alaoui MA, Bertelle C. Emergent Properties in Natural and Artificial Dynamical Systems. Springer, 2006

Bruendel M.Herstellung photonischer Komponenten durch Heißprägen und UV-induzierte Brechzahlmodifikation. Universitätsverlag Karlsruhe, 2008 (http://dx.doi.org/10.5445/KSP/1000007719)

Günther P, Huignard JP. Photorefractive Materials and Their Applications 1. Springer, 2006

Henzi P.UV-induzierte Herstellung monomodiger Wellenleiter in Polymeren. Forschungszentrum Karlsruhe, 2004 (in German)

Hu WS. Cell Culture Engineering. Springer, 2006

Kaneko K. Life: An Introduction to Complex Systems Biology. Springer, 2006

Kragl U. Technology Transfer in Biotechnology: From Lab to Industry to Production. Springer, 2005

Ichihashi Y.UV-induzierte Brechzahländerung zur Herstellung von Wellenleitern und Integration von Siziliumphotodioden. Forschungszentrum Karlsruhe, 2007(in German)

Michalzik R.VCSELs: Fundamentals, Technology and Applications of Vertical-Cavity Surface-Emitting Lasers. Springer 2013

Numai T. Fundamentals of Semiconductor Lasers. Springer, 2004

Punke M. OrganischeHalbleiterbauelemente für mikrooptische Systeme. 2008. (http://dx.doi.org/10.5445/KSP/1000007359)

Rabus DG. Integrated Ring Resonators. Springer, 2007

Schneider H, Liu HC. Quantum Well Infrared Photodetectors. Springer, 2007

Schubert EF. Light Emitting Diodes. Cambridge University Press, 2006

Stuke M. Dye Lasers: 25 Years. Springer, 1992

Träger F. Handbook of Lasers and Optics. Springer, 2012

Worgull M. Hot Embossing. Elsevier, 2009

Yariv A, Yeh P. Photonics. Oxford University Press, 2007

Yu PY, Cardona M. Fundamentals of Semiconductors. Springer, 2010

Chen LZ, Nguang SK, XD Chen. Modelling and Optimization of Biotechnological Processes. Springer, 2006

Manz A, Becker H. Microsystem Technology in Chemistry and Life Science. Springer, 1998

Renneberg R, Lisdat F. Biosensing for the 21st Century. Springer, 2008

Seger A. Nanoscale Electrical and Optical Sensing of Vertebrate and Invertebrate Cultured Neural Networks in Vitro. PhD Thesis University of California Santa Cruz, 2009

Zhong JJ. Biomanufacturing. Springer, 2004

# 5 Fluidics and fluid control systems

This chapter provides the fundamentals of fluidics, microfluidics, and fluid control systems, which includes theory and devices. Basic building blocks of fluidic circuits are fluidic channels, valves, and pumps. Nowadays microfluidic technologies have created microvalves and micropumps [247], which offer the potential of being integrated into a platform in our case a polymer platform. The chapter begins with an introduction into fluidics, which provides the basic theoretical background for designing fluidic circuits and describes the functionality of valves and other fluidic devices relevant for achieving the goal of an integrated platform.

## 5.1 Valve basics

This chapter is based on the fundamental and profound knowledge of Buerkert Fluid Control Systems, and the material presented here is used with kind permission.

The basic building block for achieving a controlled flow of fluid in a fluidic channel or fluidic system is a valve. A valve, which is actively controlled, is a pressure-containing mechanical device that is used to shut off or otherwise modify the flow of a fluid that passes through it.

The parts a valve is made out of are as follows:
- the body to contain the fluid and its pressure,
- the valve seat which is used to manipulate the fluid,
- the actuator to control the position of the valve seat.

The requirements that valves need to obey are very versatile and are extremely application dependent. The most common requirements are as follows:
- reliable, fast and safe switching
- low power consumption (switching and control power)
- response time
- temperature range
- closing force (actuator dependent)
- high chemical resistance of the materials of the valve against the fluid which is being flown through

There are numerous types of valves, which are in use today for all kinds of fluids and, therefore, there are many ways to categorize them. A common way to categorize valves is based on their initial working state: normally open (NO), normally closed (NC), and bistable. Another common way to categorize valves is by their actuation principle. A few actuation principles are as follows:
- electromagnetic
- pneumatic

https://doi.org/10.1515/9783110546156-005

- thermopneumatic
- thermomechanical
- piezoelectric
- electrostatic
- electrochemical
- capillary force
- shape memory also referred to SMA (shape memory actuator)

The authors in [165] give a good overview regarding the classification of valves and their respective description. The classification is shown in Table 5.1.

**Table 5.1:** Classification of microvalves. (Reproduced with kind permission from J. Micromech. Microeng. 16 (2006) R13–R39, Kwang W Oh and Chong H Ahn, A review of microvalves, IOP Publishing).

| Categories | | | |
|---|---|---|---|
| Active | Mechanical | Magnetic | External magnetic field |
| | | | Integrated magnetic inductor |
| | | Electric | Electrostatic |
| | | | Electrokinetic |
| | | Piezoelectric | |
| | | Thermal | Bimetallic |
| | | | Thermopneumatic |
| | | | Shape memory alloy |
| | | Bistable | |
| | Non-mechanical | Electrochemical | |
| | | Phase change | Hydrogel |
| | | | Sol-gel |
| | | | Paraffin |
| | | Rheological | Electro-rheological |
| | | | Ferrofluids |
| | External | Modular | Built-in |
| | | | Rotary |
| | | Pneumatic | Membrane |
| | | | In-line |
| Passive | Mechanical | Check-valve | Flap |
| | | | Membrane |
| | | | Ball |
| | | | In-line mobile structure |
| | Non-mechanical | | Diffuser |
| | | Capillary | Abrupt |
| | | | Liquid triggered |
| | | | Burst |
| | | | Hydrophobic valve |

One of the most commonly used valves is the solenoid valve. Solenoid valves are described as an example of the operation principle of valves. Later in this chapter, other examples of valves with their respective actuation principle are described.

The core of a solenoid consists of an electrical magnet, which is also referred to as the solenoid coil.

The well-known principle of the generation of a magnetic field is used to create a force. The magnetic field from the coil penetrates the core, also referred to as plunger. The plunger moves in the lateral position and the valve is closed or opened depending on the type of valve, for example, NC or NO as described earlier in this chapter.

Direct acting valves do not rely on a pressure differential. The function of the valve depends on the following three characteristics:
1. size of the seat
2. operating pressure
3. magnetic force of coil.

With direct-acting valves, the pressure range decreases with increasing seat diameter. This type of control is preferably used for small nominal sizes, low pressures, and vacuum. Direct-acting valves can be both NC and NO types.

Solenoid valves are also referred to as pilot valves, which enable the opening or closing of the flow of a media. Another example would be to toggle between the two media. As direct-acting valves are binary meaning either open or closed, other types of valves are required if the flow of a media is to be controlled between 0 and 100% of the flow. These types of valves are referred to as proportional valves.

Fluid control system design starts of course by choosing the appropriate material, which withstands the media that the valve is supposed to handle. The other main interest is of course the quantity of media required to flow through the valve at a given pressure differential. If the valve is a proportional valve, then additional features such as the flow/pressure curve and operating behavior are necessary for designing the fluid control system. The following section provides a basic introduction into the theory of fluid mechanics giving the necessary toolbox equations to describe the behavior of fluids.

## 5.2 Introduction into fluidic mechanics

The basis for the following equations is the Navier-Stokes equations (named after Claude-Louis Navier and George Gabriel Stokes which have derived these equations independently from one another), which are of subject in literature in several fluid dynamics publications. These equations describe how the velocity, pressure, temperature, and density of a moving fluid are related. In this book, "every day" equations are derived and no further details are provided, as they are readily available in literature.

To determine the volume and mass flow of media through a valve, the following equation is used:

$$\dot{V} = \alpha \cdot A \sqrt{\frac{2 \cdot \Delta p}{\rho}} = A \sqrt{\frac{2 \cdot \Delta p}{\zeta \cdot \rho}}$$

$$\alpha = \frac{1}{\sqrt{\zeta}}$$

where $\alpha$ is the flow coefficient, $A$ is the cross-section, $\Delta p$ is the pressure drop through the valve (pressure loss), $\rho$ is the density of the fluid, $\zeta$ is the loss index or resistance index. The pressure drop through a valve acting as a fluid resistance is calculated by:

$$\Delta p = p_1 - p_2$$

where $p_1$ and $p_2$ are the measured pressure directly before and after the valve. The mass flow is derived from the previous equation to be:

$$\dot{M} = \rho \cdot \dot{V}$$

$$\dot{M} = \alpha \cdot A \sqrt{2 \cdot \rho \cdot \Delta p} = A \sqrt{\frac{2 \cdot \rho \cdot \Delta p}{\zeta}}$$

An essential parameter of a valve is the so-called $k_v$ value, describing the flow characteristics of a valve. The $k_v$ value is the volume flow $\dot{V}$ in m³/h of water ($\rho_0 = 1 \text{ kg/dm}^3$) at a temperature between 5 and 40 °C at a pressure drop (permanent pressure loss) at the valve of $\Delta p_0 = 100$ kPa. The $k_v$ value is derived to be:

$$k_v = \alpha \cdot A \sqrt{\frac{2 \cdot \Delta p_0}{\rho_0}}$$

To determine the $k_v$ value experimentally, only the flow coefficient $\alpha$ and the cross-section $A$ of the valve need to be known. The $\rho_0$ and $\Delta p_0$ are already given in the definition of the $k_v$ value. The $k_v$ value is thus an important parameter to describe a valve and compare valves in terms of their flow dynamic behavior.

If, for example, different valves with the same diameter (derived from the cross-section A) have different $k_v$ values, the valve with the highest $k_v$ value has been engineered the best. The $k_v$ value is often wrongly referred to as "flow rate coefficient" or "valve coefficient" assuming that $k_v$ is dimensionless, which is not the case. The $k_v$

value as seen in the definition is in the metric system. Countries using the imperial system refer to the $c_v$ value.

The $c_v$ value is defined as: the volume flow (in US gallons/min) of water at a temperature of 60 °F with a pressure loss of 1 psi through the valve. Where 1 US gallon/min = 0.227 m³/h and 1 psi = 0.069 bar.

Another important parameter used for valves in pneumatics is the $Q_{Nn}$ value. The $Q_{Nn}$ value is the volume flow (in liters/min) of air at a temperature of 20 °C and at an input pressure of 6 bar and a pressure drop through the valve of 1 bar. To convert from $k_v$ to $c_v$ and $Q_{Nn}$, the following equations can be used:

$k_v = 0.86 \cdot c_v$

$k_v = 1078 \cdot Q_{Nn}$

The volume flow and the $k_v$ value have the following relationship:

$$\dot{V} = k_v \sqrt{\frac{\rho_0 \cdot \Delta p}{\rho \cdot \Delta p_0}}$$

In this way, it is possible to calculate the volume flow of any fluid when the $k_v$ value is known. From the definition of the $k_v$ value, it is possible to simplify this equation even further, thus leading to the volume flow:

$$\dot{V} = 100 \cdot k_v \sqrt{\frac{\Delta p}{\rho}}$$

Using $\dot{M} = \rho \cdot \dot{V}$ leads to the mass flow:

$$\dot{M} = 100 \cdot k_v \sqrt{\rho \cdot \Delta p}$$

Mass flow and volume flow derived for liquids can of course be adapted for gases if compressibility, sub-critical, and super-critical outflow are accounted for.

Critical, subcritical, and supercritical flow can be described with the help of the Froude number [99]. The Froude number, $F_r$, is a dimensionless value that describes different flow regimes of open channel flow. The Froude number is a ratio of inertial and gravitational forces.

$$F_r = \frac{v_w}{\sqrt{g \cdot D}}$$

where $v_w$ = water velocity, $D$ = hydraulic depth (cross-sectional area of flow/top width), $g$ = Gravity. When:

$F_r = 1$, critical flow,

$F_r > 1$, supercritical flow (fast rapid flow),

$F_r < 1$, subcritical flow (slow/tranquil flow)

The Froude number is a measurement of bulk flow characteristics such as waves, sand bedforms, and flow/depth interactions at a cross-section or between boulders.

In the following paragraphs, critical, sub-critical, and super-critical outflows are derived and explained in terms of the pressure ratio $p_2/p_1$.

In the case of gases, the mass flow and the volume flow are dependent on the ratio of the pressure $p_2$ and $p_1$ leading to an outflow function named $\psi$. The mass flow can then be calculated using:

$$\dot{M} = \alpha \cdot A \cdot \psi \sqrt{2 \cdot \rho_1 \cdot p_1}$$

Where $\alpha$ is the flow coefficient, $\psi$ is the outflow function, $A$ the cross-section, $\rho_1$ is the density of the gas upstream of the valve, $p1$ is the pressure of the gas upstream of the valve. The outflow function $\psi$ is not only dependent on the pressure ratio but also dependent on the adiabatic exponent $\kappa$, which is a constant for an ideal gas. Specific values of certain media are given in Table 5.2. The value of the outflow function $\psi_{max}$ is between 0.5 and 0.6 for common media.

**Table 5.2:** Specific values of certain media.

| Type of gas | $\kappa$ | $\left(\dfrac{p_2}{p_1}\right)_{crit}$ | $\psi_{max}$ |
|---|---|---|---|
| Diatomic gases like N2, O2, CO | 1.400 | 0.528 | 0.484 |
| Air | 1.402 | 0.53 | 0.49 |
| Triatomic gases like CO2 | 1.300 | 0.546 | 0.473 |
| Saturated steam | 1.135 | 0.577 | 0.45 |

To determine the maximum gas flow through a valve inlet/outlet or nozzle, it is essential to calculate the critical pressure. The critical pressure ratio is the pressure ratio that will accelerate the flow to a velocity equal to the local velocity of sound in the fluid [98]. The critical pressure ratio is derived as follows:

$$\left(\frac{p_2}{p_1}\right)_{crit} = \left(\frac{2}{\kappa+1}\right)^{\frac{\kappa}{\kappa-1}}$$

The outflow function is plotted against the pressure ratio. The critical pressure ratio takes on values in the range between 0.45 and 0.55. Engineers can assume a critical pressure ratio of 0.5 for practical applications.

$$\left(\frac{p_2}{p_1}\right)_{crit} \approx 0.5$$

$$p_2 = \frac{1}{2} \cdot p_1$$

The Froude number would be 1 in the critical case. The conditions for sub-critical and super-critical outflows for the pressure ratio are as follows:

$$\frac{p_2}{p_1} > \left(\frac{p_2}{p_1}\right)_{crit}$$

for sub-critical outflow and

$$\frac{p_2}{p_1} \leq \left(\frac{p_2}{p_1}\right)_{crit}$$

for super-critical outflow.

The mass flow in the supercritical outflow case can be calculated using $\psi_{max}$ leading to:

$$\dot{M} = \alpha \cdot A \cdot \psi_{max} \sqrt{2 \cdot \rho_1 \cdot p_1}$$

The value of $\psi_{max}$ can be approximated to be 0.5. The mass flow of gases in the sub-critical case depending on the $k_v$ value can be calculated using the previous equations, leading to:

$$\dot{M} = k_v \cdot \psi \sqrt{\frac{\rho_0 \cdot \rho_1 \cdot p_1}{\Delta p_0}}$$

For the super-critical outflow dependent on the $k_v$ value:

$$\dot{M} = k_v \cdot \psi_{max} \sqrt{\frac{\rho_0 \cdot \rho_1 \cdot p_1}{\Delta p_0}}$$

For engineering purposes, the following equations for the volume and mass flows of gases and steam have been derived for practical applications, where correction factors are already taken into account (Table 5.3).

Table 5.3: Everyday formulae for gases and steam (Courtesy Buerkert Fluid Control Systems).

| Everyday formulae for gases and steam | | | |
|---|---|---|---|
| | Gases | | Steam |
| | Mass flow | Volume flow | Mass flow |
| Sub-critical | $\dot{M} = 5,140 k_v \sqrt{\frac{\rho \Delta p p_2}{T}}$ | $\dot{V} = 5,140 k_v \sqrt{\frac{\Delta p p}{\rho T}}$ | $\dot{M} = 100 k_v \sqrt{\frac{\Delta p}{v_2}}$ |
| Super-critical | $\dot{M} = 2,750 k_v p_1 \sqrt{\frac{\rho}{T}}$ | $\dot{V} = 2,750 k_v p_1 \frac{1}{\sqrt{\rho T}}$ | $\dot{M} = 100 k_v \sqrt{\frac{p_1}{2 v_k}}$ |

Sections 5.1 and 5.2 provided the basic understanding of valves and the design of fluid control systems. The following section deals with the behavior of fluid control elements in systems.

## 5.3 Theory of fluid control elements in systems

One of the simplest examples of fluid control is a valve in a pipe, controlling a fluid reservoir. The valve opening and closing results in a velocity change of the fluid in the system, which in turn generates a pressure wave. This pressure wave propagates from the valve with a propagation speed $a$ against the flow direction until it reaches a fluidic resistance, which can be another valve or opening, for example. The pressure wave gets reflected back and forth at these fluidic resistances, creating undershoots and overshoots also called as suction and pressure surges ("water hammer phenomena").

The propagation speed $a$ of the pressure change is equal to the speed of sound if dealing with a compressible fluid. For fluids in thick walled, rigid pipes:

$$a = \sqrt{\frac{E}{\rho}} \ [\text{m/s}]$$

Where E is the modulus of elasticity of the medium in N/m². The speed of sound for fluids in thin-walled, flexible pipes is given by:

$$a = \sqrt{\frac{E_F}{\rho} \cdot \frac{1}{1 + \frac{D \cdot E_F}{S \cdot E_R}}} \ [\text{m/s}]$$

Where $D$ is the pipe diameter, $S$ is the wall thickness, $E_F$ is the modulus of elasticity of the fluid, $E_R$ is the modulus of elasticity of the pipe material. The speed of sound in water can be approximated for these two cases to:

Thick-walled metal pipes: $a \approx 1{,}300$ m/s

Thin-walled pipes: $a \approx 1{,}000$ m/s

The impact of the closing and opening of the valve is mainly determined by the opening/closing time of the valve. The velocity of the fluid after opening a valve in a fluidic circuit, which is pressurized at a pressure $\Delta p$ is given by:

$$w_0 = \frac{\Delta p}{a \cdot \rho}$$

Where $w_0$ is the velocity after opening of the valve at the time $t = 0$, $\Delta p$ is the pressure before opening, $\rho$ is the density of the fluid, and $a$ is the propagation speed of the pressure wave. This is an approximation to that extent, that it is assumed the valve opens and closes at time $t = 0$. This is an ideal assumption, as different influences are not taken into account such as the properties of the media, type of actuation of the valve, and other damping effects.

After a certain period of time, the velocity $w_1$ with stationary outflow is reached and is calculated as follows:

$$w_1 = \sqrt{\frac{2 \cdot \Delta p}{\rho}}$$

The opening and closing of a valve produce suction and pressure surges, whereas a suction surge, produced at the time $t = 0$ by a sudden opening of a valve, reduces the pressure of the fluid to a maximum of 0 bar, a pressure surge can be several factors higher than the idle pressure.

The maximum pressure obtained, also known as "water hammer", is calculated using:

$$\Delta p_{max} = a \cdot \rho \cdot w_1$$

The maximum value $\Delta p_{max}$ of the pressure surge will be less, if the closing time $t_{cl}$ of the valve is longer than the wave propagation time $T_L$ which is the back and forth distance $L$ between the valve and the first fluid resistance. The wave propagation time $T_L$ is calculated as follows:

$$T_L = \frac{2L}{a}$$

The ratio of the closing time to the wave propagation time is known as impact number $Z$:

$$Z = \frac{T_L}{t_{cl}}$$

The pressure surge can thus be calculated with the following formula:

$$\Delta p = Z \cdot a \cdot \rho \cdot w_1$$

where $\Delta p$ is the pressure directly after closing the valve, a is the speed of propagation/ sound, $\rho$ is the density of the fluid, $w_1$ is the velocity at stationary outflow, and $Z$ is the impact number.

The equations described in this chapter are sufficient for a daily usage to estimate engineering problems and find adequate practical solutions. Further reading and discussions are necessary if one has to tackle more sophisticated tasks.

The following section provides a more detailed insight into the topic of control theory, which is necessary when implementing optofluidic systems with, for example, feedback loops or control loops.

## 5.4 Introduction into control theory

This chapter on the introduction into control theory is based on the fundamental knowledge and know-how of Buerkert Fluid Control Systems.

Control theory is merging disciplines, engineering, and mathematics, which are the main drivers. Basically, all systems we use in our daily life are in one way or the other based on control theory and make life for us easier. The goal of control theory is to reduce complexity and to make complex systems understandable and manageable.

### 5.4.1 Open-loop and closed-loop controls

The following section deals with the function and sequence of an open-loop control system.

An open-loop control system is characterized by the fact that one or more input variables of a system influence its output variables in accordance with the system's own interrelations. One everyday example of an open-loop control system: the inside temperature of a room is to be maintained at a constant value as a function of the outside temperature.

The room temperature PV (output variable of the open-loop control system) is to be maintained at a constant value by adjusting the electrically operated mixer valve and thus, the temperature of the heating supply line or radiator. If the outside temperature changes the room temperature as well, then the outside temperature is referred to as the disturbance variable and identified with the letter $z$ ($z1$ in the example). The task of the open-loop control system is to counteract the influence of the "outside temperature" disturbance variable. For this purpose, the outside temperature is measured with an outside temperature sensor. The mixer valve is adjusted or the temperature of the radiator is varied using the control unit. The interrelationship between the outside temperature and the heating output required for maintaining a constant room temperature must be stored in the control unit (e.g., in the form of characteristic curves). Use of such a control unit allows the influence of the outside temperature on the room temperature to be eliminated. Beside the outside temperature, the example is shown in Figure 5.1 also contains other disturbance variables, which also affect the room temperature: opening a window or a door, changing wind conditions, and the presence of persons in the room. Since it is not detected by the control unit, the effect of these disturbance variables on the room temperature is not compensated for by the open-loop control system. The use of such an open-loop control system is practical only if it can be assumed that there is a low influence of (secondary) disturbance variables. The block diagram, Figure 5.2, shows the open action sequence characteristic of an open-loop control system.

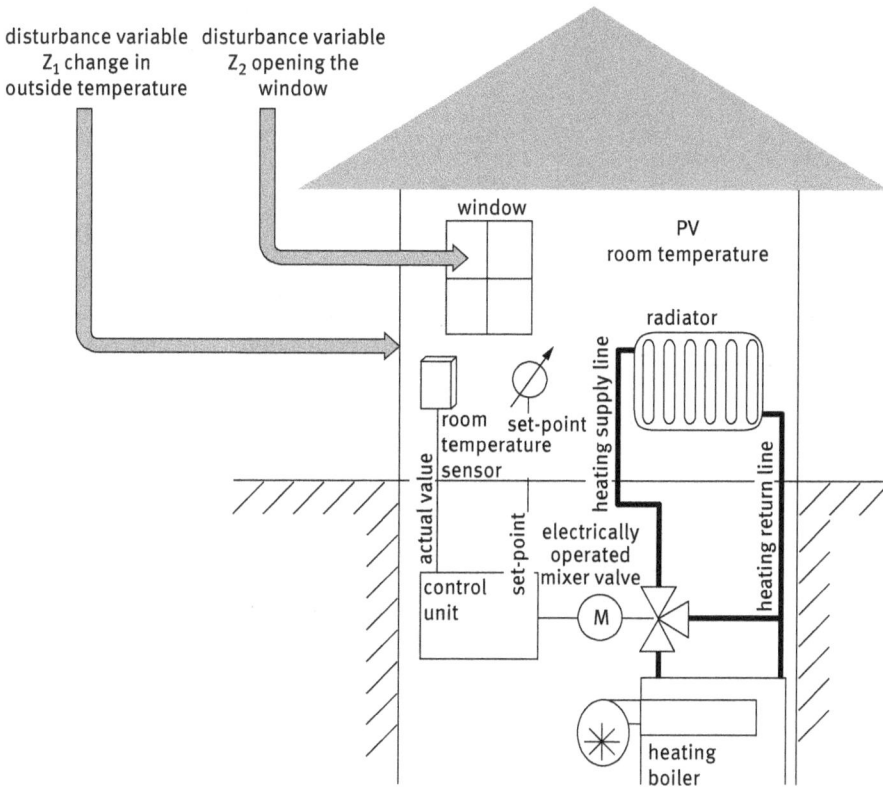

**Figure 5.1:** Open-loop control of the inside temperature of a room (Courtesy Uwe Krug, Buerkert Fluid Control Systems).

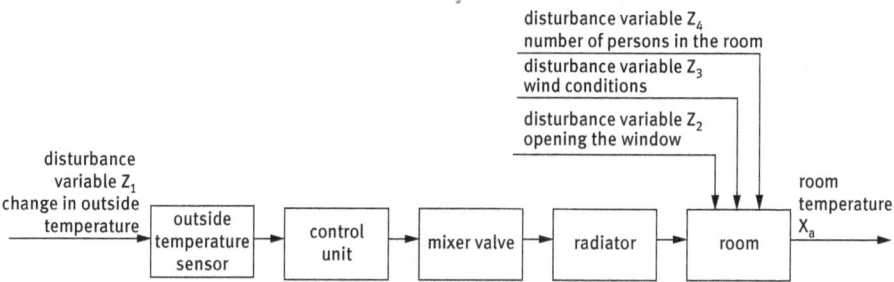

**Figure 5.2:** Block diagram of the open-loop control system for room temperature (Courtesy Uwe Krug, Buerkert Fluid Control Systems).

However, if the effect of the other disturbance variables is so strong that it also needs to be compensated for, it becomes necessary to control the room temperature on the basis of a closed-loop control system. The block diagram, Figure 5.3, shows the closed action sequence that is typical of a closed-loop control system.

disturbance variable $Z_1$
change in outside temperature

disturbance variable $Z_2$
opening the window

disturbance variable $Z_3$
wind conditions

disturbance variable $Z_4$
number of persons in the room

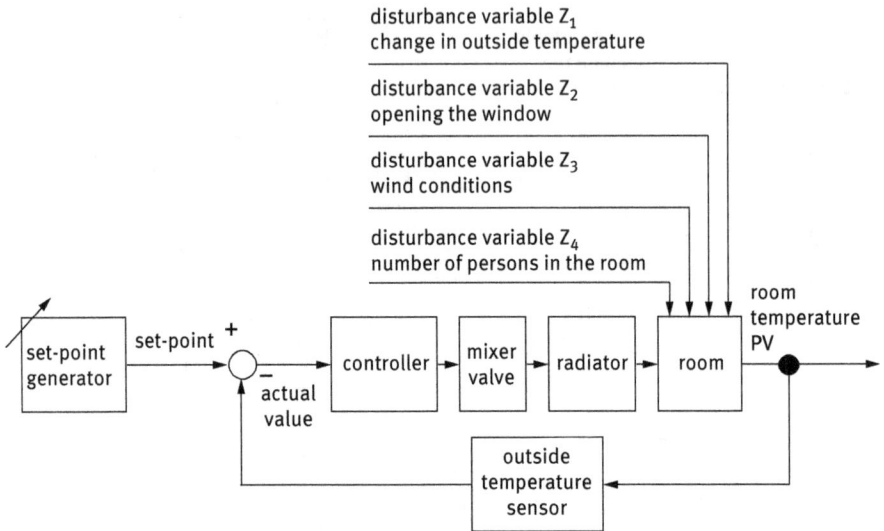

**Figure 5.3:** Block diagram of the closed-loop control system for room temperature (Courtesy Uwe Krug, Buerkert fluid control systems).

The function and sequence of a closed-loop control system is described in the following section.

The fundamental difference with respect to an open-loop control system is that the output variable of the system (the actual value) is constantly measured and compared with another variable (the set-point value or the reference variable). If the actual value is not equal to the set-point value, the controller responds to this. It changes the actual value by adjusting it to the set-point value. A common example of a closed-loop control system: the inside temperature of a room is to be maintained at a preset temperature (Figure 5.1).

In this case, the effects of disturbance variables on the room temperature should be eliminated. The effects of all disturbance variables influencing the room temperature,

– change in outside temperature
– opening a window
– changing wind conditions,

are registered by measurement of the room temperature and comparing this with the set-point value. On the basis of the comparison between set-point and actual value, the controller adjusts the temperature of the heating supply line or radiator via the mixer valve until the required room temperature is reached. The table below contains application recommendations for open-loop and closed-loop controls.

A summary on open- and closed-control loops is provided in the Table 5.4.

**Table 5.4:** Open- and closed-control loops (Courtesy Buerkert Fluid Control Systems).

|  | Open-loop control | Closed-loop control |
|---|---|---|
| **Application** | If no or only one essential measurable disturbance variable is present. | If several essential disturbandance variables are present. If disturbance variables cannot be detected or can only poorly be detected with measuring systems. If unforeseeable disturbance variables may occur. |
| **Advantages** | Low implementation effort. No stability problems due to the open action sequence. | Disturbance variables are detected and compensated for. The preset value (set-point SP) is more precisely achieved. |
| **Disadvantages** | Disturbance variables that occur are not detected automatically. Measurement is required for each disturbance variable to be compensated for. All interrelationships of the system to be controlled must be known in order to design the control unit optimally. | The equipment effort and complexity are greater than with open-loop control. A measurement is always required. |

## 5.4.2 The control loop

Closed-loop control is a process that is used in more than just technical applications. Closed-loop control systems run virtually everywhere and always. The process of setting the required water temperature when showering or complying with a speed limit when driving a car involves closed-loop control. These two examples demonstrate the task of a closed-loop control system: adjusting the specific variables such as temperature, speed, flow rate, or pressure, to a required value. In principle, a closed-loop control process appears to be a very simple one. However, when implementing technical closed-loop control systems, problems are very quickly encountered. The precondition for correct functioning of a closed-loop control system is the interplay of the individual components involved in a closed-loop control system. The totality of components of a closed-loop control system is referred to as the control loop. In the following, the control loop is explained in further detail.

A control loop consists of the following components:

- the measuring instrument or sensor for detecting the variable to be controlled
- the controller, the core of the closed-loop control system
- the system to be controlled (this part is referred to as controlled system).

One example:

The fluid level in a tank is to be maintained at or adjusted to a preset value. To implement a closed-loop control system, it is necessary to continually measure the filling level in the tank (process value [PV]). This is done here, for example, by an ultrasonic level transmitter. The PV is constantly compared with the preset target-filling level (set-point value), which is set, for example, on a control unit via buttons or selector switches. The comparison between PV and set-point value is performed by the controller. If a deviation occurs between the process and set-point value (control deviation), the controller must respond to it. The controller has to adjust a suitable final control element or actuator so that the PV adjusts to the set-point value, that is, so that the control deviation becomes zero. If the PV is higher than the set-point value, the control valve must be closed further. If the filling level is too low, the valve must be opened wider.

Control deviations in a control loop are caused by two factors:

- disturbance variables and
- changes in the set-point value.

In our example, the following two disturbance variables may occur:

- outflow from the tank, occurring abruptly due to opening of one or more ON/OFF valves and
- slow filling-level change due to evaporation of the fluid from the tank.

The elements of a control loop are discussed in the following section. Block diagrams are used to represent control loops. This mode of representation has the advantage that it concentrates on the control-engineering problem. The interplay of the individual components of the control loop is represented graphically. For example of a closed-loop filling level control system, the block diagram looks as follows (Figure 5.4):

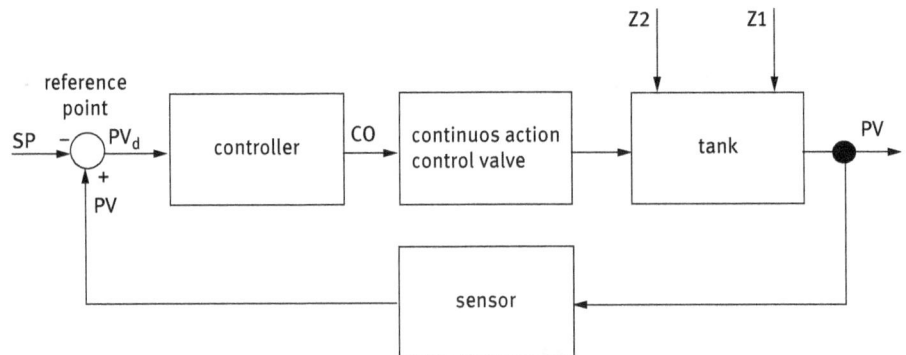

**Figure 5.4:** Block diagram of the closed-loop filling-level control system (Courtesy Uwe Krug, Buerkert Fluid Control Systems).

Where:

SP: Set-point value or reference variable (required filling level)

PV: Process value or controlled variable (measured filling level)

PVd: Control deviation (actual value – set-point value)

CO: Manipulated variable or control output (output value of the controller)

z1: Disturbance variable 1 (outflow from the tank)

z2: Disturbance variable 2 (evaporation of fluid from the tank)

The basic structure of this block diagram corresponds to that of the closed-loop room temperature control structure. Thus, the following general block diagram (shown in Figure 5.5) can be used to summarize the closed-loop control engineering.

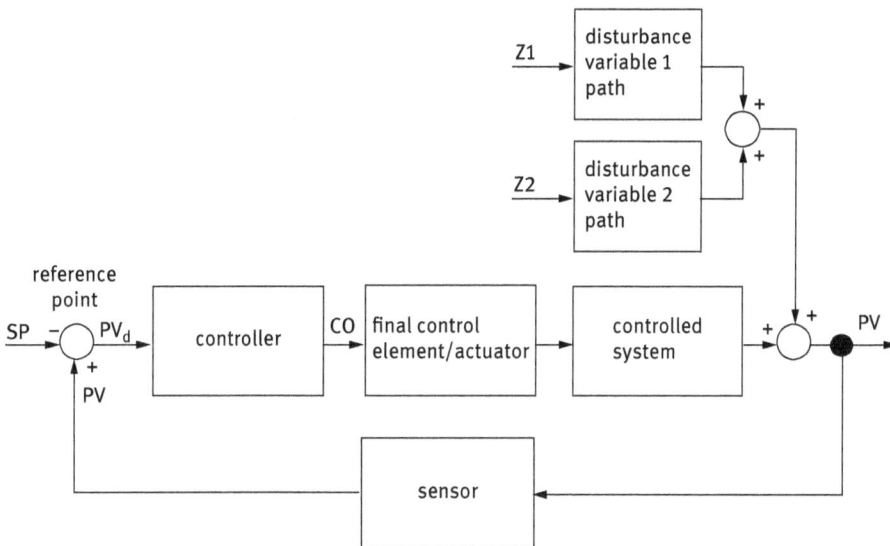

**Figure 5.5:** General block diagram of the control loop (Courtesy Uwe Krug, Buerkert Fluid Control Systems).

In this case, it is assumed that the action point of the disturbance variables does not always need to be at the output of the controlled system, but that the action of the disturbance variables can be converted to this point. Usually, however, a simplified block diagram termed "control system" is used which consists (refer Figure 5.5) of the reference point, the controller, the final control element/actuator, and the sensor.

The disturbance variables are combined and their action point is at the output of the controlled system. Block diagrams are used to create a closed-loop control engineering model of a real system. The main components of the control loop are represented by function blocks, frequently also referred to as transfer elements. Action lines show the functional relationship between the individual blocks with regard to the environment. Each function block is characterized by the dependence of its output signal on the input signal. This dependence is described by the response. There are numerous possible ways of representing the response. The most conventional way is stating the step response or transfer function. It is plotted as a simple timing diagram in the relevant function block. The step response is the characteristic of the output signal (Figure 5.6), which occurs when the input signal changes abruptly as a function of time. The transfer function (designated h(t)) is the step response standardized with respect to the magnitude of the input step or input signal ($h(t) = PV_o(t)/PV_{io}$). If, in the block diagram, we replace the individual control loop elements or function blocks by the form of representation shown in Figure 5.6, then we obtain the signal flow diagram of the control loop. Figure 5.7 shows the signal flow diagram of the closed-loop filling-level control system. The signal flow diagram or the block diagram is an important aid to designing control loops and for

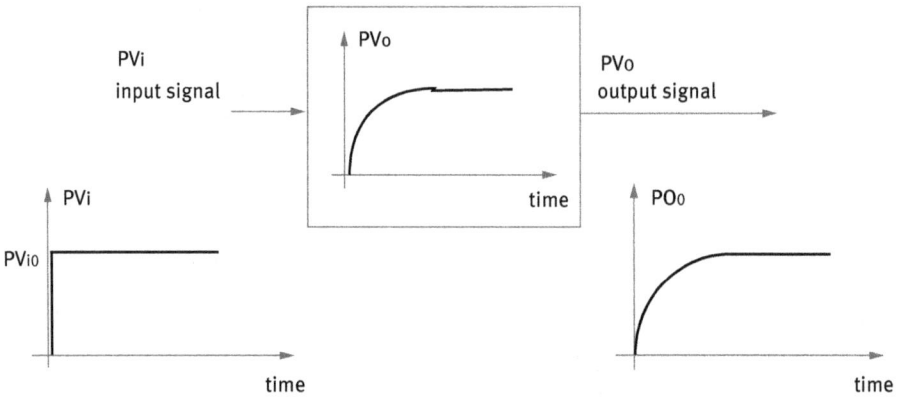

**Figure 5.6:** Step response of a transfer element (Courtesy Uwe Krug, Buerkert Fluid Control Systems).

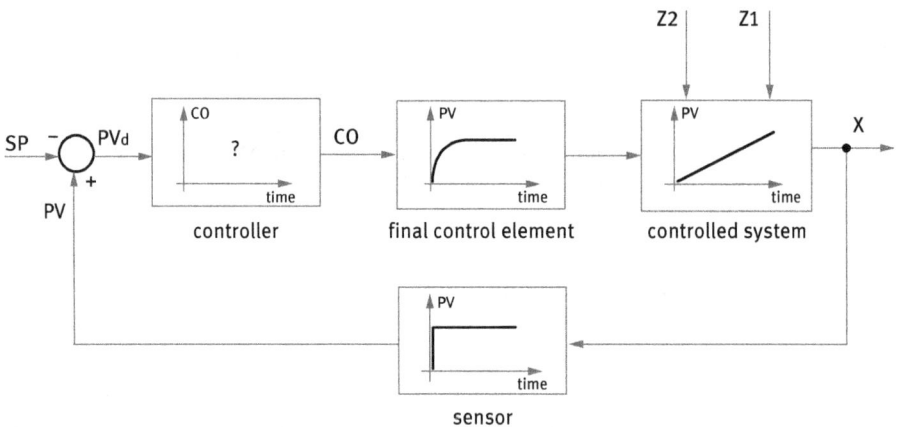

**Figure 5.7:** Signal flow diagram of the closed-loop filling-level control system (Courtesy Uwe Krug, Buerkert Fluid Control Systems).

adapting the controller to the controlled system. In many cases, adaptation of the controller is one of the most demanding tasks, and one that requires basic knowledge of controlled systems and controllers. This is covered in the following sections.

To select a suitable controller and be able to adapt it to the controlled system (the system or equipment to be controlled), it is necessary to have precise information on the behavior of the controlled system. Factors that must be known include to what extent and in what timeframe the output signal of the controlled system responds to changes of the input signal. Real transfer elements differ from ideal ones by virtue of the fact that they almost always feature a time-delayed response. This means that a certain time elapses until the output signal responds to a changing input signal. Controlled systems can be subdivided into two categories in terms of their time response or steady-state condition:

Controlled systems with compensation:

In the case of controlled systems with compensation, the output variable of the system reassumes steady-state condition within a specific period. One example of a controlled system with compensation is the flow rate in a pipe. If the degree of opening of a continuous-action control valve is changed, a constant flow rate is established after a specific period assuming constant pressure conditions. The transfer element shown in Figure 5.6 symbolizes a controlled system with compensation.

Controlled systems without compensation:

In the case of controlled systems without compensation, there is no steady-state condition. Even with a constant input variable (greater than 0), the output signal changes at a constant rate or acceleration. In the example of closed-loop filling-level control in a tank, this relates to a controlled system without compensation. If the valve for filling the tank is open and the ON/OFF valves are closed, the filling level increases continuously without assuming a steady-state condition. In the case of real controlled systems without compensation, there is generally a limitation: in the example of closed-loop filling-level control, this results from overflowing of the tank. The transfer element shown in Figure 5.8 symbolizes a controlled system without compensation.

When selecting and setting the controller, the aspect of whether the controlled system is a controlled system with or without compensation is of crucial importance. The most frequently occurring controlled systems and their transfer functions are described below in greater detail. Table 5.5 provides an initial overview.

**The P-element:**
On the P-element or proportional element, the output signal follows the input signal directly, with no time delay. Input and output signals are proportional to each other. There is no time delay. Figure 5.9 shows the behavior or step response of a P-element.

**The 1st-order time-delay element:**
On the first-order time-delay element, the output signal follows the input signal with a time delay (Figure 5.10). In this case, the output signal changes immediately, but

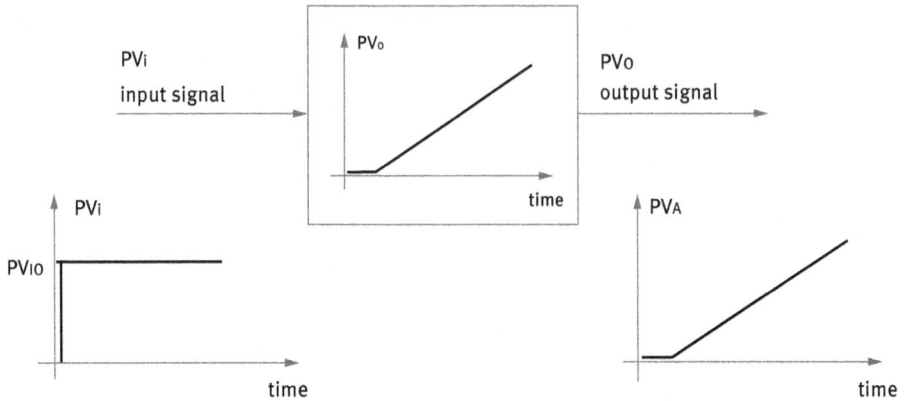

**Figure 5.8:** Transfer element with plotted step response of a controlled system without compensation (Courtesy Uwe Krug, Buerkert Fluid Control Systems).

**Table 5.5:** Overview of typical controlled systems (Courtesy Uwe Krug, Buerkert Fluid Control Systems).

| description | transfer element | application |
|---|---|---|
| Proportional system |  | Pressurecontrol with liquids flow control with liquids |
| 1st order time delayed system |  | pressure control with liquids and gases flow control with gases temperature control speed control |
| 2nd order time delayed system |  | temperature control |
| 3rd order time delayed system |  | indirect temperature (steam through heat exchanger) |
| dead time system |  | delivery rate control for belt conveyors |

**Table 5.5** (continued)

| description | transfer element | application |
|---|---|---|
| Dead time with 1st order time delayed system | PVI input signal / PVO ... PVO output signal (time) | pH-control conductivity control in-line blending of the liquid stream in one line |
| integrating system | PVI input signal / PVO ... PVO output signal (time) | level control conductivity control |

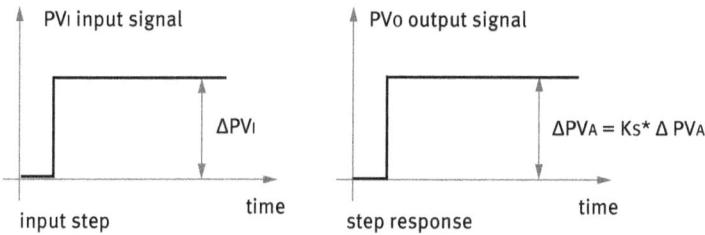

**Figure 5.9:** Step response of a P-element (Courtesy Uwe Krug, Buerkert Fluid Control Systems).

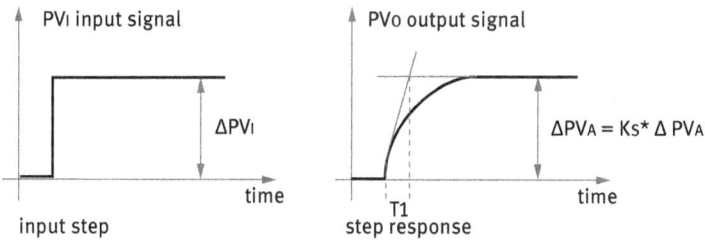

PVI input signal — $\Delta PV_I$ — time — input step

PVo output signal — $\Delta PV_A = K_S*\,\Delta PV_A$ — time — step response

**Figure 5.10:** Step response of a first-order time-delay element (Courtesy Uwe Krug, Buerkert Fluid Control Systems).

PVI input signal — $\Delta PV_I$ — time — input step

PVo output signal — $\Delta PV_A = K_S*\,\Delta PV_A$ — time — T1 — step response

increases continuously to the full-scale value with a time delay. The voltage through a capacitor when charging via a series resistor shows an analog response.

**The 2nd-order time-delay element:**
The second-order time-delay element is a controlled system with two delays (two first-order time-delay elements connected in series). The second-order time-delay systems are characterized by three parameters, the system gain Ks and the two time constants Tu and Ta (Figure 5.11). Unlike the first-order time-delay element, the step response is

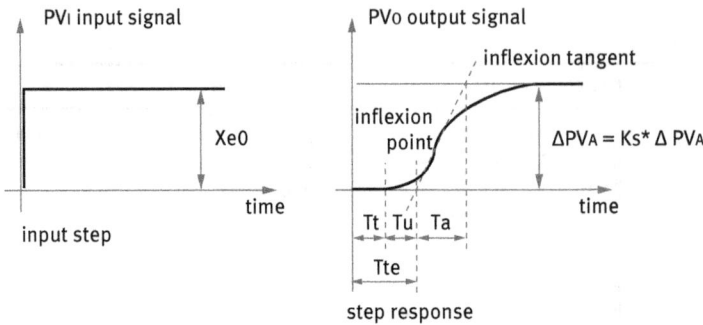

**Figure 5.11:** Model for controlled systems with compensation and dead time – approximated (Courtesy Uwe Krug, Buerkert Fluid Control Systems).

initially characterized by a horizontal tangent, features a flex point and then runs asymptotically toward the full-scale value. Real second-order time-delay elements are controlled systems with two (energy) stores, such as those occurring when tempering a tank via a heat exchanger. What follows is a consideration of the controllability of controlled systems with compensation (with time-delays/dead time). The model shown in Figure 5.11 can approximate these controlled systems.

On the basis of practical experience, it is possible to make an approximate statement on the controllability of a controlled system with compensation and equivalent dead time via the ratio Tte/Ta (Table 5.6).

**Table 5.6:** Estimation of the controllability of a system with compensation.

| Tte/Ta | Controllability | Control engineering effort |
|---|---|---|
| < 0.1 | Very well controllable | Low |
| 0.1 ... 0.2 | Well controllable | Moderate |
| 0.2 ... 0.4 | (Still) controllable | High |
| 0.4 ... 0.8 | Poorly controllable | Very high |
| > 0.8 | Barely controllable | Special measures or controller structures required |

**The I-element:**
On the I-element (integral element), the output variable is proportional to the time integral of the input variable (Figure 5.12). In the case of a constant input variable, the output variable increases continuously.

**The lag element:**
On the lag element, there is a similar behavior to that of the P-element with system gain 1 (Ks = 1). However, the lag element does not respond immediately to changes of the input value. In the case of a stepped change of the input variable PVi, the same stepped change of the output variable occurs upon expiration of time Tt (Figure 5.13).

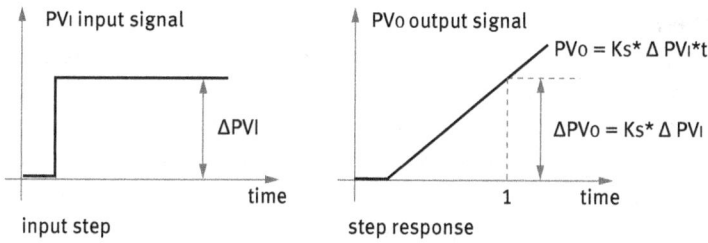

**Figure 5.12:** Step response of an I-element (Courtesy Uwe Krug, Buerkert Fluid Control Systems).

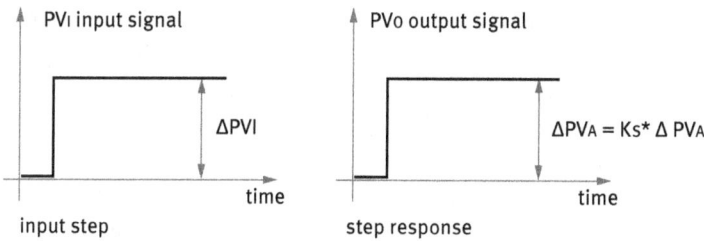

**Figure 5.13:** Step response of a lag element (Courtesy Uwe Krug, Buerkert Fluid Control Systems).

### 5.4.2.1 The controller

A closed-loop control system must ensure that the PV is equal to the set-point value or is adjusted to the set-point value under all circumstances, even under the influence of disturbance variables, that is, the control deviation must be 0. In addition, the closed-loop control system must operate stably and the PV may neither drift from the required operating point nor oscillate around it as a consequence of a change of disturbance variable or set-point value. To meet these requirements when designing a closed-loop control system or a control loop, the appropriate controller must be selected for the given controlled system and must be matched to the controlled system. In addition to knowledge of the dynamic and static behavior of the controlled system, this also necessitates knowledge of the characteristics of the individual controller versions or controller types. In the following, the individual controller types are described in greater detail. Controllers can initially be subdivided into two main groups:
–   Continuous-action controllers and
–   On/off controllers.

### 5.4.2.2 On/off controllers

On/off controllers are frequently used in temperature, filling level, pressure, pH value, and conductivity control loops. On/off controllers are also used in day-to-day applications or appliances such as automatic coffee machines, irons, refrigerators, or building heating systems.

**Two-point controller:**

A two-point controller operates in the same way as a switch. Its output can assume only two states: switched on or switched off. This means that the controlled final control element or actuator is either switched on or opened or is switched off or closed. A two-point controller can be seen as a P controller (continuous-action controller) with very high gain. A two-point controller can only be used in conjunction with time-delayed systems (first-order or second-order time delay systems) or controlled systems without compensation (I-systems). Figure 5.14 illustrates the principle of operation of a two-point controller. Ideally, the switch-on point and switch-off point of the two-point controller will coincide. In practice however, the switch-on point and switch-off point are reciprocally offset. The interval between the two switching points is referred to as (switching) hysteresis and is identified with PVh. If the PV drops below the preset set-point value SP minus half the hysteresis, the output of the controller is switched on (CO = 100%). If the PV rises above the set-point value plus the hysteresis, the output is switched off again (CO = 0%). Half of the hysteresis prevents a constant switching on and switching off at the same point owing to very minor disturbances. The described two-point controller can be used, for example, for tempering a room. The control response of the two-point controller is illustrated below on the basis of this example. The diagram in Figure 5.15 shows the principle characteristic of the PV (temperature in the room). The upper diagram in Figure 5.15 shows the PV (temperature) and the set-point value (desired temperature) as a function of time and the lower diagram shows the manipulated variable (control output). At instant t = 0 (switch-on instant of the closed-loop control system), the controller switches its output on and thus opens the heating valve. For the period of the dead time ($T_t$), the actual value initially does not change. After the dead time expires, it increases. The characteristic of the PV corresponds to the step response of the controlled system (characteristic of the PV shown with a dashed line).

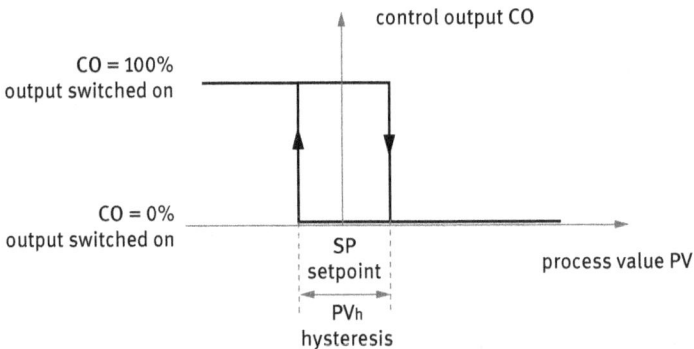

**Figure 5.14:** Principle of operation, characteristic of a two-point controller (Courtesy Uwe Krug, Buerkert Fluid Control Systems).

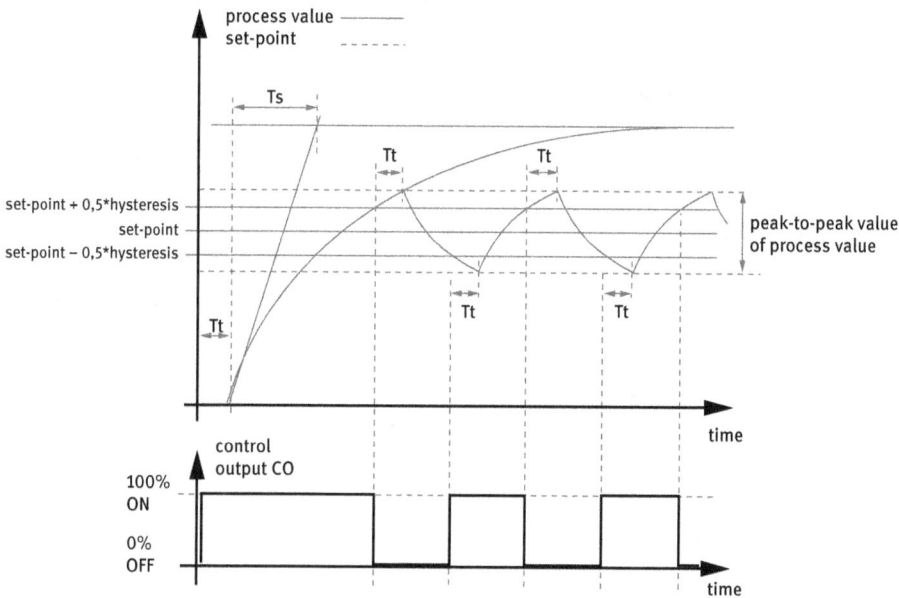

**Figure 5.15:** Process value and manipulated variable or controller output of the closed-loop temperature control system as a function of time. Tt: Dead time; Ts: System time constant (Courtesy Uwe Krug, Buerkert Fluid Control Systems).

If the PV reaches the set-point value plus half the hysteresis, the controller switches its output off and thus closes the heating valve. For the duration of the dead time, the PV initially still rises. After the dead time elapses, it drops. If the PV drops below the set-point value minus half the hysteresis, the controller switches its output back on and the heating valve opens. In the same way as with the rise in temperature, it initially still drops before it rises again, due to the time delay or dead time of the system. The control cycle then starts again. The dead time of the controlled system and the hysteresis of the controller result in periodic fluctuation of the PV about the set-point value. To assess the controllability of a controlled system with compensation and dead time using a two-point controller, it is possible to use the ratio of equivalent dead time to system time constant ($T_t/T_s$ – Table 5.7). Figure 5.15 shows how the two times are determined from the step response of a controlled system.

**Table 5.7:** Estimation of the controllability of a system with compensation and dead time using a two-point controller.

| $T_t/T_s$ | Controllability |
| --- | --- |
| <0.1 | Well controllable |
| 0.1 ... 0.3 | Controllable |
| >0.3 | Poorly controllable |

The peak-to-peak displacement of the actual value is chiefly dependent on two aspects:
- the controller hysteresis (this can generally be set on the controller). The peak-to-peak displacement increases with increasing hysteresis
- the system time constant (this generally cannot be changed and is determined by the structure of the controlled system). The lower the system time constant, the greater will be the peak-to-peak displacement.

The lower the time constant of the controlled system (system time constant) and the hysteresis of the two-point controller are, the more frequently the controller will switch. Depending on the design of the controller, the final control element/actuator and the system, increased wear on the control loop components will occur in the case of frequent switching. Consequently, a two-point controller cannot be used on controlled systems without time-delay (P-systems). The above-described response of the two-point controller is referred to as the "heating function". Besides this "heating function", two-point controllers are also used for the "cooling function". The principle of operation is similar in this case but the controller output is switched on when the set-point value is exceeded. Figure 5.16 shows this principle of operation. On most modern two-point controllers, it is possible to set the circuit function so that they can be used for both applications, depending on the setting.

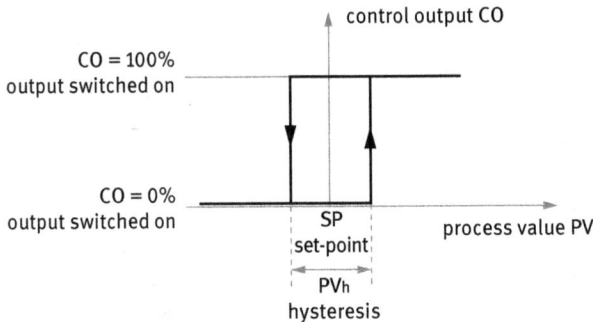

**Figure 5.16:** Principle of operation, characteristic curve of a two-point controller, cooling function (Courtesy Uwe Krug, Buerkert Fluid Control Systems).

**Three-point controller:**
A three-point controller is a switch, like a two-point controller. In contrast to the two-point controller, its output may assume three switch positions. three-point controllers are used, for example, in the following applications:
- **Closed-loop temperature control**
  Closed-loop temperature control of a room on which the disturbance variables can be counteracted by heating and cooling.

-   **Closed-loop pH control systems**
    For neutralization of media. The required pH value can be set by adding acid or lye.
-   **Control of motorized actuators**
    Motorized actuators operate valves, such as butterfly valves, ball valves, or gate valves via mechanisms. The valves can be opened, closed, or stopped at any position.
    Figure 5.17 demonstrates the principle of operation of a three-point controller.

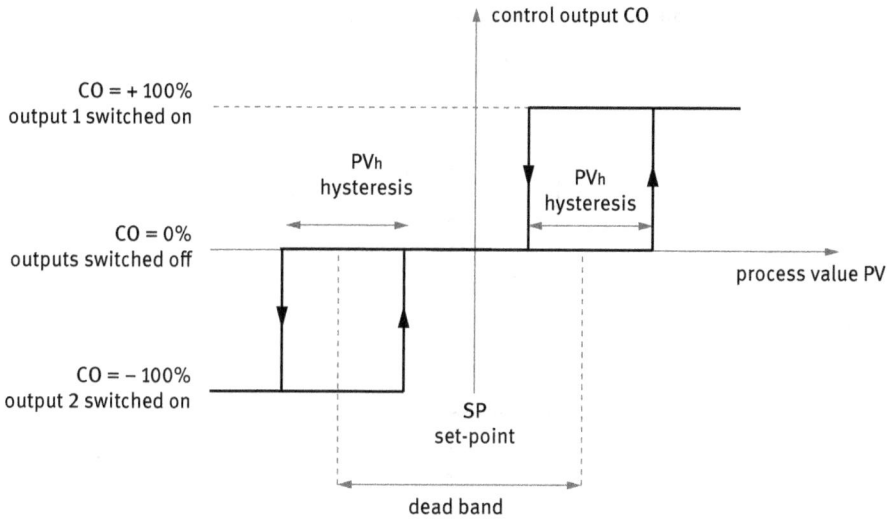

**Figure 5.17:** Principle of operation, characteristic curve of a three-point controller (Courtesy Uwe Krug, Buerkert Fluid Control Systems).

If the PV rises above the preset set-point value SP plus half the dead band plus half the hysteresis, the output of the controller is switched on (CO = +100%). If the PV drops below the set-point value plus half the dead band minus half the hysteresis, the output is switched back off again (CO = 0%). The difference between the switch-on and switch-off points is referred to as the hysteresis (as on the two-point controller). The controller displays the same principle of operation in the other direction. If the PV drops below the preset set-point value SP minus half the dead band minus half the hysteresis, the output of the controller is switched on (CO = –100%). If the PV rises again above the set point value minus half the dead band plus half the hysteresis, the output is switched back off again (CO = 0%). A three-point controller is shown by the following symbol (Figure 5.18):

In principle, the three-point controller comprises two two-point controllers whose set-point values are mutually offset (Figure 5.19). One of the controllers must be operated in "cooling" mode and the other must be operated in "heating" mode.

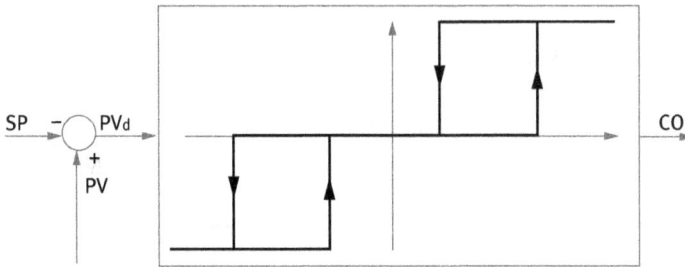

**Figure 5.18:** Symbolic representation of a three-point controller (Courtesy Uwe Krug, Buerkert Fluid Control Systems).

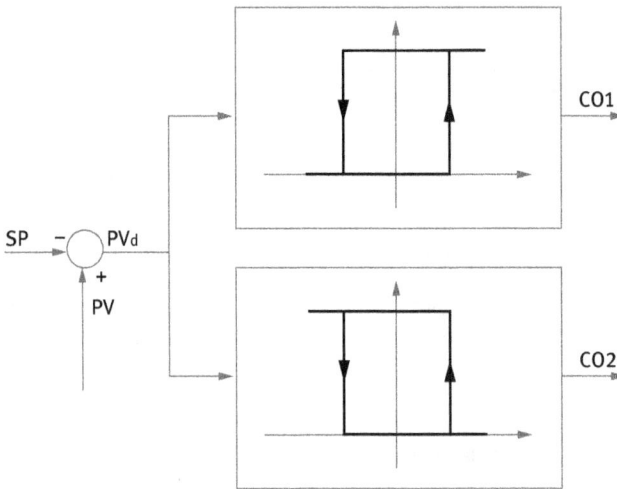

**Figure 5.19:** Symbolic representation of a three-point controller comprising two-point controllers (Courtesy Uwe Krug, Buerkert Fluid Control Systems).

## Continuous-action controllers:

Continuous-action controllers are used for demanding control engineering tasks. Unlike on/off controllers, the manipulated variable of these controllers may assume any value within the range of the manipulated variable/control output (i.e., the range between the maximum and minimum possible values of the control output, e.g., generally between the positions "OPEN" and "CLOSED" on a control valve; this then generally corresponds to a range of 0 ... 100%). These controller types are characterized by the fact that they respond to any change in control deviation (PVd = set-point value – process value) at the output. There are different types of continuous action controller:

– P controller
– PI controller
– PD controller
– PID controller

These controller types differ by virtue of their dynamics, that is, by virtue of their time response of their control output as a function of the control deviation. The various controllers are characterized by their step response, that is, by the time characteristic of their control output after an abrupt change in input variable, the control deviation PVd. The individual types of controller are explained in greater detail below.

### 5.4.2.3 P controller

The *P* controller is a purely proportional-action controller. Its control output is directly proportional to its input variable, the control deviation *PVd*, in stationary state. The control output of the *P* controller is calculated as follows:

$$CO = Kp \cdot PVd$$

$$CO = Kp \cdot (\text{Process value} - \text{Set} - \text{point value})$$

Depending on *Kp*, the control output may drop below (*Kp* < 1) or increase above (*Kp* > 1) the control deviation. *Kp* is referred to as proportional gain factor or proportional coefficient. As can be seen from the above calculation formula for the control output, the *P* controller requires a control deviation (*PVd* = PV − SP) for forming a control output (CO = *Kp* • *PVd*). For this reason, control loops with P controllers feature a permanent control deviation, which decreases with increasing *Kp*. On the basis of dynamic aspects however, it is not possible to achieve *Kp* of any arbitrary magnitude. This may lead to instabilities of the control loop.

The P controller is represented by the following symbol (Figure 5.20):

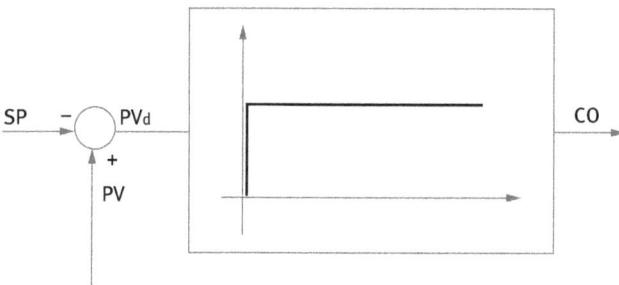

**Figure 5.20:** Symbolic representation of a P controller (Courtesy Uwe Krug, Buerkert Fluid Control Systems).

Characteristics of the P controller:
- The P controller operates without delay and very quickly; it responds immediately to changes in the control deviation.
- Control loops with P controller have a permanent control deviation.
- Setting parameter: *Kp* (proportional gain factor).

### 5.4.2.4 PD controller

On the PD controller, not only the control deviation but also its rate of change is used to form a control output. The controller thus already responds when a control deviation occurs and counteracts the occurrence of a higher control deviation. The control output increases all the faster control deviation changes. The control output of the PD controller is calculated as follows:

$$CO = Kp \cdot \left( Td \cdot \frac{d(PV_d(t))}{dt} + PV_d(t) \right)$$

where $PV_d = PV - SP$ is the control deviation, $Kp$ is the proportional gain factor and $Td$ is the derivative action time.

As can be seen from the above calculation formula for the control output, the influence of the D-component is determined via parameter $Td$. The higher Td becomes, the higher the D-component becomes when calculating the control output. As is also the case on the P controller, control loops with PD controllers have a permanent control deviation, which decreases with increasing $Kp$ (Figure 5.21). However, the D-component produces a stabilizing effect, which allows the proportional gain factor $Kp$ to be set higher than on the pure P controller. Figure 5.21 shows the step response of the PD controller. On real-world PD controllers, the D-component is time-delayed (time constant T1), which is allowed for in the transfer function shown. The time constant T1 can, however, not be set directly on most controllers.

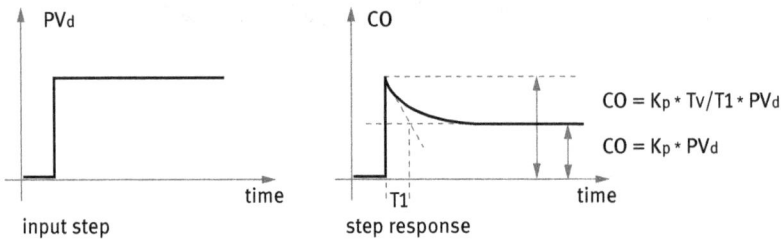

PV_d input step      CO step response

CO = Kp * Tv/T1 * PVd

CO = Kp * PVd

**Figure 5.21:** Step response of the PD controller (Courtesy Uwe Krug, Buerkert Fluid Control Systems).

The PD controller is represented by the following symbol (Figure 5.22):
Characteristics of a PD controller:
- Like the P controller, the PD controller operates without delay and responds immediately to changes in the control deviation.
- The PD controller responds to the rate of change of the control deviation and thus counteracts the buildup of a higher control deviation.
- Control loops with PD controller have a permanent control deviation.
- The D-component of the controller may lead to a situation in which minor fluctuations of the PV, and thus minor fluctuations of the control deviation, as

Kp, Tv

SP ─⊖ PVd          CO
   +
   PV

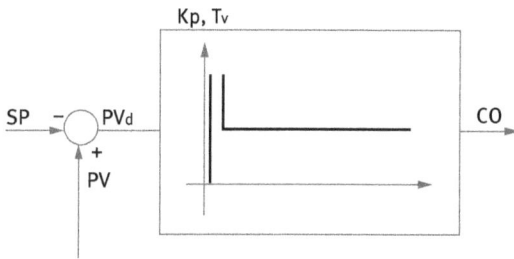

Figure 5.22: Symbolic representation of a PD controller (Courtesy Uwe Krug, Buerkert Fluid Control Systems).

caused, for example, by disturbances in electrical transfer of the PV (e.g., by standardized signals), lead to constant fluctuations of the control output.
- Setting parameters:
  Kp (proportional gain factor)
  Td (derivative-action time)

### 5.4.2.5 PI controller
The PI controller consists of a proportional component and an integral component.

The integral component calculates its share of the control output via the time integral of the control deviation. If there is a control deviation, the integral component increases the control output. This avoids a permanent control deviation as occurs on P and PD controllers. The control output of a PI controller is calculated as follows:

$$CO = Kp \cdot \left( \frac{1}{Tr} \int (PV_d(t)dt) + PV_d(t) \right)$$

where $PV_d = PV - SP$ is the control deviation, $Kp$ is the proportional gain factor or proportional coefficient, and $Tr$ is the reset time.

As can be seen from the above calculation formula for the control output, the influence of the I-component is determined by parameter $Tr$. The lower $Tr$ becomes, the larger the I-component becomes when calculating the control output. Reset time $Tr$ is the time which the controller requires to generate a control output of the same magnitude as that which occurs immediately as the result of the P-component by means of the I-component. Figure 5.23 shows the step response of the PI controller. The PI controller is represented by the following symbol (Figure 5.24):
Characteristics of the PI controller:
- The PI controller is advantageous in that it responds quickly due to its P-component and eliminates permanent control deviations owing to the I-component.
- Since two parameters can be set on the PI controller (Kp and Tr), it is possible to better adapt it to the dynamics of the controlled system.
- Setting parameters:
  Kp (proportional gain factor)
  Tr (reset time)

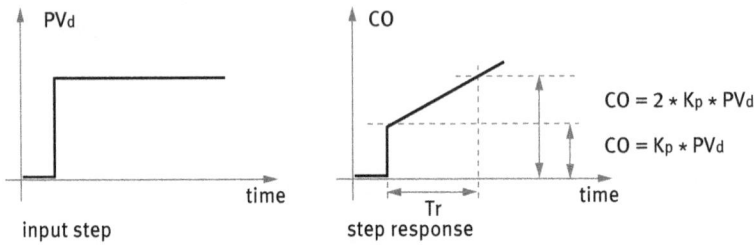

**Figure 5.23:** Step response of the PI controller (Courtesy Uwe Krug, Buerkert Fluid Control Systems).

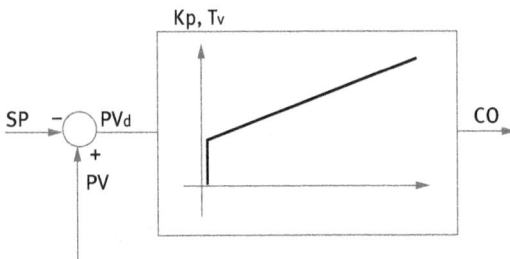

**Figure 5.24:** Symbolic representation of the PI controller (Courtesy Uwe Krug, Buerkert Fluid Control Systems).

### 5.4.2.6 PID controller

The control output of the PID controller is calculated from the proportional, integral and differential component. The control output of the PID controller is calculated as follows:

$$CO = Kp \cdot \left( \frac{1}{Tr} \int (PV_d(t)dt) + Td \cdot \frac{d(PV_d(t))}{dt} + PV_d(t) \right)$$

where $PV_d = PV - SP$ is the control deviation, $Kp$ is the proportional gain factor or proportional coefficient, $Tr$ is the reset time, and $Td$ is the derivative action time. Figure 5.25 shows the step response of the PID controller.

The PID controller is represented by the following symbol (Figure 5.26):
Characteristics of the PID controller:

– The PID controller unites the characteristics of the P controller, PD controller and PI controller.
– Setting parameters:
  Kp (proportional gain factor)
  Tr (reset time)
  Td (derivative-action time)

On the basis of practical experience, it is possible to provide the following estimation (Table 5.8) of the suitability of the various continuous-action controllers for closed-loop control of important technical-controlled variables.

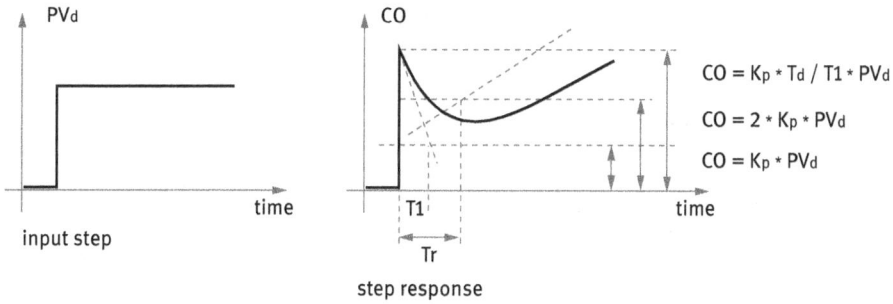

$$CO = K_p * T_d / T1 * PV_d$$

$$CO = 2 * K_p * PV_d$$

$$CO = K_p * PV_d$$

**Figure 5.25:** Step response of the PID controller (Courtesy Uwe Krug, Buerkert Fluid Control Systems).

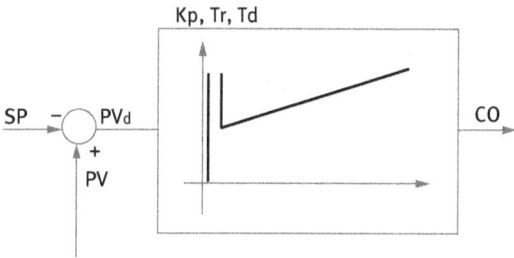

**Figure 5.26:** Symbolic representation of the PID controller Courtesy Uwe Krug, Buerkert Fluid Control Systems).

**Table 5.8:** Suitability of various continuous-action controllers for controlling important technical-controlled variables.

| | Controller Type | | | |
|---|---|---|---|---|
| | P | PD | PI | PID |
| Controlled variable | Permanent control deviation | | Non-permanent control deviation | |
| Temperature | Conditionally suitable | Conditionally suitable | Suitable | Suitable for stringent demands |
| Flow rate | Unsuitable | Unsuitable | Suitable | Over-dimensioned |
| Pressure | Suitable | Suitable | Suitable | Over-dimensioned |
| Filling level | Suitable | Suitable | Suitable | Over-dimensioned |

### 5.4.3 Adapting the controller to the controlled system

There are two requirements made on a controller or control loop.

**Variable command control:**
In the case of variable command control, the set-point value is not constant but changes over the course of time. The PV must be corrected to the set-point value.

The behavior of a closed-loop control system in the case of changing set-point value is referred to as response to set-point changes.

**Fixed command control:**
In the case of fixed command control, the set-point value is constant. In this case, the closed-loop control system has the task of maintaining the PV at the value of the set-point. Disturbance variables acting on the controlled system should be compensated for in this case. The behavior of a closed-loop control system with changing disturbance variables is referred to as disturbance response. In addition to having a good response to set-point changes, a closed-loop control system should, in most cases, feature a good disturbance response. If a disturbance occurs in a control loop, this leads to a control deviation, which the controller must compensate for. When planning a closed-loop control system, disturbance variables are of special significance. If several disturbance variables act on a controlled system, the individual disturbance variables generally have a different time response. Many disturbance variables occur abruptly and others less abruptly. Even the magnitude of the influence on the PV differs with the individual disturbance variables. When planning a closed-loop control system, there is the risk that the control loop becomes unstable due to the selected combination of controller and controlled system or owing to the selected parameters of the controller. The following behaviors may occur, for example, after occurrence of a set-point change or disturbance variable change. The control loop is at the stability limit; the PV oscillates at constant amplitude and frequency (Figure 5.27).

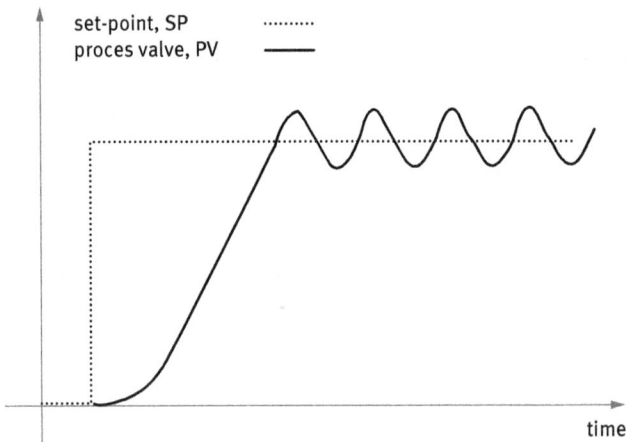

**Figure 5.27:** Process value characteristic, control loop at the stability limit (Courtesy Uwe Krug, Buerkert Fluid Control Systems).

The control loop is unstable. The PV oscillates with increasing amplitude (Figure 5.28).

The control loop is stable; the PV is corrected to the new set-point value (Figure 5.29).

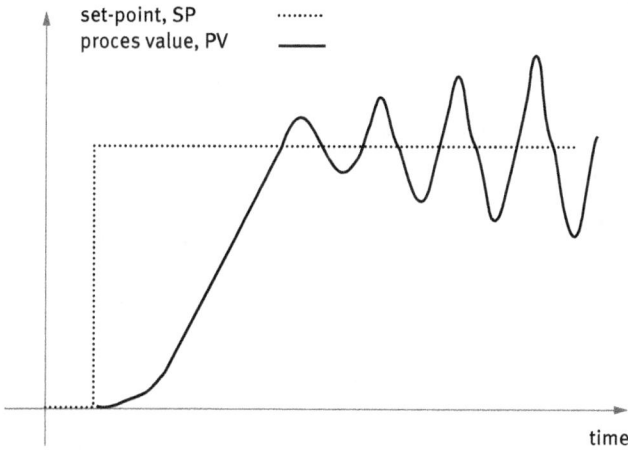

**Figure 5.28:** Process value characteristic, control loop unstable (Courtesy Uwe Krug, Buerkert Fluid Control Systems).

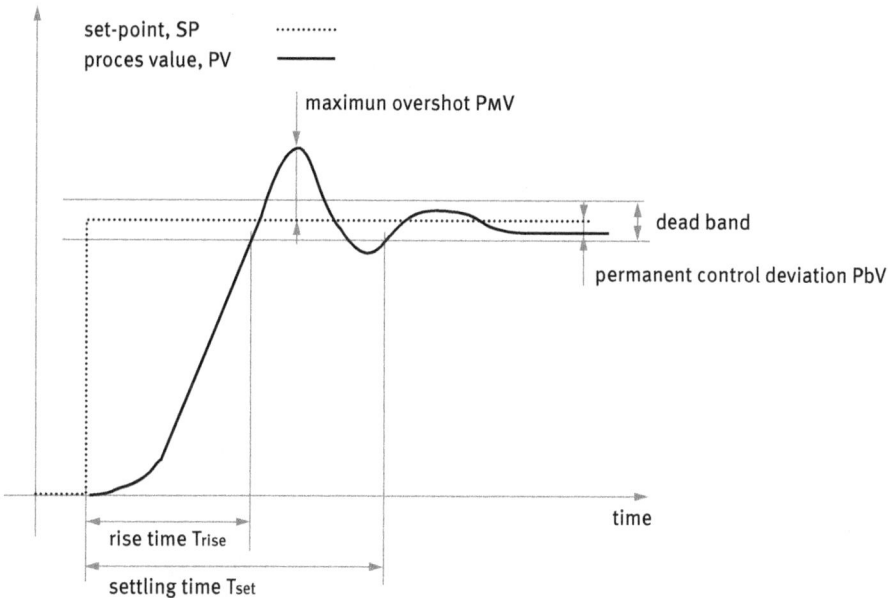

**Figure 5.29:** Process value characteristic after a set-point change in the case of a stable operating control loop (Courtesy Uwe Krug, Buerkert Fluid Control Systems).

The behavior of a well-tuned or well-set control loop after a disturbance variable change is similar (Figure 5.30).

The quality of a closed-loop control or a control loop is assessed on the basis of the following parameters.

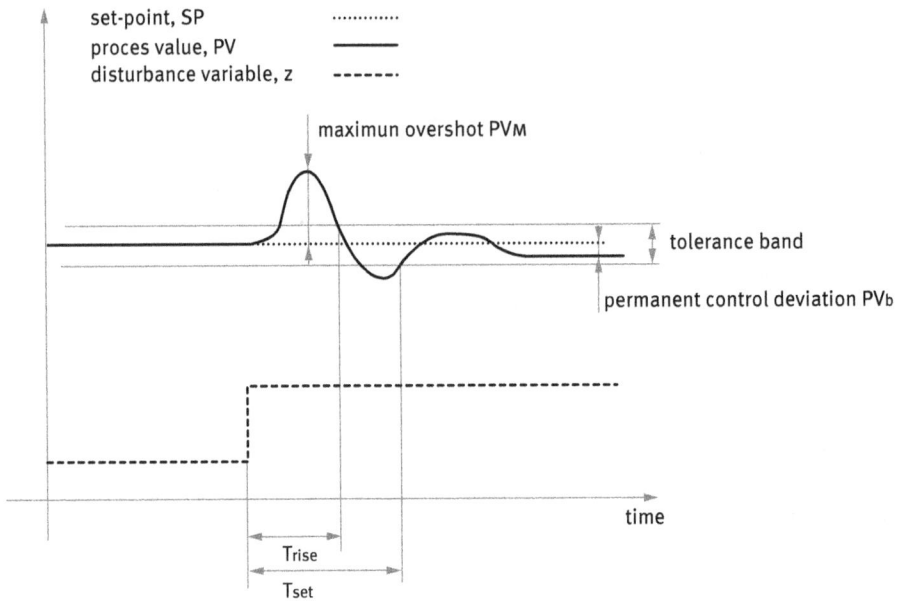

**Figure 5.30:** Process value characteristic after a disturbance variable change in the case of a stably operating control loop (Courtesy Uwe Krug, Buerkert Fluid Control Systems).

### 5.4.3.1 Permanent control deviation PVb

The permanent control deviation occurring after the adjustment process has decayed.

### 5.4.3.2 Overshoot PVm

Maximum value of the PV or of the controlled variable minus the PV in steady state.

### 5.4.3.3 Rise time Trise

The time which elapses after a set-point or disturbance variable change until the PV occurs for the first time in an agreed tolerance band (e.g., 2% or 5%) about its stationary end value.

### 5.4.3.4 Settling time Tset

The time which elapses after a set-point or disturbance variable change until the PV occurs and permanently remains in an agreed tolerance band (e.g., 2% or 5%) about its stationary end value.

On the basis of these parameters, it is possible to formulate the requirements made of an optimally tuned control loop:

- permanent control deviation $PV_b$ = 0 wherever possible
- maximum overshoot $PV_m$ as low as possible

–   settling time $T_{det}$ as low as possible
–   rise time $T_{rise}$ as low as possible.

### 5.4.3.5 Selecting the suitable controller

The controller must be matched to the controlled system in order for a control loop to operate optimally. Suitable combinations of controllers and controlled systems on which a stable control response can be achieved by appropriate setting of the controller parameters,

–   Kp (proportional gain factor)
–   Tr (reset time)
–   Td (derivative-action time),

can be established based on the dynamics and stability of control loops and allowing for empirical values. There are, of course, also control loops necessitating other combinations of controlled systems/controllers.

Table 5.9 provides an overview of suitable combinations of controllers and controlled systems.

### 5.4.3.6 Determining the controller parameters

After a suitable controller has been selected, a second step is to match the parameters of the controller to the controlled system. A number of setting guidelines with which a favorable setting of the controller parameters can be determined experimentally are cited in control engineering literature. To avoid incorrect settings, the conditions under which the relevant setting guidelines were established must always be followed. Besides the characteristics of the controlled system and of the controller themselves, other important factors include whether a disturbance variable change or a reference variable change is to be compensated for optimally.

### 5.4.3.7 Setting guidelines in-line with Ziegler and Nichols (oscillation method)

With the Ziegler and Nichols methods, the controller parameters are set on the basis of the behavior of the control loop at the stability limit. The controller parameters are initially set so that the control loop starts to oscillate. Critical characteristic values then occur which allow conclusions to be drawn in terms of the controller parameters. The precondition for using this method is that the control loop can be caused to oscillate.

**Procedure:**
–   Set controller as P controller (i.e., Tr = 9,999, Td = 0), initially select a low value for Kp.
–   Set the required set-point value.
–   Increase Kp until the PV executes an undamped sustained oscillation (refer Figure 5.31).

**Table 5.9:** Suitability of continuous-action and on/off controllers for combination with various types of controlled system (Courtesy Uwe Krug, Buerkert Fluid Control Systems).

| Controlled system | Continuous action controller | | | | NO/OFF controller | |
|---|---|---|---|---|---|---|
| | P | PI | PD | PID | 2-point | 3-point |
| Proportional | Not suitable | Well suited for control action well suited for disturbance respone | Not suitable | Not suitable | Not suitable | Not suitable |
| Proportional With dead-time | Not suitable | suited for control action suited for disturbance respone | Not suitable | Not suitable | Not suitable | Not suitable |
| Proportional With 1st order time delay | Well suited for control action | well suited for disturbance respone | Not suitable | Not suitable | suitable | suitable |
| Proportional With 1st order time delay and dead-time | Not suitable | suited for control action suited for disturbance respone | Not suitable | Well suited for control action well suited for disturbance respone | Only suitable if hysteresis is low | Only suitable if hysteresis is low |

| Process | Response | | | | | |
|---|---|---|---|---|---|---|
| **Proportional With 2nd order time delay** | | Not suitable | suited for control action / suited for disturbance respone | Well suited for control action | Well suited for control action / well suited for disturbance respone | suitable | suitable |
| **Proportional With 2nd order time delay and dead-time** | | Not suitable | suited for control action / suited for disturbance respone | Not suitable | Well suited for control action / well suited for disturbance respone | Not suitable | Not suitable |
| **Proportional With 3rd order time delay** | | Not suitable | suited for control action / suited for disturbance respone | Not suitable | Well suited for control action / well suited for disturbance respone | Not suitable | Not suitable |
| **Integrating** | | Well suited for control action | well suited for disturbance respone | suited for control action | suited for disturbance bance respone | suitable | suitable |
| **Integrating With 1st order time delay** | | Not suitable | Not suitable | Well suited for control action | well suited for disturbance respone | suitable | suitable |

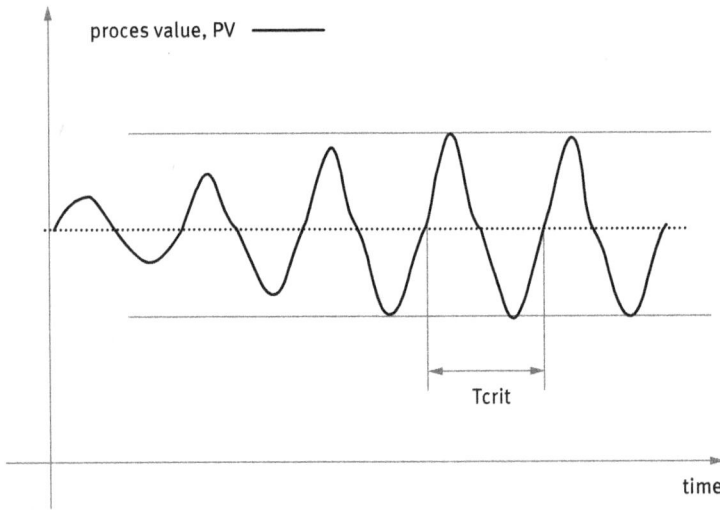

**Figure 5.31:** Process value characteristic of the control loop at the stability limit in order to determine the control parameters in line with Ziegler and Nichols (Courtesy Uwe Krug, Buerkert Fluid Control Systems).

The proportional gain factor set at the stability limit is designated as $K_{crit}$. The resultant period of oscillation is designated as $T_{crit}$.

The controller parameters can then be calculated in accordance with Table 5.10 from $K_{crit}$ and $T_{crit}$.

**Table 5.10:** Controller parameters in-line with Ziegler and Nichols (Courtesy Buerkert Fluid Control Systems).

| Setting the parameters in line with Ziegler and Nichols | | | |
|---|---|---|---|
| Controller type | Setting parameters | | |
| P controller | $Kp = 0.5\ K_{crit}$ | | |
| PI controller | $Kp = 0.45\ K_{crit}$ | $Tr = 0.85\ T_{crit}$ | |
| PID controller | $Kp = 0.6\ K_{crit}$ | $Tr = 0.5\ T_{crit}$ | $Td = 0.12\ T_{crit}$ |

The Ziegler and Nichols setting were determined for P systems with first-order time-delay and dead time. They apply only to control loops with disturbance response.

**Setting guidelines in-line with Chien, Hrones and Reswick (control output step method):**

With this method, the controller parameters are set on the basis of the transient response or step response of the controlled system. The controller emits a control output step. The times Tu and Tg are read off (refer Figure 5.32) from the characteristic of the PV

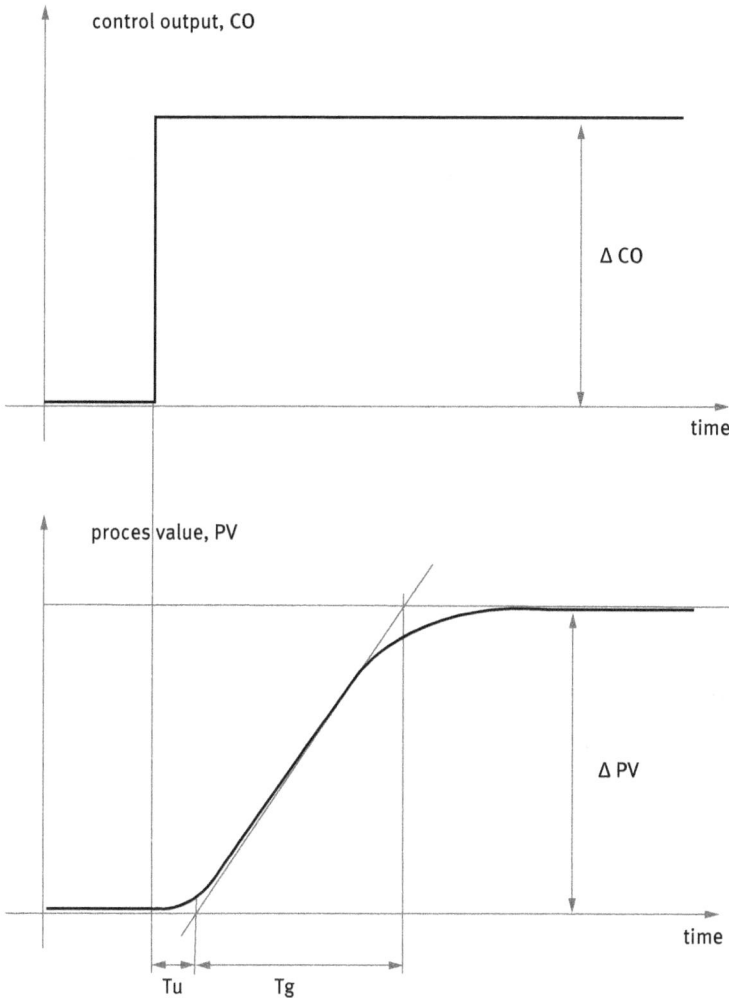

**Figure 5.32:** Step response of a controlled system for determining control parameters in-line with Chien, Hrones and Reswick (Courtesy Uwe Krug, Buerkert Fluid Control Systems).

or the controlled variable. The control output step must be selected so that a PV is produced that lies within the range of the subsequent operating point of the controlled system. *Ks* is the proportional coefficient of the controlled system. It is calculated as follows:

$$Ks = \frac{\Delta PV}{\Delta CO}$$

where *ΔPV* is the magnitude of the PV step, *ΔCO* is the magnitude of the control output step.

**Procedure for determining the step response of the controlled system:**
- Switch the controller to manual operating mode.
- Emit a control output step and record the PV with a recorder.
- In the case of critical-controlled systems (e.g., in the case of the risk of overheating), switch off in good time.

It must be noted that the PV may increase again after switch-off on thermally sluggish systems.

Table 5.11 lists the controller parameter settings as a function of Tu, Tg, and $Ks$ for response to set-point changes and disturbance response and for an aperiodic control process and a control process with 20% overshoot. The figures apply to systems with P-response, with dead time and with first-order delay-time.

Table 5.11: Controller parameters in-line with Chien, Hrones and Reswick (Courtesy Buerkert Fluid Control Systems).

**Setting the parameters in-line with Chien, Hrones and Reswick:**

| Controller Type | Parameter settings | | | |
|---|---|---|---|---|
| | **Aperiodic control process** | | **Control process with overshoot (approx. 20% overshoot)** | |
| | Resp. to set p changes | Disturbance response | Resp. to set p changes | Disturbance response |
| P controller | $Kp = 0.3\frac{Tg}{Tu \cdot Ks}$ | $Kp = 0.3\frac{Tg}{Tu \cdot Ks}$ | $Kp = 0.7\frac{Tg}{Tu \cdot Ks}$ | $Kp = 0.7\frac{Tg}{Tu \cdot Ks}$ |
| PI controller | $Kp = 0.35\frac{Tg}{Tu \cdot Ks}$ $Tr = 1.2Tg$ | $Kp = 0.6\frac{Tg}{Tu \cdot Ks}$ $Tr = 4.0Tu$ | $Kp = 0.6\frac{Tg}{Tu \cdot Ks}$ $Tr = Tg$ | $Kp = 0.7\frac{Tg}{Tu \cdot Ks}$ $Tr = 2.3Tu$ |
| PID controller | $Kp = 0.6\frac{Tg}{Tu \cdot Ks}$ $Tr = Tg$ $Td = 0.5Tu$ | $Kp = 0.95\frac{Tg}{Tu \cdot Ks}$ $Tr = 2.4Tu$ $Td = 0.42Tu$ | $Kp = 0.95\frac{Tg}{Tu \cdot Ks}$ $Tr = 1.35Tg$ $Td = 0.47Tu$ | $Kp = 1.2\frac{Tg}{Tu \cdot Ks}$ $Tr = 2.0Tu$ $Td = 0.42Tu$ |

## 5.4.4 Rating and selection of control valves

In addition to controllers and sensors, actuators or final control elements which intervene in the process to be controlled as a function of the signals preset by the controller and which change the process variable to be controlled are required for constructing closed-loop control systems.

### 5.4.4.1 Introduction and definition of terms
Valves are final control elements or actuators for influencing fluid streams in fluid control systems. In accordance with DIN IEC 534, a positioning valve is a device,

operated with auxiliary energy, which varies the flow rate in a process. It consists of a valve fitting, connected to the actuator, which is able to change the position of the restrictor in the valve as a function of the controller signal (control output). Generally, a control system is required between the actuator and controller to act as a signal transducer and/or amplifier. In the case of many positioning valves, the control system is integrated as far as a field bus interface in the actuator. In accordance with DIN IEC 534, positioning valves are subdivided on the basis of the following types (Table 5.12):

**Table 5.12:** Classification of positioning valves in accordance with DIN IEC 534.

| Valve type | Restrictor |
| --- | --- |
| Lift-type valve<br>Through-way valve<br>3-way valve<br>Angle valve | The restrictor is generally designed as a cone. It moves perpendicular to the seat plane. |
| Gate valve | The restrictor is a flat or wedge-shaped plate. |
| Diaphragm valve | A flexible restrictor performs the function of isolation and sealing. |
| Ball valve | The restrictor is a ball with a cylindrical bore or a segment of a ball. |
| Butterfly valve | A disc mounted in such a way as to allow it to rotate. |
| Plug valve | The restrictor may be a cylindrical, conical, or eccentrically mounted ball segment. |

Valves can also be classified in accordance with the distinction between the main functions of final control elements/actuators in compliance with DIN 19226, dividing them into CONTROL-final control elements and ON/OFF-final control elements. ON/OFF valves having only two or a few circuit states are used for open-loop control tasks. Control valves which are able to continuously set the fluid stream are used for closed-loop process control tasks. ON/OFF valves and control valves have extremely different tasks in some cases, so that the rating and selection of both valve types necessitate greatly different procedures.

### 5.4.4.2 Rating and selection of ON/OFF valves

This kind of valves can either open or close a line (ON/OFF valve) or switch over a material stream from one line to another. An important criterion for the valve to be selected is initially that the required fluid quantity be able to flow through the valve at a given pressure differential, that is, the valve cross-section must be adequately large. The following rule of thumb frequently applies: the line cross-section is equal to the valve (fluidic connection) cross-section. The next requirement is that the valve be able to switch against the maximum pressure differential, that is, that the valve actuator be adequately powerful. The maximum switchable pressure differential is specified in the data sheet. If the type of auxiliary energy has been defined and the material suitability has been checked, it is already possible to define a specific valve type and to select the specific valve, if it is available on the market.

### 5.4.4.3 Rating and selection of control valves

Control valves are able to constantly change their opening cross-section and thus continuously influence fluid streams. They, thus, represent variable flow resistors.

### 5.4.4.4 Fluidics fundamentals related to Valves in Control Loops

Flow resistances occur in fluid control systems in various forms as follows:
- as resistances in capillaries, gaps, nozzles, diaphragms, and valves;
- as line resistances in pipes, hoses, and ducts; and
- as leakage resistances in gaps and porous components.

In general, the ratio of pressure drops $\Delta p$ to fluid flow $Q$ can be defined as the flow resistance $R$ of a component.

$$R = \frac{\Delta p}{Q}$$

Basically, a distinction must be made between two types of resistance on the basis of the physical causes:
- frictional resistances due to flow involving friction and
- cross-sectional resistances owing to variations in the flow cross-section.

The following distinction between cases for dependence between $\Delta p$ and $Q$ must be made for frictional resistances in non-compressible fluids as a function of the Reynolds number $Re$.

$$Re = \frac{\bar{u} \cdot D_h}{v}$$

where $u$ is the mean flow velocity, $D_h$ the hydraulic diameter, $D_h = kA/U$, and $v$ is the kinematic viscosity. (Table 5.13)

**Table 5.13:** Range of Reynolds numbers with respect to the flow form.

| Range of the Reynolds number | Flow form | Interrelationship between $\Delta p$ and $Q$ |
|---|---|---|
| Re low | Laminar | $\Delta p \approx Q$ |
| Re high | Turbulent | $\Delta p \approx Q^{7/4}$ |
| $Re \approx Re_{critical}$ | Transitional form | To be determined experimental |

From this, we can conclude that the flow resistance $R$ is constant only in the case of laminar flow owing to $\Delta p \approx Q$. Otherwise, a non-linear relationship always applies between pressure drop $\Delta p$ and fluid flow $Q$. The following applies to fluid resistances in the case of cross-sectional variation in non-compressible fluids and with turbulent flow:

The permanent pressure loss $\Delta p_{loss}$ is taken as the basis for the flow resistance $R$. The flow resistance coefficient $\zeta$ is introduced as a "non-dimensional pressure loss" by referring the permanent pressure loss to the dynamic pressure.

$$\zeta = \frac{\Delta p_{loss}}{\left(\frac{\rho \cdot u^2}{2}\right)}$$

where $\rho$ is the density of the fluid.

The following applies to the model case of flow through an orifice plate:

$$\zeta = \left[\left(\frac{A_{pipe}}{\mu \cdot A_{orifice}}\right) - 1\right]^2$$

where $\mu$ is the contraction coefficient and

$$\Delta p_{loss} = (1 - m_B) \cdot \Delta p_B$$

where $\Delta p_B$ is the effective pressure through the orifice plate and $m_B = A_{orifice}/A_{pipe} =$ the opening ratio. After introduction of a flow coefficient $\alpha$, we obtain the flow-rate equation:

$$Q = \alpha \cdot A_{orifice} \sqrt{\frac{2 \cdot \Delta p}{\rho}}$$

In the case of a high Reynolds number, that is, in turbulent flow, the following applies to cross-sectional resistances in non-compressible fluids: $\Delta p \approx Q^2$.

The flow-rate variable $kv$ which was introduced in an earlier chapter is used to identify valves as orifice-type fluidic components: the $kv$ value (in $m^3/h$) is the volume flow of water at +5 to +30 °C passing through the valve at the relevant stroke s with a pressure loss $\Delta p_{ovalve} = 100$ kPa. (1 bar; 14.5 psi). Analogous to this, the flow-rate coefficient, cv, is described in the American literature and defined as follows: the cv value (in US gal/min) is the volume flow of water at 60 °F, which passes through at a pressure loss of 1 psi with the relevant stroke s.

Moreover, the $Q_{Nn}$ value (in l/min) is specified as the flow characteristic for compressed air under standardized conditions for pneumatic valves. The following conversion factors apply:

Kv ⟷ cv: kv = 0.86 cv
Kv ⟷ ζ: kv = 4 · d²/(ζ)^{1/2}
Kv ⟷ $Q_{Nn}$: kv = 1,078 $Q_{Nn}$

### 5.4.4.5 Valve characteristic

The valve characteristic represents the dependence of the aperture cross-section $A$ on a stroke $s$ of the valve spindle: A = f(s). In the simplest case, the valve characteristic is linear, that is, A = $K_1$ s. Note: Despite the linearity of the valve characteristic, there is no linear interrelationship between the volumetric flow rate Q through the valve and the valve stroke s due to non-constant pressure drops through the valve. The equal-percentage valve characteristic is described by a constant percentage increase in the aperture cross-section A with stroke s (referred to the relevant aperture cross-section A present).

$$\frac{dA(s)}{ds} \cdot \frac{1}{A(s)} = K_2$$

$$A = A_0 \cdot e^{K_2 s}$$

The equal-percentage valve characteristic approximates practical requirements to a greater extent than the linear characteristic, since
- low variations in stroke $\Delta s$ cause low $\Delta A$, that is, fine in feed movements
- high variations in stroke $\Delta s$ cause high $\Delta A$, that is, coarse in feed movements

In the case of s = 0, a minimum aperture cross-section $A_0$ is present. The valve closes only with an additional sealing edge.

Various valve characteristics are implemented by the contour of the closure elements, the valve cones. Conventional designs include, for example, parabolic cone, lantern cone, perforated cone, V-port cone, and many others.

### 5.4.4.6 Flow characteristic and range ability

The most conventional characteristic curve required for valve selection is the flow characteristic. The flow characteristic represents the dependence of the standardized flow rate $kv$ on the stroke s: kv = f(s).

### 5.4.4.7 ON/OFF flow characteristic

Due to their plate cone, ON/OFF valves initially have a linear flow characteristic in the range of small stroke (up to approx. 30% stroke). At an opening angle of 30 to 40% stroke (s), such valves already achieve approximately 90% flow rate (kv). As the aperture opens even further, the flow rate rises only very slowly through to the maximum, at full stroke. Since the total stroke of the ON/OFF valves is low in relation to the stroke of control valves, the actual task of producing a high change in flow rate with low stroke is performed.

### 5.4.4.8 Linear flow characteristic

In the simplest case, the flow characteristic is linear, that is, $kv = K1 \cdot s$.

Note: Unlike the linear valve characteristic, the interrelationship between the volumetric flow rate $Q$ through the valve and the valve stroke s is linear in the case of the linear flow characteristic.

### 5.4.4.9 Equal-percentage flow characteristic

The equal-percentage flow characteristic is described by a constant percentage increase in the flow rate kv with the stroke s (referred to the relevant flow rate kv present).

$$\frac{dkv(s)}{ds} \cdot \frac{1}{kv(s)} = K_2$$

$$kv = kv_0 \cdot e^{K_2 s}$$

At $s \approx 0$ a minimum aperture cross-section $A_0$ is present, causing a minimum flow $kv_0$. The valve closes only with an additional sealing edge.

The maximum aperture cross-section $A_{max}$ is reached at maximum stroke s. The related $kv$ value is referred to as $kv_s$ value. The range between $kv_0$ and $kv_s$ is the total range of the manipulated variable of the valve. The ratio of $kv_0$ to $kv_s$ is referred to as the range ability and defines a valve characteristic value:

$$\frac{1}{\alpha} = \frac{kv_0}{kv_s}$$

Conventional values are as follows:

$$\frac{1}{\alpha} = \frac{1}{25}; \frac{1}{30}; \frac{1}{50}$$

### 5.4.4.10 Operating characteristic and pressure ratio

The operating characteristic identifies the flow behavior of the valve under operating conditions in the fluid control system. It represents the dependence of the volume flow $\dot{V}$ on the stroke s of the valve spindle.

$$\dot{V} = f(s)$$

The following main elements of the installation influence the operating behavior:
- the pump; the pressure generated by the pump drops as $\Delta p$ over the entire fluid control system
- the tubes with the pressure drops $\Delta p_{Li}$
- and other resistances $\Delta p'_i$ in the installation (shut-off valves, heat exchangers, pipe elbows, branches, changes in cross-section, and other installed fittings).

The operating characteristic curve ($\dot{V} = f(s)$) thus differs from the flow characteristic curve $kv = f(s)$ of the valve considered in isolation.

The magnitude of the difference (the degree of characteristic distortion) is represented by the pressure ratio. The pressure ratio $\psi$ is stated for the fully open valve:

$$\psi = \frac{\Delta p_{v0}}{\Delta p_0} = \frac{\Delta p_{v0}}{\Delta p_{v0} + \Delta p_{L0} + \Delta \dot{p}_0}$$

where
$\Delta p_0$: Pressure drop over entire installation
$\Delta p_{v0}$: Pressure drop at fully opened valve (max. flow)
$\Delta p_{L0}$: Pressure drop at tubes, fittings ...
$\Delta \dot{p}_0$: Pressure loss at pump (at max. flow rate)

The behavior of the system as an interplay between source (pump) characteristic and load (valve) characteristic can be described with the help of a characteristic map.

The following standardized equation applies to the operating characteristic:

$$\frac{\dot{V}}{\dot{V}_{max}} = \frac{1}{\sqrt{1 + \psi\left[\left(\frac{kv_s}{kv}\right)^2 - 1\right]}}$$

that is, the operating characteristic depends on the pressure ratio and the flow characteristic:

$$\frac{kv_s}{kv} = \int\left(\frac{s}{s_{max}}\right)$$

An approximation of a linear operating characteristic can be achieved
- in the case of linear flow characteristic with high pressure ratio $\psi$ and
- in the case of equal-percentage flow characteristic with low pressure ratio $\psi$.
  The non-linarites for both valve types have approximately the same magnitude at $\psi \approx 0.3$.

A linear operating characteristic is achieved only if the valve features an optimum flow characteristic as the result of a special valve contour.

### 5.4.4.11 Rating and selection of control valves for fluid control systems

Control valves must be rated and selected with a view to their specific task to be able to ensure a faultless control function. Initially, the connection nominal diameter must be defined in accordance with the medium and the related, efficient flow velocity. The following guideline values apply:

- 2 m/s for fluids
- 20 m/s for gases
- 45 m/s for steam

The following formulae are helpful for practical application:

**Fluids:**

$$NW = 0.42 \cdot \sqrt{Q}$$

where $NW$ is the connection nominal diameter, $Q$ is the volumetric flow rate in l/h.

**Gases:**

$$NW = 4.2 \cdot \sqrt{\frac{Q_N}{p_1}}$$

where $Q_N$ is the volumetric flow rate in $Nm^3/h$ and $p_1$ is the pressure upstream of the valve in absolute bar.

**Steam:**

$$NW = 2.8 \cdot \sqrt{G \cdot v''}$$

where $G$ is the mass flow rate in kg/h and $v''$ is the specific volume in $m^3/kg$.

**In general:**

$$NW = 18.8 \cdot \sqrt{\frac{Q_B}{c}}$$

where $Q_B$ is the volumetric flow rate in $m^3/h$ and $c$ is the flow velocity in m/s.

In the case of simple control valves on which a connection nominal diameter is assigned directly to a $k_{vs}$ value, the anticipated flow velocity from a practical point of view should at a minimum be checked. The nominal pressure stage results from knowledge of the valve material, the operating temperature and the maximum operating pressure, for example, from DIN 2401, or from the specific valve data sheet. The actual closed-loop control function, that is, setting the fluid flow rate of a given temperature and given pressure, while simultaneously producing a defined pressure loss, is determined by the flow characteristic, the $k_v$ value (described also earlier in Chapter 4). The $k_v$ value is a reference variable and is defined as follows: $k_v$ value = quantity in $m^3/h$ of cold water (+5··+30 °C), which flows through the valve at 1 bar differential pressure across the valve and at stroke s. The $k_{vs}$ value, in analogy, is the quantity at stroke $s = 100\%$.

In analogy to the $k_v$ value, the flow rate coefficient $c_v$ is described in the American literature. The following conversion factor applies: $k_v = 0.86\ c_v$. The $k_v$ value must be calculated for the current operating data. A distinction must be made between maximum load (maximum quantity $Q_{max}$, minimum $\Delta p_{min} \gg k_{vmax}$) and minimum load (minimum quantity $Q_{min}$, maximum $\Delta p_{max} \gg k_{vmin}$). Both load cases must be calculated individually and be adjusted on the basis of the valve range ability.

**The following applies to cold water:**

$$k_v = Q \cdot \sqrt{\frac{1}{\Delta p}}$$

where $Q$ is the volumetric flow rate in m3/h and $\Delta p$ is the pressure differential at the valve in bar.

**The following applies in general to fluids (sub-critical):**

$$p_2 > ps_2$$

where $ps_2$ is the saturated steam pressure in bar of absolute pressure, related to the temperature downstream of the valve.

$$k_v = Q \cdot 0.032 \cdot \sqrt{\frac{\rho_1}{\Delta p}}$$

where $\rho_1$ is the density of the medium in $kg/m^3$ and $\Delta p$ is the pressure differential at the valve in bar.

$$k_v = G \cdot 0.032 \cdot \sqrt{\frac{1}{\rho_1 \cdot \Delta p}}$$

where $G$ is the mass flow rate in kg/h and $\Delta p$ is the pressure differential at the valve in bar.

**The following applies to fluids in general (super-critical):**

$$p_2 < ps_2$$

where $ps_2$ is the saturated steam pressure, in bar of absolute pressure, related to the temperature downstream of the valve.

The $k_v$ value is calculated here in two steps: the $k_v$ value for the evaporating steam quantity $k_{vD}$ and the $k_v$ value for the fluid $k_{vF}$ are calculated separately and both values are added.

$$k_v = k_{vF} + k_{vD}$$

The evaporating quantity is calculated from the mass and energy balance around the valve, assuming isenthalpic relaxation [50].

**The following applies to saturated steam (sub-critical, i.e., $p_2 > \frac{p_1}{2}$):**

$$k_v = \frac{G_S}{22.4 \cdot \sqrt{\Delta p \cdot p_2}}$$

where $G_S$ is the saturated steam quantity in kg/h, $p_1$ is the pressure upstream of the valve in bar of absolute pressure, and $p_2$ is the pressure downstream of the valve in bar of absolute pressure.

**The following applies to saturated steam (super-critical, i.e., $p_2 < \frac{p_1}{2}$):**

$$k_v = \frac{G_S}{11.2 \cdot p_1}$$

where $p1$ is the pressure upstream of the valve in bar of absolute pressure.

**The following applies to superheated steam (sub-critical, i.e., $p_2 > \frac{p_1}{2}$):**

$$k_v = \frac{G_S}{31.7 \cdot \sqrt{\frac{\Delta p}{v''_2}}}$$

where $v''_2$ is the specific volume at $p_2$ and $t_1$ in m$^3$/kg.

**The following applies to superheated steam (super-critical, i.e., $p_2 < \frac{p_1}{2}$):**

$$k_v = \frac{G_S}{22.4 \cdot \sqrt{\frac{p_1}{v^*}}}$$

where $v^*$ is the specific volume at $\frac{p_1}{2}$ and $t_1$ in m$^3$/kg.

The following applies to gases (sub-critical, i.e., $p_2 > \frac{p_1}{2}$):

$$k_v = \frac{Q_N}{514} \cdot \sqrt{\frac{\rho_N \cdot T_1}{\Delta p \cdot p_2}}$$

where $Q_N$ is the volumetric flow rate in $Nm^3/h$ and $\rho_N$ is the standard density in $kg/m^3$ (standard state: 0 °C and 1013 mbar).

The following applies to gases (super-critical, i.e., $p_2 < \frac{p_1}{2}$):

$$k_v = \frac{Q_N}{257 \cdot p_1} \cdot \sqrt{\rho_N \cdot T_1}$$

where $T_1 = 273 + t_1$

After calculating the $k_v$ values, the $k_{vs}$ value is determined with the aid of tables in data sheets. The $k_{vs}$ value should only be slightly higher than the $k_{vmax}$ value. Excessive $k_{vs}$ values diminish the usable ability range and thus the control response when subject to weak load. The $k_{vmin}$ value must be able to be reached with the selected control valve, that is, it must lie within the ability range. If $k_{vmin}$ lies below this limit, a consideration should be made as to whether to split the quantity over two differently sized valves, whereby the $k_{vs}$ value of the smaller valve should be approximately 10% of the $k_{vs}$ value of the larger valve.

- **Critical pressure gradient with water or fluid**
  If the outlet pressure $p_2$ is less than the saturation pressure $p_S$ at the inlet of the control valve, evaporation will occur. This leads to a much higher stressing of the control valve. To prevent premature failure of the valve, special measures will be required, such as use of perforated cones, armoring of the sealing edges, and multi-stage pressure reduction. In this case, consulting the manufacturer is recommended.

- **Critical pressure gradient with gases and vapors**
  Super-critical relaxation causes noise problems. Special noise-reduction measures such as perforated cones, restrictor plates, flow straighteners, and sound-control insulation will be required. Individual restriction steps larger than 0.56 $p_1$ under full load should not be selected.

- **$k_v$ value for valve groups connected in series**
  The total $k_v$ value can be calculated as follows for series connection of several valves:

$$k_{v_{tot.}} = \frac{1}{\sqrt{\frac{1}{kv_1^2} + \frac{1}{kv_2^2} + \frac{1}{kv_3^2} + \cdots + \frac{1}{kv_n^2}}}$$

- **k$_v$ value for valve groups connected in parallel**
  The total $k_v$ value can be calculated as follows for several valves connected in parallel:

$$k_{v_{tot.}} = k_{v_1} + k_{v2} + k_{v3} + \cdots + k_{vn}$$

Section 5.4 provided the necessary basis for implementing control loops in optofluidic systems. The following section describes the current state of the art on microvalves and derivatives.

## 5.5 Microvalves

Microvalves and micropumps are well known in the field of microfluidics and especially lab-on-chip systems. Several technologies exist side by side, and their implementation into fluidic system depends on the valve characteristics. This is one of the major drawbacks in implementing microvalves and micropumps as one has to exactly know the behavior and the technology behind the microvalve to have a best fit for the system. This is in contrast to solenoid valves where the actuation principle is well known and only the fluidic part is adapted to the requirements of the fluid.

Modular active microvalves are constructed in a way that the fluidic and membrane part are basically the same and only the actuating part is different. The schematic of a modular microvalve is shown in Figure 5.33 [20].

The microvalve consists (from bottom to top) of a cavity in a passive layer for the fluid to be handled, the membrane layer and the actuator layer, which includes the electrical contacts. The passive layer and the membrane layer can remain while the actuator layer is adapted to the different actuating principles.

Common actuating principles are sketched in the following Figure 5.34. Electromagnetic microvalves have naturally evolved to be one of the first valves to be miniaturized due to the fact that as said in a paragraph before, the actuation principle is well known and downscaling does not really impose a hurdle to overcome, as technology is readily available, especially nowadays. The goal as usual is to have a platform, which can be used for various actuation principles. This is perfectly shown in Figure 5.34. Here, only the actuation principle is changed.

A perfect overview of microvalve development during the last 20 years is presented by Oh et al. [188]. Detailed information with references to fabricated mechanical active microvalves is provided.

In the following paragraph, the focus lies on the description of monostable and bistable microvalves as well as their integration to microfluidic systems. The information described is provided with kind permission by C. Megnin.

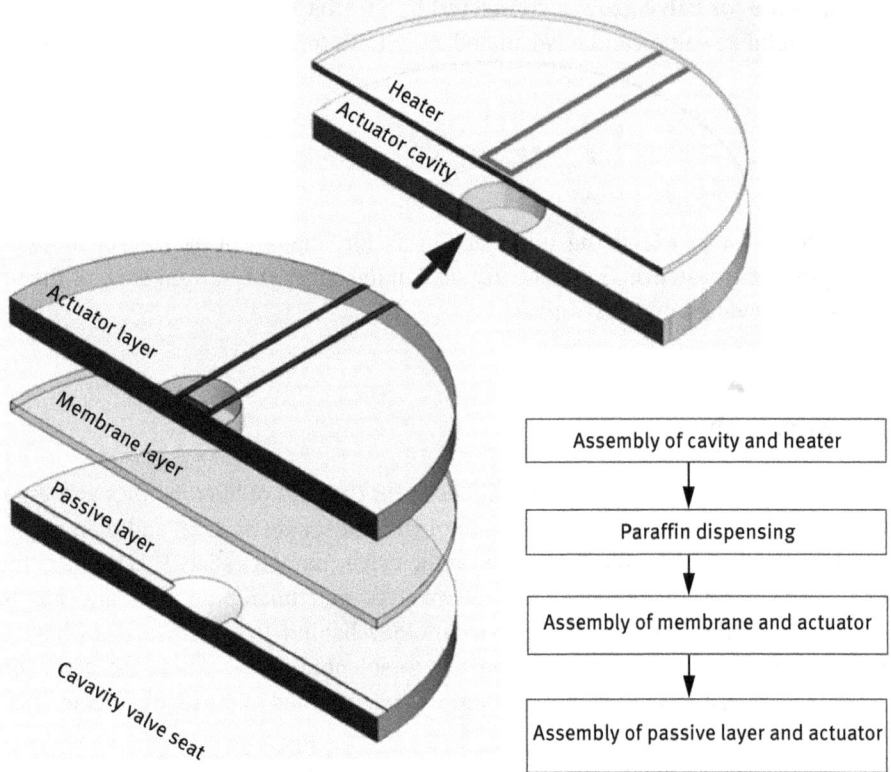

**Figure 5.33:** Schematic of the microvalve, built up from three main module layers. The actuator layer is made up of two sub-layers containing the actuation material and a microheater. The assembly sequence is illustrated as a flow diagram. (Reprinted from Microelectronic Engineering, Volume 98, October 2012, Pages 638–641, Allwyn Boustheena, F.G.A. Homburg, Andreas Dietzel, A modular microvalve suitable for lab on a foil: Manufacturing and assembly concepts, with permission from Elsevier)

To enable a compact design at preferably high fluidic power, a structured foil of a shape memory alloy (SMA) is used for actuation. The monostable microvalves are designed in a modular way to reduce the number of components. This enables economic fabrication of NO and (NC) microvalves as well as fast adjustment to required pressure and flow range. For fabrication in an industrial environment, packaging and interconnection technologies are systematically investigated and assessed with respect to quality and manufacturing criteria. The simultaneous fluidic and electric interconnections of several microvalves are achieved for fluidic systems by introducing a novel plug-in interconnection technology.

The components used for fabricating the NO microvalve are shown in Figure 5.35. The difference between the NO and the NC versions is that the NC version uses an integrated spring to enable the normally closed functionality and thus possesses a slightly larger height than the NO version.

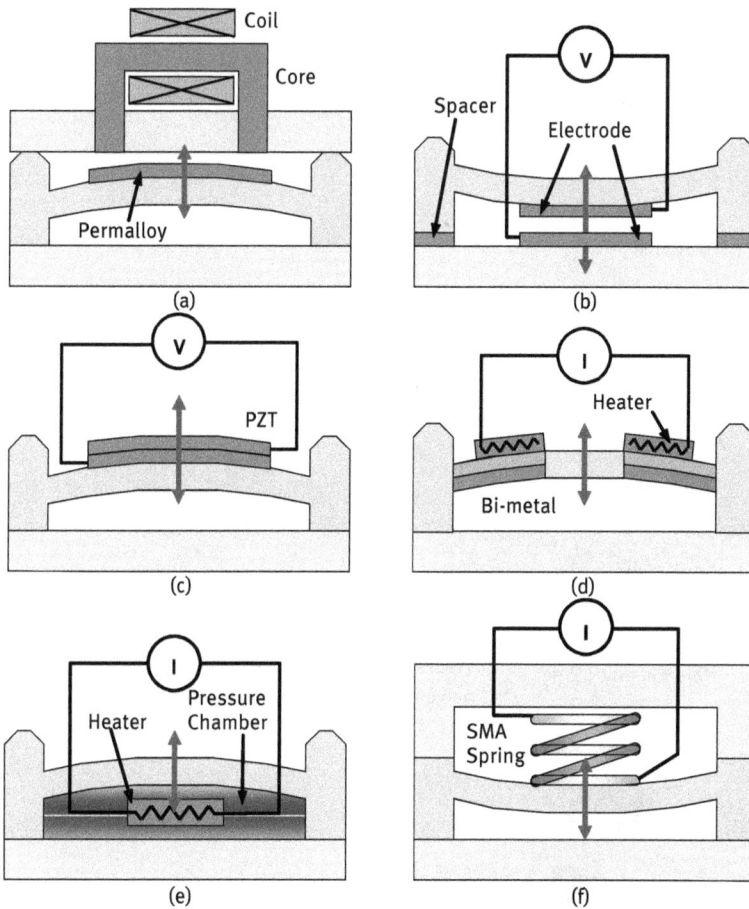

**Figure 5.34:** Illustrations of actuation principles of active microvalves with mechanical moving parts: (a) electromagnetic; (b) electrostatic; (c) piezoelectric; (d) bimetallic; (e) thermopneumatic; and (f) shape memory alloy actuation. (Reproduced with kind permission from J. Micromech. Microeng. 16 (2006) R13–R39, Kwang W Oh and Chong H Ahn, A review of microvalves, IOP Publishing).

The fabricated NO microvalve is shown in Figure 5.36.

The NO microvalve has been tested using nitrogen gas at a pressure difference between 50 kPa and 200 kPa (Figure 4.61). At a pressure difference of 200 kPa, the flow rate is measured to be approximately 2,000 sccm for the case where no switching energy is applied. The microvalve starts to close at approximately 70 mW and is completely closed at 80 mW. As can be seen from Figure 5.37, the opening and closing of the valve follow the typical SMA hysteresis, which needs to be taken into account when driving the microvalve.

The NO microvalve has been also tested for water at a pressure difference between 50 and 200 kPa (Figure 5.38). The flow rate in the case for no switching

Screws

Spring Contacts

Cap

Heat Sink

Shape Memory Actuator

Actuator Holder

Plunger

Membrane

Valve seat

Hard Magnetic Ring

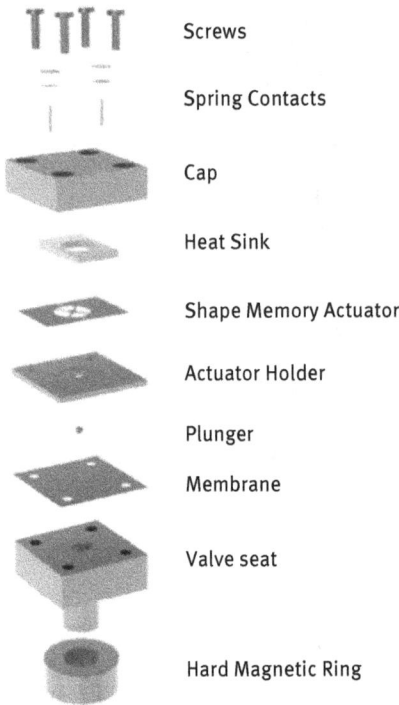

Figure 5.35: NO microvalve layout. The size of the valve is 10 × 10 mm (Courtesy C. Megnin).

Figure 5.36: NO microvalve – 10 × 10 mm (Courtesy C. Megnin).

power applied is measured to be 12.5 ml/min. As can be seen in the diagram, the applied energy is three times higher than in the case for the nitrogen gas measurement.

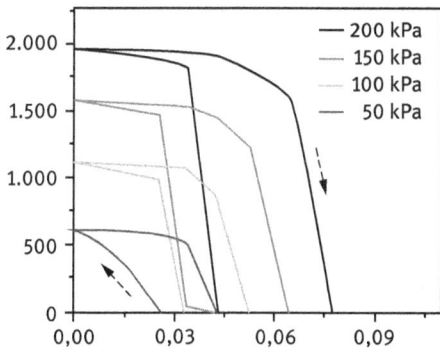

**Figure 5.37:** NO microvalve at different Nitrogen differential pressures. X-axis in W and Y-Axis in sccm (Courtesy C. Megnin).

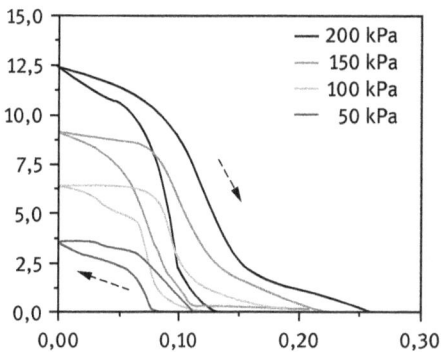

**Figure 5.38:** NO microvalve at different Water differential pressures. X-axis in W and Y-axis in ml/min (Courtesy C. Megnin).

An image of the normally closed (NC) microvalve is shown in Figure 5.39. The center screw is used for applying a preset condition for the valve.

The NC microvalve is closed at a differential pressure of 200 kPa. The valve is completely open at an applied energy of 200 mW for the nitrogen measurement (Figure 5.40) and 500 mW for the water measurement (Figure 5.41). At these applied energies, the force of the SMA actuator is equal to the force of the integrated spring.

The flow rate achieved with a completely open microvalve is measured to be 880 sccm for nitrogen and 7.7 ml/min for water. This is lower than compared with the NO microvalve. This can be explained by the integrated spring, which applies a force onto the membrane and thus onto the valve seat, resulting in a smaller gap for realizing the flow through the valve. A screw in the center of the valve can adjust this applied force of the integrated spring (refer Figure 5.39). The measurement (nitrogen gas) in dependence of the applied rotation angle is shown in Figure 5.42.

The NC valve with integrated spring is measured for an applied energy of 0 mW and 250 mW.

The integration of valves on a microfluidic backplane and integrated electronic for addressing each valve individually is shown in Figure 5.43.

**Figure 5.39:** NC microvalve (Courtesy C. Megnin).

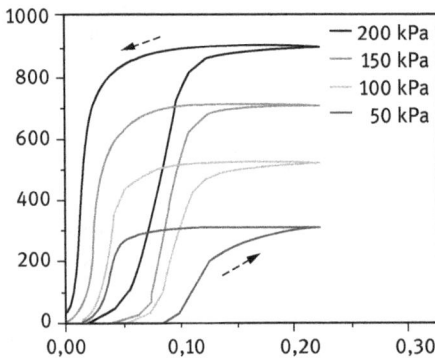

**Figure 5.40:** NC microvalve at different Nitrogen differential pressures. X-axis in W and Y-Axis in sccm (Courtesy C. Megnin).

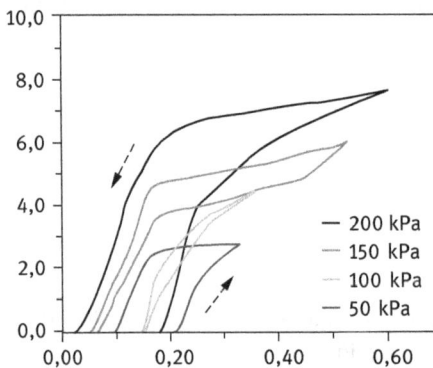

**Figure 5.41:** NC microvalve at different Water differential pressures. X-axis in W and Y-Axis in sccm (Courtesy C. Megnin).

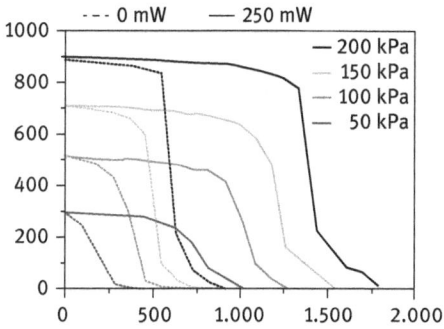

Figure 5.42: NC microvalve at a differential Nitrogen pressure between 50 and 200 kPa in dependence of the rotation angle of the integrated screw. X-axis rotation angle of the screw in degrees and Y-axis flow rate in sccm (Courtesy C. Megnin).

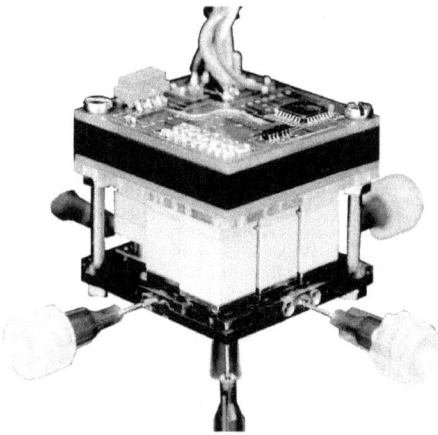

Figure 5.43: NC microvalves assembled on a microfluidic circuit (refer also Section 8.2) (Courtesy C. Megnin).

The following paragraph shows two images (Figures 5.44 and 5.45) of bistable 3/2 microvalves consisting of two counteracting SMA devices, which enable bidirectional actuation. Stable positions are maintained in power-off condition by magnetostatic forces in between magnetic components. As electrical power is only required for switching, the bistable microvalves are suitable for fluidic systems operating at low average power consumption and low duty cycles. The switching and latching structures are analyzed separately and their coupled behavior is optimized to improve fluidic power. By design variation of these components, the microvalve can be tailored to a given pressure and flow range. The bistable microvalve design is described in great detail in [156].

The presented microvalve based on shape memory actuators (miniature actuators on the basis of shape memory alloys are commercialized by for example www.memetis.com) can serve as building blocks for optofluidic systems, as they are easy to integrate and offer the potential of being fabricated in an industrial environment. A further example of the integration of these types of valves is given in the following chapters.

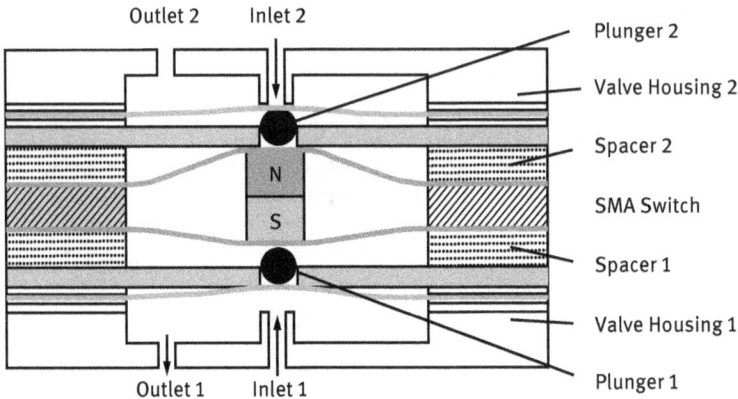

**Figure 5.44:** Bistable 3/2-way microvalve based on magnetic shape memory actuators – State I (Courtesy C. Megnin).

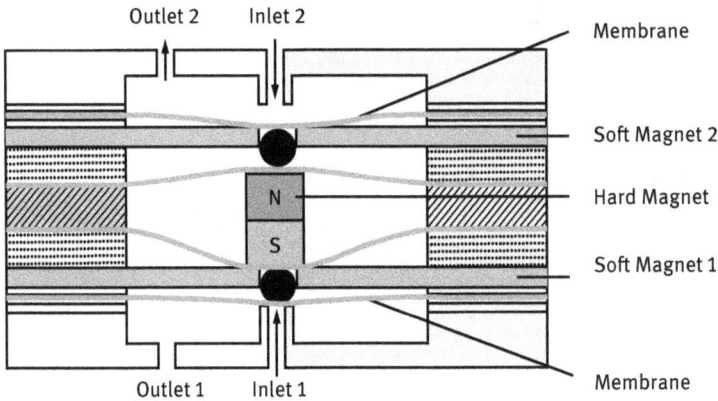

**Figure 5.45:** Bistable 3/2-way microvalve based on magnetic shape memory actuators – State II (Courtesy C. Megnin).

## 5.6 Fabrication of fluidic channels: Bonding

To integrate optics and fluidics, it would be advantageous to use a similar technology to create an optofluidic device. The deep UV technique is one method of combining optics which includes waveguides and light sources and fluidics as has been described in earlier chapters.

Two types of polymers have been investigated: PMMA (Hesa@Glas, a homopolymer from Notz-Plastic, Switzerland) and alicyclic methacrylate copolymers which were obtained from Hitachi Chemical as OPTOREZ-series (OZ-series). For deep UV modification, a commercial UV-exposure systems is used, a mask aligner EVG620 having a DUV lamp combined with a cold mirror with reflectance in the wavelength range of 200–240 nm in the exposure system.

Using the DUV process, it is possible to fabricate fluidic channels and reservoirs. There are several possibilities of realizing fluidic channels with this method. The PMMA can be spin-coated onto a glass wafer and then be exposed and developed. The PMMA is used in this case like a conventional photo-resist. The DUV fabrication method can also be used to realize a Ni-Shim, which can then be used for hot embossing of fluidic channels. Another possibility is to directly expose a PMMA bulk substrate and develop the exposed regions. The penetration depth of the DUV light is only a few μm, which defines the maximum height of the channels. The advantage of using spin-coated substrates is a defined height structure for realizing a flat and smooth bottom of the channels. A photograph of an unsealed T-junction is shown in Figure 5.46. The fluidic channels have a width of about 5 μm.

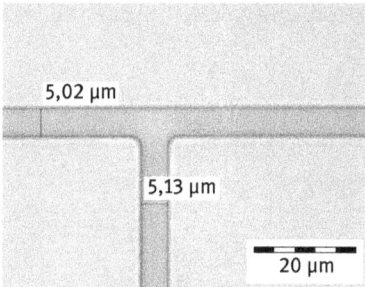

Figure 5.46: Unsealed fluidic channels in PMMA.

As the fluidic channels are fabricated in PMMA, it takes only another DUV aligned exposure to integrate the waveguides (Figure 5.47).

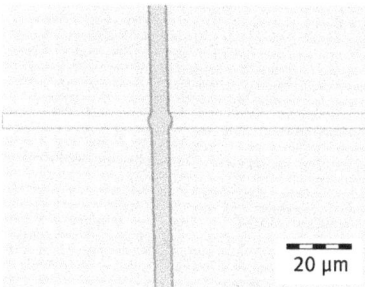

Figure 5.47: Fluidic channel and waveguide integrated on the same chip.

In a next step, cover plates are heat-sealed onto the fluidic channel. The DUV flood exposure is applied to both the substrate containing fluidic channel and waveguides and the cover plates. The DUV radiation lowers the glass transition temperature, which is used advantageously in this bonding process [219] making welding at lower temperatures possible.

## 5.7 Conclusion

This chapter on fluidics and fluid control systems has provided the necessary theoretical and practical background for designing optofluidic systems. When designing optofluidic systems, it is extremely helpful to rely on a dedicated system platform. This is an advantage as the designer can solely focus on solving the need for the application and does not need to worry on applied technologies. In this way, applications can be solved more easily and products can be fabricated in a shorter development cycle in the end.

The following chapter provides a brief introduction into the vast area of biology, focusing only on certain technologies needed for integration with optical waveguides and fluidic channels. This chapter is purely written from an engineering perspective.

## Further reading

Anandharamakrishnan C. Computational Fluid Dynamics Applications in Food Processing. Springer, 2013

Dongqing L. Encyclopedia of Microfluidics and Nanofluidics. Springer, 2008

Kohl M. Shape Memory Microactuators. Springer, 2010

Megnin C.Formgedächtnis-Mikroventile für eine fluidische Plattform. PhD Thesis KIT 2012

Petrila T, Trif D. Basics of Fluid Mechanics and Introduction to Computational Fluid Dynamics. Springer, 2005

Wendt J. Computational Fluid Dynamics. Springer, 2009

Yamaguchi H. Engineering Fluid Mechanics. Springer, 2008

# 6 Sensors for Optofluidic systems

Sensors are the eyes and ears of every system not only of Optofluidic systems. They play an ever-increasing role and their implementation into Optofluidic systems is the key for obtaining the wanted results. In creating a true Optofluidic system platform, several sensors need to be integrated which are vital for the performance, for example temperature, pressure, flow sensors, and also specific analytical sensors. This chapter is organized as follows: first, standard process sensors are introduced followed by process analytical sensors with focusing on spectroscopy, multivariate data analysis for data handling, and a conclusion.

By the end of this section, you should be able to:
– differentiate classical sensors from process analytical sensors.
– understand how radiation interacts with molecules in different regions of the electromagnetic spectrum.
– describe the operation of optical spectrometers.
– explain advantages and disadvantages of working in the ultraviolet-visible, fluorescent, near-IR, mid-IR, and Raman region.

## 6.1 Process sensors

Optofluidic system technology comprises of several manual and automatic processes including fluids, their transport, conditioning, or disposal. A wide range of developments such as sensors, valves, and actuators that find their applications in the different domains of optofluidic system technology have been conducted in recent years and some have already led to products. In order to monitor and control these fluids, sensors and actuators play a very important role. Basic parameters like temperature, pressure, humidity, etc., need to be controlled and analyzed while working with fluids. The chapter on temperature, flow, and level sensors are based on the profound knowledge and documentation, which has been gathered and derived over several years by Uwe Krug and colleagues at Buerkert Fluid Control Systems and used with kind permission. For further information, see further reading numbers [1–3].

### 6.1.1 Temperature sensors

The closed-loop optofluidic control system, and thus the detection or measurement of temperatures, is of major significance in this large range of processes, for example, in chemical and pharmaceutical fluid control systems, in medical devices, or in vehicles and conditioning systems. The variations and physical measuring principles of

https://doi.org/10.1515/9783110546156-006

relevant sensors used are just as diverse as the fields of application of temperature measurement. There is no such thing as a general purpose measuring instrument or sensor. The requirements differ depending on the application and as this book is about system technology, some aspects to consider are:

- measuring range
- accuracy
- dynamic behavior/response behavior
- overall size, especially for microfluidic applications
- measurement with/without physical contact to fluid
- mechanical/chemical resistance
- and finally the price

Depending on the relevant and specific structure, temperature sensors have the function of converting a measurable temperature variable to an electrical signal (current, voltage, or resistance). Most sensors used today can be equipped with an additional electronic module (transmitter), which converts the original output signal of the sensors to a standard electrical signal. The following section provides an overview of the most popular sensors for temperature measurement and their basic characteristics and fields of application.

### 6.1.1.1 Definition
The SI unit of temperature is kelvin K, in which one kelvin is conveniently equal to one Celsius degree. To get an exact conversion, you have to add 273.15 to the Celsius value.

### 6.1.1.2 NTC (Negative temperature coefficient) temperature sensors
The output variable of an NTC temperature sensor is electrical resistance. The electrical resistance of the NTC decreases with increasing temperature (Figure 6.1).

The resistance at 25 °C provides a convenient reference point for thermistors. The tolerances in $R_{25}$ are mainly due to variations in ceramic material manufacturing and tolerances in chip dimensions. Due to the use of highly homogeneous material compositions, the tolerance limits for $R_{25}$ are usually less than 1%. B is a material constant (obtained from manufactures datasheet) that controls the slope of the RT characteristic. By knowing the B value of a thermistor, it is possible to produce a table of temperature versus resistance to construct a suitable graph using the following normalized equation:

$$R_T = R_{25}\; e^{B\left(\frac{1}{T} - \frac{1}{298.15}\right)}$$

In practice, B varies somewhat with temperature and is therefore defined between two temperatures 25 °C and 85 °C by the formula:

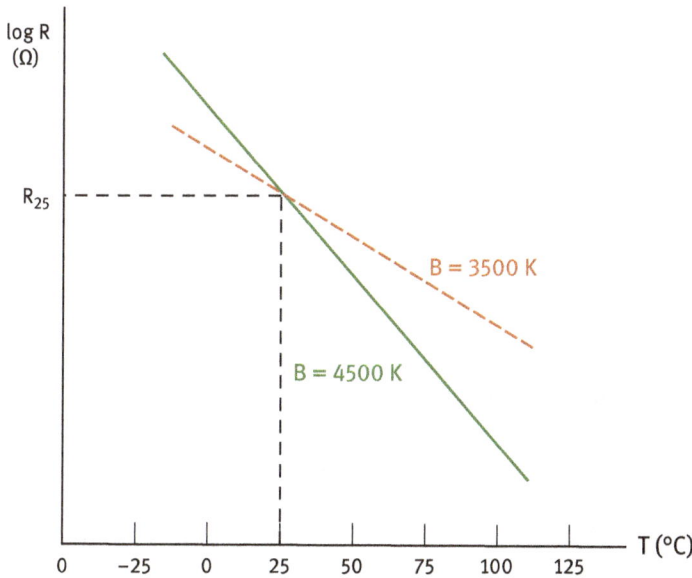

**Figure 6.1:** Typical resistance as a function of temperature for an NTC temperature sensor.

$$B_{25/85} = \ln\left(\frac{R_{85}}{R_{25}}\right) \Big/ \left(\frac{1}{385.15} - \frac{1}{298.15}\right)$$

The *total tolerances* of the NTC temperature sensor over its operating temperature range is a combination of the tolerances on $R_{25}$ and on B-value given by the formula:

$$\frac{\Delta R}{R} = \frac{\Delta R_{25}}{R_{25}} + \Delta B\left|\frac{1}{T} - \frac{1}{298.15}\right|$$

The sensitivity of such a sensor to temperature changes is defined over the temperature coefficient of resistance $\alpha$:

$$\alpha = \frac{1}{R} \times \frac{\Delta R}{\Delta T} = \frac{\Delta B}{T^2}$$

NTC thermistors are made of polycrystalline mixed oxide ceramics. The typical temperature range is between $-50\ °C$ ... and $+100\ °C$. The NTC possesses a high offset coefficient, i.e. great change in resistance with temperature providing a good control response. NTC sensors usually come with a moderate accuracy and thus the sensors often are not suitable for very precise measurements. Nevertheless, if high accuracy is required, small diameter nickel leads are recommended. Their low heat conductivity effectively isolates the component from the outside world, enabling it to accurately monitor the temperature. In addition, NTC

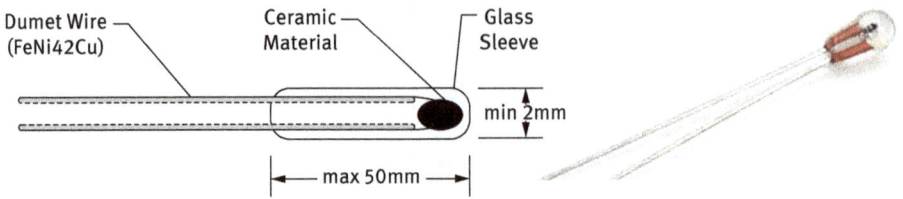

**Figure 6.2:** Glass sealed encapsulated NTC thermistor.

sensors possess only a moderate reproducibility. This greatly impairs interchangeability of the sensors. Figure 6.2 shows a schematic construction of a glass sealed encapsulated NTC thermistor.

Typical applications can be classified into two groups depending on their physical properties; see figure 6.3:
- **Temperature sensor:** Application in which the sensitive change of the resistance versus the temperature is used.
- **Time delay thermistor:** Application in which the time dependence is decisive, when the temperature is considered as a parameter.

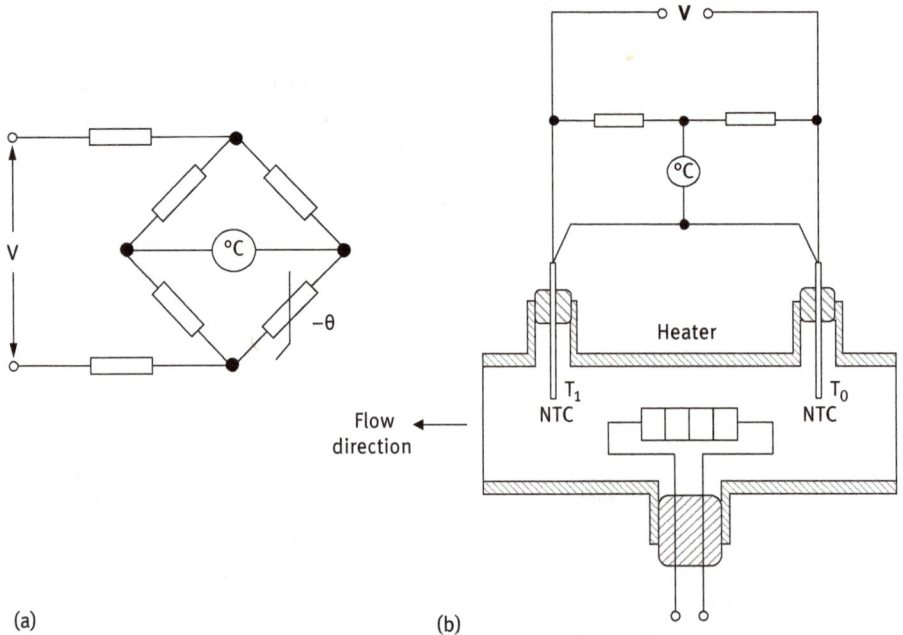

(a)                                    (b)

**Figure 6.3:** Examples of applications. (a) Temperature measurement with a Wheatstone bridge; (b) The temperature difference between $T_1$ and $T_0$ is a measure for the velocity of the fluid.

Some examples of the fields of applications of NTC sensors are the plastics industry, automotive engineering, mobile measuring instruments, and medical devices. In addition to measurement tasks, NTC sensors are also used in the following applications:

- Temperature compensation of coils
- Working point stabilization of transistors
- Over-temperature protection and cut-outs

### 6.1.1.3 PTC (Positive temperature coefficient) temperature sensors

PTC thermistors can be classified into two major categories. The first category consists of thermally sensitive silicon resistors. An example is a KTY silicon sensor. The output variable is as was the case with NTC temperature sensors, electrical resistance. The silicon sensors have a positive temperature coefficient. Their electrical resistance increases with temperature. They have approximately a linear characteristic (Figure 6.4). These devices are most often used for temperature compensation of silicon semiconducting devices in the range of −60 °C to +150 °C.

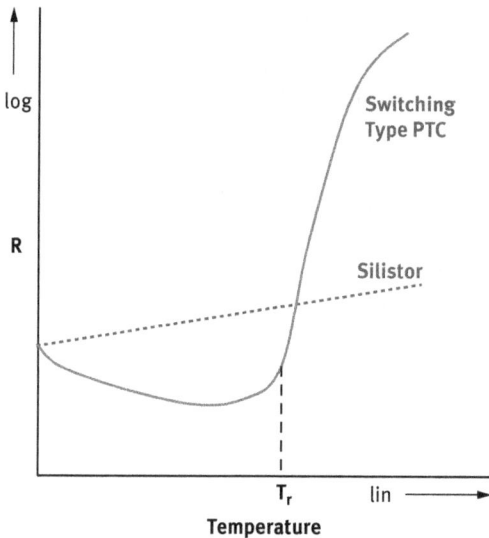

**Figure 6.4:** R–T characteristics of PTC thermistors.

The second category are the switching PTC thermistors. These devices are polycrystalline ceramic materials that are normally highly resistive but are made semiconductive by the addition of dopants. They are most often manufactured using compositions of barium, lead, and strontium titanates with additives such as yttrium, manganese, tantalum, and silica. These devices have a resistance-temperature characteristic that exhibits a very small negative temperature coefficient until the device reaches a critical temperature $T_c$. Above this point, the device begins to exhibit a rising positive temperature coefficient of resistance as well as a large increase in resistance. The resistance change can be as much as several orders of magnitude within a temperature span of a few degrees.

PTC sensors have a high offset coefficient, i.e., a large change in the resistance with temperature providing good control response. In comparison to NTC sensors, PTC sensors posses a higher accuracy and reproducibility. However, PTC sensors are also not suitable for precision measurements. Some examples of the fields of applications of PTC sensors are industrial measurement and control systems, automotive engineering, medical technology, building/facility services management, and over-temperature protection for motors, generators, and transformers.

### 6.1.1.4 Resistance temperature detector (RTD)

The output variable of these types of sensors is electrical resistance. Nickel (Ni) and platinum (Pt) resistance thermometers have a positive temperature coefficient. To measure the change, the sensor output is fed into a Wheatstone bridge with a reference voltage. Their resistance increases with increasing temperature. Depending on the nominal value (resistance value at 0 °C), the sensors are referred to as Pt/Ni 100, Pt/Ni 500, or Pt/Ni 1,000 sensors. The numerical value indicates the resistance value in Ohm ($\Omega$).

Figure 6.5 shows the resistance–temperature characteristics curve of the three different metals. For Platinum, its resistance changes by approximately 0.4 ohms per degree Celsius of temperature.

While nickel sensors are less expensive and have a higher measurable variable sensitivity than platinum sensors, they also have the following disadvantages:
- lower temperature range
- poorer corrosion resistance
- poorer accuracy or linearity
- are more complex to produce.

Typical measuring range of platinum thermometers is between −200 °C and +850 °C and for nickel thermometer between −60 °C and +240 °C.

Resistance thermometers can be used universally and virtually in all fields of application, specifically in applications requiring high accuracy and reproducibility. Resistance thermometers are available in a wide variety of enclosure designs. Some of them are described or stipulated by standards. In order to minimize the effects of the line resistances and their fluctuation with temperature, it is usual practice to employ a three-wire circuit. It consists of running an additional wire to one contact of the RTD. This results in two measuring circuits of which one is used as reference.

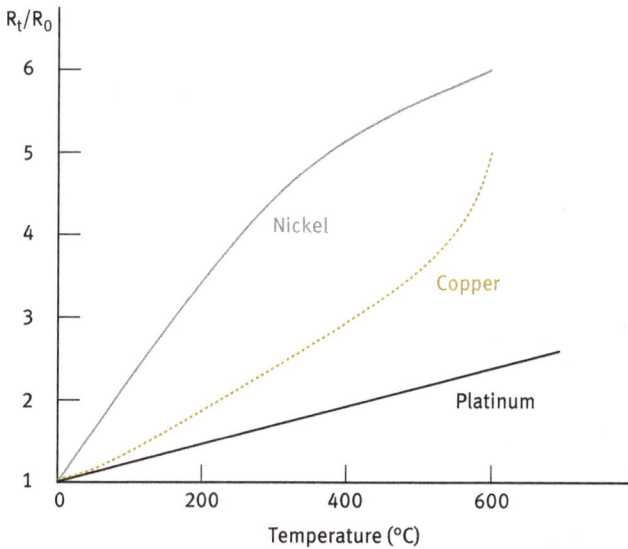

**Figure 6.5:** Resistance–temperature characteristics curve for three metals.

### 6.1.1.5 Thermocouples

Thermocouples are based on the Seebeck effect. The Seebeck effect states that a voltage, which depends on the temperature of the contact point, is produced at the point of contact of two different metals. In Figure 6.6, thermocouple consists of two metal wires welded together at specific points. When this junction is heated, a voltage called the thermal electromotive force (thermal emf) is produced at the open ends. In general, the Seebeck effect is described locally by the creation of an electromotive field

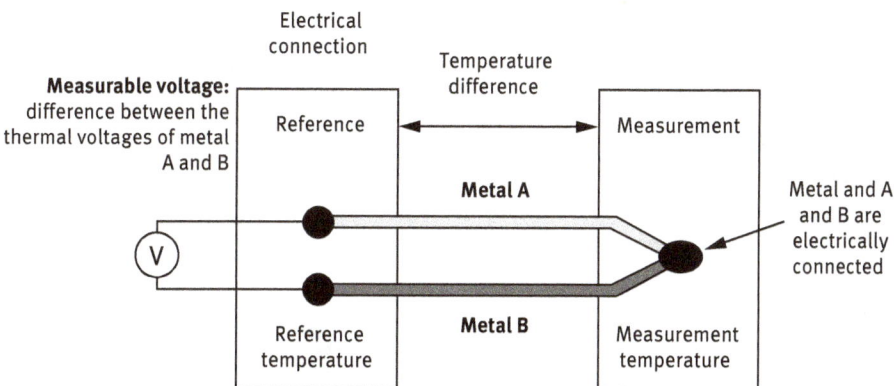

**Figure 6.6:** Standard thermocouple measurement configuration. The measured voltage can be used to calculate temperature, provided that the temperature is known.

$$E_{emf} = -S\Delta T$$

where S is the Seebeck coefficient, meaning a property of the local material, and ΔT is the difference in temperature T.

Thermocouples are connected to the evaluation unit or controller either directly or via equalizing conductors. This connecting point is referred to as reference junction. The temperature measured by the thermocouple or its output voltage is not directly proportional to the temperature of the measuring junction, but is proportional to the temperature difference between the measuring junction and reference junction. So, the temperature of the reference junction must be constant or has to be measured and compensated for accordingly. The reference junction compensation circuit is generally incorporated in the evaluation and control unit.

Thermocouples have the following characteristics:
- good accuracy
- good reproducibility
- very high dynamics, depending on design.

An extremely wide variety of thermocouples (metal combinations) is used depending on the application. Some examples are:

*Type J, material: Fe–CuNi*
Temperature range: −40 °C to +750 °C
- high thermal EMF coefficient, i.e. high resolution
- iron (Fe) is at high risk of corrosion
- very low-cost.

*Type T, material: Cu–CuNi*
Temperature range: −40 °C to +350 °C
- high thermal EMF coefficient, i.e. high resolution
- copper (Cu) is at risk of corrosion at high temperatures
- very low-cost.

*Type K, Material: NiCr–Ni*
Temperature range: −200 °C to +1,200 °C
- low thermal EMF coefficient, i.e. low resolution, used preferably for high temperatures
- scaling at high temperatures.

*Type E, material: NiCr–CuNi*
- Temperature range: −200 °C to +900 °C
- low thermal EMF coefficient, i.e. low resolution, used preferably at high temperatures
- good chemical resistance

*Type R, material: Pt10Rh–Pt*
*Type S, material: Pt13Rh–Pt*
    Temperature range: −40 °C to +1,600 °C
- highly accurate, very good reproducibility
- low thermal e.m.f. coefficient, i.e. low resolution, used preferably at high temperatures
- good chemical resistance
- very expensive.

If operating conditions allow, thermocouples can be used directly as unsheathed components. In the simplest case, the contact point is the measuring sensor directly. In aggressive environments, sheathing protects the elements. Thermocouples can be used in all kinds and fields of applications. Specifically in applications that requires a good dynamic behavior of the sensor. Thermocouples as can be seen from the past paragraph are readily available in a large variety of metal combinations, sizes, and designs.

### 6.1.1.6 Pyrometers

Unlike the measuring methods or sensors discussed in this chapter, which have to come into contact with the body to be measured, pyrometers operate without physical contact. The pyrometer measures the temperature via the heat dissipated by the bodies. Besides depending on its temperature, the thermal radiation of a body depends on its ability to radiate heat (emissivity), i.e. on its surface characteristics and color. Since the emissivity differs from measured object to measured object, a pyrometer must be calibrated for the relevant application. Ideal prerequisites for measurement with pyrometers include constant conditions concerning color, shape, surface, condition of the measured object, and the measurement distance. In addition, it must be ensured that the pyrometer does not detect extraneous radiation, i.e. thermal radiation from other bodies or reflected radiation from the body to be measured. The typical measuring range is between −50 °C to +3,000 °C.

    Pyrometers are used whenever measuring objects or fluids with direct contact is difficult; for example:
- sticky, adhering media
- very hot surfaces
- moving objects
- parts which are not easily accessible
- aggressive media and environments
- very small objects with low mass
- sensitive surfaces.

## 6.1.2 Pressure sensors

Pressure is one of the standard parameters used today not only in Optofluidic control systems, but due to the availability of pressure sensors as MEMS (micro-electro-mechanical systems) sensors, these sensors are integrated into a multitude of applications ranging from smartphones to cars to process monitoring. The following paragraph provides a brief introduction to some of the commonly used pressure sensors in optofluidic controls systems.

### 6.1.2.1 Definition

The pascal (Pa) is the SI unit of pressure, equal to the force in newton (N) divided by the area in square meters (m$^2$). The standard atmosphere (atm) is a unit of pressure defined as 101,325 Pa. Bar is also a convenient unit, which equal 1.01325 bar regarding to 1 atm. Most pressure transmitters measure the gauge pressure (barg), not the absolute pressure (bara). For calculations, we have to use always the absolute pressure with x barg = (x+1) bara. In U.S., the most common unit for measuring pressure is pounds per square inch (psi); 1 bar = 14.503 psi. If you get data in psig, first convert it to psia using x psig = (x+14.7) psia.

### 6.1.2.2 Thin-film strain gauge

A thin-film Wheatstone bridge, as a resistive sensor element, is bonded directly to a stainless steel diaphragm; see figure 6.7. Flexure of the diaphragm as the result of the external pressure causes a change in the resistances of the Wheatstone bridge, which is then converted to a pressure-proportional signal. Very high burst pressures and highly precise pressure measurements can be implemented by using this measuring method. Even applications in environments subject to shock and vibration can easily be implemented.

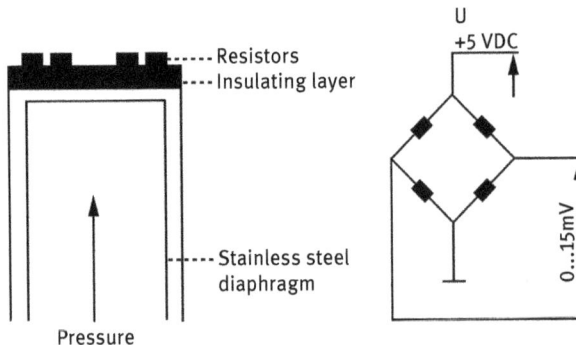

**Figure 6.7:** Principle of thin-film strain gauge.

### 6.1.2.3 Piezoresistive sensor

For protective purposes, the piezoresistive sensor is flushed with a hydraulic fluid. The medium is separated by means of a stainless steel diaphragm. Flexure of the diaphragm as the result of the external pressure causes a change in the hydraulic pressure of the fluid around the piezoresistive sensor. The sensor, like in figure 6.8 emits a pressure proportional signal, which is then converted for example to a standard 4–20 mA output signal. The measuring method is very well suited for the detection of low pressures. High overload factors can be achieved with this type of sensor as well.

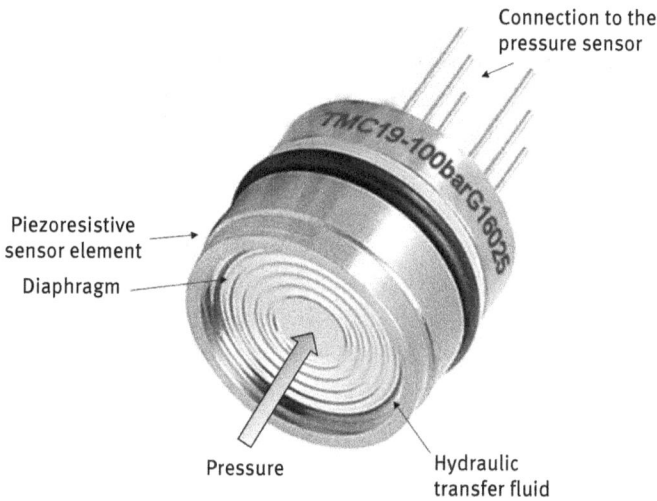

**Figure 6.8:** Principle of piezoresistive sensor.

### 6.1.2.4 Thick-film ceramic measuring cell

Unlike the thin-film strain gauge method, the Wheatstone bridge is bonded directly to a ceramic diaphragm in this case. Flexure of the diaphragm as the result of the external pressure causes a change in the resistances of this Wheatstone bridge, which is converted to a pressure-proportional signal. Using ceramics achieves a higher chemical resistance to aggressive media. The measuring range is higher than with the thin-film strain gauge method. The measuring accuracy however, is not as high.

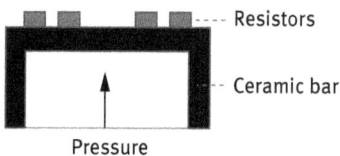

**Figure 6.9:** Principle of thick-film ceramic measuring cell.

### 6.1.3 Flow sensors

Flow measurement is another standard parameter of interest in Optofluidic control systems. Depending on the required measuring range, properties of the medium, pressures and temperatures obtained, it may be necessary to use different measuring instruments or measurement principles. The following section provides an overview of the most conventional measuring methods and when they can be used. When not otherwise stated, the measurement can be obtained only for fluids and not for gases.

#### 6.1.3.1 Definition
The SI unit of volumetric flow rate is the cubic meter per second. Based on one cubic meter that is equal to 1,000 liters, we use in this context volume in liter (L) and flow in liter per minute (L/min).

#### 6.1.3.2 Paddle wheel and turbine
This measuring method requires contact with the medium. It can be used in electrically non-conductive fluids. It is suitable for media without large solids proportions, for example sand, granulates, and fibers. Due to the materials used to fabricate the flow sensor, it can also be made suitable for use in very aggressive media and in special cases for high viscosity or contaminated media, or media containing solids.

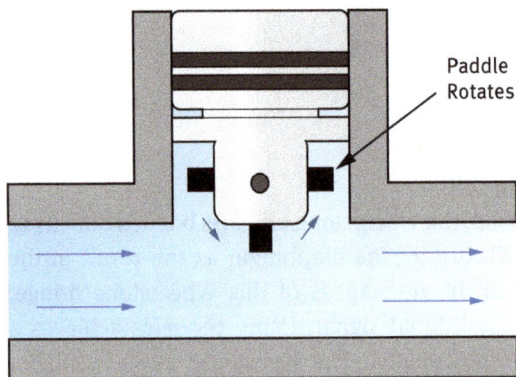

Paddle
Rotates

**Figure 6.10:** Principle of a paddle wheel flowmeter.

Typical sensor's characteristics are:
- Nominal diameters: DN 1 to DN 300
- Measuring range: up to 2,000 m$^3$/h
- Temperature: up to 100 °C
- Pressure: up to 16 bar

### 6.1.3.3 Magnetic inductive flow (MID) sensor

Magnetic inductive flow sensor is used universally as a measuring method without moving parts in electrically conductive measured media. It can be used with or without wetted electrodes, depending on the application. The MID sensor has a broad range of applications from clean to contaminated to pasty media. Depending on the design, this measurement method is also suitable for sterile applications in the pharmaceutical industry and food sector.

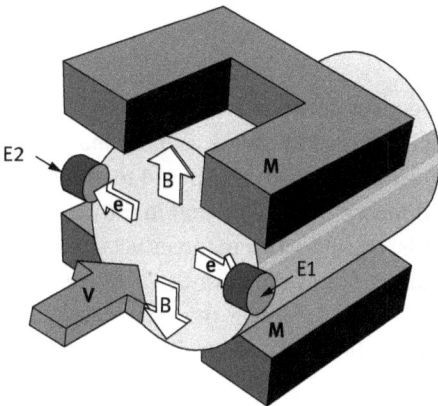

**Figure 6.11:** Principle of full MID. B = field strength (constant due to design); L = conductor length (distance between electrodes E1 and E2); K = constant; v = flow velocity.
Source: Bürkert GmbH & Co. KG

The liquid, a conductive measured medium flows through a pipe section with flow velocity v, in which a constant magnetic field B is created via two solenoid coils (M). Each of these solenoid coils are arranged on the outside of the pipe, opposite to one another. A voltage, in accordance with Faraday's law, is induced between electrodes E1 and E2 on account of magnetic field M and the flow of a conductive medium. The voltage is proportional to the flow velocity and is thus proportional to the flow rate. A measured value transducer amplifies the signal and converts it to a standard signal (e.g., 4–20 mA).

Typical sensor characteristics are:

- Nominal diameters: DN 4 to DN 300 and above
- Measuring range: up to 6,000 $m^3/h$
- Temperature: up to 130 °C
- Pressure: up to 40 bar

These are suitable only for gas formation or foaming media and when strong vibrating pipes are used.

### 6.1.3.4 Vortex

It is a universal measuring method with wetted shedder bar. It can be used for low flow velocities. This measuring method is relatively insensitive to pipe vibrations. It is also suitable for steam flow rate measurements.

Typical characteristics are:
- Nominal diameters: up to DN 150
- Measuring range: up to 700 m³/h
- Temperature: up to 400 °C
- Pressure: up to 40 bar

It is conditionally suitable in the case of highly viscous media, media forming coatings, media containing solids, or contaminated media.

### 6.1.3.5 Coriolis

This measurement method is regarded as the most accurate and most reliable flow measurement method, which uses the Coriolis effect to provide a direct mass flow measurement. The fluid flows through a curved tube, which is exposed to vibration. The Coriolis force acts perpendicular to the vibration of the fluid and in the direction of the flow. While the tube vibrates downwards, the fluid flow in forces acts up along the tube. When the fluid flows out of the tube, it forces downwards. This creates a torque and twists the tube. The reverse process occurs when the tube swings up. These opposing forces cause the tube to twist as the amplitude is directly related to the mass flow of liquid through the tube.

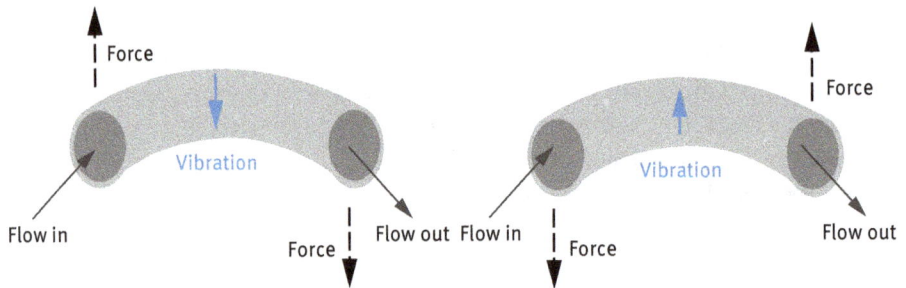

**Figure 6.12:** Principle of the Coriolis flow meter.

The volume flow rate is accessible due to the simultaneous measurement of the medium density. It is also suitable for measurement with steam. It is easy to use with media containing solids and media tending to outgas. Depending on the design of the sensor, it is also suitable for sterile applications in the pharmaceutical industry and food sector.

Typical sensor characteristics are:
- Nominal diameters: up to DN 100
- Measuring range: up to 250 metric ton/h
- Temperature: up to 150 °C
- Pressure: up to 40 bar

This method is conditionally suitable in the case of media forming coatings and in strongly vibrating pipe systems. In some cases, vibration-absorbing assembly sets are required for installation. Generally, this method is used at high gas densities (medium pressure > 3 to 4 bar).

### 6.1.3.6 Thermal flow sensors

It is a measuring method for low flow rates in liquids when there are no moving parts in the measured medium. All wetted parts in traditional and non-semiconductor based thermal flow sensors are commonly made of stainless steel. The measuring range is up to 1 l/h, the temperature range is up to 150 °C, and the pressure can be as high as 40 bar. There are two types of measurements. The first type measures the current required to maintain a fixed temperature across a heated element. During the flow, the fluid touches the element and dissipates or carry away the heat. As the flow rate increases, more current is required to maintain the element at a fixed temperature. The power requirement is proportional to the mass flow rate. The second type of measurement involves measuring the temperature at two points on a hot element (see also Figure 6.14). When the fluid flows over the element, it consumes heat. The upstream side of the element is hotter than the downstream side. The change in temperature is related to the mass flow of the liquid.

In gases, the thermal flow measurement principle is used to directly determine the mass flow rate. It possesses a high measuring dynamic in the case of the main flow measuring principle. It has a broad measuring span (100:1). It is usually used in nominal diameters up to DN 1000, for temperatures up to 150 °C, and pressures up to 40 bar. It is conditionally suitable for contaminated media.

Thermal flow sensors have been miniaturized in semiconductors and polymer foils and are still of high interest to both industry and research institutes due to their ability to be integrated in fluid and Optofluidic control systems.

### 6.1.3.7 Oval gear

Another classical flow measuring method, involving contact with the medium is the oval gear. It can be used in fluids containing no solids or with very small solid particles. It is particularly suitable for highly viscous and aggressive media.
Typical characteristics are:
- Nominal diameters: up to DN 400
- Measuring range: up to 1,200 m³/h
- Temperature: up to 290 °C

It is conditionally suitable at low upstream pressures (liquid columns) and with contaminated media or very thin-bodied media. Various aspects for ensuring trouble-free operation must be noted when designing a flow measuring Optofluidic control system: flow rates, viscosity, pressure, and measurement error considerations. Flow rates stipulated as a function of the nominal diameter are possibly depending on the measuring method and device type. The higher the flow velocity decreases the measurement error but increases the pressure loss. Viscosity describes the degree of internal friction (the interaction between the atoms and molecules) in the case of real fluids. It can be distinguished between the term "dynamic viscosity η" and "kinematic viscosity ν".

The interrelationship between these two is based on multiplication of the relevant densities of the substances.

$$\eta = \upsilon \cdot \rho$$

where the dynamic viscosity is $[\eta] = 1\,N/m^2 \cdot s = 1\,Pa \cdot s = 10^3\,mPa \cdot s = 10\,Poise = 10^3\,cP$ (centipoise) $\rightarrow 1\,mPa \cdot s = 1\,cP$ and the kinematic viscosity is $[\nu] = 1\,m^2/s = 10^6\,mm^2/s = 10^6$ cSt (centistoke) $\rightarrow 1\,mm^2/s = 1\,cSt$.

Table 6.1 provides a general overview of conventional media. Viscosity has a major influence on the channel, tube, and pipe design of Optofluidic control systems. At constant flow velocity, with a rise of viscosity (media becoming more viscous) the pressure loss in a pipe also rises. In order to maintain a constant flow velocity, either the flow velocity drops or the upstream pressure needs to be increased. The medium temperature also influences the viscosity. While with water, the change in viscosity can frequently be ignored; it is essential to consider it in the case of oils.

**Table 6.1:** Viscosity values of conventional media (Courtesy: Buerkert Fluid Control Systems).

| Medium/temp. [°C] | Dyn. viscos. $\eta$ [cP] | Density $\rho$ [kg/m³] | Kinem. viscosity $\nu$ [cSt] |
|---|---|---|---|
| Water/20 °C | 1.01 | 1,000 | 1.01 |
| Ethanol/20 °C | 1.19 | 1,580 | 0.75 |
| Turpentine/20 °C | 1.46 | 860 | 1.70 |
| Juice | 2–5 | – | – |
| Milk | 5–10 | – | – |
| Glycol/20 °C | 19.90 | – | – |
| Cream (body lotion) | 70–150 | – | – |
| Olive oil/20 °C | 107.50 | 919 | 117 |
| Detergent/20 °C | 360 | 1,028 | 350 |
| Transformer oil/20 °C | 986 | 860 | 1,146.50 |
| Thin honey | 1,000–2,000 | | |
| Ketchup | 5,000 | | |

In the case of pressure with respect to designing a flow control loop, an additional pressure loss occurs depending on average flow velocity in the case of fittings in channel, tubes, or pipes. In order to be able to estimate the total pressure loss in an Optofluidic control system, it is frequently necessary to be aware of the individual

pressure losses. An important aspect in the design of a flow control loop is the measurement error consideration in terms of required linearity, measurement error, and repeat accuracy. A decision to opt for a specific measuring method may also depend on the required accuracy. Basically, percentages refer either to the measured value or to the full-scale value. The maximum measurement error refers to the full-scale value and describes the sum of all possibly occurring individual deviations which is frequently shown graphically as a bell-shaped curve. This includes:
-   Linearity over the entire measuring range
-   Repeat accuracy (referred to the measured value)
-   Production-related tolerances
-   Installation tolerances as the result of installation in the pipe system.

The production-related tolerances and installation tolerances can be eliminated by calibration (teach-in), which greatly reduces the measurement error. Another important issue to consider is that the pressure resistance of plastics drops with increasing medium temperature.

## 6.1.4 Level sensors

Level sensors are next to temperature and flow sensors which are the most needed and sought after sensors in standard fluid control systems. Some of the commonly used measurement principles are described in the following section.

### 6.1.4.1 Radar and microwave
It is a non-contact and continuous measuring method and it is applicable in virtually all fluids and pasty media. It can even be used under difficult conditions such as the formation of hazes, mists, vapors, and gases. Typical characteristics of this type of sensor:
-   Measuring range: up to 30 m
-   Temperature: up to 300 °C
-   Pressure: up to 64 bar

This method cannot be used in applications with media with a low dielectric constant $(\varepsilon_r < 1.4)$.

### 6.1.4.2 Ultrasonic level measurement
The transmitter emits an ultrasonic wave and determines the propagation time of the signal reflected at a surface. On the basis of this propagation time, the device calculates the distance between the lower edge of the sensor and the surface. The influence of the sound velocity depends on the surrounding atmosphere and

can be automatically compensated for by entering specific values and the measurement of the ambient temperature by the transmitter.

If the distance between the lower edge of the sensor and the bottom of a tank is known, the device is able to indicate the filling level. If the tank geometry is known, the volume inside the tank can be indicated. Various disturbance echo filters even enable use in containers with built-in fixtures generating a disturbance echo. This measuring method allows very precise level measurements even with contaminated, aggressive fluids and in the case of bulk goods. The flow rate can be calculated by measuring the filling level in open channels and gutters. In principle, all circumstances involving attenuation of the signal or the reflected signal not being reflected back to the sensor restrict the application. This includes frothing on the fluid surface (attenuation), very damp atmosphere (attenuation), and condensate formation on the sensor (very great attenuation). Coarse-grained bulk goods reflect only a low share of the signal directly back, meaning that the maximum measurable distance or spacing is greatly reduced. In addition, formation of bulk goods cone must be taken into account when measuring. The gas properties influence the measured value of an ultrasonic level measuring system as well and have to be considered. Depending on temperature and the gas to be measured, the propagation speed of the sound wave will vary (see the Table 6.2 below for examples).

**Table 6.2:** Density, sound velocity, and temperature gradient of various gases (Courtesy: Buerkert Fluid Control Systems).

| | Density at 0 °C [kg/m$^3$] | Sound velocity at 0 °C [m/s] | dv/dt [m/s K] |
|---|---|---|---|
| Dry air | 1.293 | 331.45 | 0.59 |
| Ammonia | 0.771 | 415.00 | |
| Carbon monoxide | 1.250 | 338.00 | 0.56 |
| Carbon dioxide | 1.977 | 259.00 | 0.60 |
| Chlorine | 3.214 | 206.00 | |
| Ethylene | 1.260 | 317.00 | |
| Helium | 0.178 | 965.00 | 0.80 |
| Hydrogen | 0.090 | 1,284.00 | 2.20 |
| Hydrogen chloride | 1.639 | 296.00 | |
| Hydrogen sulfide | 1.539 | 289.00 | |
| Methane | 0.717 | 430.00 | |
| Nitrogen oxide | 1.340 | 324.00 | |
| Nitrogen | 1.251 | 334.00 | 0.60 |
| Oxygen | 1.429 | 316.00 | 0.56 |
| Sulfur dioxide | 2.927 | 213.00 | 0.47 |

Since there may be a wide variety of gas mixtures present between the fluid and the sensor, these physical characteristics of the gases must be known or determined by two-point calibration. A temperature change in the tank can be detected, for example by integrating a temperature sensor into the ultrasonic level transmitter resulting in the measured value to be corrected automatically.

In summary: It is a non-contact measuring method for continuous measurement and for limit level detection in fluids. No physical contact with the fluid is required. This measurement method can be used universally in a multitude of fluid control systems applications, even with aggressive fluids. The measuring range is usually up to 50 m with temperatures of the fluid up to 90 °C and pressures ranging up to 2 bar. This measurement principle can only be used restrictedly in measurements dealing with the formation of hazes, mists, gases, or foaming media. It cannot be used in measurements involving a high tank pressure or vacuum.

### 6.1.4.3 Hydrostatic pressure

A fluid column generates a specific hydrostatic pressure as a function of density and filling level. A pressure sensor attached to the bottom of a tank measures this pressure with respect to a reference pressure (generally ambient pressure). Conclusions are then drawn as to the filling level with the aid of the known fluid density. Hydrostatic level measurement is suitable for virtually all types of fluids and produces very precise measured values depending on the accuracy of the pressure transmitter. Restrictions apply to applications in pressurized tanks. In such cases, it is then necessary to measure the gauge pressure. This can be done by using a second pressure sensor that detects the pressure above the filling level.

A corresponding evaluation unit corrects the measured value of the first pressure sensor on the tank bottom based on this value. The higher the internal pressure of the tank, the lower the share of the hydrostatic pressure in the overall pressure, and the level measurement error increases. The measuring accuracy also drops further as the use of two pressure sensors adds the measurement errors. When a differential pressure transmitter is used, the absolute pressure at the bottom of the tank is applied to the front side of the pressure diaphragm and the absolute pressure above the filling level is applied to the rear side. This means that the measuring accuracy of the transmitter remains unchanged even in the case of increasing internal pressure. However, these measuring methods involve very complex designs and are thus extremely costly.

In summary: It is a measuring method, which involves physical contact with the medium. It can be used with fluids and pasty media in a continuous manner. This measurement method is insensitive to foaming.

Typical sensor characteristics are:

- Measuring range: up to over 40 m (special versions up to 200 m)
- Temperature: up to 100 °C

The measurement result is dependent on the density of the medium. In the case of pressurized tanks, two measuring instruments or a differential pressure-measuring instrument must be used.

### 6.1.4.4 Capacitive

It is a universal measuring method for continuous and level limit detection in fluids, involving physical contact with the medium. It can even be used in aggressive fluids. The measurement principle is largely insensitive to adhering media. Typical sensor characteristics are:

-   Measuring range: up to 25 m
-   Temperature: up to 400 °C
-   Pressure: up to over 100 bar

The dielectric constant of the medium should be greater than 1.4. Fluctuations in the dielectric constant of the medium (e.g., resulting from temperature changes or a change in composition) affect the measurement results. Special designs or installations are required for measurement in plastic tanks.

### 6.1.4.5 Vibration

It is a measuring method for limit level detection, involving physical contact with the medium. It can be used in all types of fluids and pasty media. The measurement is independent of turbulence, contamination of the medium, and the electrical properties of the medium.
Typical sensor characteristics are:

-   Temperature: up to 150 °C
-   Pressure: up to 30 bar

The medium must meet the following requirements with respect to density and viscosity:

-   Density: > 0.7 kg/dm$^3$
-   Viscosity: < 10,000 mPas

### 6.1.4.6 Conductive

The measuring method for limit level detection involves physical contact with the medium. It is a simple, low-cost measuring method for virtually all conductive fluids. Sensor characteristics are:

-   Temperature: up to 150 °C
-   Pressure: up to 60 bar

The measuring method is sensitive to adhering media and cannot be used in non-conductive media (e.g., hydrocarbons).

## Further reading

Lipták, BG., (ed.). *Process Control: instrument engineers' handbook*, volume one: process measurement and analysis. Butterworth-Heinemann, 2013.

Fraden, J. *Handbook of modern sensors: physics, designs, and applications*. Springer Science & Business Media, 2016.

Mukhopadhyay, SC. *Intelligent sensing, instrumentation and measurements*. Berlin, Germany: Springer, 2013.

## Self-Assessment Questions

1.  What is (a) the difference between a thermistor, RTD, and a thermocouple and (b) what are the two factors on which the thermocouple effect depends?
2.  In a resistance thermometer, a metal wire shows a resistance of 500 Ω at ice point and 700 Ω at steam point. Temperature when resistance is 630 Ω would be (a) 60 °C; (b) 65 °C; (c) 70 °C; and (d) 75 °C.
3.  Sensitivity of a thermometer refers to
    a)  how quickly thermometer can register change in temperature
    b)  amount of change in thermometric property for a unit change in temperature
    c)  min and max temperatures that thermometer can measure
4.  A 10 kΩ NTC thermistor has a B value of 3,455 between the temperature range of 25 °C to 100 °C. Calculate its resistive value for both temperatures and plot the graph.
5.  Express the following pressures as absolute pressures in pascal or kilopascal:
    a)  A process pressure of 15 barg
    b)  A pressure gauge reading of 14.5 psig
    c)  A pressure differential of 4.0-inch water column
6.  Which statement represents the correct relation between flow rate and area of pipe?
    a)  Direct proportionality
    b)  Inverse proportionality
    c)  Equal
7.  A fluid sample must travel over a micro mixer 1 to the analyzer location in the micro mixer 2 in less than 14 seconds. The volume of mixer 1 is 1.1 mL with a channel length of 0.8 m and the volume of mixer 2 is 1.7 mL with a channel length of 1.2 m. Find the necessary flow rate and the velocity in the transport line.

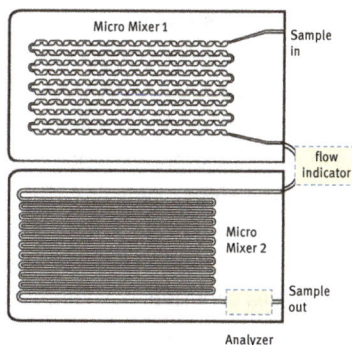

8. In _____, velocity of fluid is constant on every point at a specific time.
   a) steady flow
   b) rotational flow
   c) non-steady flow

## 6.2 Process analytical sensors

### 6.2.1 The added-value concept

The goal of modern production techniques is to produce materials with well-defined and tailor-made properties. In order to meet these requirements, the process has to be controlled in such a way that both the various qualities of the raw materials and the different requirements of the final user are taken into account. Measurements of process variables are therefore important in controlling a process and represent a dynamic feature of the process that may change rapidly, such as due to changes in feedstock composition or temperature profile. Our particular interest in measurement is to gain a better understanding of the process and to quantify specific (bio)chemical components [315, 316]. Actually, the majority of instruments used for sensing in process control are classical sensors, utilized for the determination of temperature, pressure, level, and flow. These types of sensors are indispensable for a basic control of processes.

The field of Process Analytical Technology (PAT) adds valuable analytical sensors to this required controlling and analyzing of the fluid of interest. Analytical instruments are quite different from the classical process sensors because they measure the quality of those fluids. They provide information about the composition of a fluid, its chemical and physical properties, and changes in chemical structures. With analytical instruments, the operator is able to control critical quality parameter rather than measuring unspecific conditions of a fluid. Figure 6.13 illustrates the added value concept for batch processing.

For example, flexible recipe management in batch processing becomes increasingly important as the market continues to evolve to favor smaller batch sizes. The recipe identifies raw materials, their relative quantities, the required processing, and the order of processing. Process parameters specify the variables, such as temperature, pressure, and time, which are set points and comparison values that. Traditional automated recipe management systems (level 1) are primarily data bases which are downloaded to a programmable logic controller (PLC) that sequences these to initiate the process steps. Typically, the steps are not easily configurable by the operator. Traditional management systems do not allow recipe optimization while the process is being run. Recipe optimization based on the development of first principle models is in many cases very time consuming and cost intensive or not feasible. Analyzing historical process data from the Production Information Management System (PIMS) and Enterprise-Resource-Planning (ERP) is the first step for improvements. The use of

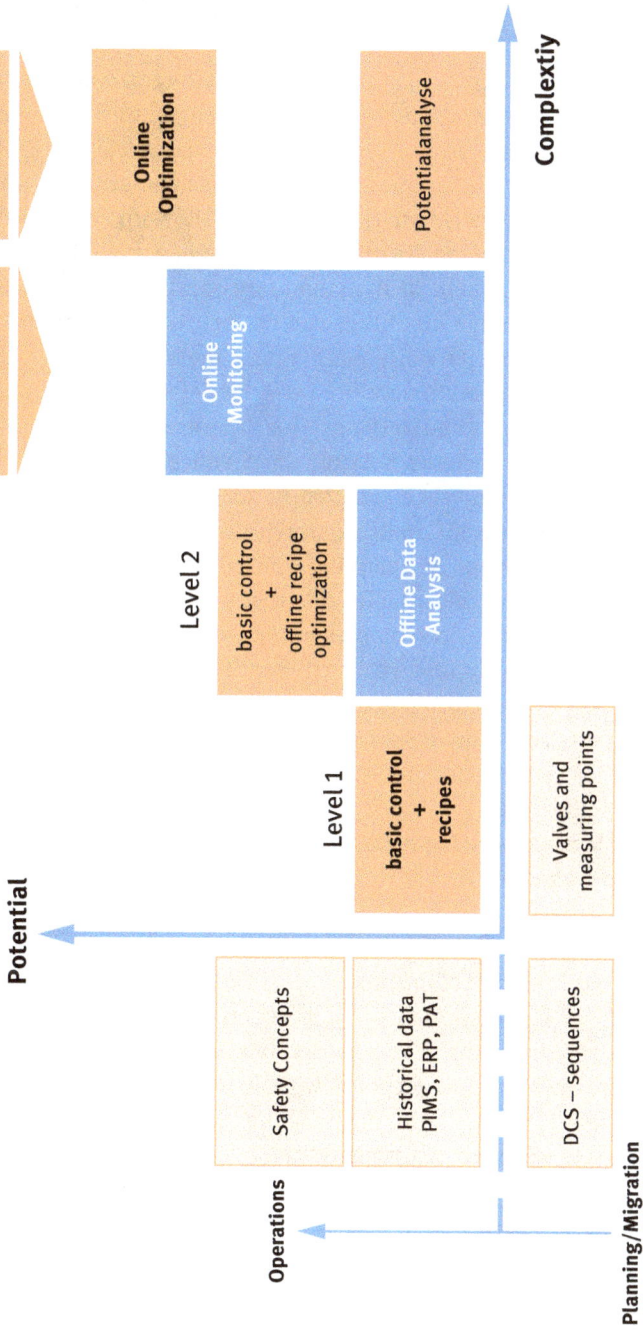

**Figure 6.13:** Advanced process control for batch processes.

product quality data from the offline laboratory (level 2) is an important step toward recipe optimization.

Online monitoring (level 3), as an "assistant," of tracking parameters that are relevant to the product quality is a fundamental step to understanding the process dynamics and is a prerequisite for knowledge based control [313]. The preparation of samples in level 2 for quality control at the end of the process can sometimes be extremely time-consuming; as a result, follow-up checks can thus be omitted by this concept. The results obtained with online analytical measurements during product development in the laboratory can be easily up-scaled to production mode. The production quantities can be rapidly and flexibly adapted to changing market requirements while maintaining the same level of quality. Online monitoring can be applied to obtain a better understanding in the chemical and physical mechanisms, find out unstable intermediates, and therefore improve the process safety. It includes not only the chemical or physical analysis, but also statistical and risk analysis conducted in an integrated manner. Online monitoring can lead to the prevention of batch rejection, decrease in production cycle times, significant decrease in sampling and conventional analyses (e.g., HPLC, GC, etc.), and thereby a meaningful decrease in costs. Furthermore, PAT tools can improve hygiene conditions by preventing sampling of extremely active substances and they can be applied to better monitor the processes and take decisions. However, continual learning and information organization will perhaps be the most important factor for the success of PAT. The utilization of process analytical sensors for quality assurance is the most usual utilization within knowledge based production.

The "Autopilot" level (level 4) is a highly sophisticated system centered around a control loop function with feed-forward control strategies based on the analytical measurements. It is proactive in that the control variable adjustment is not error-based, instead it is based on molecular knowledge of the process that takes the form of a multivariate statistical model of the process, and knowledge about, or analytical measurements of, the process disturbances. In this context, Quality by Design (QbD) is the overarching paradigm and PAT is a toolbox to achieve this [339, 345]. Obviously, this approach should be prepared with an active response; such as, by monitoring the process that allows the manufacturer to make production adjustments on this basis. It relies on the primary point of PAT which is to produce information on product quality in real-time and proceed with a smart control strategy.

### 6.2.2 Taxonomy

Process control with the help of process analytical sensors can be categorized into four different methodologies:

- **In-line:** The analytical instrument is directly located in the process line and can perform the required analysis in real time, thus providing the wanted parameters (e.g., amount of solvent in the sample, percentage of ingredients, etc.) immediately.
- **On-line:** This method automated sampling and analysis of the sample, and if necessary, transport the sample through a by-pass to an automated analytical instrument.
- **At-line:** This is a method of manual sampling with transport of the sample to an analytical instrument near the manufacturing or production line.
- **Off-line:** This is also a manual sampling method which transports the sample to a specialized laboratory.

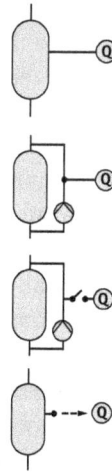

The purpose of sampling for at-line and on-line systems is to extract and transport a representative sample that is compatible to the analyzer. The analyzer itself is only one part of the system. Results can be accurate only to that extent so that the sample received by the analyzer represents the source. Over 80% of problems with analytical instrumentation can be attributed to the sample conditioning system [353].

### 6.2.2.1 Selection of analytical techniques

To design a robust process analysis tool, there are many decisions to make:
- The type of the reactor (batch or continuous)
- The scale of operation
- The sampling frequency and location
- Calibration and signal interpretation requirements
- Use of the data for a feed-back or feed-forward control loop
- Safety and environmental concerns
- Cost

Specific data manifests a property and makes it possible to gain qualitative and quantitative information about individual components of a sample stream. Therefore, for real-time analysis, the following should be considered:
- The type and number of analytes to be measured
- The number of components in the process stream
- The physical state of the sample
- Cycle time and response time required
- Accuracy, precision, and robustness required

Depending on the branch of industry, the requirements for process analytical sensors differ. In general, all analytical sensors must have a high degree of accuracy and reproducibility, low cross-sensitivity, high reliability, and should be non-invasive to the process. Additionally, analytical sensors must meet requirements specific to the application. These include robustness against pressure, temperature, vibration, and shock; high chemical corrosion resistance; no contamination effect on the process; rapid response; minimal maintenance; and system integration by standardized mountings and standardized output signals (4–20 mA, Profibus, Modbus). In addition, they should conform with the hygienic standards of the food and biotechnology industries and the standards for electrical apparatus in potentially explosive atmospheres. A process sensor has to resist temperatures ranging from −40 °C to + 180 °C and pressures of 40 bar. However, requirements can be much higher such as, when parameters related to the properties of gases are to be measured. When applications in process industries are considered, a sensor has to resist aggressive media, such as hot acids. Measurements in this harsh environment can be done using optoelectronic sensors which is also non-invasive.

### 6.2.3 Process analyzer categories

Process analzyer covers over hundred different analytical techniques to measure a wide variety of varbiables of interesst to the processing industries. We simplify the analytical techniques to six main categories of process analytical sensors. Further readings for chemical and biochemical sensors can be found in [290, 291].

#### 6.2.3.1 Physical property analyzer

These analyzers measure a quality property of the process fluid, in which the set-up is complex as for analytical instruments. In many cases, ASTM (American Society for Testing and Materials) laboraty test procedures define the measurement set-up for the process sensors. Typical physical property analyzers measure

- Optical properties: turbidity, refractive index
- Flow properties: viscosity, freezing point
- Volatility properties: boling point, kinetic vapour pressure

An example of an optical physical process analyzer is a turbidimeter to accurately measure the concentration of light-scattering components. Turbidity is a measure of cloudiness or the lack of clarity of a fluid caused by undissolved particles or droplets. Turbidity causes the light to scatter rather than transmit in straight direction through the sample. Turbidity measurements are found in many areas, such as in water analysis (both drinking water and wastewater) and in the beverage industry (beer, juice, and wine production), for their application. The turbidity range from 0 to 10,000 NTU (Nephelometric Turbidity Units) is of particular interest for this application and process turbidimeter are on the market. Forward-scattering signals, for example, are inversely proportional to increased turbidity up to 4,000 NTU and are suitable for lower turbid samples. Above this value, no more light reach the detector and the instrument's noise becomes a major interfering factor. As the light scattered in the forward direction, the measurement of light transmitted through the sample will yield variable results depending upon the particles' size. This problem can be eliminated when scattering is measured at right angle to the incident light beam because the angle is very sensitive to light scattered by particles. The signal for back-scattering increase is proportional to the increase in turbidity for measurements of high turbidity between 1,000 and 10,000 NTU.

Miniaturized and cost-effective sensors that can detect 0 to 40 NTU are currently rarer and therefore the subject of current research. A prototype of Arduino control is shown in Figure 6.14. The 90° set-up is used together with a GRIN lens. The flat surface allows the GRIN lens to be easily fused to an optical fiber to produce collimated output. The design of such lenses involves detailed calculations of aberrations as well as efficient manufacturing of the lenses. Additionally, antireflection coatings are typically effective for narrow ranges of frequency or angle of incidence.

### 6.2.3.2 Molecule property analyzer

This type of analyzer measures characteristic properties, such as paramagnetic force of the analyte molecules, and uses of it to derive the concentration of those molecules. This analyzer requires a non-condensing sample gas flow at ambient temperature and pressure. Most of them require a sample without suspended solids, so that an upstream filer must be used frequently.

This group includes the following measurement methods:
-  Paramagnetic force: Oxygen
-  Gas thermal conductiviy: Hydrogen in air
-  Flame or photoionization: Hydrocarbons, volatile organic compounds
-  Catalytic oxidation: Combustabile gases
-  Quartz frequency: Moisture

For example, a photoionization detector (PID) measures a broad range of volatile organic compounds (VOCs) such as methane, formaldehyde, benzene, or hydrocarbons. The detector uses high-energy photons from the vacuum ultraviolet region to break molecules into positively charged ions. Compounds entering the detector are bombarded with high-energy UV photons and ionized as they absorb the UV light. This leads to the ejection of electrons and the formation of positively charged ions. The ions can undergo numerous reactions including reaction with oxygen or water vapor, rearrangement, and fragmentation. Because only a small fraction of the analyte molecules are actually ionized, this method is considered nondestructive,

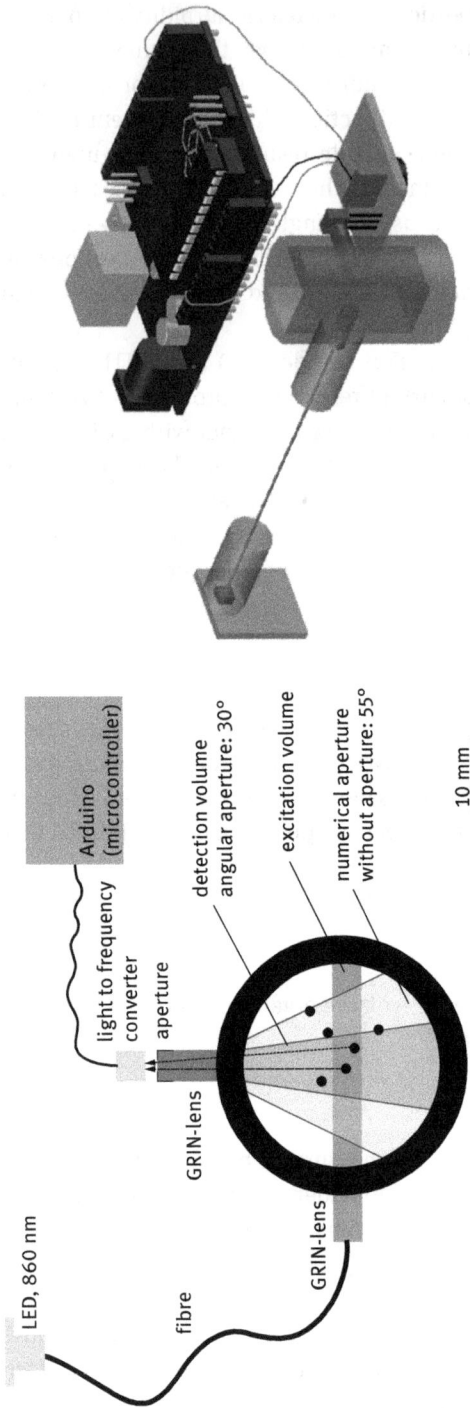

**Figure 6.14:** Design of a miniaturized low cost turbidimeter. Courtesy of Marc Brecht [328].

allowing it to be used in conjunction with another detector to confirm analytical results. In addition, PIDs are available in handheld portable devices [308] or are the subject of research in miniaturized gas chromatographs (micro-GCs) [310]. A critical parameter is the response time, which can be high for large ionization chambers and thus large dead volumes. Hongbo Zhu et al. presented a microfluidic PID with a significantly reduced ionization chamber volume which is nearly 100 times smaller than commercial detector systems [356]. The microfluidic PID consists of a conductive silicon wafer with a spiral microchannel structure. The body was anodically bonded to a Pyrex glass wafer; see Figure 6.15.

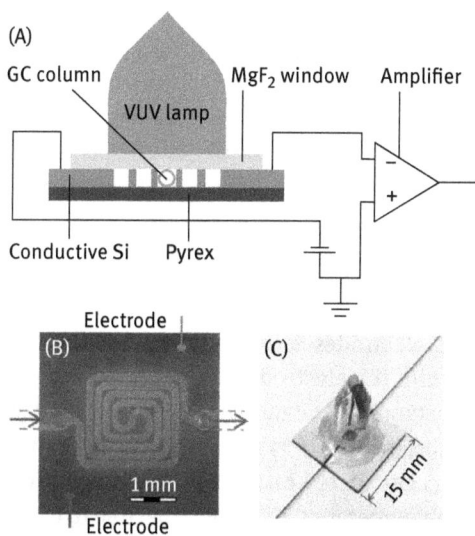

**Figure 6.15:** Flow-through microfluidic photoionization detector for rapid and highly sensitive vapor detection [356].

The channel structure has a cross-section of 150 μm in width and 380 μm in depth with a total length of 2.3 cm. In order to reduce the dead volume at the interconnection between the GC column and the microfluidic channel, the terminal of the microfluidic channel has a trapezoidal shape. Figure 6.15(b) shows a microscopy image of the microfluidic channel portion of the PID. Two electrodes were wire-bonded to the aluminum layer on the microchannel and connected to the voltage supply and the amplifier, respectively. On the top of the microchannel, a VUV Kr lamp together with an MgF$_2$ crystal window is mounted and sealed with an optical adhesive. In Figure 6.15(c), the lamp source covers the entire microfluidic PID area with an illumination diameter of 3.5 mm. The inlet and outlet ports are connected with the column and sealed with an optical adhesive. The ions produce an electric current, which is the signal output of the detector. The greater the concentration of the component, the more ions are produced, and the greater the current. The current is amplified and displayed on an ammeter or on a digital concentration display.

### 6.2.3.3 Electrochemical analyzer

These analyzers comprise the large family of devices that use an electrochemical cell to measure the concentration or activation of ions in solution.

- Conductivty: Ion concentraton
- Glass electrode: pH value
- Galvanic membrane cell: Dissolved oxygen
- Potentiometric titration: Acid–base, oxidation–reduction
- Amperimetric cell: Oxygen, residual chlorine

Conductivity sensors, as the name suggests, measure the ability of a solution to conduct a current. The higher the concentration of ions in the solution, the higher the conductivity. The unit of conductivity is microsiemens ($\mu m$ S). Unlike pH sensors, which only measure the concentration of hydrogen ions, conductivity measurements take all the ionic species present into account. Conductivity measurements can be used to determine the concentration of a solution provided the dissolved species are known. The conductivity measurement although will not differentiate between an acid and a base.

### 6.2.3.4 Conductive conductivity

The measuring cell consists of two open electrodes to which an AC voltage is applied. The medium is in direct contact with the electrodes. The applied voltage generates a current depending on the resistance of the medium (Ohm's law). The geometry of the measuring cell (area S and distance d) is defined by its quotients $K = d/S$. The conductivity of the solution is calculated on the basis of this known cell constant K and by measuring the current generated. In order to be able to cover a broad conductivity range, measuring cells with various cell constants are used. The lower the conductivity, the lower the cell constant must be. The conductivity measurement of the ultra-pure water depends on the selection of the cell constant up to concentrated solutions. A PT1000 temperature sensor is integrated for temperature compensation and its use in coating-forming media is recommended only if the measuring electrodes are cleaned regularly, otherwise the insulating effect of the coating would mean that the measured value does no longer corresponds to the actual value.

### 6.2.3.5 Inductive conductivity

An inductive conductivity cell consists of two coils: a transmitter coil and a receiver coil. The coils are integrated in the measuring cell. A bore is routed through the finger and the coil is integrated into it. The fluid encloses the finger and is also present in the bore. A sinusoidal AC voltage is applied to the transmitter coil. This produces a current in the fluid, which is proportional to the conductivity. This current in turn generates a voltage in the receiver coil. By measuring this voltage and knowing the cell constant, it is possible to determine the conductivity. A temperature sensor can be integrated for temperature compensation. The measuring method also allows to be used in very aggressive and hazardous fluids. Owing to the separation of the medium and the measurement method, all that needs to be ensured is that the housing has adequate resistance if used in such media. Since the measuring electrode has a very broad measuring range, different cell constants are not required. Use of the device is, however, not possible in very pure media, since no measured value can be detected below a specific conductivity. The calibration procedure of conductivity sensors consists merely of a check of the cell constant, which may possibly have changed as a result of deposition or chemical attack.

A reference that may either be a buffer solution or a reference measurement is required for this. The new cell constant is then calculated on the basis of the following equation:

$$K_{new} = K_{old} \cdot \frac{Cond_{ref.}}{Cond_{meas.}}$$

where $K_{new}$ is the new value of the sensor coefficient; $K_{old}$ is the former value of the sensor coefficient; $Cond_{ref.}$ is the conductivity measured with a reference instrument; $Cond_{meas.}$ is the conductivity measured with the actual sensor (former value of sensor coefficient). In the case of conductivity measurement, the measuring cell has no dependence on temperature but the temperature dependence of the solution must be compensated for in order to allow a comparison between various measurements. In certain cases, linear compensation is accurate enough to monitor and control processes. Linear temperature compensation merely requires an input value, i.e. the average compensation both for the temperature range and the conductivity range.

### 6.2.3.6 pH sensors
One of the most common pH measuring instruments is the pH glass electrode. The hydrogen ion concentration (actually referred to as pH value) in an aqueous solution generates a potential difference at a measuring electrode (e.g., pH-sensitive glass diaphragm) with respect to a reference electrode (e.g., Ag/AgCl). This voltage is measured by a high-impedance pH-measuring instrument and converted to a pH value. The relationship between pH value and the voltage is linear with a slope of, for example, 59.16 mV/pH value. The slope is temperature-dependent and must be compensated for by an integrated temperature sensor.

A review on pH monitoring is provided in [351] where also the historical developments on pH sensors based on field effect transistors (FET) are provided. Nowadays, also the pH ion selective field effect transistors (ISFETs) are used to measure the pH value. The advantage of pH ISFETs is the ability to integrate them into small housings and thus present a nice way for integration into Optofluidic systems. References [293, 294] provide one of the first review papers on ISFETs, Theory and practice.

An open access review on the technology of ISFETs including fabrication technologies is presented in [311]. ISFETs are commercially available and several companies have integrated them into their standard sensor portfolio.

Recently, optical pH sensors are gaining more and more attention and several companies offer optical pH probes for use in different environments. An open access review on optical chemical pH sensors is given in [354]. The oxydo-reduction potential electrode measures the potential of a solution on the basis of the presence of specific ions. This potential occurs between a metallic measuring electrode (platinum or gold) and a reference electrode (Ag or AgCl). It provides information on the oxidizing or reducing capability of the solution. The following paragraph provides a more closer look at oxygen sensors.

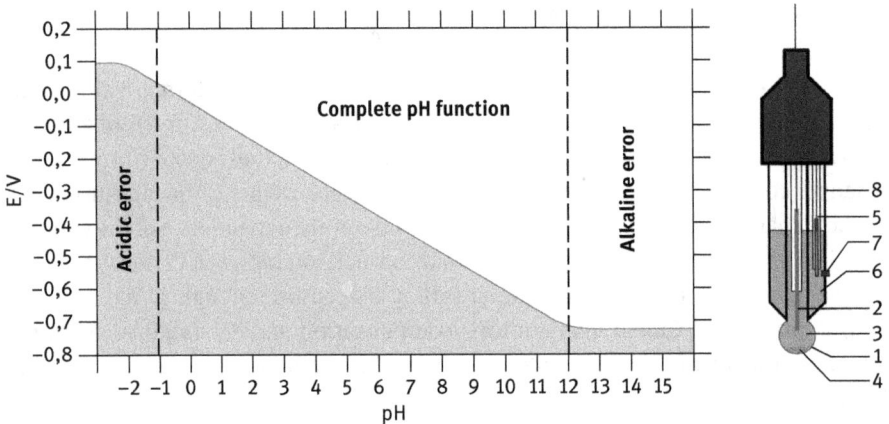

**Figure 6.16:** Scheme of the typical dependence E (Volt) – pH for ion-selective electrode (right). 1: glass bulb; 2: Ag/AgCl electrode; 3: KCl buffer; 4: AgCl precipitate; 5: reference electrode; 6: reference internal solution; 7: ceramics; 8: electrode body.

### 6.2.3.7 Electrochemical oxygen sensor

With technological advancement, the electrochemical sensors used to measure oxygen are gradually being replaced with the optical oxygen sensors, which utilize modern technology and are more user friendly. Oxygen plays an important role in the monitoring and control of fluids. The parameter "oxygen concentration" is of great interest in many industries like waste water treatment and Food & Beverage.

Oxygen is an immensely important chemical species – essential for life. The need to determine levels of oxygen occurs in many diverse fields. Oxygen measurement provides an indispensable guide to overall condition and functioning of various processes. The presence or absence of oxygen can play a vital role in many fields and such an important parameter should ideally be monitored continuously. Here, the principle of determining oxygen content optically is described.

The traditional way of measuring oxygen concentration in a water sample is wet chemistry techniques, such as the Winkler Titration method. However, there are commercially available oxygen sensors that measure the oxygen concentration in liquids and gases with great accuracy. There are mainly two kinds of oxygen sensors, which work on two different measuring principles: eectrochemical oxygen sensors and optical oxygen sensors.

For measuring dissolved oxygen, Clark-type electrodes are most commonly used. The basic principle is that cathode and anode are submersed in an electrolyte. Oxygen enters the sensor through a permeable membrane by diffusion and is reduced at the cathode creating a measurable electrical current (see the chemical equation below). There is a linear relationship between the oxygen concentration and the electrical current. With a two-point calibration, it is possible to measure oxygen in a sample.

$$O_2 + 4e^- + 2H_2O \rightarrow 4OH^-$$

Electrochemical oxygen sensors are designed either for use in gas or liquids. They have a response time of approximately 1–1.5 minutes. Oxygen is consumed by the electrodes at nearly 0.1 micrograms/hour. Calibration is required on a frequent basis. Optical sensors vary depending on the field of application; see Section 6.3.5. They could be either invasive or non-invasive. Invasive sensors are completely part of the measurement setup, whereas non-invasive sensors influence the medium.

### 6.2.3.8 Separation analyzer

These analyzers separate selected analytes from a liquid or gaseous mixture. The mixture is dissolved in a fluid called the mobile phase, which carries it through a structure holding another material called the stationary phase. The various constituents of the mixture travel at different speeds, causing them to separate. Samples for these analyzers must be very clean. Any solids in the sample will soon damage injection valve. It's recommended to install a guard filter (0.5 μm) just before the sample injection valve. Gas samples must be completely vapor with no trace of liquid mist and must stay at constant pressure. Often atmospheric pressure is used, but its normal variations can cause noticeable measurement error.

- Gas chromatography: Almost any selected substance in a gas or liquid
- Liquid chromatography: More complex and thermolabile liquid substances
- Mass spectrometry: Any substance in a gas or vapor sample

Combination of optofluidic systems and separation analyzers are possible. Shopova et al. [344] reported a concept and experimental set-up for an on-column micro gas chromatography detector using capillary-based optical ring resonators (CBORR). The optical ring resonator consists of a thin-walled fused silica capillary with an inner diameter of only few hundreds of micrometers. The circular cross-section forms a ring resonator and supports whispering gallery modes (WGMs) that circulate along the ring resonator circumference hundreds of times. The evanescent field extends into the core and is sensitive to refractive index changes induced by the interaction between the gas sample and the stationary phase. The WGM can be excited and monitored at any location along the CBORR by placing a tapered optical fiber against the CBORR, thus enabling on-column real-time detection. Figure 6.17(a) shows set-up in which the sample is introduced by a short pulse in the gas injector and travels via high purity hydrogen gas to the CBORR. Figure 6.17(b) shows the cross-section of the CBORR with the propagation of a WGM confined in the coated capillary wall.

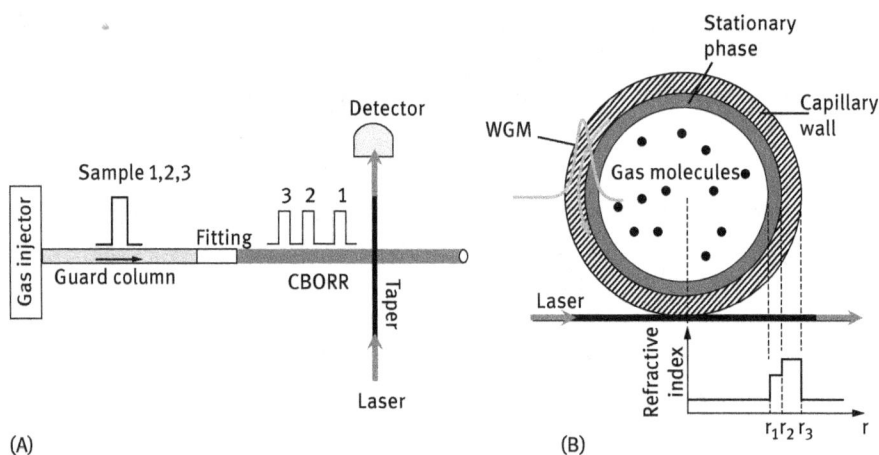

**Figure 6.17:** On-column micro gas chromatography detection with capillary-based optical ring reso-nators. $r_1$, inner radius; $r_2$, radius at the polymer coating layer; and r3, outer radius [344].

Optofluidic devices can also be used in High-Performance Liquid Chromatography (HPLC) or Capillary Electrophoresis (CE) systems [332] to detect small liquid sample quantities to be injected into the HPLC separation column.

### 6.2.3.9 Chemical analyzer

Chemical analyzers rely on a chemical reaction to measure an analyte. The idea is to convert the analyte to a new chemical species that is easier to measure, often in the visible wavelength range. Samples are usually liquid and are often aqueous solutions such as effluent water. Some dirty samples may need grinding and homogenizing before analysis. Standard analytical methods have known interferences.

- Titration by chemical reagent: Chloride in water acidity
- Colormetric analysis by photometer: ppb of silica in boiler water
- Flow injection analysis: Any analyte amenable to wet chemical analysis

A common method with flow injection analysis is the "stop flow method" in which the sample and reagent are introduced in flowing streams and then stop the flow with the reaction mixtrue in the microfluidic cell. In this technqiue, drive syringe plungers press streams of reagent and sample rapidly into the mixer and the flow of mixed solution is stopped suddenly. Array-based spectrometers allow entire spectra to be taken at different time intervals and multiple wavelengths can be used for measuring the absorbance. The reaction product should be stable for at least 10 seconds. In comparison, photometer can be attached that allow rates to be obtained at a single wavelength by measuring the absorbance as a function of time. The set-up shown in Figure 6.18 is designed to stopped-flow mixing and to allow measurements within milliseconds.

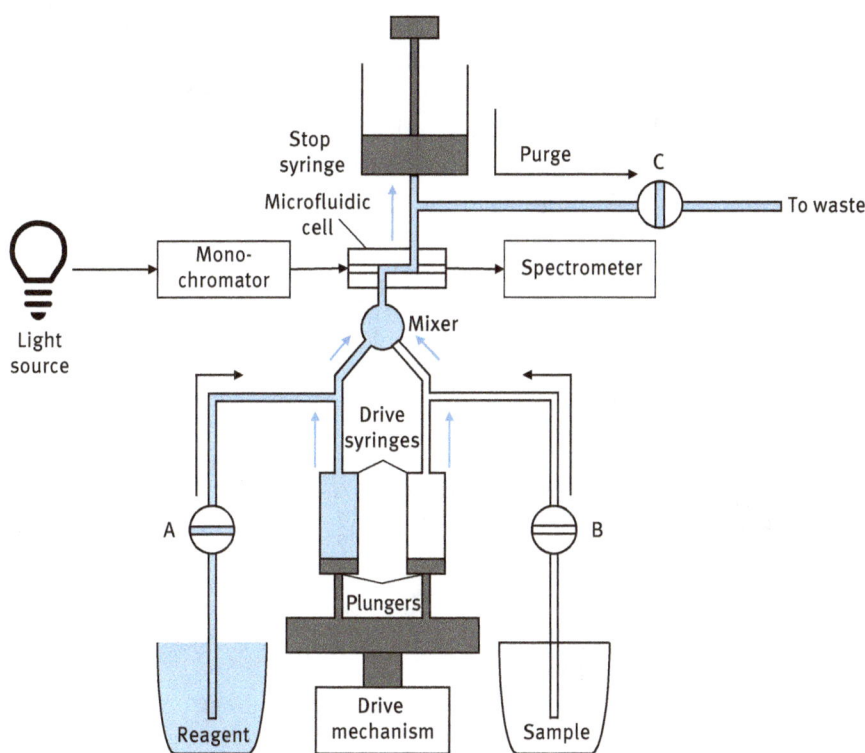

**Figure 6.18:** Spectrometric stopped-flow mixing method.

### 6.2.3.10 Spectroscopic analyzer

These analyzers will be discussed in detail in Section 6.3 and measure the amount of light absorbed, scattered, or emitted by molecules in the sample. A photometer measures the light energy absorbed or emitted at a few selected wavelength bands, whereas a spectrometer splits the light into a full spectrum of wavelengths and may muse multiple wavelengths to determine analyte concentration.

- X-ray fluorescence: metal ions in aqueous solution
- Ultraviolet absorption: many liquids, such as benzene residual in water
- Visible light absorption or scatter: opacity, color, turbidiy
- Fluorescence: ppm levels of toxic elements, chlorophyll in water
- Near-infrared spectroscopy: protein, fat, and water content in milk products
- Infrared-spectroscopy: gases with more than one element in their molecule
- Microwave absorption: moisture content in sheet materials

Recently, a new approach using light spectroscopy has been applied to measure CHO suspension cells at three different angles in parallel. As light passes through a sample containing suspended CHO cells, the cells absorb the light energy and re-radiate the energy in all directions in the form of scattering. In general, particle shape, size, and refractive index determine the spatial distribution of the scattered light by the cells. Cell fragments smaller than the wavelength of light scatter light at equal intensities in all directions, while cells larger than the wavelength of light cause greater forward scattering. Figure 6.19 shows a set-up with corresponding measurements for back-, side-, and forward scattering spectroscopy.

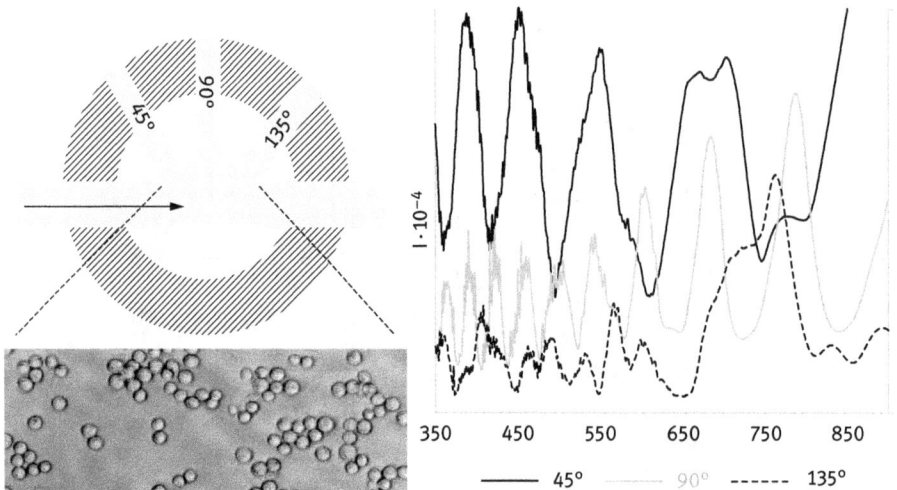

Figure 6.19: Elastic light scattering spectroscopy of CHO cells in parallel for 45° backscattering, 90° side scattering, and 125° forward scattering in the visible range.

The design allows for the accurate determination of cell sizes when the refractive index of the cells and the medium is known. The modulations result from Mie scattering patterns [306], which can be very specific for a given particle size. Especially at very low concentrations, single scattering is present and this type of measurement is well suited.

Since different scattering angle generates different spectral pattern, each spectral characteristic is necessary for a reproducible and evaluable result. If only the turbidity needs to be determined, and no information about particle sizes is required, the favored angle to measure light scattering is 90° at one single wavelength; see also "Physical property analyzer."

## Self-Assessment Questions

1. Instruments that measure process quality might measure physical conditions/ physical property/process composition (choose all that apply).
2. What are the gains of analytical sensors over classical process sensors?
3. Prepare a table with pros and cons for in-line, on-line, at-line, and off-line. What factors affect the choice of technique?

## 6.3 Spectroscopic in-line monitoring

This chapter provides an introduction into the vast area of spectroscopy, especially the area of micro spectrometers, which has gained high attention recently and enables highly integrated optofluidic systems. Spectroscopy has found its way into a multiple of applications, which are still growing as the technology, especially the miniaturization and robustness, on spectrometers progresses.

### 6.3.1 Industry relevance

Figure 6.20 illustrates important spectroscopic methods, which are used in industry and that have great potential for new applications in the future. This figure is a part of the technology roadmap for process analytical sensors in the chemical and pharmaceutical industries [337, 347]. The "benefit" and "time line" axes are used to classify the most frequently used methods. The symbol size distinguishes the expected frequency of its use. It is thus possible to gain an impression of the market-related importance of a method. Methods with the greatest benefit and the shortest development horizon can be found in the diagram's top left quadrant.

The techniques of widest applicability and greatest interest in this industry sector are near infrared (NIR) spectroscopy, mid-infrared (MIR) spectroscopy, and light-scattering techniques. Ultraviolet-visible (UV-VIS) spectroscopy, fluorescence, and Raman spectroscopy have also been used successfully for a range of industrial applications involving on-line, in-line, and non-invasive analysis [289]. Of particular interest is the compositional analysis of small quantities of liquids and low-concentration ranges [336], or the integration of new methods [338] in optofluidic systems. An insight into the applied spectroscopy world can also be gained through, for example, the Society for

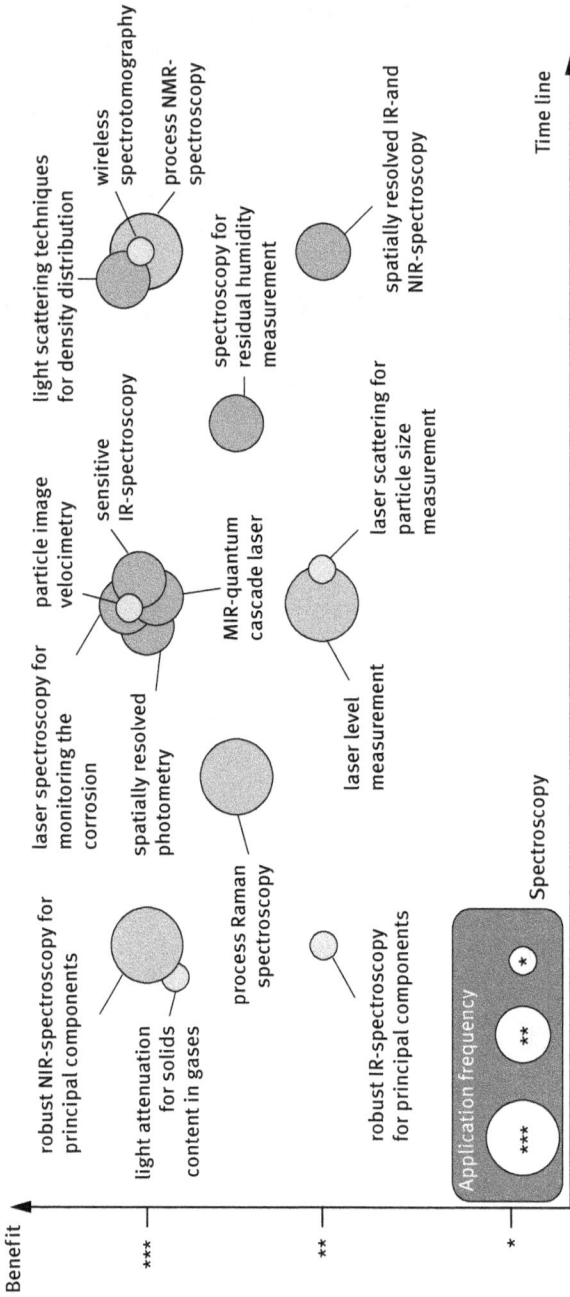

**Figure 6.20:** Industrially relevant spectroscopic measurement techniques [337].

Applied Spectroscopy (s-a-s.org) or the conference from the Optical Society of America, Applied Industrial Optics: Spectroscopy, Imaging, and Metrology (AIO), which has devoted itself to utilize photonics and optical technology to solve real world problems (osa.org).

### 6.3.2 The interaction of radiation with matter

On the molecular level, the interaction of radiation with matter results in a change in the state of the molecule. This is normally discussed in terms of energy levels and transitions between them since the molecules are quantum objects. Depending on which part of the molecular subsystem is involved, the energy levels are divided into electronic, vibrational, and rotational levels. Figure 6.21 shows the transition states for ultraviolet (UV), visible (Vis), and IR (infrared) regions. The UV region covers the wavelengths from 200 to 400 nm and the visible region extends from 400 to 800 nm. The near-infrared (NIR) region covers from 0.8 to 2.5 μm.

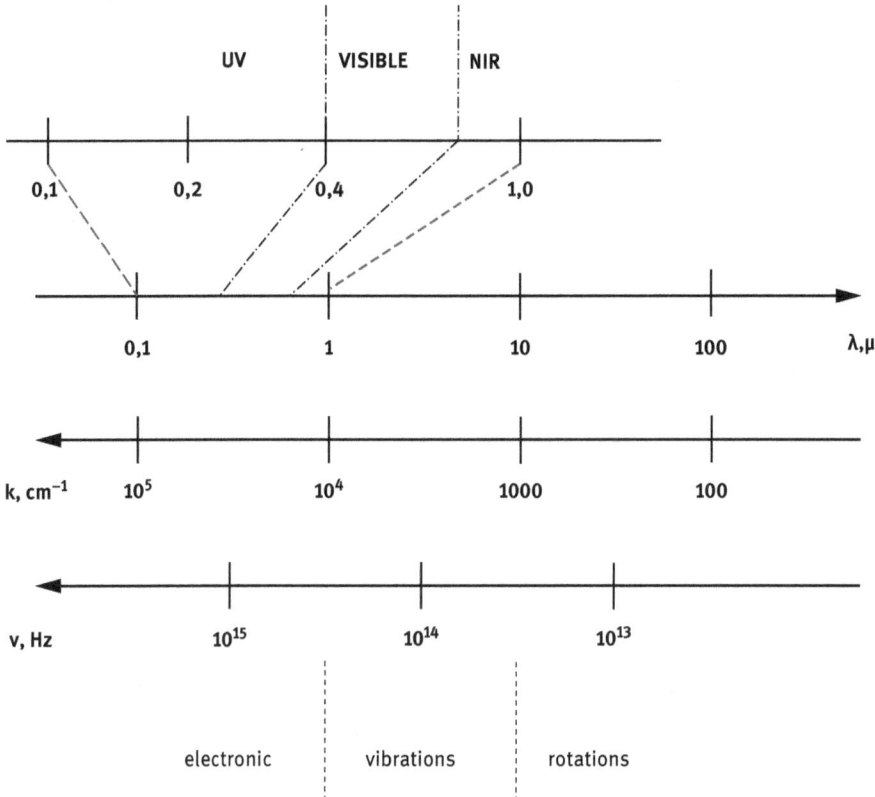

**Figure 6.21:** Relations between different scales used in spectroscopy.

Each of the electronic levels is related to the energy of the electron subsystem and corresponds to the transition from one electronic level, or orbital, to another. The vibrational levels are linked with the vibrational motions of the molecules and the transitions between them have almost no effect with regard to the electronic level. The rotational levels arise from rotations of molecules as a whole. Therefore, only electronic transitions fall into the optical range. Fluorescence spectroscopy, for example, uses ultraviolet light to excite the electrons in the molecules of certain compounds and emits light typically in the visible range. The vibrational and rotational transitions correspond to the infrared and far infrared wavelength regions. The energies of the transitions decrease in the following order: electronic > vibrational > rotational. The UV-VIS-NIR is a relatively small part of the electromagnetic radiation spectrum. Here, as the wavelength becomes shorter, the radiation can penetrate further into matter. Raman spectroscopy typically uses lasers in the near UV-VIS-NIR range, where the energy of the laser photons becomes shifted up or down.

The shift in energy gives information about the vibrational modes in the system. Infrared spectroscopy yields similar but complementary information. The wavelength $\lambda$, frequency $v$, and photon energy $E$ are equivalent measures in spectroscopy. One practical difficulty in this is that numerous units are used to characterize the same parameter – the transition energy. The most commonly used values are documented in Table 6.3, where the respective values for red light are also noted.

**Table 6.3:** Units commonly used in spectroscopy and the respective values of the parameters of red light.

|  |  | Units | Values for red light |
|---|---|---|---|
| Wavelength | $\lambda$ | nm | 650 |
| Wave number | $k = \lambda^{-1}$ | $cm^{-1}$ | 15,385 |
| Frequency | $v = c/\lambda$ | Hz | $4.61 \times 10^{14}$ |
| Photon energy | $E = hv$ | J | $3.06 \times 10^{-19}$ |
| Electron energy | $U = E/e$ | eV | 1.90 |

Constants and conversion factors: Speed of light in vacuum $c = 2.99 \times 10^8$ m/s; Planck constant, $h = 6.62 \times 10^{-34}$ Js; 1 eV $= 1.602 \times 10^{-19}$ J

The wavelength in nanometers is probably the most used value for current spectroscopy applications. The wave number is the typical measure taken in IR spectroscopy and here the historical unit is the reciprocal centimeter ($cm^{-1}$). As the wave number is directly proportional to the frequency and the energy, it is convenient to use the energy or frequency dependency. The electron energy is related to the photon energy in an electric field and is useful when a photon shows a transition of an electron from one energy level to another. High-energy electronic transitions are observed in the shorter wavelength UV-Vis regions. Moderate-energy (vibrational and rotational) transitions are observed in the longer wavelength IR region.

Molecules are able to absorb, emit, or scatter radiation at various frequencies depending on the energy of their molecular transitions. In the following section, various measuring principles based on these phenomena are explained.

### 6.3.3 UV-Vis spectroscopy

The absorption of UV-Vis light generally occurs following the excitation of bonding electrons. As a result, the wavelengths of absorption bands can be correlated with the types of bonds in the species under study. Molecular absorption spectroscopy is therefore valuable for identifying functional groups in a molecule. There are three main groups of compounds that can be characterized by UV-Vis spectroscopy: transition metal ions, charge transfer complexes, and organic compounds with a high degree of conjugation. Organic compounds can absorb electromagnetic radiation because they contain valence electrons that can be excited to higher energy levels. In a UV-Vis spectrum, the absorption bands typically consist of many individual narrow lines, which are affected by numerous "line broadening" mechanisms. The wavelengths, at which absorption lines appear and their relative intensities, depend primarily on the electronic and molecular structure of the sample. Figure 6.22 shows the structures and absorption spectra of benzene, anthracene, and perylene dissolved in cyclohexane. Benzene shows a specific fine structure that is due to the numerous vibrational levels associated with the excited electronic states of this aromatic molecule. In the condensed phase or in solution, molecules have less freedom to rotate than when they are in the

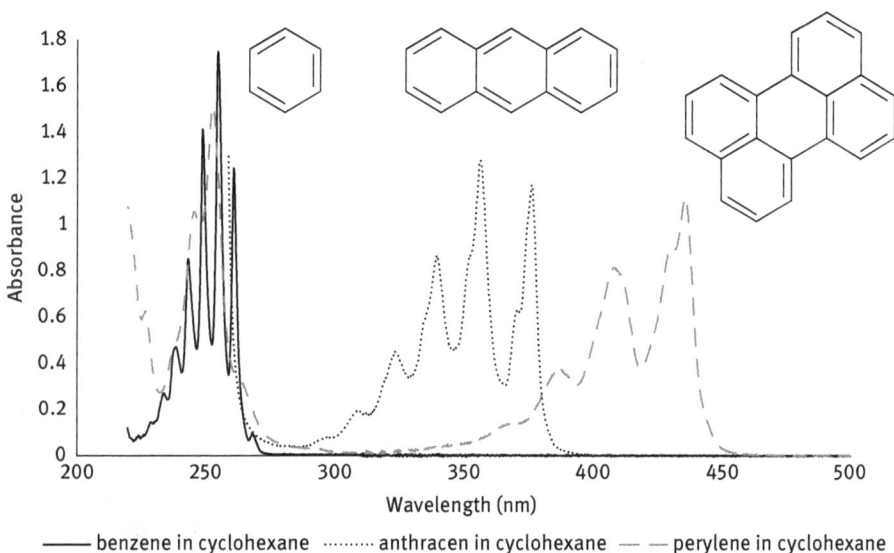

**Figure 6.22:** Ultraviolet absorption spectra for benzene. Anthracen and perylene in cyclohexane.

gaseous state. As such, the lines due to the differences in rotational energy level are not resolved in the spectra. For the anthracene and perylene, it is observed that peak wavelengths tend to be shifted toward longer wavelengths as the conjugated system gets larger.

Excitation of $\pi$ electrons in double bonds and in aromatic systems requires less energy. Highly conjugated double bonds can result in absorption in the UV and even in the visible range. Table 6.4 lists typical organic chromophores and the approximate wavelengths at which they absorb.

**Table 6.4:** Some chromophores and the corresponding transitions

| Chromophores with examples | Transition | $\lambda_{max}$ (nm) |
|---|---|---|
| $H_2O$ | $\sigma \rightarrow \sigma*$ | 183 |
| C–C, C–H, $H_3C$–$CH_3$ | $\sigma \rightarrow \sigma*$ | 170, 171 |
| C–X, $CH_3OH$, $CH_3$–$NH_2$, $CH_3$–I | $n \rightarrow \sigma*$ | 180–260, 187, 215, 258 |
| C≡C; $H_2$–C=$CH_2$ | $\pi \rightarrow \pi*$ | 160–190, 162 |
| $H_2C$=CH–CH=$CH_2$ | $\pi \rightarrow \pi*$ | 217 |
| C=O, H–CH=O | $n \rightarrow \pi*, \pi \rightarrow \pi*$ | 270, 170–200, 270, 185 |
| $H_2C$=CH–CH=O | $n \rightarrow \pi*, \pi \rightarrow \pi*$ | 328, 208 |
| C=N | $n \rightarrow \pi*$ | 190, 300 |
| N=N | $n \rightarrow \pi*$ | 340 |
| C=S | $n \rightarrow \pi*$ | 500 |
| N=O | $n \rightarrow \pi*$ | 630–700 |

The data for both position ($\lambda_{max}$) and peak intensity ($\varepsilon_{max}$) are influenced by structural details and solvent effects. The value is a rough estimation for identification. Moreover, conjugation between chromophores tends to cause shifts in absorption maxima to longer wavelengths; see Table 6.5.

**Table 6.5:** The effect of conjugation

| n | H(CH=CH)$_n$–H | | $CH_3$-(CH=CH)$_n$–$CH_3$ | |
|---|---|---|---|---|
| | $\lambda max$ (nm) | log $\varepsilon$ | $\lambda max$ (nm) | log $\varepsilon$ |
| 2 | 217 | 4.3 | 223 | 4.4 |
| 3 | 268 | 4.7 | 275 | 4.5 |
| 4 | 304 | 4.8 | 310 | 4.9 |
| 5 | 334 | 5.1 | 341 | 5.1 |

Finally, vibrational effects broaden absorption peaks. The molar absorptivities for $n - \pi*$ transitions are normally low and usually range from 10 to 100 L $\cdot$ mol$^{-1}$ $\cdot$ cm$^{-1}$. On the other hand, values for $\pi$ ->$\pi*$ transitions generally range between 1,000 and 15,000 L $\cdot$ mol$^{-1}$ $\cdot$ cm$^{-1}$. Absorption bands in aqueous solution tend to be very broad and overlapped, resulting from multiple electronic-vibrational-rotational transitions. Nevertheless, useful structural information can be derived from a spectrum. A spectrum can be characterized in a univariate manner, in terms of the position of the maximum absorbance or in a multivariate manner, in terms of the shapes and slopes of peaks. Electronic transitions occur for sigma (σ) electrons as they do in simple alkenes; however, these are only accessible to a limited extent because excitation requires high energy UV radiation to occur.

### 6.3.3.1 Quantification

One of the primary tasks in optical spectroscopy is the determination of the wavelength dependencies which is measured in the absorption spectrum. More important, however, are the applications to the quantitative determination of compounds containing absorbing groups. As shown in Figure 6.23, not only absorption occurs in the sample cell but also reflection at interfaces and attenuation of a beam may occur as a result of scattering by large molecules and sometimes from absorption by container walls. To compensate these effects, the power of the beam transmitted by the analyte solution is compared with a reference without the analyte.

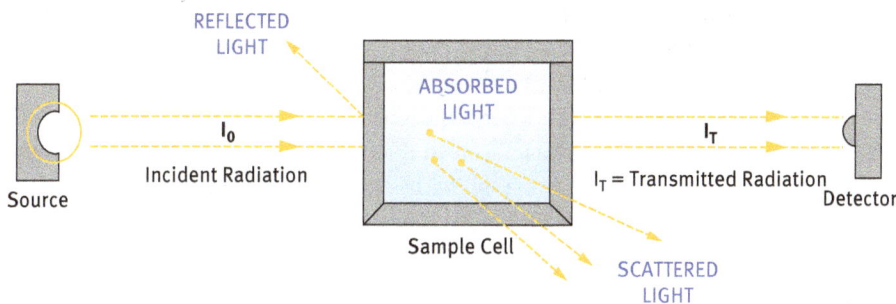

**Figure 6.23:** Radiation of initial intensity is attenuated to transmitted intensity by a solution containing c moles per liter of absorbing solution with a path length of d centimeters.

The transmission or transmittance is given by the following equation:

$$T = \frac{I_1}{I_0}$$

where $I_0$ is the intensity of the light before passing through the sample and $I_1$ is the intensity of the transmitted light which has passed through the sample. Instead of the transmittance, the terms light absorbance $A_\lambda$ or the optical density $OD_\lambda$ is used, which are given by:

$$A_\lambda = OD_\lambda = -log_{10}\frac{I_1}{I_0}$$

The higher the absorption of a sample, the higher the optical density and the lower the transmission. The transmission is dimensionless and usually expressed in percentage. For example, an optical density of 2.0 corresponds to a transmission of 1%. The optical density is dependent on the wavelength $\lambda$ used for measuring and the path length $d$ of the light path travelling in the medium. In order to compare and share measurement data, it has become a standard to use cuvettes with a path length of 10 mm.

The amount of absorbed light depends on the concentration $c$ of the substance, the thickness of the liquid layer $d$, and a specific molar absorption or extinction coefficient $\alpha_\lambda$ at a defined wavelength $\lambda$. If the extinction or absorption coefficient and the path length are known, then the concentration of a sample can be calculated using the famous Lambert-Beer equation:

$$A_\lambda = -log_{10}\frac{I_1}{I_0} = \alpha_\lambda \cdot c \cdot d$$

For quantification of components in complex mixture, the linearity of the Beers law is presupposed. However, under certain circumstances the relationship breaks down and gives a non-linear relationship that is based on three reasons.

First, *real deviations*. These are fundamental deviations due to the limitation of the law itself, e.g. deviations in absorptivity coefficients at high concentrations (>0.01 M) due to electrostatic interactions between molecules in close proximity. Second, *chemical deviations*. These are deviations observed due to specific chemical species of the sample which is being analyzed, e.g. shifts in chemical equilibria as a function of concentration. Third, *instrument deviations*. These are deviations which occur when the absorbance measurements are made, e.g. polychromatic illumination of the sample [298].

## Self-Assessment Questions

1. What is the wavelength of a photon that has three times as much energy as that of a photon whose wavelength is 785 nm?
2. A photometer with a linear response to radiation measures 83.4 pA with a reference in the light path. Replacement of the reference with an absorbing solution yielded a response of 39.6 pA. Calculate
   a) the percent transmittance of the sample solution.
   b) the absorbance of the sample solution.
   c) the transmittance to be expected for a solution in which the concentration of the absorber is one-third that of the original sample solution.
   d) the transmittance to be expected for a solution that has twice the concentration of the sample solution.

3. A tryptophan solution with $4.21 \times 10^{-3}$ M showed a transmittance of 0.232 when measured in a 2.00 cm cell. What concentration of tryptophan would be required for the transmittance to be increased by a factor of 3 when a 1.00 cm cell was used?

4. A solution containing 5.23 ppm $KMnO_4$ had a transmittance of 0.1095 in a 1.00 cm cell at 522 nm. Calculate the molar absorptivity of $KMnO_4$ at 520 nm.

5. Which molecule absorbs at a longer wavelength, A or B?

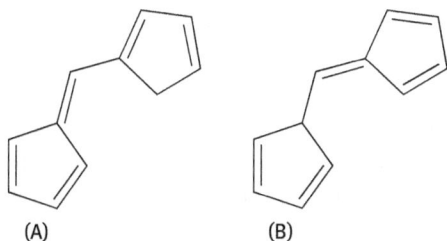

(A)                    (B)

6. The black lines in the energy diagram represent $\pi$ bonding orbitals, and the dashed ones $\pi$ antibonding orbitals. (a) Why are there more of each of these as you go from ethene to buta-1,3-diene to hexa-1,3,5-triene? (b) Explain the relationship between this diagrams and the absorption maximum of these substances.

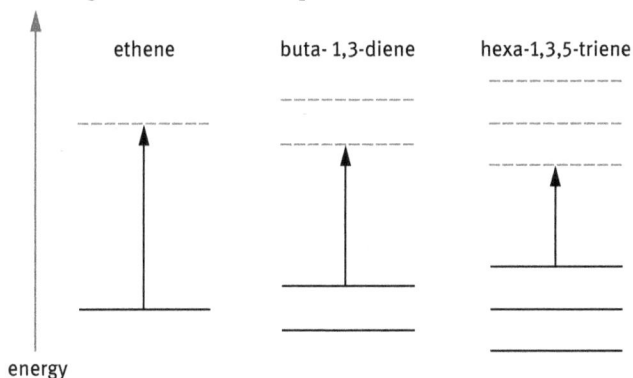

## 6.3.3.2 Color measurement

As explained in the previous section, visible light can be absorbed, reflected, or transmitted as it falls onto an object. Color is perceived in relation to the used light source. Therefore, when quantitatively measuring color, one always has to take the light source into account. Without light, there exists no color. Examples of natural light sources are sunlight, atmosphere, stars, and galaxies. Examples of artificial light sources are black body radiator, incandescent lamp, and arc lamp. These light sources can be categorized as temperature radiators. Examples of artificial lumines- cence radiators are gas discharge lamps, fluorescent lamps, light emitting diodes (LEDs), lasers, etc.

The human color perception is characterized mainly by three properties:

- Saturation: It defines how color is perceived, quality wise. Saturation defines a range from pure color (100%) to gray (0%) at a constant lightness level. A pure color is fully saturated.
- Hue: Hue defines pure color in terms of the fundamental colors, such as red, green, blue, yellow.
- Brightness: Brightness is an attribute of our perception, which is mainly influenced by a color's lightness. The difference between the colorimetric brightness and the photometric brightness is that the colorimetric brightness is compared to a color standard or standard solution.

A color value can only be compared with another one in very limited circumstances as different color spectra can produce the same color perception for the human eye. Colorimetry combines human color perception with fundamental physics in order to provide a mean to distinguish and measure color. The following paragraph deals with the color stimulus function $\Phi(\lambda)$, which is dependent on the wavelength. The color stimulus is equal to the product of the spectral distribution of the light source $S(\lambda)$ and the transmission T or reflection R spectrum $\Gamma \in \{T, R\}$ of the illuminated layer.

$$\Phi(\lambda_i) = S(\lambda_i)\Gamma(\lambda_i)$$

The color stimulus function $\Phi(\lambda)$ is valid only for colors which are non-self-luminous. The stimulus function is equal to the emission spectrum in the case of light sources. T and R provide for each spectrum the part of the intensity, which is either absorbed or transmitted. This is why the calculations for transmission and reflection are basically identical. In order to derive the color stimulus function, it is required to introduce the color stimulus specification and the laws of Grassmann.

One of the first researchers who dealt with color physics was Hermann Grassmann (1853). He postulated and proved that by using monochromatic light sources of red, green, and blue wavelengths, one could additively mix colors and generate basically all visible colors. The three color values are combined in the color stimulus specification C that is given by:

$$C = \begin{pmatrix} R \\ G \\ B \end{pmatrix}$$

The primaries **R**, **G**, and **B** are defined to be the unit vectors in the three dimensional space. The direction of vector C provides information on the chromaticity. According to the first Grassman law, each color stimulus C can be generated as a result of linear combinations of the three distribution functions. This is called the color equation, which is given by:

$$C = R\boldsymbol{R} + G\boldsymbol{G} + B\boldsymbol{B}$$

For identification of a color stimulus specification, three independent primaries, which cannot be matched by additive mixture of the other stimuli, are necessary and sufficient. The second Grassman law states that the result of the additive color mixture is influenced only by the color stimulus specifications and not by the spectral compositions. For instance, if there are two stimulus

$$C_1 = R_1\boldsymbol{R} + G_1\boldsymbol{G} + B_1\boldsymbol{B}$$

and

$$C_2 = R_2\boldsymbol{R} + G_2\boldsymbol{G} + B_2\boldsymbol{B},$$

the color addition is equal to:

$$C = C_1 + C_2 = (R_1 + R_2)\boldsymbol{R} + (G_1 + G_2)\boldsymbol{G} + (B_1 + B_2)\boldsymbol{B}$$

A new color stimulus is produced from which it is not possible to discern whether or not it consists of several individual color values.

The third law of additive color mixing implies that if one or more components of the mixture are gradually changed, then the resulting color values also change gradually, that is, as opposed to changing in a discontinuous fashion. Therefore, if we assign the color stimulus specification to a point in three-dimensional space, all additively mixed color stimuli generate an continuous and coherent color solid, the color space.

The laws of additive color mixing emphasize the fact that any given color is characterized by the three primaries Red, Green, and Blue with the following conditions:
- the primaries have to be specific, reproducible, and calculable from the color measurements.
- the primaries must have a positive sign.
- the primaries must refer to the photopic vision, meaning under well-lit conditions.
- the numerical color differences must relate to the visually perceived color difference.

The color system introduced largely the listed first three criteria from the CIE (Commission Internationale de l´Eclairage/International Commission on Illumination) in 1931 meets. The basis of the CIE system is a result of the so-called tristimulus color matching experiment, which was conducted by Wright and Guild.

In order to measure color, discrete wavelengths are required. Therefore, the color stimulus produced at a given wavelength $\lambda_i$ is termed as $C_i$. Each color stimulus $\Phi_i$ is represented by a corresponding color stimulus specification $C_i$. If there are simultaneously i = 1, 2, ..., N stimuli $C_i$, the total color stimulus specification is given by the additive mixture:

$$C = \sum_{i=1}^{N} C_i$$

Applying the additive law of Grassmann, with $N \geq 2$, we obtain:

$$C = \sum_{i=1}^{N} C_i = \left( \sum_{i=1}^{N} R_i \right) R + \left( \sum_{i=1}^{N} G_i \right) G + \left( \sum_{i=1}^{N} B_i \right) B$$

The color values $R_i$, $G_i$, and $B_i$ are added up leading to the vector components that can be interpreted as the red, green, and blue component of a color tint, which are given by:

$$R = \sum_{i=1}^{N} \Phi_i \bar{r}_i \Delta \lambda$$

$$G = \sum_{i=1}^{N} \Phi_i \bar{g}_i \Delta \lambda$$

$$B = \sum_{i=1}^{N} \Phi_i \bar{b}_i \Delta \lambda$$

where $\bar{r}(\lambda_i)$, $\bar{g}(\lambda_i)$, $\bar{b}(\lambda_i)$ are the distribution functions; also known as color matching functions (CMFs). These equations can also take the integral form if the wavelength interval is reduced. Then the CMF becomes continuous. For normal computation and implementations in software, the discrete form is generally preferred due to the ease of use. Therefore using the equations above, the R, G, and B values of any given color sample can be calculated. The values of $\Phi_i$ are generally calculated by using a spectrometer. This leads to the determination of R, G, and B values that are representatives of any color if the accompanying reflection or transmission is known from measurement.

For colorimetric applications, the CIE introduced the color system with the so-called virtual primaries X, Y, Z, which are derived from the previously described real primaries R, G, B. The virtual primaries X, Y, Z are chosen in such a way that the following criteria are fulfilled:

- X, Y, Z are by definition independent of each another; therefore, they cannot be produced by mixing (the same requirement as for the real primaries R, G, B).
- The new CMFs $\bar{x}(\lambda)$, $\bar{y}(\lambda)$, $\bar{z}(\lambda)$ always take values $\geq 0$ and follow from the previous CMFs by an appropriate transformation.
- The virtual primaries X, Y, Z are adjusted in such a way that the corresponding tristimulus values X, Y, Z are equally valued for the chromaticity of the equi-energy spectrum: X = Y = Z.

   –   The two virtual primaries X and Z are chosen in such a way that only the color value Y is proportional to the lightness of a color.

The new color quantities X, Y, Z are called standard color values. The new CMFs $\bar{x}(\lambda)$, $\bar{y}(\lambda)$, $\bar{z}(\lambda)$ are named standard color-matching functions (SCMFs) and are given with the previous CMFs $\bar{r}(\lambda)$, $\bar{g}(\lambda)$, $\bar{b}(\lambda)$ by the empirical matrix equation:

$$
\begin{pmatrix} \bar{x}(\lambda) \\ \bar{y}(\lambda) \\ \bar{z}(\lambda) \end{pmatrix} = \begin{pmatrix} 2.768892 & 1.751748 & 1.130160 \\ 1 & 4.590700 & 0.060100 \\ 0 & 0.056508 & 5.594292 \end{pmatrix} \cdot \begin{pmatrix} \bar{r}(\lambda) \\ \bar{g}(\lambda) \\ \bar{b}(\lambda) \end{pmatrix}
$$

These color-matching functions are also referred to as CIE 1931 two-degree observer standard color matching functions.

As time progressed, it was shown through applied research that the visual area for sample assessments, which was thought to be $2^0$, was slightly larger. It was in fact 10 degree since this incorporated all the rods and cones that extended beyond the highest density area covered in 2 degree known as fovea. This was why the CIE introduced the so-called 10 degree observer which is also known as CIE 1964. The SCMFs of the 10 degree observer are distinguished from those of the 2 degree observer by the subscript 10, and are given by the following matrix:

$$
\begin{pmatrix} \bar{x}_{10}(\lambda) \\ \bar{y}_{10}(\lambda) \\ \bar{z}_{10}(\lambda) \end{pmatrix} = \begin{pmatrix} 0.341080 & 0.189145 & 0.387529 \\ 0.139058 & 0.837460 & 0.073160 \\ 0 & 0.039553 & 1.026200 \end{pmatrix} \cdot \begin{pmatrix} \bar{r}_{10}(\lambda) \\ \bar{g}_{10}(\lambda) \\ \bar{b}_{10}(\lambda) \end{pmatrix}
$$

The CIE recommended the 2 degree observer for visual angles between 1 and 4 degrees, and the 10 degree observer for visual angles larger than 4 degree observer. The 4 degree limit was arbitrarily established although there exists no discontinuity of color perception at this angle.

The CIE standard color values X, Y, Z (here for the 2 degree observer) are given by:

$$
X = \sum_{i=1}^{N} \Phi_i \bar{x}_i \Delta\lambda
$$

$$
Y = \sum_{i=1}^{N} \Phi_i \bar{y}_i \Delta\lambda
$$

$$
Z = \sum_{i=1}^{N} \Phi_i \bar{z}_i \Delta\lambda
$$

The tristimulus values are applicable for opaque, translucent, and transparent non-self luminous colors. The formulas for three tristimulus values change if the object under observation is opaque or translucent. For the opaque samples spectral reflectance is used, whereas for the translucent samples spectral transmittance is used. It is important to note that while specifying color values, the observer function and the underlying illuminant function are stated. Both have a drastic effect on the values calculated. Generally, the illuminant function and the observer function are tabulated in lookup tables. It is the spectral reflectance and the transmittance that has to be calculated for the values to be computed. Table 6.6 shows the standardized tristimulus values of different illuminants under different observer functions. The value of $Y_n$, which is the lightness of the illuminant, is kept fixed at 100. The mentioned illuminants are described in the paragraph below.

**Table 6.6:** CIE standardized color values of different illuminants.

| Illuminant | 2° Standard Observer | 2° Standard Observer | 10° Standard Observer | 10° Standard Observer |
|---|---|---|---|---|
| | $X_n$ | $Z_n$ | $X_n$ | $Z_n$ |
| D 65 | 95.047 | 108.883 | 94.811 | 107.304 |
| A | 109.850 | 35.585 | 111.144 | 35.200 |
| C | 98.074 | 118.232 | 97.285 | 116.145 |
| FL 2 | 99.186 | 67.393 | 103.279 | 69.027 |
| FL 7 | 95.041 | 108.747 | 95.792 | 107.686 |
| FL 11 | 100.962 | 64.350 | 103.863 | 65.607 |

A standard illuminant is a theoretical light source, which has got a published spectral power distribution curve for reference. Standard illuminants become useful while comparing images or colors rendered under different lighting conditions. While discussing various illuminants, it's important to note the following terms:
- CIE illuminant: This is a theoretical light source having a tabulated spectral power distribution curve.
- CIE standard illuminant: Illuminant A and illuminant D65 have been standardized by the CIE.
- CIE source: It refers to the physical source that is in accordance with the CIE illuminant.
- CIE simulator: This is a technical source that simulates the CIE illuminant within an approximation.

The CIE has specified certain illuminants that are being used in the color industry and are characterized on the basis of color temperature and the emission spectra. Some of the most common illuminants and there salient features are:
a) Illuminant A: This illuminant is defined to represent tungsten filament lighting that is used domestically. The spectral power distribution that it holds is that of the Planckian radiator at a temperature of 2,856 K. CIE color illuminant A has to be used

in all of the colorimetric applications involving the use of incandescent lighting. The emission at a particular wavelength and temperature follows the Planck's law of radiation.

b) Illuminants B and C: This category corresponds to the daylight simulators. They are obtained by using liquid filters in front of the illuminant A. B represents sunlight during afternoon with a correlated color temperature of 4,874 K while C represents average daylight at a color temperature of 6,774 K. These illuminants are poor approximations of any common light source and are generally not preferred in favor of illuminant D. The liquid color filters have a high absorbance at the red end of the spectrum, thereby increasing the color temperature to daylight levels. The liquid filters are composed of distilled water, copper sulphate, mannite, pyridine, sulfuric acid, cobalt, and ammonium sulphate added in specific proportions to different separated glass sheets.

c) Illuminant D: This standard is created to represent natural daylight. They are difficult to produce physically but have an easy mathematical form. The chromaticity coordinates (chromaticity coordinates are derived from the XYZ tristimulus values and are explained in the following paragraph) are given by the equation:

$$y = 2.870x - 3x^2 - 0.275$$

The characteristic vector analysis also gives the equation of the spectral power distribution (SPD) in terms of vectors.

$$S(\lambda) = S_0(\lambda) + M_1 S_1(\lambda) + M_2 S_2(\lambda)$$

So, as seen from the above equation, the SPD of the daylight sample can be calculated by the linear combination of three fixed SPDs. The first vector $S_0$ is the mean of all SPD samples. The second vector $S_1$ corresponds to a yellow blue variation that is due to the absence of clouds or direct sunlight. The third vector $S_2$ corresponds to the pink green variation present in the atmosphere due to water vapors and haze. So, one needs the coefficients M1 and M2 of the characteristic vectors in order to generate the spectrum. Expressing the chromaticities x and y as:

$$x = \frac{X_0 + M_1 X_1 + M_2 X_2}{S_0 + M_1 S_1 + M_2 S_2}$$

and

$$y = \frac{Y_0 + M_1 Y_1 + M_2 Y_2}{S_0 + M_1 S_1 + M_2 S_2}$$

Making use of the known tristimulus values, $M_1$ and $M_2$ can be expressed as:

$$M_1 = \frac{-1.3515 - 1.7703x + 5.9114y}{0.0241 + 0.2562x - 0.7341y}$$

$$M_2 = \frac{0.0300 - 31.4424x + 30.0717y}{0.0241 + 0.2562x - 0.7341y}$$

Initially, the trichromaticities of the illuminants has to be calculated which are given as a function of color temperature. After obtaining them, the values of $M_1$ and $M_2$ can be calculated using the equations above. Then, finally, the SPD values can be obtained.

There are different variants of the D series primarily depending upon the color temperature used. D65 refers to the color temperature of 6,504 K, which is the noon light.

d) Illuminant E: This illuminant has been defined keeping a constant SPD in the visible spectrum. It is mostly used as a theoretical reference, and gives equal weights to all the wavelengths thus producing an even color. It can also be approximated by D series having a color temperature of 5,455 K.

e) Illuminant F: The CIE illuminant F2 represents cool white fluorescent source. In the SPD graph, the spikes are due to the principle mercury emission lines. The CIE F7 represents the broadband daylight fluorescent source. Next is F11 that represents narrow band white fluorescent source.

In Summary, light sources need to be taken into account when deriving color values in a specific color system. So far, the CIE XYZ color space has been explained. In the following paragraph, the chromaticity coordinates will be introduced which also have their origin with the CIE in 1931.

The chromaticity coordinates xyz are calculated from the tristimulus values:

$$x = \frac{X}{X+Y+Z}$$

$$y = \frac{Y}{X+Y+Z}$$

$$z = \frac{Z}{X+Y+Z}$$

Each coordinate is given by the ratio of that particular tristimulus value divided by the sum of all the values. Hence, the sum of the three coordinates give a value of 1 that means, x+y+z=1. Knowing two coordinates gives the third coordinate. Likewise, the coordinates for the 10° standard observer can also be computed. It is also impossible to calculate the three tristimulus values just knowing the two coordinates x and y.

To calculate three values, one of the value has to be known in advance. The chromaticities x and y combines both hue and the chroma of the color.

From these coordinates, it is possible to have a graphical representation of this system. The value Y is the measure of lightness of the color and the primaries X and Z contain no lightness component. The quantities x and y can be set as coordinates of a plane triangle. The figure obtained is the well-known chromaticity diagram, which can be regarded as the work horse of colorimetry. Non-self-luminous colors are situated inside the sole shaped area. The bordering areas are spectrum locus and the straight line of purples. The x and y coordinates of the visible spectral colors are represented by the spectrum locus. The purple boundary joins the end points of the spectrum locus. The colors on the line of purples do not exist naturally, but can be produced spectrally by the additive color mixing. In the case of equi-energy spectrum, the chromaticity coordinates take the same values x=y=z=1/3 and is also known as achromatic point. Every point, from the achromatic point to the spectrum locus, has the same hue. The saturation, however, decreases while coming to the center of the locus. Increasing lightness also decreases the spectrum locus and the line of purple. The chromaticity diagram is also standardized for certain illuminant and standard observer.

Another important color space is the so-called CIE 1976 Lab color space. This color space is based on the Munsell color space which used value V, chroma C, and hue H to represent colors in 3-D space. The CIE Lab tends to present data equidistantly on the color space as perceived in reality. This is the advantage of using this space as compared to the CIE XYZ color space described earlier. However, the syntax is different as it uses $L$ for the lightness or black to white contribution, $a$ for the red to green spectrum, and $b$ for yellow to blue contribution. These color assignments are based on the color opponent theory. The values $L$, $a$ and $b$ are calculated from the XYZ tristimulus values and the illuminant used. The color space is based on three orthogonal axes, where L stretches from 0 to 100. The L, a, and b values are calculated as follows:

$$L = 116 \cdot f\left(\frac{Y}{Y_n}\right) - 16$$

$$a = 500 \cdot \left[f\left(\frac{X}{X_n}\right) - f\left(\frac{Y}{Y_n}\right)\right]$$

$$b = 200 \cdot \left[f\left(\frac{Y}{Y_n}\right) - f\left(\frac{Z}{Z_n}\right)\right]$$

$$C_{ab} = \sqrt{a^2 + b^2}$$

$$h_{ab} = \frac{180}{\pi \cdot arctan\left(\frac{b}{a}\right)}$$

This color space graphically characterizes the properties of the industrial color standards such as lightness, chroma, hue, and hue angle. This can be shown by using the equation for chroma ($C$), whose length serves as a measure for the saturation degree. Moreover, the angle starting from positive $a$ axis serves as the hue measure. There is an interesting feature that the CIE Lab color space offers. Along the axis of $L$, $a$, $b$, opponent colors change; the color difference between a sample color with values $L_s$, $a_s$, $b_s$ and a reference color with values $L_r$, $a_r$, $b_r$ is given by the following equation:

$$\Delta L = L_S - L_r$$

$$\Delta a = a_S - a_r$$

$$\Delta b = b_S - b_r$$

If $\Delta L > 0$, then the color is lighter; if $\Delta L < 0$ then the color is darker. If $\Delta a > 0$, then the color is redder; if $\Delta a < 0$, then the color is greener. If $\Delta b > 0$, then the sample is yellower; if $< 0$, then the color is bluer.

The entire color difference can be summed over the following equation. The formula is a Pythagorean that describes the length of a vector between the two color points, the sample and the reference.

$$\Delta E_{ab} = \sqrt{\Delta L^2 + \Delta a^2 + \Delta b^2}$$

In summary, the described color spaces provide the basis for comparing colors and industrial standards are derived from these values. Color measurement is one of the main measurements of Vis spectroscopy. For further reading, see [320].

### 6.3.4 Flow chemistry applications with UV/Vis spectroscopy

In flow chemistry, reagents are continuously pumped into a microreactor system where they are mixed and subsequently react in microchannels under thermal control. Such systems have the advantage that mixing can be achieved in seconds because of the larger surface area to volume ratio, which results in faster reactions. Based on this, exothermic reactions can be controlled very effectively. Due to small amounts of hazardous substances, microreactors contribute to improved overall safety [303]. For process control, parameters such as flow, temperature, and pressure can all be rapidly altered and varied. One reaction can follow another separated by solvent, each cleaning out the previous reaction, which enables quick reaction optimization. Figure 6.24 shows a microreactor unit with seven single microreactor plates. The reactants can be pre-heated before the fluid enters the mixing system. After mixing, the reaction takes place in the resident time channel. To stop the reaction, a quenching substance can be pumped into a specific port. The main

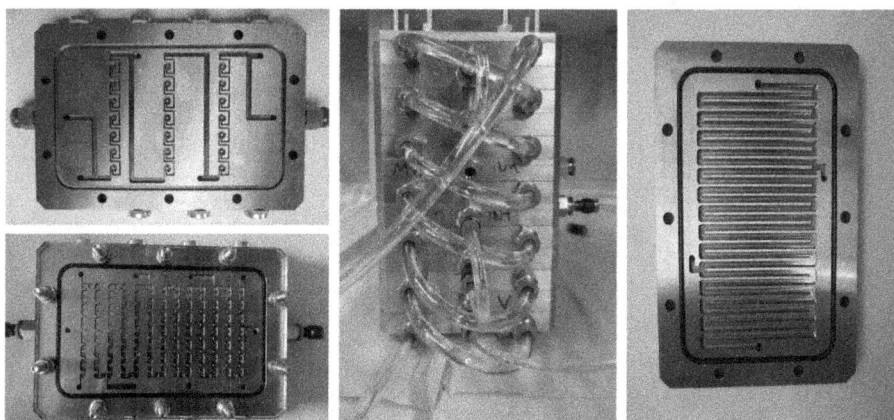

**Figure 6.24:** Flow chemistry reactor with different intervention options.

advantage of this technology is its scalability. The central image shows the complete module that has been developed to allow operation over a wide range of flow rates. During process development, the operation of only a few single reactors with avoidance of parallelization is favored. This allows total control of the chemical systems because they are often meta-stable. For example, they can form deposits that are stable over time and such phenomena create unpredictable pressure drops over the reactor. One single plate is suitable to enable process development under small flow rates, when reagent availability is still limited or a visual control is necessary.

For each size step up, the plate area is doubled which means that the heat exchange area and reactor volume is doubled. For viscous system and low temperature applications, the pressure drop becomes very important at high flow rates. In addition, the mixing zone within the reactor causes the largest pressure drop. Therefore, an enlargement of the mixer elements at higher flow rates drastically reduces the overall pressure drop. To handle higher output ratios, higher flow low rates, larger reactors or numbering up, enable the production of kilogram quantities using a microreactor.

With fixed fiber optics at critical points, in-line spectroscopy can be carried out to analyze the reaction progress and the spectra can be used to control the flow and temperature through the microreactor unit. For analysis of different points along the reactor, a line-scanning (push-broom) spectrometer can be used [296]. In such systems, the full spectral information for each line is acquired. The system is a direct sight imaging spectrograph that can be quickly combined with a broad range of matrix detectors to form a spectral imaging device. An imaging lens, or multiple optical fibres, like those in Figure 6.25, focuses the light onto an entrance slit; the light is collimated, dispersed by a grating, and focused on a matrix detector. One dimension of the matrix detector corresponds to a spatial line image, and the other dimension corresponds to the spectrum. In this case, the parallel acquisition of

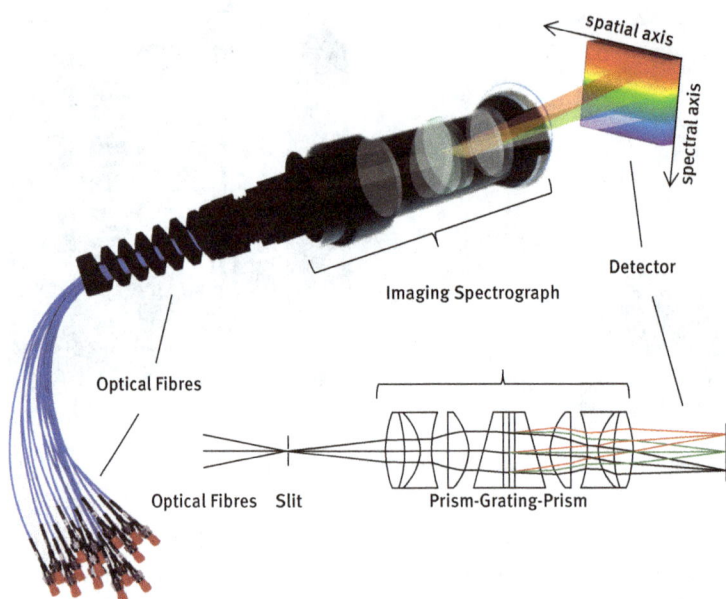

**Figure 6.25:** Multipoint spectrometer: Push-broom imaging system with multiple optical fibers.

thousands of spectra can be used to compare differences between samples, rather than the more common implementation of exploring spatial heterogeneity within a single sample. The reactor technology, together with the spectroscopy, can contribute to the understanding of the process and ensures a measure of speed and flexibility in the process development and keeps the chemical-technical aspects to a minimum. Reactor technology, together with spectroscopy, can contribute to the understanding of processes and ensures speed and flexibility in process development and also keeps the chemical-technical aspects to a minimum.

After process development, the use of parallel microreactor modules provides access to kilogram or even ton of quantities, as does the use of advanced microreaction technology as depicted in Figure 6.26.

Numbering up can be performed in different ways. One possibility is the external numbering up with parallel connection of similar reactor elements. Second possibility is the internal numbering up with a parallelization of microchannel within one microreactor element. Depending on the microchannel's geometry and material and physical properties of the solvents, the contact between two phases can create different flow patterns. The main advantage of the external strategy is that the plates can be easily replaced with no effect on the other parallel units if one of the units fail during production. Industrial production has confirmed that microreactor technology is applicable for performing liquid phase reactions. They can be used for different single and multiple phase reactions and even for explosive and flammable reactions or those that use highly toxic components. If the reactor is made of (quartz) glass like

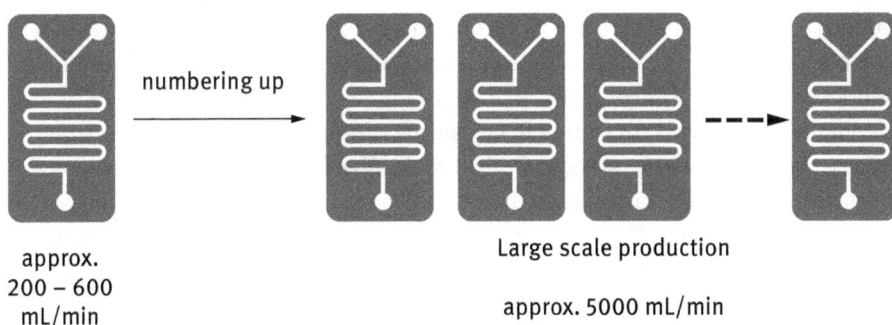

numbering up

approx.
200 – 600
mL/min

Large scale production

approx. 5000 mL/min

**Figure 6.26:** Numbering up with microreactors and microfluidic cells.

in laboratory application, flexible points for spectroscopy can be selected and even spectroscopy in transmission mode can be realized. The transmitted light from different local points of the rector can be measured via the multiple fibers on the entrance slit of the spectrograph. To demonstrate the functionality of this design, the reaction of 2,6-Dichlorophenolindophenol (DCIP) with ascorbic acid in respect to Trefz et al. [349] is used. DCIP is used as a redox dye and shows a maximum absorption at 614 nm when it is oxidized, while the reduced form is colorless. DCIP can be used as an indicator for ascorbic acid because of its ability to be reduced. If ascorbic acid is present in the system, the blue DCIP turns pink in acidic conditions and then it is

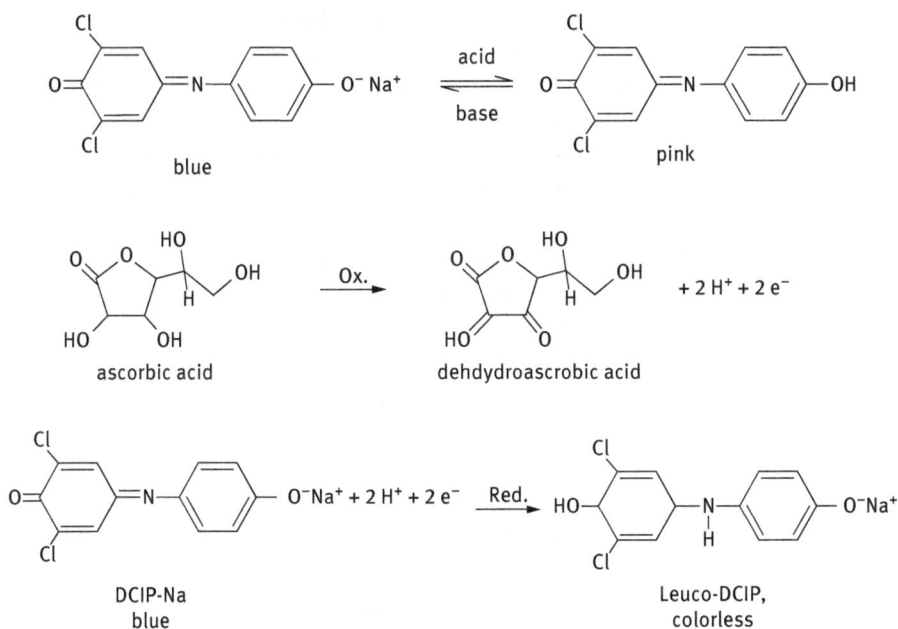

**Figure 6.27:** Reaction scheme showing the reaction of vitamin C with the DCIP.

reduced to a colorless compound by ascorbic acid. This reaction is a redox reaction in which ascorbic acid is oxidized to dehydroascorbic acid, and DCIP is reduced to the colorless compound Leuco-DCIP.

Figure 6.28 shows the on-line measuring set-up for characterization of the process along the mixing system. Optical multi-fibers are used for spectroscopy in the visible wavelength range. Each fiber bundle is placed over a specific point along the reactor. The decrease in absorbance at 614 nm at different flow rates is monitored. In the entrance area of the microreactor, mixing of the reagents superimposes the reaction. Only when complete mixing is achieved, the reaction process can be tracked. This effect decreases with increasing flow rate so that full mixing is achieved at the beginning of the microchannel at higher flow rates. In general, no loss of performance is observed as long as the same energy dissipation rate in the mixing zone is maintained. Thus, the mixing zone becomes the only scaled factor that is considered in this technology and it must be properly designed and adapted. The reaction rate is strongly pH-dependent.

### 6.3.5 Fluorescence spectroscopy

Fluorescence spectroscopy is a sensitive method to detect the smallest changes in the molecular environment. The method describes the emission of light at a defined wavelength by a substance that has absorbed light. It occurs when an orbital electron of a molecule, atom, or nanostructure relaxes to its ground state by emitting a photon of light after being excited to a higher quantum state by some type of energy. The characteristics of fluorescence spectra can be rationalized by the Jabolonski diagram that describes the electronic states of a molecule and the transitions between them. An exited molecule can return to its ground state by a combination of several steps. Fluorescence and phosphorescence involve the emission of a photon or radiation. On the other hand, there are several deactivation steps, which are radiationless processes. These are vibrational relaxation, internal conversion, external conversion, and intersystem crossing. Relaxation of a higher quantum state can also occur through interaction with a second molecule through fluorescence quenching.

Figure 6.29 shows two-dimensional excitation-emission spectra of milk with 3.5% fat, which mainly shows the riboflavin fluorescence between 450 and 600 nm. For measuring the fluorescence spectra, the excitation wavelength is held constant and the emission intensity is measured as function of the emission wavelength. In the excitation spectra, the emission is measured at one wavelength while excitation wavelengths are scanned. The excitation spectrum closely resembles an absorption spectrum since the emission intensity is usually proportional to the absorbance of the molecule. The rate of photon absorption is very rapid and takes place in about $10^{-15}$ s while fluorescence emission typically occurs in $10^{-5}$ s. The total fluorescence spectrum is either a

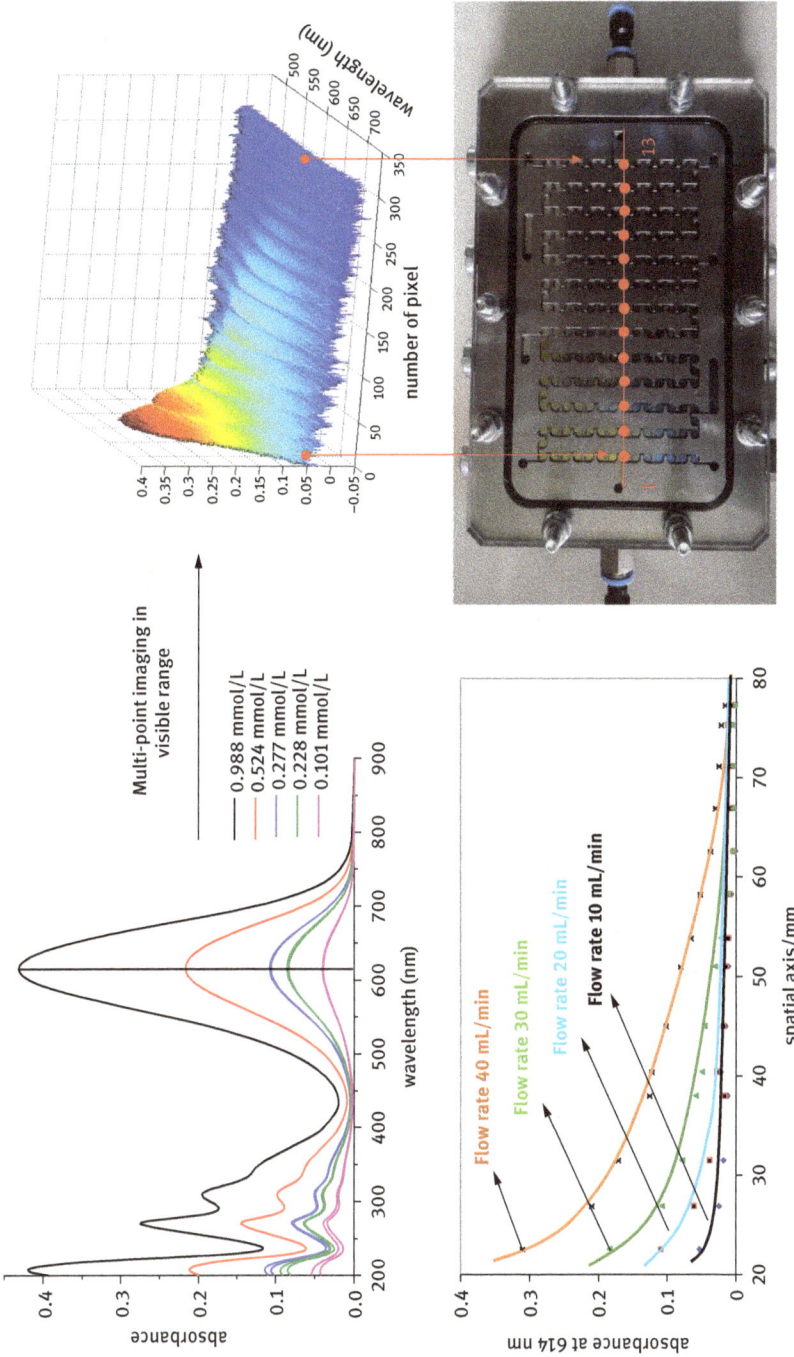

**Figure 6.28:** Reaction monitoring with multipoint Push-broom imaging device of the reduction of DCIP by ascorbic acid. Absorbance of DCIP @ 614 nm. Reaction conditions: c(DCIP) = 1 mmol/L; c(Ascorbic acid) = 1 mmol/L; pH = 7; T = 35 °C; Flow rates: 10 mL/min–40 mL/min; Spatial resolution: 64 µm; Integration time: 300 msec.

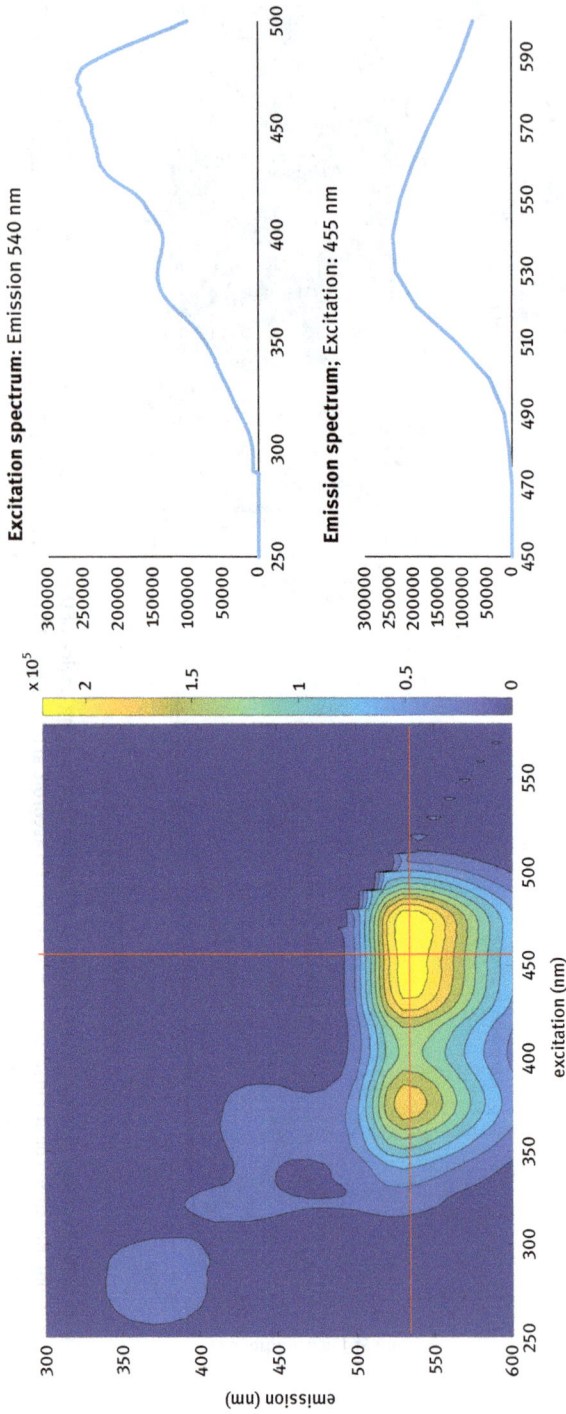

**Figure 6.29:** Excitation-emission matrix of milk with 3.5% fat.

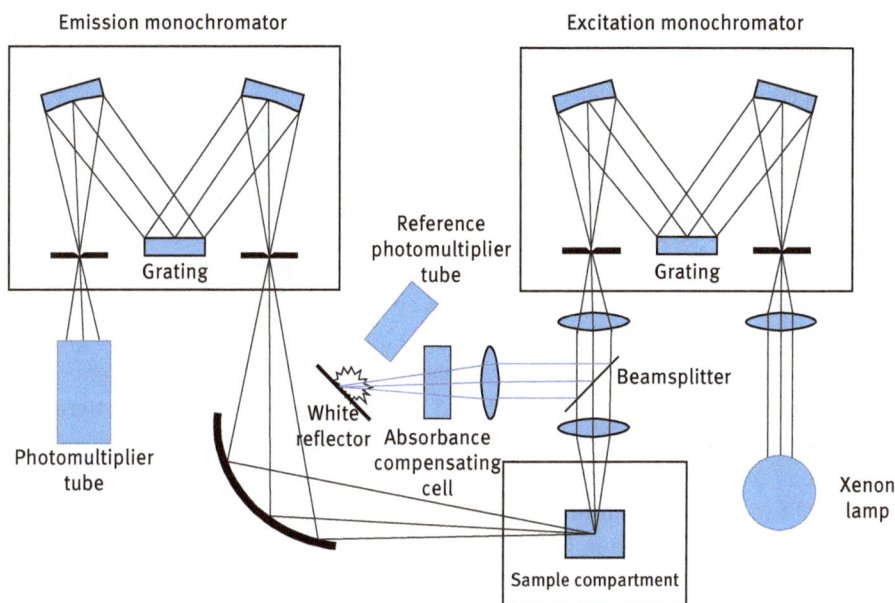

**Emission monochromator**

**Excitation monochromator**

Reference
photomultiplier
tube

Grating

Grating

Beamsplitter

White
reflector   Absorbance
compensating
cell

Photomultiplier
tube

Xenon
lamp

Sample compartment

**Figure 6.30:** Set-up of a desktop fluorescence spectrometer.

three-dimensional representation or a contour plot. Such data are called an excitation-emission matrix.

### 6.3.5.1 Instruments for measuring fluorescence

The components of instruments are very similar to those in UV/Vis spectroscopy. For spectroscopy, xenon lamps with a continuum band from about 250 nm to 1,300 nm are often used. Typical radiation sources for fluorophotometer are low-pressure mercury lamps with sharp peaks for excitation at 254, 302, 313, 546 and 578, 691, 773 nm. The sample beam passes through the excitation wavelength selector to the sample. The emission wavelength selector isolates the emitted fluorescence before striking the transducer. For wavelength selection, the most spectrometers are equipped with at least one and often two grating monochromator. In fluorophotometer, interference and absorption filter have been used. Emission signals are typically quite low in intensity. Therefore, transducers like photomultiplier or Peltier cooled charge-coupled-devices (CCDs) to improve signal-to-noise ratios are used.

### 6.3.5.2 On-line excitation-emission spectroscopy

Two-dimensional (2D) fluorescence spectroscopy has been used to control some food products [341] or biotechnology applications [305]. However, until now, because of the high cost of fluorescence spectrometers and the difficulty in interpreting fluorescence spectra, this often remains an off-line method of quality control. However, on-line excitation-emission spectroscopy offers the possibility of the real-time differentiation of mixtures, often without the use of a chromatographic separation method. Normally, fluorescence spectrometers detect the emission spectrum at a fixed excitation wavelength, or different excitation wavelengths are switched in succession with a filter wheel. The acquisition of a complete 2D spectrum takes several minutes at a spectral resolution of approximately 20 nm [324]. The spectrometer design in Figure 6.31 can record a complete 2D fluorescence spectrum with a spectral resolution of 2 nm in 100 ms [327].

The basic element is a grating monochromator. An external xenon lamp serves as an excitation light source, which is coupled by means of mirrors in the monochromator. A grating splits the light spectrally in the horizontal plane. Part of the spectral region is coupled in a mirror system and imaged on a flow-through fluorescence cell. In the beam path, a filter changer is installed to filter out the 2nd order light. Here, the length of a flow cell is irradiated with an excitation beam that has been dispersed along the xy plane by a monochromator. A pushbroom imager was adapted at 90° to the exit slit of the monochromator. The transducer is a CCD device that image the dispersed excitation radiation in the xy-plane and the dispersed radiation from the emission monochromator in the yz-plane. The calibration of the CCD camera for 2D-fluorescence spectroscopy can be carried out with a line spectrum of a Cd/Hg-EDL in n-decane (quartz envelope; 0.7 kPa argon atmosphere; 5 mg of cadmium; Hg vapors), see Figures 6.31 and 6.32.

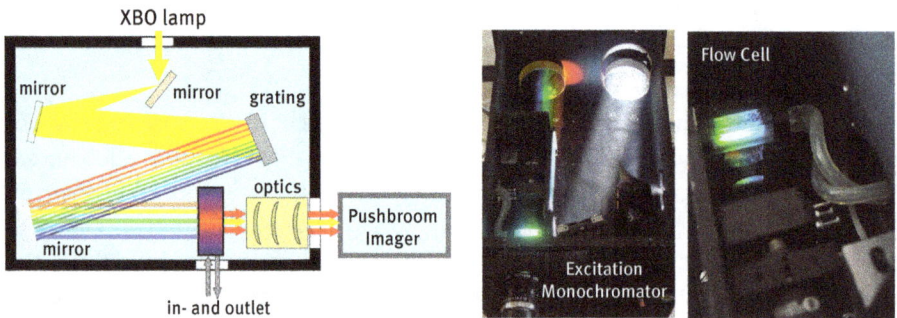

**Figure 6.31:** Set-up of a 2D-fluorescence spectrometer with a pushbroom imager for simultaneous measurements of excitation and emission spectra.

These lines can be used to calibrate the emission wavelength. With test samples, the excitation wavelengths can be correlated to the monochromator wavelengths. After the calibration, the excitation wavelengths can be assigned to the pixels on the x axis and the emission spectra to the pixels on the y axis. Typical integration times are appropriate in the millisecond to second range. The integration times could be further shortened by more sensitive cameras. For calibration, image acquisition and evaluation, software for the system was developed based on LabView. In addition to its online capabilities, the software also has the option of online evaluation

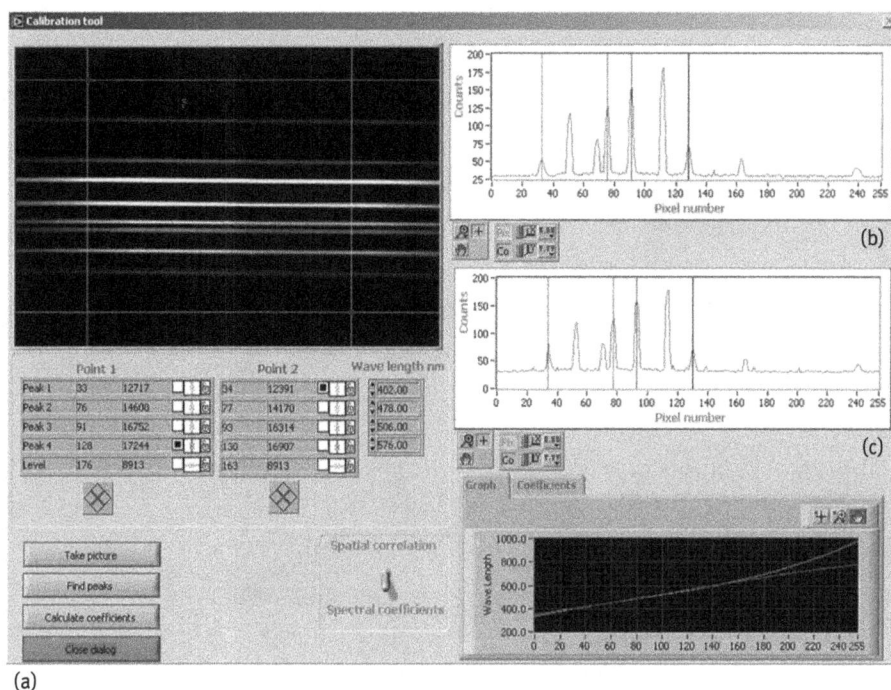

**Figure 6.32:** Calibration tool with (A) image section of the CCD; (B) line spectra of the "calibration cube," and (C) calculation of the correlation coefficients.

with a multivariate model. The change in the main components over time can thus also be displayed online. Figure 6.33 shows three test substances used to adjust the excitation wavelength of the monochromator.

For this purpose, the flow cell is exchanged for a fluorescence standard (Starna® Brand) and the desired excitation range is selected via the monochromator control. A second example shows the yeast cultivation of saccharomyces cerevisiae by the 2D-fluorescence spectrometer; see Figure 6.34. For this purpose, the intrinsic fluorescent component NADH at 460 nm emission is detected.

On-line control in industry by means of fluorescence is often only the observation of changes in fluorescence intensity at one single wavelength. But, chemical or biological reactions can cause the quenching of the fluorescence, peak shifts, or the emergence of new species and their fluorescence. Therefore, the observation of intensities is not sufficient and requires further spectral parameters. In addition, an understanding of the influencing factors is necessary.

### 6.3.5.3 Variables affecting fluorescence
The intensity of fluorescence emission F is proportional to the power of the exictation beam that is absorbed by the system.

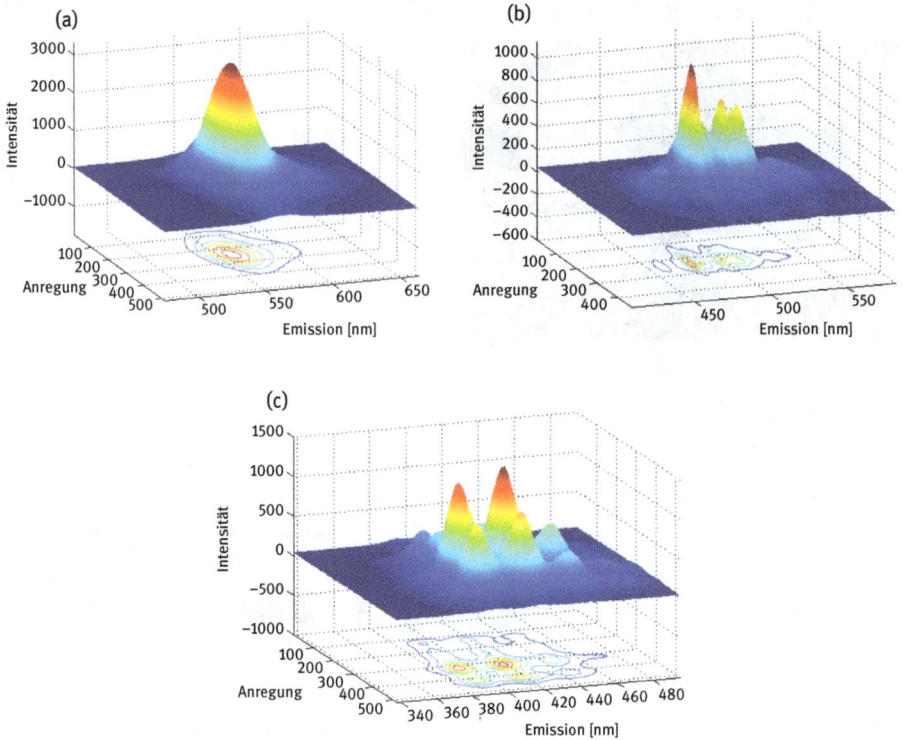

**Figure 6.33:** 2D-Fluoresence spectra of (a) Rhodamine ($2 \times 10^{-7}$ mol/L) integration time: 0.25 s; (b) Ovalene ($2 \times 10^{-7}$ mol/L) integration time: 1 s; and (c) Anthracene ($2 \times 10^{-3}$ mol/L) integration time: 1 s.

$$F = 2.303\phi K' \varepsilon b c I_o$$

$I_0$ is the intensity of the beam incident in the fluid, b is path length where the light travels through, $\varepsilon$ is the molary aborbivity of the fluorescing molecule, and c is the concentration. A plot of the fluorescence intensity against concentration should be linear at low concentations. As concentrations increase and absorbance exceed over 0.05, the linearity is lost due to "primary absorption," (Figure 6.35). Additional negative effect for linearity at high concenration is the "secondary absorption" that occurs when the wavelength of emission overlaps an absorption band. The fluorescence intensity is then decreased as the emission pass the solution and is reabsorbed by other molecules in solution.

### Fluorescence and structure

Both molecular structure and chemical environment influence wheater a substance will or will not show flouresence. These factors also determine the intensity of the emission when fluorescence occur. The quantum efficiency $\Phi$ is an important

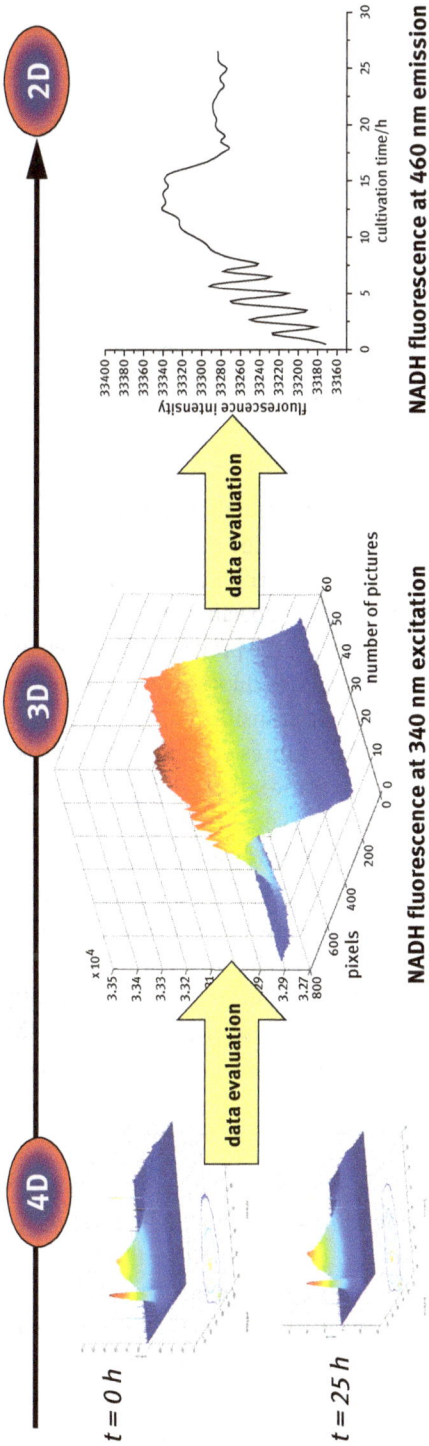

**Figure 6.34:** Saccharomyces cerevisiae batch cultivation.

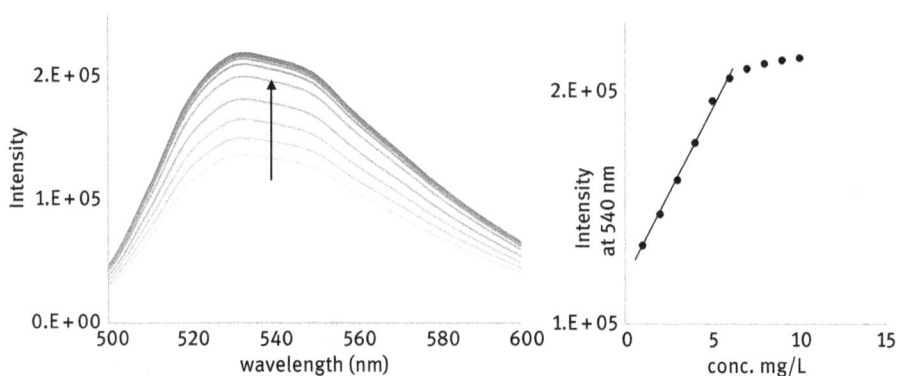

**Figure 6.35:** Changes in emission spectra with increasing concentration of the fluorophore. The inset shows the concentration profile at emission maximum 540 nm.

parameter in fluorescence and is defined as the ratio of the number of molecules that fluorescence to the total number of excited molecules. Molecules do not fluoresce very well when the efficiencies go to zero and vice versa. Highly fluorescent molecules as fluorescein have high quantum efficiency. More specifically, the quantum efficiency is determined by the relative rate constants k for the processes by which the lowest are exited singlet state is deactivated. High intense fluorescence is found in compounds with aromatic functional groups with $\pi-\pi^*$ transitions.

Structures with aliphatic and alicyclic carbonyl structures or conjugated double-bond structures may exhibit fluorescence but most of these structures does not show a fluorescence effect. In addition, simple heterocyclic molecules like furan, pyrrole, or pyridine do not exhibit fluorescence. In comparison, fused aromatic ring structures with nitrogen heterocyclics involve $n-\pi^*$ transitions that convert to the triplet state and prevent fluorescence. However, the fusion of such structures results in higher molar absorptivity and the lifetime is shorter than in pure structures like quinolone, isoquinoline, or indole. An additional effect to increase the fluorescence is the rigidity of the molecule structure. So, the molecules fuorene and biphenyl show under the conditions a 1:0.2 ratio regarding the fluorescence intensity. A non-rigid molecule can undergo vibrations in which energy is lost.

### Examples

Quinoline-based fluorescence sensors are widely used to measure intensity changes of fluorescence emission signals or shifts in fluorescence wavelength. The main mechanisms for signal transduction are photo-induced electron transfer (PET), intermolecular charge transfer (ICT), and fluorescence resonance energy transfer (FRET). In the PET mechanisms, atomic spacers are often used to connect a fluorescence group to a receptor containing a high-energy non-bonding electron pair, such as sulfur or nitrogen, which can transfer an electron to the excited

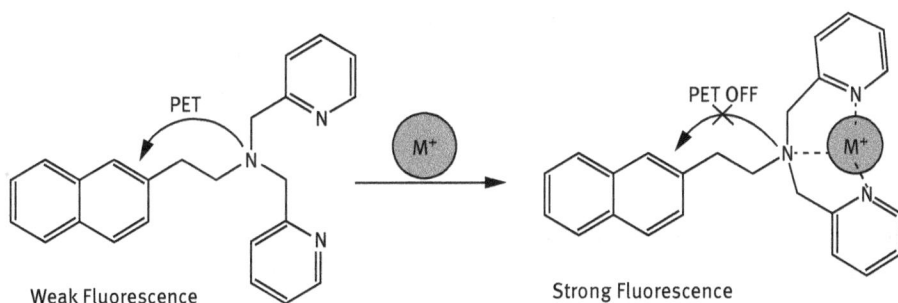

**Figure 6.36:** The intensity of fluorescence will increase after combination of ions, according to PET mechanism [35].

fluorescence group and result in fluorescence quenching. In such cases, the electron pair is coordinated with a metal ion, the electron transfer will be prevented, and fluorescence is switched on [326].

Ou, Shengui, et al. [333] introduced a new strategy for a selective and sensitive optical sensor based on fluorescence analyzing of $Hg^{2+}$ in natural water. Here, the core molecule quninoline group, which is normally insoluble in water, is coupled with a water-soluble D-glucosamine group. The experimental results show a weak emission band at 480 nm following 315 nm excitation. In this example, the quantum yield of the unbounded molecule is 0.005 which suggests the presence of a PET-quenching process of the quinolone group by the lone pair electron of the imine nitrogen atom. After adding the $Hg^{2+}$ ion, the blue emission band shifts to 415 nm with the luminescence intensity enhanced by an order of magnitude. This signal enhancement suggests a photo-induced charge transfer (PCT) signaling mechanism from the deprotonated hydroxyl group to the core quinoline molecule. The deprotonation increases the electronic density on the quinoline ring, which leads to the charge transfer state responsible for the luminescence toward higher energy. Generally, the influence of rigidity can account for the increase in fluorescence of certain organic agents when they are combined with a metal ion. Their lack of rigidity, when not combined, normally causes an enhanced internal conversion rate with decreased fluorescence.

## Fluorescence and pH effects

Most of the aromatic structures with acidic and basic ring substituents are usually pH- dependent. Both the emission intensity and wavelength are different for the protonated and unprotonated form of the molecule. Fluorescein, for example, has an absorption maximum at 494 nm and an emission maximum of 512 nm. The isosbestic point, that is equal absorption for all pH values, is at 460 nm. The pKa is 6.4 and its ionization equilibrium leads to pH-dependent absorption and emission over the range of 4 to 9. In cellular biology, the isothiocyanate derivative of fluorescein is often used to label and track cells in fluorescence microscopy applications or flow cytometry. Figure 6.37 shows pH dependency of fluorescein between 4 and 8. The smaller images below show the whole CCD images of pH-dependent fluorescence measurement. Not only the intensities, but also the wavelengths and peak shapes change drastically.

**Fluorescence and Dynamic Quenching**

The presence of oxygen often reduces the intensity of fluorescence in solution. This effect may be a consequence of the paramagnetic properties of molecular oxygen which promotes intersystem crossing and conversion of excited molecules to the triplet state. The optical oxygen sensors work on the principle of fluorescence quenching. The fluorescence effect is directly correlated to the partial pressure of oxygen. Oxygen molecule quenches efficiently the fluorescence and phosphorescence of certain luminophores. This effect is called "dynamic fluorescence quenching." In the excited state of a fluorophore if a collision with an oxygen molecule occurs, a transfer of energy takes place. The degree of fluorescence quenching relates to the frequency of collisions, and therefore to the concentration, pressure, and temperature of the oxygen containing media. With the advent of optical oxygen sensors, the principle of fluorescence quenching has gained popularity in terms of use. With this method, a chemical formulation layer, which is permeable to oxygen, forms a junction with the process (usually a coating or a patch). The number of oxygen molecules exposed to this coating is directly proportional to the amount of oxygen molecules in the medium. The light from an LED excites the formulation and it emits energy at a fixed wavelength. If the excited chemical complex encounters an oxygen molecule, the excess energy is transferred to the oxygen molecule, decreasing or quenching the fluorescence signal. The degree of quenching correlates to the level of oxygen concentration to the oxygen partial pressure.

Optical oxygen sensors can apply various optical measurement schemes such as reflection, absorption, luminescence (fluorescence and phosphorescence). In recent times, the most popular methodology is fluorescence quenching. Fluorescence in solution obeys the linear Stern–Volmer relationship. Using the Stern–Volmer equation, the sensor electronics calculates either the phase shift between the excited and emitted state or the fluorescence intensity. These values are then converted into a known unit of concentration of oxygen. Fluorescence of a molecule is quenched by specific analytes, e.g., ruthenium complexes are quenched by oxygen. A non-linear relationship between the fluorescence and the quencher (analyte) is created. The signal (fluorescence) to oxygen ratio is not linear, and an optical oxygen sensor is most sensitive at low oxygen concentration, i.e., the sensitivity decreases as oxygen concentration increases.

The Stern–Volmer algorithm requires at least two standards of known oxygen concentration. The first standard must have 0% oxygen concentration and the last standard must have a concentration in the high end of the working concentration

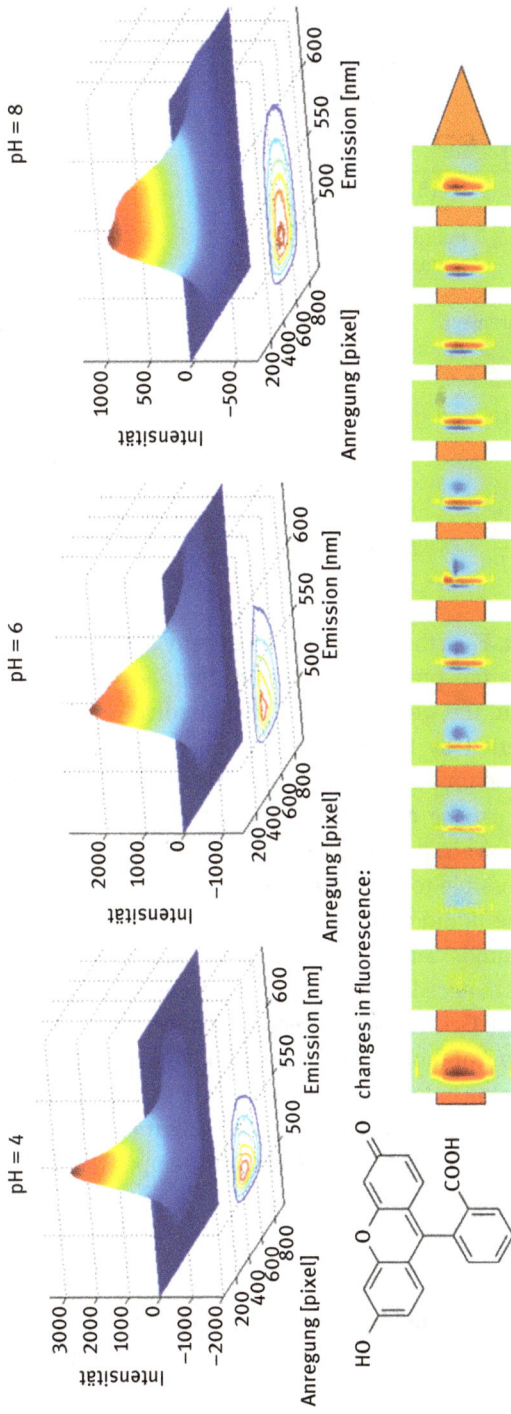

**Figure 6.37:** pH-Dependency of fluorescein.

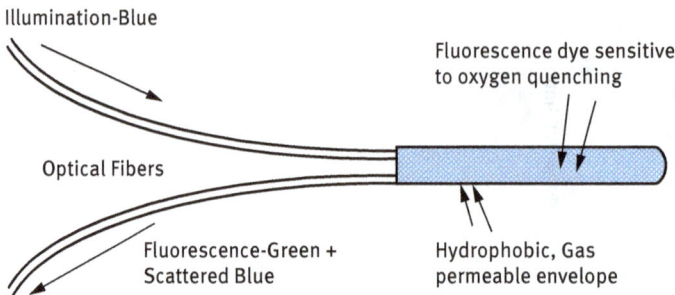

**Figure 6.38:** Optical fiber for oxygen measurement.

range. The fluorescence intensity is related quantitatively to the partial pressure of oxygen:

$$\frac{F_0}{F} = 1 + kp_{O2}$$

where k is the Stern-Volmer constant, $F_0$ is the fluorescence intensity at 0%, F is the fluorescence intensity, $p_{O2}$ is the partial pressure of oxygen. Henry's Law-Liquid Phase: When temperature is constant, the weight of a gas that dissolves in a liquid is proportional to the pressure exerted by the gas on the liquid. Therefore, the pressure of the gas above a solution is proportional to the concentration of the gas in the solution. The concentration can be calculated if the absolute pressure is known. Henry's law is used to convert partial pressure to concentration and Henry's law only proves that it is possible to calibrate the sensor in the gas phase and then make measurements in the liquid phase and vice versa. Optical oxygen sensors measure both in gases and liquids. They have fast response times, which are below 1 s. Frequent calibration is not required. Some examples of companies providing optical oxygen sensor solutions are Ocean Optics, Presens, and Pyroscience. The presented sensors are regarded as the workhorses for future Optofluidic systems and technologies for integration are available today. An early example of a multisensor in thick film technology for water quality control is presented in [325]. It consists of a multisensor designed to measure the following parameters: pH, temperature, dissolved oxygen, conductivity, redox potential, and turbidity.

**Example:**
The fluorescence of a component is quenched by a high concentration of oxygen. The following signals F were obtained as the function of the oxygen concentration. With the help of a spreadsheet in Excel, you can determine the quenching constant $k = 305 \pm 12 \text{ M}^{-1}$.

| F | [$O_2$] | $F_0/F$ |
|---|---------|---------|
| 175.0 | 0.000 | 1.000 |
| 82.5 | 0.005 | 2.121 |
| 53.0 | 0.010 | 3.302 |
| 38.2 | 0.015 | 4.581 |
| 30.0 | 0.020 | 5.833 |
| 23.5 | 0.025 | 7.447 |
| 19.5 | 0.030 | 8.974 |
| 14.1 | 0.040 | 12.411 |
| 10.7 | 0.050 | 16.355 |

| | | | |
|---|---|---|---|
| **Slope** | 304.531 | 0.294 | **Intercept** |
| **SD Slope** | 11.800 | 0.314 | **SD intercpt** |
| **R²** | 0.990 | 0.547 | **Sr** |
| **F** | 666.036 | 7 | **df** |
| **SSregress** | 199.388 | 2.096 | **Ssresid** |

**Spreadsheet Documentation**
Cell D2 = $B$2/B2
Cells B11:C15 = RGP(D2:D10;C2:C10;1;1)

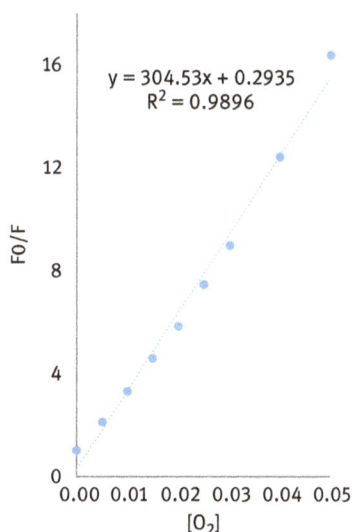

y = 304.53x + 0.2935
R² = 0.9896

**Figure 6.39:** Determine quenching constant k.

## Self-Assessment Questions

1. Many molecules do not fluoresce, even though they have absorbance. Explain why.
2. What kind of information is obtained by scanning the excitation wavelength and fixing the emission wavelength? How is it different from absorbance?
3. Why is fluorescence spectroscopy more sensitive than fluorophotometry?
4. Common fluorimeters have two monochromators, but these are normally not double monochromators. Explain their uses.
5. Many fluorimeters have two detectors, one for emission signal, and another placed at some point in the light path before the sample. Explain its use.
6. The quantum effienciy of flourescence $\phi_f$ can be written as

$$\phi_f = \frac{\tau}{\tau_0}$$

where $\tau$ is the observed lifetime of the exited state in the presence of a quenching agent and $\tau_0$ is the natural lifetime in the absence of a quencher. Derive an equation to show that the F–$\tau$ ratio is independent of collisional quenching and directly related to concentration [307].

7. Quinoline yellow is representative of a large class of quinophthalone pigments. The structure is given below. Predict the part of the molecule that is most likely to behave as the chromophore and fluorescent center.

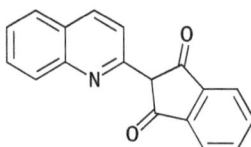

Quinoline yellow, $C_{18}H_{11}NO_2$

8. Fluorescence emission signals
   a) increase with quantum yield
   b) increase with molar absorptivity
   c) increase with sample pathlength
   d) all of the above

### 6.3.6 Infrared spectroscopy

Infrared spectroscopy has gained much popularity for improving the monitoring and control of industrial processes. To ensure that the most appropriate analysis strategy is implemented and robust IR methods are developed, the optical characteristics of the sample, the analyte's sensitivity and selectivity, and the production and control requirements must be considered.

#### 6.3.6.1 Absorption in MIR and NIR wavelength range

Infrared spectroscopy can be divided into the short-near-infrared region (sNIR) from 780–1,000 nm, the near-infrared region (NIR) from 1,000–2,500 nm, and the mid-infrared region MIR from 2,500–25,000 nm (4,000 $cm^{-1}$–4,00 $cm^{-1}$) (Figure 6.44). The region beyond 25,000 nm is regarded as the far-infrared region (FIR). MIR wavelength bands are used to detect absorption regions of fundamental modes of functional groups within molecules. The frequencies range from $10^{13}$–$10^{14}$ Hz. Conventional wavenumber units from 4,000 $cm^{-1}$–400 $cm^{-1}$ are used to cover the spectral range.

The radiation is not energetic enough for electronic transitions but enough for measuring differences in vibrational and rotational states. In comparison, NIR absorption bands are overtones and combinations of C–H, N–H, and OH of fundamental vibrations which normally occur between 3,000 and 1,700 $cm^{-1}$. Such transitions are forbidden by the selection rules of quantum mechanics.

The characteristic of an atomic vibration can be simplified by a harmonic diatomic oscillator model where the vibrating masses $m_1$ and $m_2$ lead to a change of the internuclear distance. Combing Hook's law with Newton's force law, the vibrational frequency corresponds to:

$$\upsilon_{vib} = \frac{1}{2\pi} \sqrt{\frac{k(m_1 + m_2)}{m_1 m_2}}$$

**FREQUENCY** (CM⁻¹)

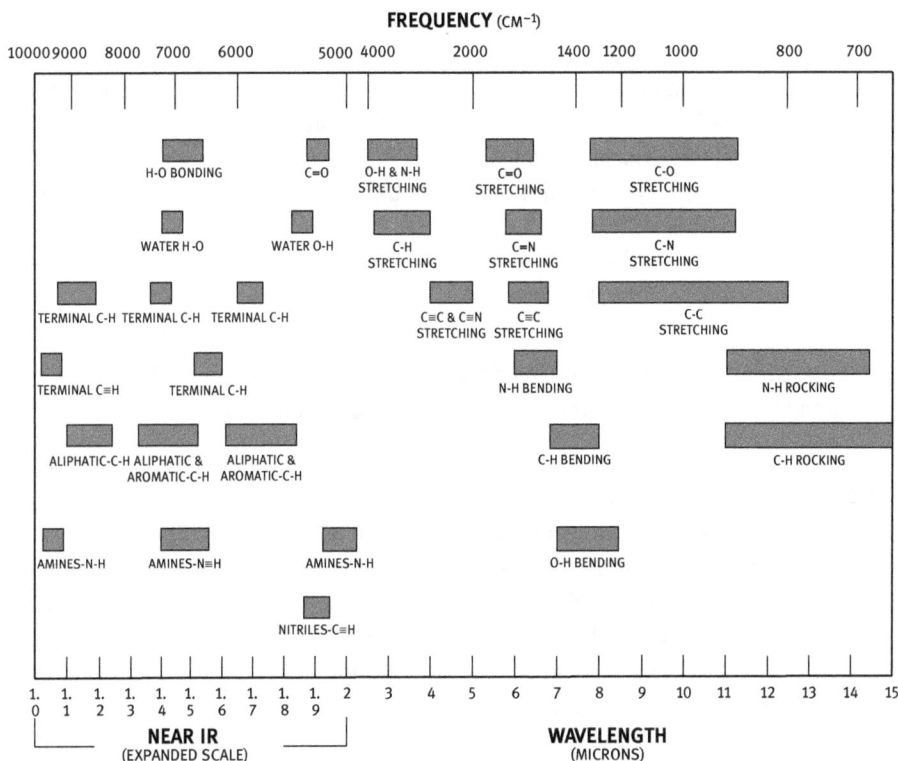

**Figure 6.40:** Functional group frequency chart. Fundamental vibrations absorb in the mid-IR; overtones and combination bonds absorb in the NIR.

Where $v_{vib}$ is the frequency of the absorption band; k is the force constant of the bond; and $m_1$ and $m_2$ are the single masses of the bonded atoms. Quantum mechanical considerations using the Schrödinger equation show that the vibrational energy for the harmonic oscillator has no continuum for vibrational energy levels, but only certain discrete energy levels.

$$E_{vib} = hv\left( v + \frac{1}{2} \right)$$

Where h is Planck's constant and v is the vibrational quantum number which can only have integer values. Combining the two equations gives the vibrational energy

$$E_{vib} = \left( v + \frac{1}{2} \right) \frac{h}{2\pi} \sqrt{\frac{k(m_1 + m_2)}{m_1 m_2}}$$

According to this equation, only transitions between neighboring and equidistant energy levels are allowed in the harmonic oscillator. The distribution of the molecules among the allowed vibrational levels follows the Boltzmann distribution. At room temperature most molecules reside at the ground vibrational level = 0. Hence, the first allowed transition to the energy level = 1, also called the fundamental transition, dominates the spectrum. The further allowed transitions derived from vibrationally excited levels and the associated bands are much weaker. To express the radiation in wavenumbers, we can use $\bar{v} = kv$ and rearrange the equation:

$$\bar{v} = \frac{1}{2\pi c}\sqrt{\frac{k}{\mu}} = 5.3 \cdot 10^{-12}\sqrt{\frac{k}{\mu}}$$

Generally, k has been found for single bonds around $5 \cdot 10^2$ N/m while double and triple bonds have force constants about two and three times higher in values.

### 6.3.6.2 Mid-infrared spectroscopy

To absorb MIR radiation, a molecule must undergo a net change in dipole moment as a consequence of its vibrational motion. In practice, any molecule having an asymmetric charge distribution will absorb IR radiation to some extent. The dipole moment is determined by the magnitude of charge differences and the distance between the two centers of charge. The longitudinal vibration of polar diatomic molecules results in the regular fluctuation of dipole moment. The resultant electric field can interact with the electric field associated with IR radiation. If the frequencies match a natural molecular vibration, then a net transfer of energy occurs, namely absorption, leading to a change in the amplitude of the molecular vibration. Dipole moment changes do not occur on the vibration of homonuclear diatomic molecules. Dipole fluctuations also occur on the rotation of asymmetric molecules and again, the interaction of radiation is possible. For each vibrational state, there are several rotational energy levels. Rotation is limited in liquids and solids which results in a broadening of the vibrational peaks. The harmonic oscillator model cannot be adhered to larger amplitudes of vibration because the repulsive forces between the vibrating atoms and the possibility of dissociation when the vibrating bond is strongly extended is not considered in this model. Accordingly, the allowed energy levels for an anharmonic have to be modified:

$$E_{vib} = hv\left(v + \frac{1}{2}\right) - y\left(v + \frac{1}{2}\right)^2$$

where y is the anharmonicity constant. Anharmonicity means that $\Delta E$ becomes smaller and the energy levels are no longer equidistant and the strict selection rules are expanded to transitions that span more than one energy level. Furthermore, the potential energy curve is represented by an asymmetric Morse function as shown in the Figure 6.41.

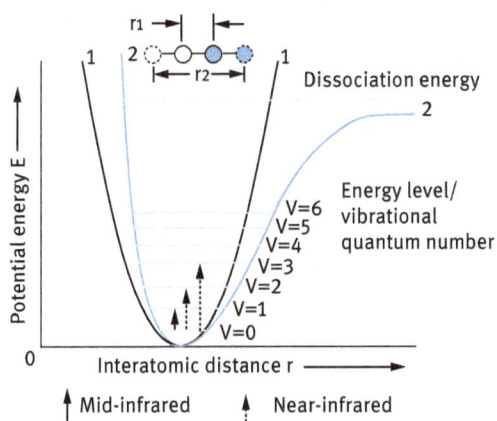

**Figure 6.41:** Potential energy diagrams of the (1) harmonic and (2) anharmonic oscillators.

As a result, weaker transitions called overtones are observed. These transitions correspond to $\Delta v = \pm 2$ or 3. The first overtone vibration is expected at twice the frequency of the fundamental. The simultaneous excitation of two different vibrational modes is possible and the combination of two individual vibrational frequencies occurs. As a result, NIR absorption bands are typically 10 to 1,000 times weaker than the corresponding fundamental bands.

Figure 6.42 shows a typical spectrum obtained with a process FT-IR spectrometer. The identification of an organic compound from a spectrum starts by examining what kind of functional groups are present in the range between 4,000 cm$^{-1}$ and 1,450 cm$^{-1}$. Bond order and the types of atoms joined by the bond are the two most important factors in determining where a chemical bond will absorb. The higher the bond order, the higher the frequency needed to stretch. Triple bonds have higher stretching frequencies than corresponding double bonds, especially bonds to hydrogen have higher stretching frequencies than those to heavier atoms. Normally, the unknown spectra is compared with known spectra of pure components from literature tables [307, 346]. In the fingerprint region (1,450–600 cm$^{-1}$), minor structural differences between molecules result in significant changes in their IR absorption pattern.

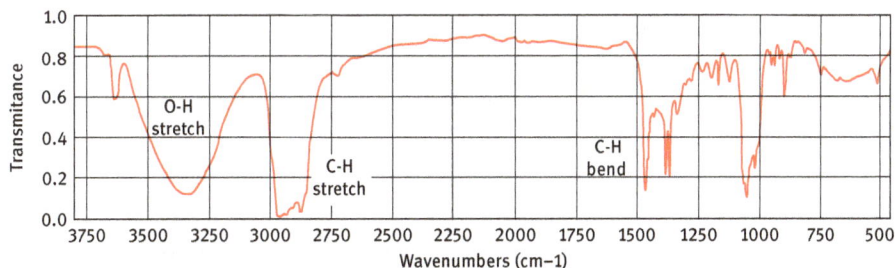

**Figure 6.42:** 4-Methyl-1-pentanol infrared spectrum.

Most of the bending frequencies appear in that region. This region is therefore excellent for the identification of compounds against standard spectra. The spectra contains an absorption band at 2,850–2,970 cm$^{-1}$ which corresponds to a C–H stretching vibration and the peak shape indicates that more than one alkane group is present. The two peaks at about 1,340 cm$^{-1}$ and 1,470 cm$^{-1}$ are also characteristic group frequencies for C–H groups and result from bending vibrations in the molecule. Stretching vibrations occur at higher frequencies than bending vibrations. Generally, bonds more easily bend than stretch or compress. Furthermore, the O–H stretching occurs at about 3,600 cm$^{-1}$, where the typical range of hydrogen-bonded alcohols are between 3,200 and 3,600 cm$^{-1}$.

The uniqueness of an IR spectrum leads to a degree of specificity that is matched or exceeded by relatively few other analytical methods. This specificity has found particular application in the analysis of mixtures of closely related organic compounds. As the IR peak positions are determined by the atomic mass and the bond strength of the functional group when it undergoes a vibration, structural isomers with a unique set of masses and bond strengths can be identified by IR spectroscopy. Figure 6.43 shows spectra of o-xylene, m-xylene, p-xylene, and ethylbenzene with individual peaks in the range of 12 and 15 μm.

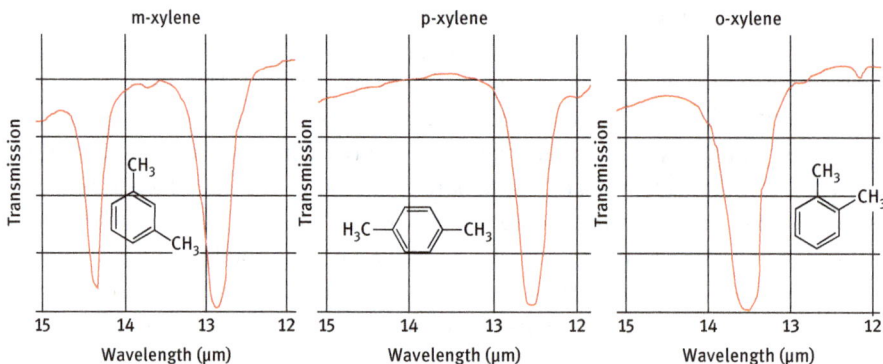

**Figure 6.43:** IR spectra of xylene (C8H10) isomers of $C_8H_{10}$ isomers.

In mixtures, however, the concentrations of components cannot be determined due to overlap of the bands. Therefore, the molar extinction coefficients of the four compounds must be determined at all four wavelengths. To determine individual components over the entire spectral range, multivariate tools like principal component analysis or multivariate calibration can be applied.

### 6.3.6.3 Near infrared spectroscopy
In NIR range, spectroscopic techniques are sensitive to C–H, N–H, and S–H absorbance and therefore well suited for process analysis of organic reactions.

Typical chemical applications include analysis of polymers, alcohol, residual solvent, melt index, hydroxyl number, or acid value. Also, strongly absorbing and highly light scattering matrices such as suspensions or slurries can be directly analyzed [321]. Historically, NIR has also been used to measure physical properties like viscosity. This is an intrinsic chemical characteristic as a result of cross-linking behavior or chain length and is measured indirectly via a spectral disturbance in the spectrum. Even, fast-growing fields of biotechnology use NIR for analysis of key analytes in real-time. The optimization of nutrient levels as well as ammonia and lactate concentrations are becoming more important [300]. Based on the cost of spectrometer devices, the measurements generally cover as many parameters as possible. However, peaks are broad and overlapping (low selectivity) since the number of combinations is large; consequently, NIR is a poor primary characterization tool. Therefore, a detailed qualitative analysis of NIR spectra is quite difficult.

Quantitative analysis is done using multivariate data analysis; see Section 6.4. As the calibration and validation of measured NIR spectral data is correlated through statistical methods to reference data, NIR spectroscopy is a secondary analytical method. Both diffused reflection and transmission measurement are used; although, diffuse reflectance is by far the more widely used. One advantage is that NIR can typically penetrate much farther into a sample than mid-infrared radiation. NIR absorption bands occurring at longer wavelengths are stronger, sharper, and better resolved than their corresponding higher overtone bands, which occur at shorter wavelengths. Figure 6.44 shows this behavior for three different organic compounds.

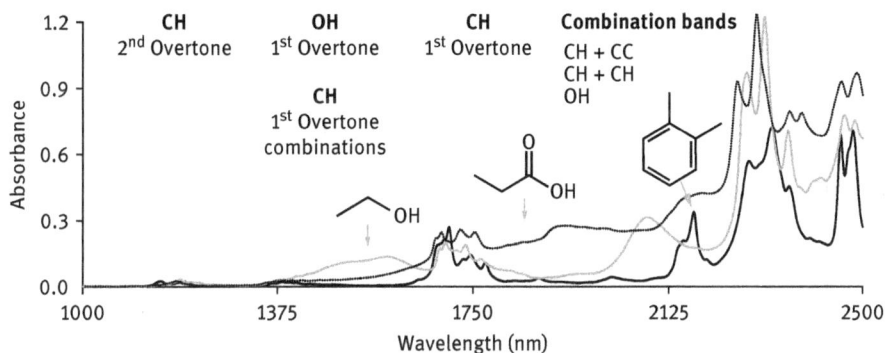

Figure 6.44: NIR spectra of three different organic compounds.

In the NIR spectral region, the apparent absorptivity increases with wavelength, whereas band overlap and penetration depth decrease with wavelength. Hence, the optimal spectral region needed for a particular analysis is determined by matching the spectral properties of the NIR region with the required analytical performance, the sampling requirements, and the physical properties of the sample. For example, measurement of a component whose bands are not well resolved from bands of other components in the matrix, or is present at low concentrations, is most easily accomplished in the longer wavelength region, which requires the use of a shorter optical path length.

NIR is extremely sensitive to OH absorbance and is therefore often used for moisture detection [301]. Water absorption features can be found at approximately 970, 1,200, 1,450, and 1,950 nm. Figure 6.45 shows a comparison of water absorption for different path lengths together with a comparison to the mid-IR. The absorption intensity is weaker than for the fundamental vibrations. However, this is not important as longer path-length cuvettes are used. The features at 1,450 and 1,950 nm are the most pronounced until the absorption band at 698 nm which is a 3rd overtone (n=4). The spectrum tails off in the visible region and this is responsible for the intrinsic blue color of water. This can be observed with a standard UV-Vis-NIR spectrophotometer, using a 5 cm path length cuvette. The color can be seen by eye when looking through a column of water about 10 m in length; the water must be passed through an ultrafilter to eliminate any color due to Rayleigh scattering which can also make water appear blue.

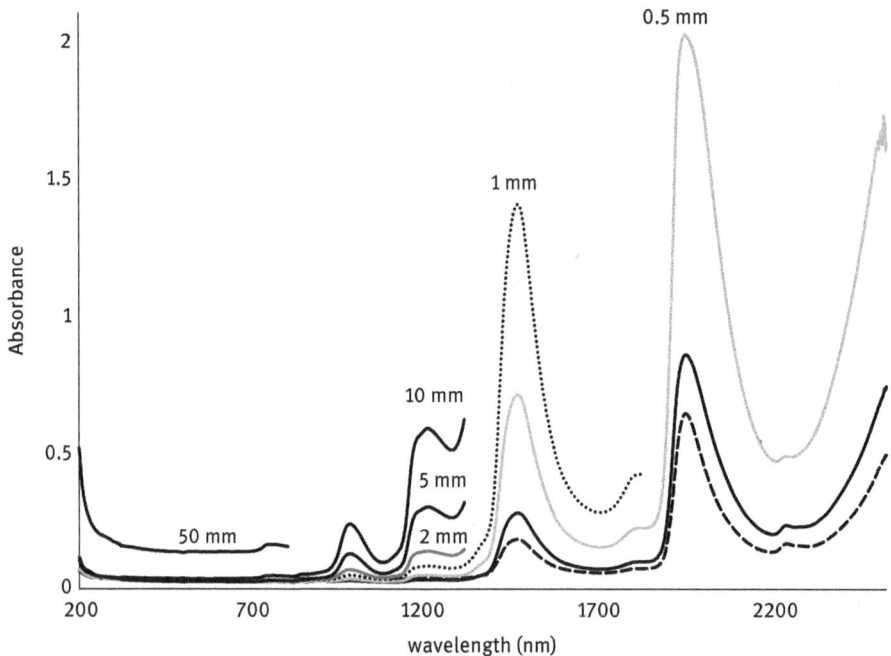

Figure 6.45: Comparison of water absorption spectra at different path lengths.

In most processes, liquid samples has to be conditioned with respect to temperature, pressure, phase, particle content, and flow characteristics. Sample conditioning means that the sample fluid is brought to a condition that does not adversely affect the analyzer and that allows reproducible results. Conditioning includes reduction and control of pressure to protect the analyzer and to maintain a steady optical path length. At long wavelengths, where the path length is very small, even the smallest dimensional change represents a large relative path length error. Other conditioning modules are the temperature controller and the filter. In NIR analysis, the samples must be maintained at a stable temperature in order to achieve precise concentration data. Slight temperature variations are known to change the hydrogen bonding structure of water in most cases, but the spectral characteristics of other materials through changing the hydrogen bonding or other chemical equilibria have also been observed. Figure 6.46 shows the spectra of water and ethanol at different temperatures ranging from 30 °C to 70 °C.

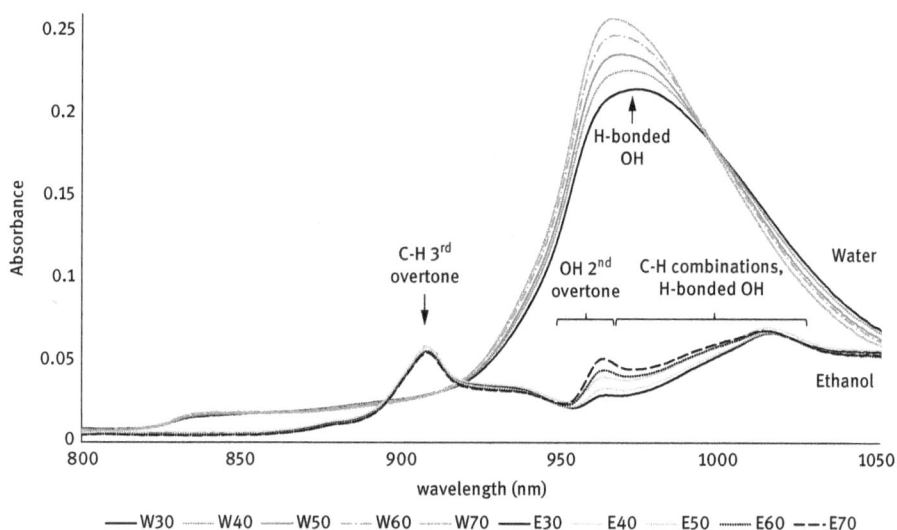

**Figure 6.46:** Spectra of water and ethanol at different temperatures ranges from 30 °C to 70 °C.

The water spectrum shows a band shift to lower wavelengths as the temperature increases which is combined with an absorption increase and band narrowing. Increasing the temperature reduces the cluster size of hydrogen-bonded molecules and therefore increases the proportion of free hydroxyl groups. The ethanol spectrum shows the strongest changes for the second overtone which is based on the free OH; only the overtone of the third CH slightly decreases and there is some increase in the CH combination bands.

### Chemical versus morphological information

In addition to the chemical information that is stored in the spectra, physical, or morphological information related to the sample can also be determined. For example, milk of various origins or dilutions, may shows nonlinear concentration dependence. Milk is an emulsion with fat globules dispersed in an aqueous environment. The fat globules do not coalesce and form a separate layer because they are protected by a membrane layer which keeps the fat particles separate from the water phase. The principal group of milk proteins, the caseins, is not soluble in water and exists in milk as small micelles with diameters smaller than 300 nm.

Liquid streams containing particulate matter are the most difficult to measure accurately. A number of models have been developed to quantitatively describe the diffused reflected and transmitted radiation for such systems [299, 330]. The most widely used model was developed by Kubelka and Munk (K–M) $f(R'_\infty)$ and is given by:

$$f(R'_\infty) = \frac{(1 - R'_\infty)^2}{2R'_\infty} = \frac{k}{s}$$

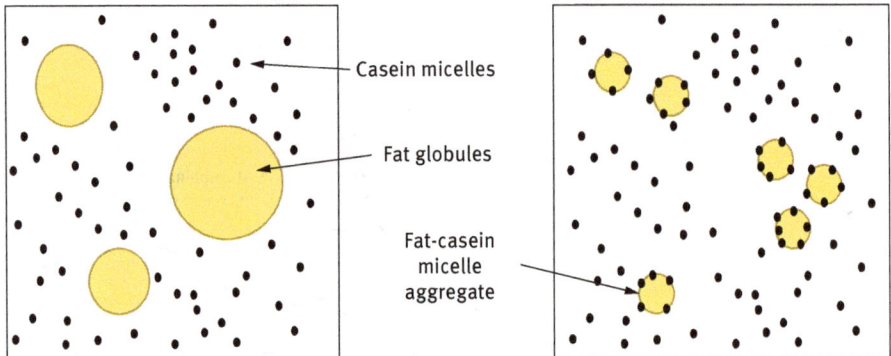

**Figure 6.47:** Schematic distribution of fat globules and casein micelles in unhomogenized milk (left) and homogenized milk (right).

where $R'_\infty$ is the ratio of the reflected intensity of the sample to that of a nonabsorbing standard. The quantity k is the molar absorption coefficient of the analyte and s is the scattering coefficient. For very diluted sample, k is related to the molar absorptivity $\varepsilon$ and the molar concentration of the analyte c is represented by the relationship

$$k = 2.303\varepsilon c$$

From this equation, it can be seen that the reflected intensity is not separated from the absorption or scattering but depends on the relationship between k and s. To determine the individual parts of scattering and absorption, at least two independent measurements must be carried out. One possibility is by using the *exponential solution* to calculate the value of $k$ and $s$ by taking the measurement of the diffused reflection of a probe with finite and infinite layers of thickness. A further possibility for the determination of $k$ and $s$ is the use of the hyperbolic solution in which a sample with finite thickness in the diffused reflection and the diffused transmission are measured. By using the following sizes of a and b obtained from the measurement,

$$a = \frac{1 + R_0^2 - T^2}{2R_0} = \frac{K+S}{S}$$

$$b = (a^2 - 1)^{1/2}$$

one can get the product of the scattering coefficient S and the layer thickness $d$, as well as the product of the absorption coefficient $K$ and the layer thickness $d$:

$$Sd = \frac{\sinh^{-1}\left(\frac{b}{T}\right) - \sinh^{-1}(b)}{b}$$

$$Kd = Sd(a - 1)$$

By using the separately determined coefficients, the penetration depth of the light in the forward direction for the measurement in transmission and the backscattered information for the reflection measurement are determined. Figure 6.52(a and b) shows the diffused reflectance (log 1/R) and diffused transmittance (log 1/T) spectra of milk with 1.5% and 3.5% fat content. Different fat content in the milk causes different light scattering and absorption behavior and therefore different penetration depth of the radiation. Samples with 1.5% fat show higher absorbance than samples with 3.8% fat content because K is small and the light can penetrate deeper into the sample and increase the absorbance. The application of the K–M formulas allows a calculation of separate coefficients for scattering and absorption of fat molecules. The milk with 3.8% fat content shows overall higher scattering coefficients, as there are more fat emulsion beads that can scatter. The scatter increases with smaller wavelength. In the range of 1,460 and 1,950 nm, the calculations are no longer valid because the water absorption of the milk is too high. Moreover, the measurements below 250 nm were out of the model. The calculated absorption coefficients show only the chemical information of the water, fat, and protein bands of the milk and can be used for quantitative investigations.

The dependence of the K–M function or the apparent absorbance based on the emulsion droplet size is rather complex, particularly in heterogeneous mixtures such as raw milk where both emulsion droplet size range and the differences in absorption coefficient among milk products are wide. Though K–M is a turbid-medium theory, it

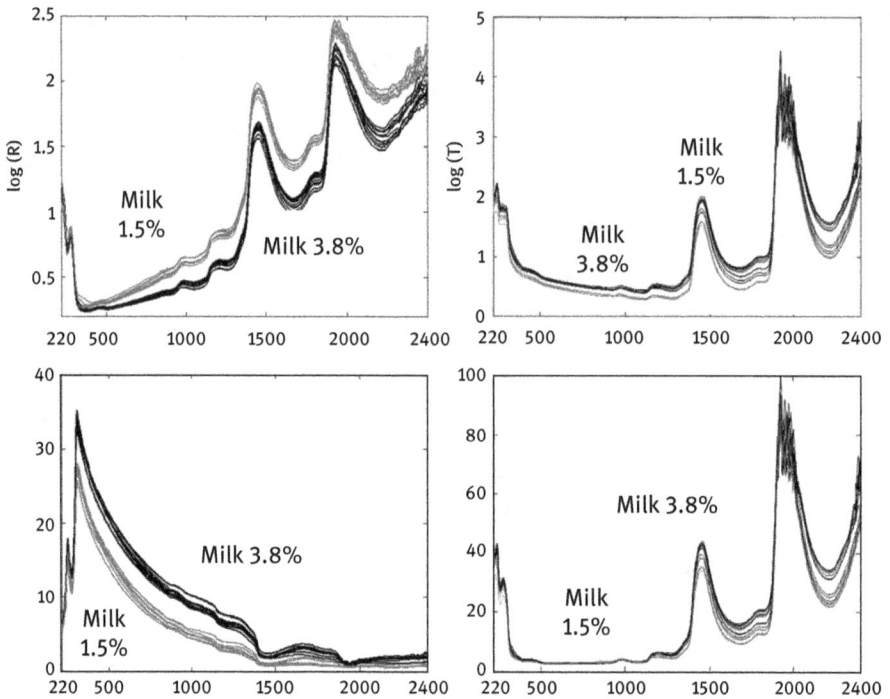

**Figure 6.48:** (Top) Diffuse reflectance (log 1/R) absorbance spectra (left) and diffuse transmission (log T) absorbance spectra of milk with 3.8% and 1.5% fat. (Below) Scattering coefficient (left) und absorbance coefficient (right).

is limited in its application. It is important to note the advancements that have been made in other areas of turbid-media theory. It is also important to note that the K–M theory is only valid for specific circumstances like a perfectly diffused and homogenous illumination incident on the surface or the medium is considered isotropic and homogenous. The effect of fat and protein globules for light scattering in Vis and sNIR range was studied by Bogomolov and Kessler [331].

## Self-Assessment Questions

1. How can infrared radiation be used to excite a molecule?
2. Which of the following statements describes the molecular vibration characteristics of IR spectroscopy?
   a) Stretching frequencies are lower than corresponding bending frequencies
   b) Triple bonds have lower stretching frequencies than corresponding double bonds, which in turn have lower frequencies than single bonds

c) Bonding with hydrogen have higher stretching frequencies than bonding with heavier atoms

d) Stretching frequencies are lower than corresponding bending frequencies

3. Calculate the approximate wavenumber and wavelength of the fundamental absorption due to the stretching vibration of a carbonyl group C=O.

4. With the help of the literature table which of the following absorption bands would be most useful for identifying the compound diethylamine?

a) Two medium peaks, one at 3,300 cm$^{-1}$ and another at 3,400 cm$^{-1}$

b) Medium stretch from 3,500 to 3,300 cm$^{-1}$

c) Very broad stretch from 3,500 to 2,500 cm$^{-1}$

d) Sharp peak at 1,750 cm$^{-1}$

5. What is the name of the compound based on the absorption spectrum?

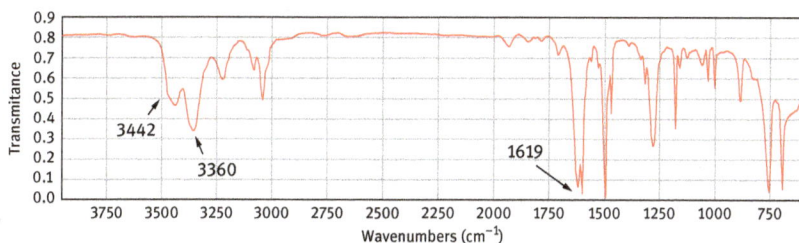

(a) aniline, (b) diethylamine, (c) pentanol, (d) butanoic acid

## 6.3.7 Spectrometer components and designs

A spectrometer can be defined as an instrument that disperses radiation into a spectrum and measures the radiation within a specific wavelength region. Although spectroscopic setups differ in configuration, most of the basic components are very similar. Typical instruments contain five components: (1) a light source, (2) a wavelength selector, (3) a sample cell, (4) an input transducer, and (5) a signal process and readout. Moreover, the requirements of these components are the same regardless of whether they are applied to the UV, Vis, or IR region of the spectrum. Sometimes, the placement of the wavelength selector and sample cell are reversed.

*Light Sources:* The purpose of the light source is to provide an output that is both intense and stable for reasonable periods to make the measurement in the region of spectral interest (Figure 6.49).

Light sources are classified as either continuum or line sources. Continuum sources are useful in several types of spectroscopy including absorption and fluorescence that emits radiation over a broad range of wavelengths, with a relatively smooth variation in intensity. For UV spectroscopy, the common continuum source is the deuterium lamp while in fluorescence spectroscopy

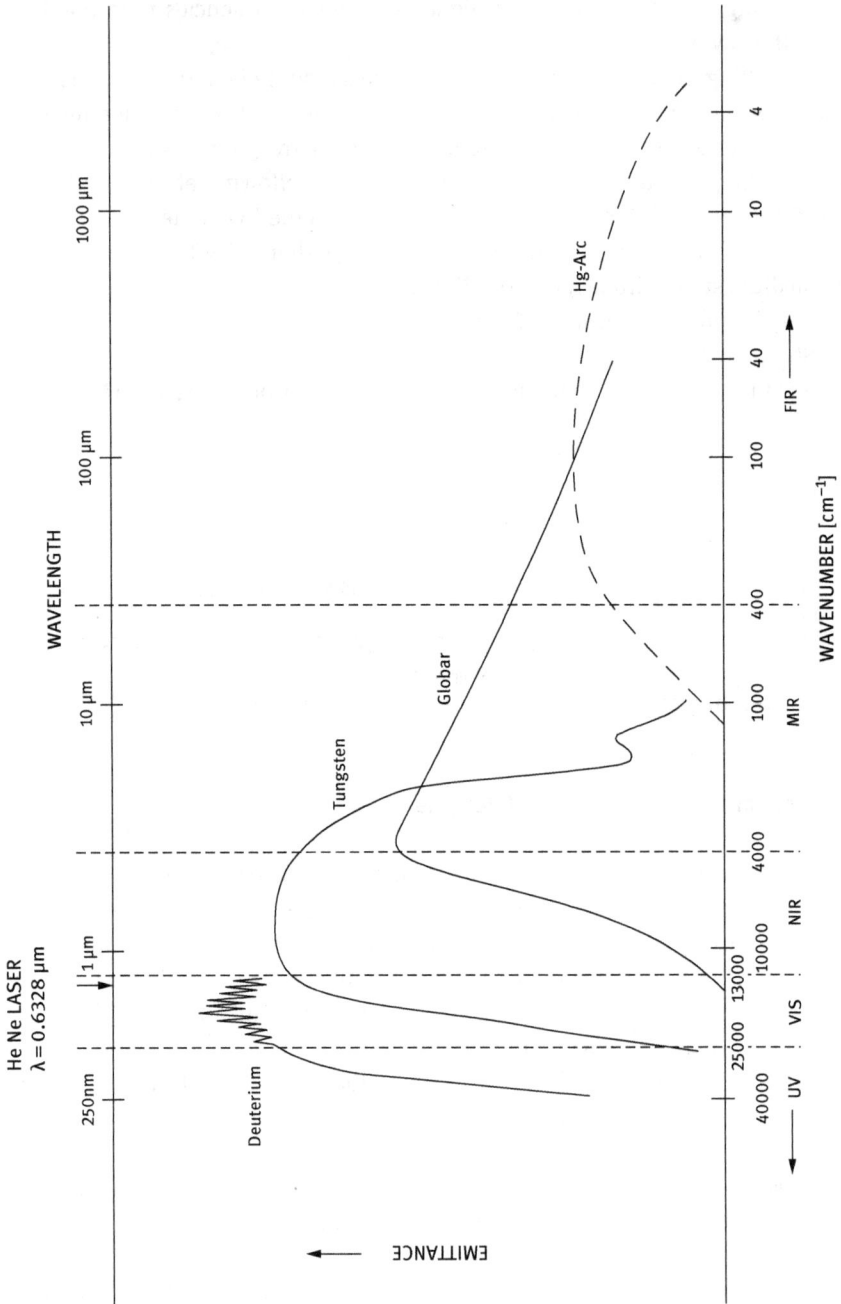

**Figure 6.49:** Sources for spectroscopic instruments. Source: Bruker [295].

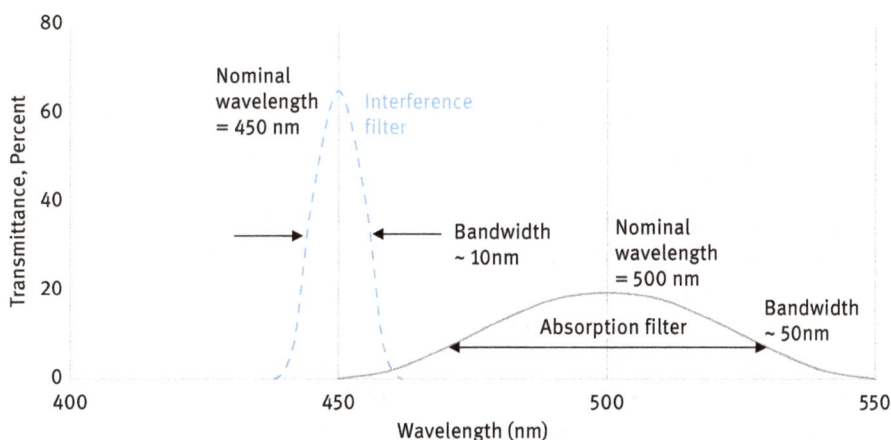

**Figure 6.50:** Comparison of effective bandwidths for two different filters.

high-pressure gas-filled lamps that contain xenon are used when a particularly intense source is required. For Vis-spectroscopy, a tungsten filament lamp is used for many applications. For common IR sources, inert solids are heated to 1,500–2,000 K, a temperature at which the maximum radiant output occurs at wavelengths of 1.5 to 1.9 μm. A line source, on the other hand, emits a few discrete lines. The familiar mercury and sodium vapor lamps provide a relatively few sharp lines in the UV and visible regions and are used in several spectroscopic instruments. Typically, the radiant power of a source varies exponentially with the voltage of its power supply. Thus, a regulated power source is crucial in spectroscopy to provide the required stability. Lasers are highly useful sources in analytical setups because of their high intensities, their narrow bandwidths, and coherent nature of their outputs.

*Wavelength selectors:* For spectroscopy, there are two types of wavelength selectors: filters and monochromators. Both systems are used to select a limited, narrow, continuous wavelength bands to provide selectivity in methods and obtaining a linear relationship between the optical signal and concentration. Practically, the output from a wavelength selector is a narrow band such as the black shown in Figure 6.55. The effective bandwidth is an inverse measure of the quality of the spectrometer; so a narrower bandwidth represents a better spectral performance. The effective bandwidth is of glass filers range from 30 to 200 nm with a transmittance often under 20% at their band peaks. This filter type is restricted to the visible spectrum, while interference of the filter covers the range from UV to IR. Filters are generally characterized by the wavelengths of their transmittance peaks, the percentage of transmittance, and the effective bandwidth.

An interference filter consists of a transparent dielectric layer in which the transmitted wavelength by the filter is

$$\lambda = \frac{2dn}{n}$$

where d is the thickness of the effective dielectric filter layer, n is the refractive index of the filter, and the integer **n** is the order of interference. A second device based on interference filter is the *Fabry–Perot etalon*. The etalon consists of spectral transparent material, with highly parallel faces coated with a reflective material. The spacer between the reflective materials is normally air, quartz, or invar. The wedged etalon consists of a pair mirrored, partially transparent plated separated by a wedge profile of a dielectric layer. The length is typically between 50 and 200 mm. The transmitted wavelengths vary continuously from one end to the other end of the wedge. Interference wedges are available for Vis, NIR, and MIR range.

Filter systems are normally used for process photometer as shown in Figure 6.52. The optical path consist of an UV or IR source, a chopper motor, a filter wheel with

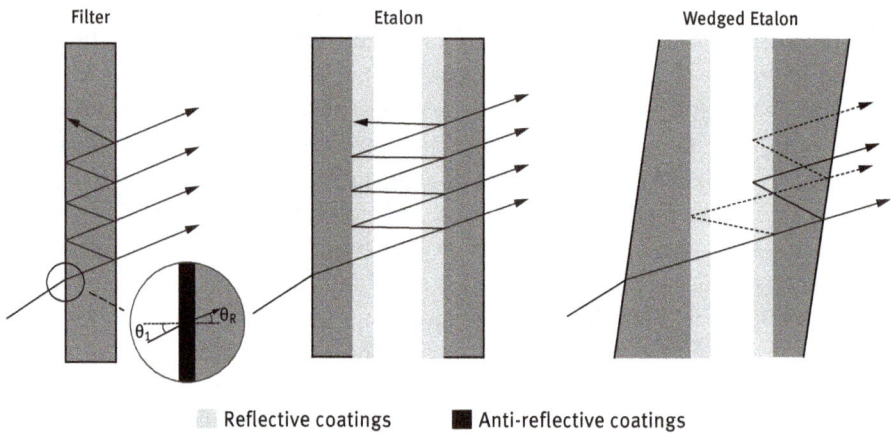

**Figure 6.51:** Different filter configurations.

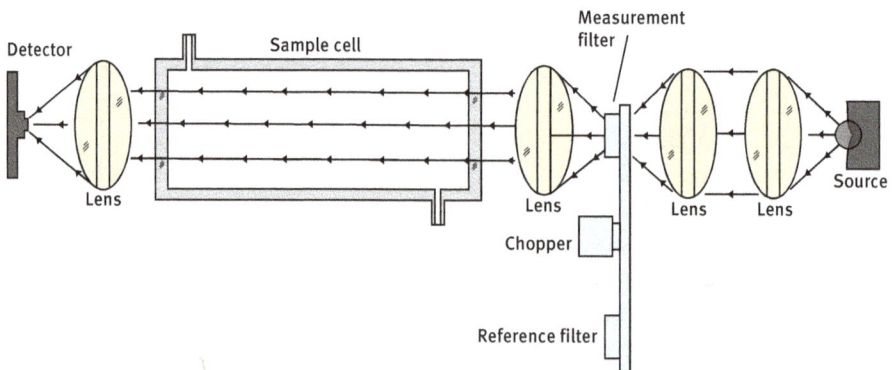

**Figure 6.52:** Principle set-up of an online photometer.

multiple interference filter, lenses, sample cell, and a solid state UV or IUR detector. The chopper motor rotates the filter wheel to the reference or measure filter position. The lenses focus and collimate the source radiation through the sample cell path to the detector. A reference wavelength is chosen where the stream components have little or no absorbance. The measure wavelength or wavelengths are chosen where the measured components have absorbance.

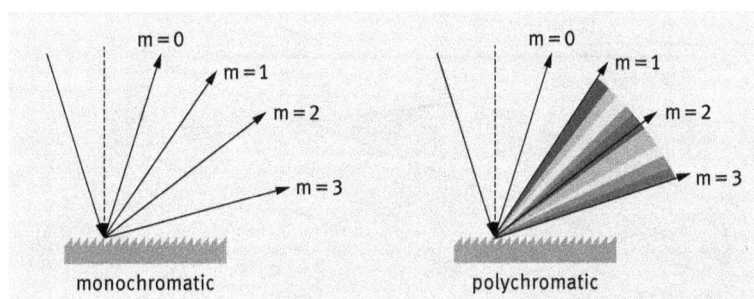

**Figure 6.53:** Diffraction gratings and their principles of operation. (Courtesy: Ocean Optics).

For many application, it is often necessary to record a complete spectrum over a broad spectral range to characterize a sample. This scanning mechanism can be carried out with a monochromator. Monochromators for UV/Vis and IR are similar in the mechanical construction regarding diffractions gratings, lenses, mirrors, and silts.

Diffraction gratings can be fabricated in several materials depending on the wavelength required. The separation of the lines in a grating is in the order of the incident wavelength. As in a prism, the reflected light from a diffraction grating is dispersed into its constituent wavelengths. It was in 1821, Josef Fraunhofer conducted the first experiment on the diffraction of light using optical gratings. He rediscovered the optical grating as David Rittenhouse, an American astronomer, fabricated the first grating. The precision of gratings was further developed and engineered by H. A. Rowland in 1882. The Italian physicist Francesco Maria Grimaldi (1618–1663) coined the word "diffraction," as he discovered that light diffracts around objects by holding a post into a light beam. The shadow of the post was larger than the original dimensions of the post.

Gratings are made up of several 2D/3D objects that are aligned with precise distance to each other. This distance influences the phase and/or the amplitude of the incident light. The effect of a grating can easily be seen in the visible region of the spectrum; if an incident light hits a transmission or reflection grating, a rainbow colored image is obtained. This image is known as the optical spectrum. In this way, gratings can be used to analyze the wavelength composition of incoming light.

Traditional bench-top spectrometers use a scanning monochromator and a single photodetector enabling a slow build up of the spectrum with a usually high accuracy and sensitivity. The common setup of a spectrometer is the so-called Czerny Turner monochromator shown in Figure 6.54.

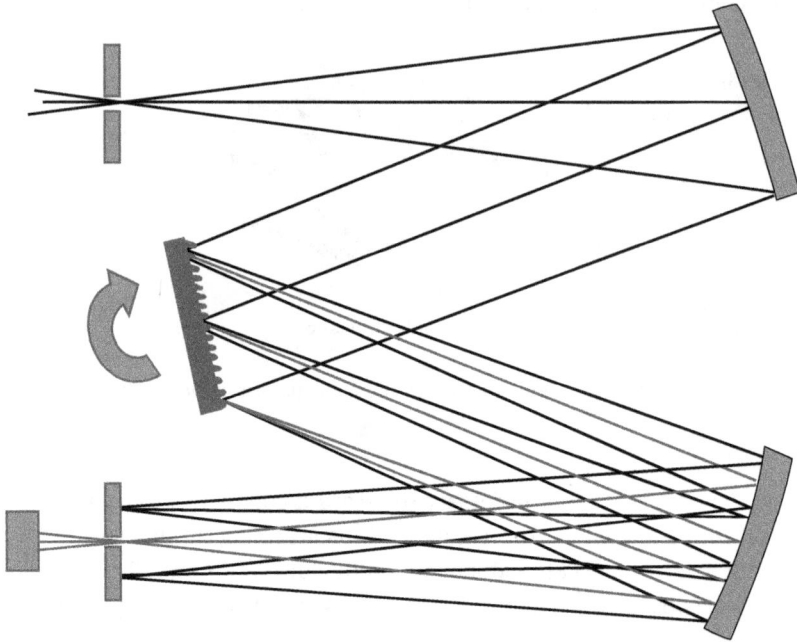

**Figure 6.54:** Czerny-Turner monochromator (Courtesy: Ocean Optics.).

The Czerny Turner principle states that the light coming from the entrance slit is collimated through the first mirror, then dispersed at the grating, and with the help of the second mirror focused onto the exit slit. The exiting light has a certain wavelength; by turning the grating, appropriate wavelengths are obtained. If the grating is fixed, then a photodiode array is needed to capture all wavelengths at the same time. These relations can be expressed by the grating equation

$$n\lambda = d(\sin\alpha + \sin\beta)$$

which governs the incident angle $\alpha$ and diffraction angle $\beta$ from a grating of groove spacing d. Here n is the spectral order, which is an integer.

This principle is now widely used in miniature spectrometers. An example of a miniature spectrometer is shown in Figure 6.61. The first miniature fiber optics spectrometer was introduced by Ocean Optics in 1992. As there are no moving parts, it is truly a portable device. Today, a large variety of microspectrometers are commercially available on the market, for example, B&W Tec., Hamamatsu, Bayspec, Serstech, Avantes, Zeiss, etc.

In order to design or specify a CCD spectrometer, certain parameters need to be taken into account like:
- wavelength range
- resolution
- sensitivity
- signal-to-noise ratio
- second order maximum
- stray light

One of the most important parameters of a spectrometer is the *resolution*. In practical environment, a high resolution is not always the best choice as the required resolution depending on the application. For example, the resolution of a NIR spectrometer can be sufficient with a 10 nm resolution, as the spectra of NIR absorbing substances consists of very broad peaks. On the other hand, the resolution of a Raman spectrometer is required to be below 1 nm or sometimes even below 0.1 nm in order to distinguish Raman peaks from certain molecules. The resolution is used in conjunction with the Rayleigh criterion, which is the criterion for the minimum resolvable detail of two adjacent peaks of the same intensity.

The Rayleigh criterion is reached when the valley between these two adjacent peaks is at 81% of the total intensity of the peak value. A more practical criterion for the resolution of a spectrometer is the full-width at half-maximum (FWHM) or spectral half-width value. This is the spectral width at 50% of the peak value. The spectral resolution, FWHM, is approximately 80% of the Rayleigh criterion.

The resolution of a spectrometer depends on the mechanical design of the spectrometer as well as the choice of the suitable components. The obvious resolution defining criteria is the size of the spectrometer. From lab spectrometers, it is known that these types of spectrometers usually have a higher resolution than those which are classified as Mini- or Microspectrometers. A further obvious criterion is the size of a pixel on the detector, resulting in a higher resolution on a smaller pixel size. The entrance slit of a spectrometer is also a means of defining the resolution of a spectrometer; a smaller width results in a higher resolution. The entrance slit also has an impact on the sensitivity as a smaller width prevents more light from entering the spectrometer. The final resolution is defined by the choice of the grating, where a greater line density results in a higher resolution but a smaller wavelength range. In summary, there are three factors in dispersive array spectrometers that determine the resolution – the slit, the diffraction grating, and the detector.

When measuring the spectral resolution, it is important to note that the observed signal $S_0$ not only depends on the resolution of the spectrometer, but also on the line width of the signal $S_r$. The resolution can then be written as a convolution

$$S(\lambda)_o = S(\lambda)_r * R(\lambda)$$

This dependency can be neglected if the signal line width is significantly greater than the spectral resolution. Therefore when measuring the spectral resolution of a spectrometer, one has to assure that the bandwidth of the measured signal is significantly narrow so that the measurement becomes resolution limited.

The spectral resolution can be calculated by the following equation:

$$\delta\lambda = \frac{RF \cdot \Delta\lambda \cdot W_S}{n \cdot W_p}$$

where Ws is the slit width, $\Delta\lambda$ is the spectral range of the spectrometer, Wp is the pixel width, n is the number of pixels of the detector, and RF is the resolution factor determined by the relationship between the slit width and the pixel width.

In order to determine the FWHM of a peak, a minimum of 3 pixels is required. When Ws is approximately Wp, then the resolution factor is 3. When Ws is approximately 2 times of Wp, the resolution factor is 2.5. The resolution factor can be approximated to 1.5 if Ws > 4Wp.

Another important feature of a spectrometer is the *sensitivity*. As with the resolution, the sensitivity required is defined by the application for which the spectrometer is used. This is especially true for low-light-level sensor applications where stray light is detected or light at different angles compared to the light incident angle is detected; for example, it is used in Raman and fluorescent measurements. The sensitivity, technically speaking, is the relationship of the number of photons collected to the number of electrons generated at the detector resulting in the electronic voltage needed to be processed further. Note that the detector is wavelength dependent and a calibration is always needed. The other key factors having an impact on the sensitivity are the slit size, the choice of the grating, and the diameter of the fiber used to couple light into the spectrometer.

The parameter of interest to the electronic engineers in processing the obtained spectrometer electronic response is the *signal-to-noise* ratio. The signal-to-noise ratio depends on the well depth of the detector pixels and not on the optical configuration discussed in the previous paragraphs of this chapter.

When designing a spectrometer, it is important to know the number of orders of the grating used. The zero order corresponds to the specular reflection. The first order is the one used usually by the spectrometer. This leads to the necessity to suppress the influence of the zero order and higher orders to obtain the wanted performance of the spectrometer.

So far when designing a spectrometer, we have dealt with the wavelength range, the resolution, the sensitivity, the signal-to-noise ratio, and second orders from the gratings. An important parameter is the suppression of stray light, as this parameter is directly related to all the other parameters. Stray light can be defined as any light which reaches the detector by a path other than the intended optical path. A detailed analysis of stray light in microspectrometers is presented in [309, 322]. Stray light can have several origins. The most occurring causes are imperfections in the grating, reflections within the spectrometer bench, and external light entering the spectrometer. Stray light is typically measured with a broadband illumination with long pass filters. When referring to the unit "stray light," it is defined as the ratio of the transmission in the blocked wavelength region to the transmission in the non-blocked region. Example parameters are: < 0.05% at 600 nm, < 0.1% at 400 nm, etc.

The performance criteria discussed in the previous paragraphs of a spectrometer are directly related to the choice of the spectrometer components: optical bench, detector, grating, slit/aperture, filters, lenses, and mirrors. An example of a typical optical bench is shown in Figure 6.55.

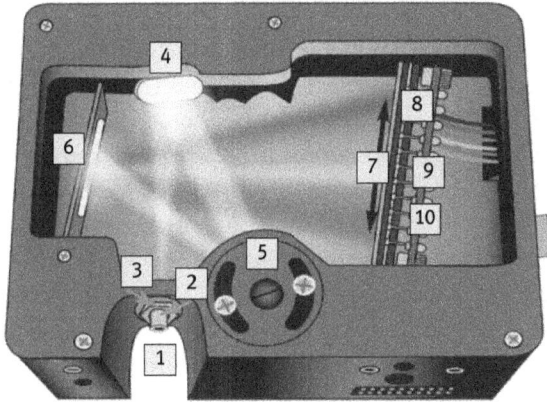

**Figure 6.55:** Optical bench. (Courtesy: Ocean Optics.).

Where:
1. SMA connector: Light from a fiber enters the optical bench through the SMA 905 connector. The SMA 905 bulkhead provides a precise locus for the end of the optical fiber, fixed slit, absorbing filter, and fiber -clad mode aperture.
2. Fixed entrance slit: Light passes through the installed slit, which acts as the entrance aperture. Slits are available in various widths from 5 µm to 200 µm. The slit is fixed in the SMA 905 bulkhead to sit against the end of a fiber.
3. Longpass absorbing filter: If selected, an absorbance filter is installed between the slit and the clad mode aperture in the SMA 905 bulkhead. The filter is used to block second- and third-order effects or to balance color.
4. Collimating mirror: The collimating mirror is matched to the 0.22 numerical aperture of the optical fiber. Light reflects from this mirror, as a collimated beam, toward the grating.

5. Grating and wavelength range: The grating is installed on a platform that is rotated to select the starting wavelength specified. Then the grating is permanently fixed in place to eliminate mechanical shifts or drifts.
6. Focusing mirror: This mirror focuses first-order spectra on the detector plane.
7. Detector collection lenses: One of these cylindrical lenses, made to ensure aberration-free performance, is fixed to the detector to focus the light from the tall slit onto the shorter detector elements. It increases the light-collection efficiency.
8. Detector: A linear CCD array detector is used. Each pixel responds to the wavelength of light that strikes it. Electronics bring the complete spectrum to the software.
9. Variable longpass order-sorting filter: This filter precisely blocks second- and third-order light from reaching specific detector elements.
10. Detector with UV2 or UV4 window: When specified, the detector's standard BK7 window is replaced with a quartz window to enhance the spectrometer's performance from 200–340 nm.

The choice of the appropriate detector specifies the operating wavelength and the overall performance of a spectrometer next to the optical parameters. There is a large variety of detector choices available (see also chapter on image sensors), ranging from silicon linear CCD arrays or silicon linear photodiode arrays, to back thinned area CCDs or InGaAs linear arrays or CMOS arrays. The main criteria for choosing a detector are the wavelength range and the sensitivity. Silicon arrays are sensitive in the wavelength range 180 nm to 1,100 nm. InGaAs detectors are used for longer wavelength typically up to 2,500–2,600 nm. By selecting the corresponding grating, slit/aperture, filters, lenses, and mirrors complement the fabrication of a spectrometer.

Once a spectrometer is assembled, the calibration is performed, linking the pixels of a detector to the measured wavelength. The output data of a spectrometer is intensity counts versus pixel number. The wavelength calibration is performed by using, for example, a Hg/Ar lamp with well-defined emission lines and deriving a polynomial fit to obtain the calibration coefficients.

Miniature spectrometers, those are readily available on the market for numerous applications, are still rising. A very nice example of how to easily create a spectrometer with available components is described by Alexander Scheeline and Kathleen Kelley in Cell Phone Spectrometer: Learning Spectrophotometry by Building and Characterizing an Instrument, which can also be found online at: http://scheeline.scs.illinois.edu/~asweb/CPS/. Another form of miniaturized spectrometers are so called MEMS based spectrometers which are described in the following chapter.

### 6.3.7.1 MEMS based spectrometers
MEMS based spectrometers are distinguished from miniaturized spectrometers by the fact that they are monolithically fabricated from silicon or polymer without the necessity to assemble optical parts on an optical bench. An overview on MEMS

based spectrometers is given in [297]. A few MEMS based spectrometers are already commercially available. One example of the MEMS spectrometer is from Hamamatsu. Hamamatsu has recently developed ultra-compact mini-spectrometers using the state-of-the-art MOEMS (micro-opto-electro-mechanical systems) technology that combines nanotechnology and MEMS technology into image sensors.

The entrance slit is precisely positioned and formed using the same photomask as the image sensor. The major specifications for the MOEMS spectrometer is listed in Table 6.7.

**Table 6.7:** Specification of MOEMS spectrometer (Courtesy: Hamamatsu)

| Parameter | Specification | Unit |
|---|---|---|
| Spectral response range | 340 to 750 | nm |
| Spectral resolution (FWHM) *2 | 12 | nm |
| Wavelength reproducibility *3 | ±0.5 | nm |
| Spectral stray light *2 *4 | −25 | dB |
| Entrance slit size | 75 (H) × 750 (V) | μm |
| Optical NA (numerical aperture) | 0.22 | – |
| Number of pixels | 256 | pixels |
| Pixel size | 12.5 (H) × 1,000 (V) | μm |
| Operating temperature | +5 to +40 | °C |
| Storage temperature | − 20 to + 70 | °C |
| Weight | 9 | g |

*2: Measured with standard slit; *3: Measured under constant conditions; *4: While a line spectrum is input, the spectral stray light is defined as the ratio of the count measured at the input wavelength to the count measured at a wavelength of 40 nm longer or shorter than the input wavelength. The layout of the optical system is shown in Figure 6.56.

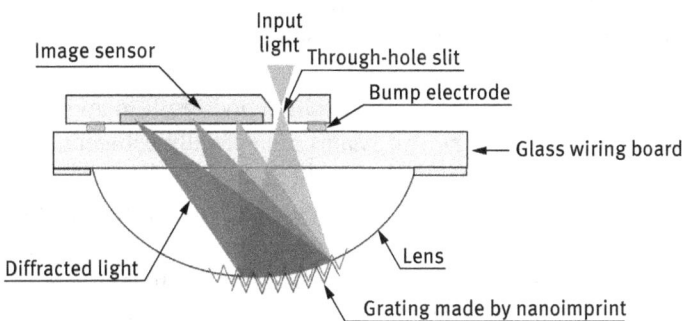

**Figure 6.56:** Optical system layout of ultra-compact minispectrometer. (Courtesy Hamamatsu.).

The technique for replicating a grating makes use of nanoimprint, which transfers an engraved grating pattern onto a glass body. Ultraviolet-curing resin (replica resin) is attached near the top of a condenser lens, and the grating is replicated on the lens by pressing the grating pattern against the resin while simultaneously irradiating it with ultraviolet light.

Another commercially available MEMS spectrometer has been presented by the company Si-Ware Systems (SWS). It is a Fourier-Transform-IR (FT-IR) spectrometer. The FT-IR spectrometer consists of a MEMS chip that measures $1 \times 1$ cm$^2$ and contains all of the optical components. The electronics interface is a separate ASIC that is designed and produced by SWS's ASIC Solutions Division. The additional components of the system are a photodetector, optical fiber, and software. The initial system is contained in a module that is powered through USB interface and consumes less than 150 mA. The module measures $8 \times 6 \times 3$ cm$^3$ and weighs less than 150 g. This is the smallest and lowest power consuming FT-IR spectrometer in the world.

FT based spectrometers offer significant advantages over dispersive spectrometers [355]. All wavelengths act simultaneously throughout the entire sample measurement. This reduces the integration time and the total number of photons observed. It has aperture advantages, is much more sensitive, and has better signal to noise ratio than dispersive spectrometer. The core of an FT-IR spectrometer is a Michelson interferometer that uses a series of mirrors and other optical components to create interference at the detector that can be measured as a function of the position of the movable mirror. The interferometer produces a light pattern that corresponds to the spectral content of the input light (the interferogram). The photo detector captures the interferogram and generates electrical signal corresponding to light intensity. Conventional spectrometers can be more complex and bulky because they contain optical components, motors, electrical interfaces, and drivers. The miniaturization of spectrometers has significant advantages in size for high portability and cost savings. It will also allow prolific installation in process lines and for environmental sensing. SWS has developed a fully monolithic MEMS based Michelson interferometer where the fixed mirror, the moving mirror, the beam splitter, and the micro-actuator are all integrated.

In Figure 6.57, the half-plane splitter separates the light beam; so one beam is transmitted through the fixed mirror while the other is reflected in the direction of the moving mirror. Both beams recombine at the beam splitter and result in an optical path length difference of $\delta = 2x$. Since the two beams are spatially coherent, they interfere on recombination. For monochromatic radiation, the two beams interfere constructively if the optical retardation is a multiple of the wavelength $\lambda$ and interfere destructively if $\delta$ is an odd multiple of $\lambda/2$. The signal on the detector is thus the radiation intensity of the combined beams as a function of the retardation $\delta$. For a source of monochromatic radiation with frequency $v$ or wavenumber $\tilde{v}$, the intensity at the detector is given by

$$I_D(\delta) = 0.5S(\tilde{v})[1 + \cos(2\pi\tilde{v}\delta)]$$

In which $S(\tilde{v})$ is the intensity of the monochromatic beam. It can be seen that $I_D(\delta)$ consists of a constant and a modulated component. Only the modulated component

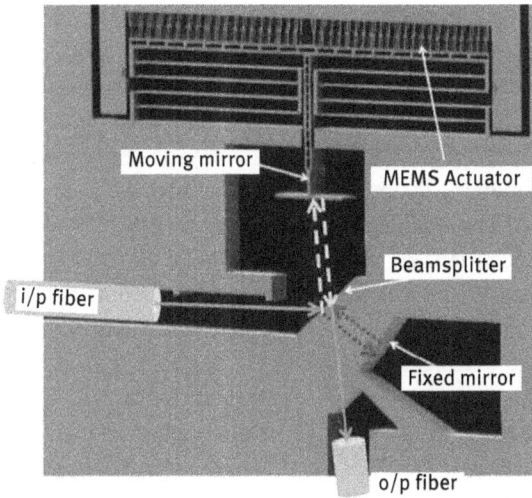

**Figure 6.57:** Michelson interferometer in silicon. (Courtesy: Si-Ware Systems.).

is important for spectroscopic measurements, while the modulated form is referred to the interferogram:

$$I(\delta) = S(\tilde{v})\cos(2\pi\tilde{v}\delta)$$

The retardation is a function of time and the interferogram is a function within the time domain. If the movable mirror is scanned at a constant velocity v, the retardation is $\delta = 2t$ and the interferogram $I(t)$ is given by

$$I(t) = S(\tilde{v})\cos(2\pi\tilde{v}2t)$$

Therefore, the signal at the detector varies sinusoidal with the frequency

$$f_v = 2v\tilde{v}$$

In comparison with a source of polychromatic radiation, the interference pattern is given by

$$I(\delta) = \int_0^\infty S(\tilde{v})\cos(2\pi\tilde{v}\delta)d\tilde{v}$$

where $S(\tilde{v})$ is the spectral power of the source.

The cleanliness of the mirror surfaces and the alignment of the moving mirror are critical to the functioning and quality of the interferometer. SWS's prototype enabled <5 nm resolution with a MEMS travel range of more than 400 $\mu$m. The overall size of the die is 8 mm × 9.5 mm. A dedicated electronic interface was used for the interferometer's MEMS actuation and control. SWS developed an application-specific integrated circuit

(ASIC) that drives the MEMS chip along with a range of supporting functions including data processing. A patented capacitive sensing technique is used to detect the mirror position with an error in the low picometer range, replacing bulky positioning solutions for conventional He–Ne lasers. An analog-to-digital converter (ADC) translates the analog signals of the interferogram into digital values that are processed through an onboard Fast Fourier Transformation (FFT) unit and into an actual spectrum.

Using the combination of the silicon Michelson interferometer, the ASIC's and SWS's software analysis package SpectroMOST, spectra from various materials have been tested and analyzed. The SpectroMOST software operates the spectrometer with the appropriate signal processing to generate and display the measured interferogram and obtained spectrum, and it is designed to enable customers to create their own application-specific software package.

The curves shown in Figures 6.58 and 6.59 depict spectra measured by the SiMOST (technology platform developed by Si-Ware) based FT-IR spectrometer including the emission of different light sources, as well as the light transmitted through different sample materials.

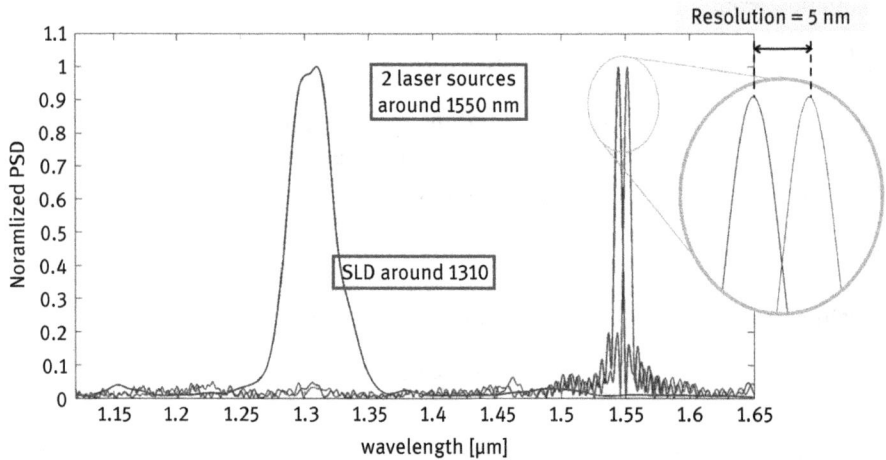

**Figure 6.58:** Emission spectra of different super luminescent diodes (SLDs) and laser sources. (Courtesy: Si-Ware Systems.).

The use of FT-IR has several advantages. First, the throughput or *Jacquinot advantage* is higher because there is few optical elements and no linear slits to attenuate radiation. This enables that radiant power that reaches the detector is much greater than in dispersive elements and also the signal-to-noise ratio is much greater. Second, the high wavelength accuracy or *Connes' advantage* is based on the fact that the wavelength scale is calibrated by using a laser beam of known wavelength that passes through the interferometer. This is much more stable and accurate than in dispersive instruments where the scale depends on the mechanical movement of

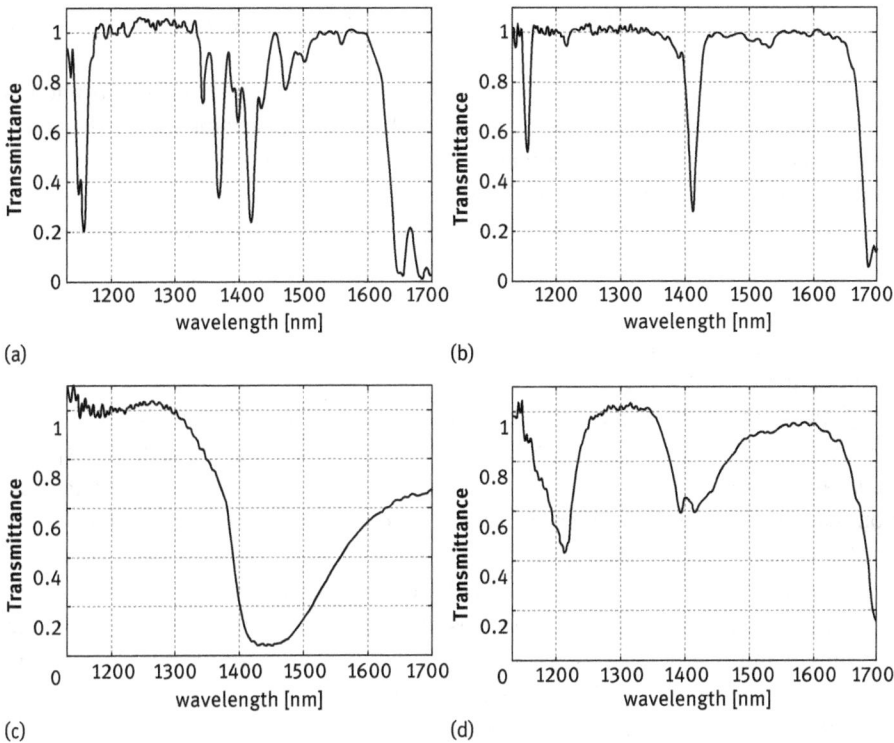

**Figure 6.59:** Transmission spectra of different sample materials: (a) Methylene Chloride, (b) Chloroform, (c) Water, (d) Olive Oil. (Courtesy Si-Ware Systems.).

diffraction gratings. In practice, the accuracy is limited by the divergence of the beam in the interferometer which depends on the resolution.

Third, the multiplex or *Fellgett's advantage* arises from the fact that information from all wavelengths is collected simultaneously. In conventional spectrometer, the spectrum $S(\tilde{v})$ is measured directly by recording the intensity. It results in a higher signal-to-noise ratio (S/N) for a given scan-time for observations limited by a fixed detector noise contribution. The S/N for the average of $n$ measurement is given by

$$\left(\frac{S}{N}\right)_n = \sqrt{n}\left(\frac{S}{N}\right)_i$$

where $(S/N)_i$ is for one measurement.

Another type of MEMS-based spectrometer gaining attention lately is the Fabry-Perot spectrometer. Using electrostatic actuation, it is possible to tune the Fabry-Perot resonator and thus scan the wavelengths and realize a very compact spectrometer. The fabrication of such a device can be found, for example, in Onny Setyawati,

MEMS-based Tunable Optical Filter Arrays for Nano-Spectrometer in the Visible Spectral Range (available online).

The authors in [314, 343] present a paper on MOEMS miniature spectrometers using tuneable Fabry-Perot interferometers. These are spectrometers covering the visible spectrum. Similar development can be found for tuneable Fabry-Perot spectrometers for the infrared part of the spectrum, which are already commercially available (see for example, Spectral Engines – www.spectralengines.com or InfraTec – www.infratec.de).

### 6.3.7.2 The Comparison of MEMS Microspectrometers
**Introduction**
The field of spectroscopy has developed fast from the 1980s laboratory work through 1990s in-line and 2000s portable devices to real mobile phones and widespread applications during 2010s. In this development, there are many technical and application-related challenges of which the most governing is the cost and the size of the hardware. To overcome this, many trials have been made to develop the next-generation spectroscopic devices, which would at the end become spectral sensors rather than spectrometers: small and low-cost devices that are seamlessly connected, providing meaningful information to the user. The most prominent wavelength range concerning the information content, availability of detectors and light sources and ease of sample preparation is the near infrared (NIR) wavelength range depicted to be between 750 and 2,500 nm. In this region, various new techniques have been developed to meet the future demand, most of them concentrating in the development of microelectromechanical (MEMS) manufacturing technologies. This chapter will summarize the current status of the main technologies and solutions around miniaturized NIR spectrometers on their way to become true spectral sensors. It gives practical guidelines on how to choose a technology for the envisaged application.

**NIR technology and market development**
NIR spectroscopy is a field of analyzing and identifying material and material content by illuminating an object with infrared light, gathering transmitted or reflected light by the spectrometer and using various mathematical data analysis methods to deliver results. The first very extensive commercial applications of NIR were developed in the agriculture and food industries, one example being the development led by Canadian Grain Commission to use NIR technology as a basis for buying and selling wheat based on analyses of protein, oil and moisture. Analytical applications of NIR spectroscopy developed rapidly in the 1980s; thanks to progress in key enabling technologies: the personal computer, optoelectronic technology and multivariate modeling or chemometrics (statistical and mathematical methods applied to the field of chemistry). Currently, analytical measurements are the most significant area of application for NIR spectrometers in laboratory studies, process monitoring as well as handheld and field

measurements. There are thousands of papers so far published on applications of NIR spectroscopy for varying analytical measurements in agriculture and food industries, forestry, environmental, medicine, pharmaceutical, chemical, textile, cosmetics and many other applications. The advantages of NIR analysis are often mentioned as follows: (1) minimum sample preparation, (2) rapid analysis, (3) no reagents needed, (4) non-invasive and nondestructive, (5) simultaneous possible multicomponent analysis, (6) sample size from very small to very large, depending on the equipment and (7) compatibility with fiber optics, in which case the sample may be separated from the instrument.

The types of spectrometers used are as follows:
- Fourier infrared spectrometers (FTS, FTIR or FT-NIR)
- Grating (dispersive) spectrometers (monochromators and compact multipixel spectrometers)
- Fabry–Perot interferometer spectrometers (FPI)
- Fixed filter arrays and filter wheels

The benchtop FTIR has many benefits, such as high throughput, wide wavelength range and good resolution and it is therefore the most used high-end NIR analyzer, manufactured by companies such as Bruker. Grating spectrometers have become compact in the recent years, especially in the so-called very near infra-red (VNIR) range. Compact NIR grating spectrometers are currently the most common devices when going for process control applications due to smaller size, higher robustness and lower cost than the FTIR. The drawback of using compact grating spectrometers is that in the sensitive wavelength ranges, the multipixel detector technology needed is expensive and often produces modest signal-to-noise ratio (SNR). Later, this obstacle has been tried to overcome by Texas Instruments by providing their digital micromirror device (DMD) technology to replace the multipixel detector with a scanning mirror and a single pixel detector. As the computation power and connectivity has been significantly improved over the years via digitalization of societies, new possibilities for expanding the possibilities for NIR spectroscopy have risen. Therefore, another type of spectrometer, earlier used mainly in astronomy and telecommunications, has been introduced to markets in the recent years, being the first that can really catch the high-quality, meaningful applications of the true NIR wavelength range, while being capable of scaling-up in volumes to hundreds of thousands or millions, while dropping the price tag to sub-100€ level. This technology is called the Fabry–Perot interferometer.

Tematys, a France-based market research company, envisions in their recent marketing report from December 2016 that the market for miniaturized spectrometers will be doubled in the next 5 years, and the trend in the 2010s is going to be the point-of-care and consumer applications. Previously, the problem preventing this significant growth (the NIR market has been growing only at a few % CAGR in the recent years) has been the size and cost of the instrumentation, but now new approaches have emerged to overcome this.

Usually, it really is the application that decides which technology suits the best. There are some basic criteria with which you can select the right platform for you without even going to the technical details of wavelength resolution. This selection logic is described below in Figure 6.60, where you end up to the right platform by asking a few application and business-related questions. Once you have gone through this simple selection tool, it is much easier to narrow down your final selection, which then depends on the final specifications, cost targets and the level of solution (sub-component vs. full solution). This graph does not consider your business model, which naturally is an important question. Some companies provide only components, leaving system development, wavelength calibration and temperature stabilization development, along with the reference library and algorithm development to the customer, whereas others offer full end-solutions only. Most companies work between these two ends: for example, Spectral Engines provides full wavelength and temperature calibrated spectrometers, including light sources (NIRONE Sensors), and for customers wanting to privately label a solution and move to the market quickly, the NIRONE Scanner platform provides the complete instrumentation, communication methods, cloud platform, reference library creation and algorithm implementation.

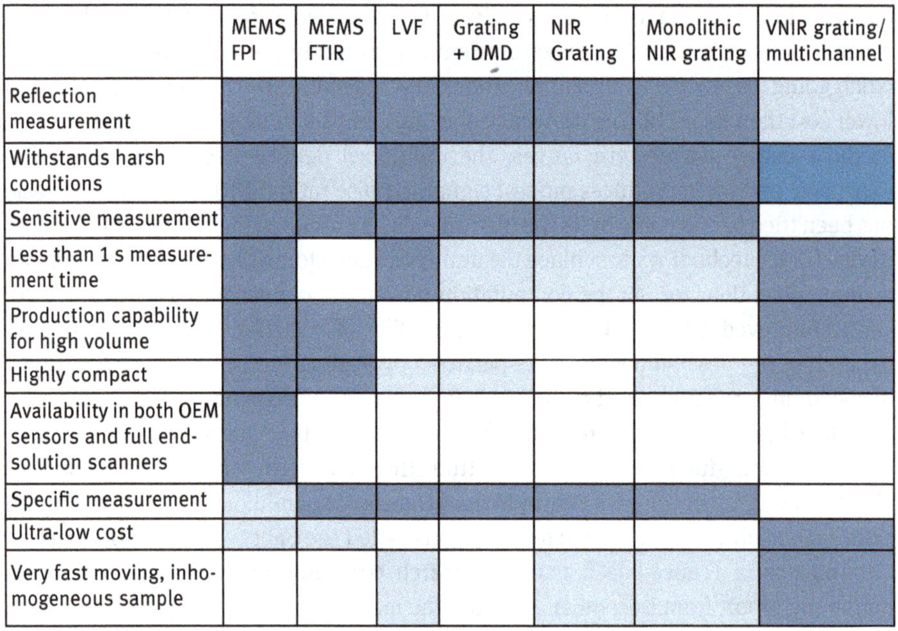

| | MEMS FPI | MEMS FTIR | LVF | Grating + DMD | NIR Grating | Monolithic NIR grating | VNIR grating/ multichannel |
|---|---|---|---|---|---|---|---|
| Reflection measurement | | | | | | | |
| Withstands harsh conditions | | | | | | | |
| Sensitive measurement | | | | | | | |
| Less than 1 s measurement time | | | | | | | |
| Production capability for high volume | | | | | | | |
| Highly compact | | | | | | | |
| Availability in both OEM sensors and full end-solution scanners | | | | | | | |
| Specific measurement | | | | | | | |
| Ultra-low cost | | | | | | | |
| Very fast moving, inhomogeneous sample | | | | | | | |

**Figure 6.60:** High-level decision tree for an application and its business case.

## Miniaturization principles and challenges

There are many limitations concerning miniaturization, most relating to manufacturing techniques. Taking separate, macroscopic parts and trying to make them smaller only gets one that far. Limitations of manufacturing techniques of mechanics and optical parts come into play, destroying tolerances, but leaving assembly and calibration costs high. In order to reach the next level, new philosophies are needed.

Currently, the dominant manufacturing technology used for miniaturizing spectrometers is MEMS, or microelectromechanical systems. This enables much smaller features and more integrated structures than was previously possible. MEMS technologies employ deposition, patterning and etching techniques, taking place in a clean room, to form mechanical and electrical structures in a microscale. Eventually, one can get hundreds of chips from a single silicon wafer, resulting in highly scalable and low-cost devices. The main philosophy is to take a traditional spectrometer type, such as Fourier Transform Infrared (FTIR), Fabry–Perot (FPI) or grating-based spectrometer that can retain the working principle but scale down the size with this new manufacturing technique. Micromechanical sensor devices and techniques have become widespread after the A-class Mercedes failure in the "moose test" in 1997. There accelerometers were quickly needed to produce active stabilization to the car that was too easy to flip over with a quick steering motion. Since then MEMS sensors of various kinds have become a commodity in cars, but later in mobile phones. A mass production example of optical MEMS is the digital light projection (DLP) technology by Texas Instruments.

Other common methods of breaking the barrier of common mechanics manufacturing tolerance limitations are nanoimprinting and Lithographie Galvanik Abformung (LIGA). Both these methods are used to manufacture and replicate very fine grating structures, and also to create monolithic optical benches for grating spectrometers.

One philosophy in the miniaturization and lowering the cost of instrumentation is to limit oneself to working below 1,000 nm in the wavelengths (the "VNIR" range) as this is a cutoff for the very inexpensive silicon detector technology. Here the price of a multipixel sensor is not an issue and LED light sources are also readily available. Later, companies such as Consumer Physics and AMS AG have demonstrated this route. The selection of this wavelength range presents a major obstacle: the molecules do not react well and all the spectral fingerprints start to get mixed, because one is actually measuring the third or fourth overtones of molecular vibrations. The sensitivity and the specificity (how well can you distinguish materials from each other) can be anywhere between 10 and 1,000 times worse in the VNIR than in the "true NIR" range from 1,000–2,500 nm. This is tried to be compensated by setting up large databases and performing analysis in the cloud using machine learning and other sophisticated methods but one can only get to a very indicative level even with the smartest of algorithms. For some applications, though, where, for example, one needs to measure through a thick sample, the VNIR range is applicable, but the vast majority of applications lie in the true NIR range. Figure 6.61 illustrates this: to get to mass applications, one needs to get from the top part to the lower boxes, and to start creating a digital ecosystem with some indicative first applications, one can function in the lower left box. But then to eventually create meaningful, sustainable applications, one must reach the green box on the lower right side. The vast majority of the examples of miniaturization presented in this chapter are concerning specifically with this green box.

**Performance, price**

| Low performance, high price | High performance, high price |
|---|---|
| Low performance, low price | High performance, low price |

Wavelength

Silicon ('VNIR') detector range

(Ext-)InGaAs ('True NIR') detector range

**Figure 6.61:** Illustration of instrument division per performance and price.

In the case of realizing a spectrometer using MEMS technologies, one usually concentrates in producing a mirror movement of some sort to create an *active interferometer*. At the same time, however, one should keep the optical area large to sustain usable SNR of the system. Movement can be achieved in multiple ways, such as using piezoelectric, electrostatic or thermal actuation. Piezoelectric and thermal actuations are very susceptible to hysteresis and ambient temperature changes, and therefore they always need an additional measurement circuitry to repeatably control the motion. For this reason, the vast majority uses electrostatic actuation, because it draws almost no power and it can be stable enough to avoid external stabilization circuitry, thus making the overall system simpler, easier to calibrate and therefore cheaper. The general benefit of silicon-based MEMS compared to macroscopic solutions – in addition to the small size and cost – is that even though there is motion, there is virtually no wear and one easily achieves billions of repetitions without a notable effect.

Many have been only concentrating on the MEMS chip itself, creating only the single moving mirror, but one should not forget that this is only one part of the entire system: you also need the light source, focusing optics, band-limiting filters and a detector, along with electronics. In addition, you need to attach the MEMS somewhere without destroying or distorting it, package it hermetically while keeping the optical path available and maybe align some multimode fibers with it. In spectroscopy, the movement usually needs to be rather large while the controlling needs to be very accurate and repeatable, as errors in the wavelength axis of even 0.5 nm might destroy an application. In addition, one practical parameter to consider is the mass that is moving: as the mirror size often needs to be large, the mass of the moving part is often rather high compared to spring forces holding the mass, which makes the system susceptible to vibrations and to directional offsets due to gravity. You can imagine a weight hanging on a spring. So, it does not help if one manages to make a large MEMS mirror, moving hundreds of micrometers, but finding it impossible to control without an external reference laser, the movement being very slow, needing a cooled detector and a shoebox worth of sophisticated electronics and optics. If the MEMS component costs a few euros to manufacture, the end result is a low performing box at 20,000€, then the business case does not really match. In other words, it is very important to design the MEMS and the whole system integration so that the entirety becomes small, easy to assemble and robust, otherwise the end result will not be usable in large scale.

### Recent main types of realizations and main actors

The commercial miniaturization of NIR spectrometers has started from making traditional grating spectrometers as small as possible, and by introducing a new type of compact spectrometer in the 2000s by JDSU (MicroNIR) where they utilized special thin film technology to create a linear variable filter, a type of Fabry–Perot interferometer (FPI). In addition, the first commercial MEMS solution came out with Axsun's NIR engines. Simultaneously, lots of work to develop new MEMS solutions were on its way at VTT (Finland), Fraunhofer (Germany), Si-Ware (Egypt), University of Western Australia (Australia) and Hamamatsu (Japan), to name a few. Coming out to the 2010s, these solutions began their commercial try-outs one-by-one with Hiperscan, Si-Ware, Hamamatsu, Texas Instruments, TellSpec and Spectral Engines. People are already looking toward consumer applications, and more work to further decrease the price and size of the spectrometers are continuously going on. Consumer Physics (Israel), for example, targets consumer markets directly with successful crowd funding campaigns and the first mobile phone integration concept. There are naturally trade-offs when going small and cheap: SNR levels suffer, device-to-device calibrations become more challenging, selecting low-cost silicon as detector material means low detectivity of materials and so on. If the performance is too low, then meaningful, sustainable applications will not carry the business – the same thing happens if the price is too high. All the companies and technologies are balancing between these topics, and often the right answer depends on the application.

Table 6.8 below summarizes current miniaturization technologies and related spectrometer solutions by type. The table also sums up the current main actors and general pros and cons of the solutions. Categorization of the technologies is performed first based on their spectral functioning principle, and then by how the optical axis of the system is organized concerning the movement of the MEMS mirror/plate. In MEMS, one always has a chip, which is rather thin (0.5 mm) and the top area is rather large (5–50mm$^2$). "In-plane" means that the light is coming into the system from the small "side" of the chip, resulting in small optical area, but easier to organize large movements, whereas "off-plane" motion means that the light is coming perpendicular to the surface. There the optical throughput can be high but large movement is more difficult to organize. The third category is "fixed," which means that there are no moving parts at all and the spectral separation is organized spatially: a multipixel detector is necessary. Figure 6.62 below explains these two main types, categorization of the tilting mirror being within the off-plane motion.

One way to categorize these different technologies would be "scanning" or "spatial", but this is in fact represented in the detector type category, as all the scanning systems use a single pixel detector and spatial systems use a multipixel detector. Spatial systems are usually faster and require no active motion control, but the cost of multipixel detectors in the true NIR is too high to be taken to masses: the detector element price can be several hundreds of euros when moving away from silicon-based detector technologies. The silicon-based detectors limit the usable wavelength range to below 1,000 nm, this being a serious drawback of the low-cost detectors, because there is very limited amount of molecular information below 1,000 nm (the so-called fourth and fifth overtones of the actual molecular vibration modes). The price of NIR pixelized detectors is not the only drawback of related solutions: they experience higher noise levels and usually need to be cooled (added complexity and cost) and they experience a noise source called pixel-to-pixel variation, which naturally is nonexistent in the case of a single pixel system. In addition, depending on the geometry of the sample and the optical setup, different pixels might "look at" different physical location on the sample, producing errors in the interpretation of spectral in the case of nonhomogeneous samples. Finally, multipixel solutions are usually too large to be really miniaturized. As a general benefit, the multipixel detectors are often the right choice with fast moving targets. Because of the price level, size and performance challenges of the multipixel systems, the main commercial effort in the recent years have been concentrating on scanning systems and using NIR single pixel detectors (InGaAs and extended InGaAs).

One obvious thing to note on the different actors is that their business model varies. Companies such as Hamamatsu sell quite low-level components, leaving the instrument realization and calibrations to the customer, whereas, for example, Thermo Fisher sells an entire portable analyzer, which includes their MEMS

**Table 6.8:** Summary table of commercial types of MEMS-based NIR spectrometers.

| Main technology category | Motion category | Description of the system principle | Detector type | Main Actor(s) | Main Pros | Main Cons |
|---|---|---|---|---|---|---|
| FTIR | In-plane | Integrated interferometer with plate comb drive | Single pixel | Si-Ware, Hamamatsu | Wide wavelenght range, whole optical bench integrated | Low optical through-put, susceptible to shocks and vibrations |
| | In-plane | Discrete moving mirror component in macroscopic optical bench | Single pixel | Thermo Fisher | Wide wavelength range | System price due to complex assembly and large chip |
| FPI | Off-plane | Surface micromachined FPI, stacked with detector in a hermetic metal can | Single pixel | Spectral engines, Hamamatsu | Very compact, high mass-producibility, robust, fast, high throughput | Limited wavelength range for single component |
| | Off-plane | Several high-order FPIs | Single pixel | Axsun | High throughput, wide range, high resolution | Very complex system, high cost, large |
| | No motion | Linear variable filter on a linear arary detector | Multipixel | Viavi | Compact, robust, no motion | Expensive, size-limited |

(continued)

**Table 6.8:** (continued)

| Main technology category | Motion category | Description of the system principle | Detector type | Main Actor(s) | Main Pros | Main Cons |
|---|---|---|---|---|---|---|
| Grating | Off-plane | Scanning grating | Single pixel | Hiperscan | Wide wavelength range, stable wavelengths | System price due to complex assembly and large chip, low throughput |
| | Off-plane | Scanning micromirror + discrete grating | Single pixel | Texas Instruments, TellSpec | Programmable wavelength selection, wide spectral range, fast | Complex optical bench, size and cost-limited |
| | No motion | Integrated grating | Multipixel | Insion, Zeiss | Integrated optical bench, robust | High price, size-limited |
| | No motion | Discrete grating | Multipixel | Horiba, Ocean Optics, Avantes, Hamamatsu | Robust, well-known | Expensive, size-restricted, high SNR needs cooling |
| Multichannel | No motion | Multiple small band-pass filters on a multipixel detector | Multipixel | Hamamatsu, AMS, Consumer Physics | Compact, low-cost if silicon, no actuation | Expensive if InGaAs, |

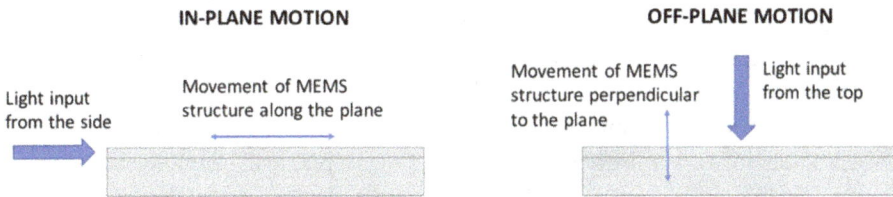

**Figure 6.62:** The two main types of MEMS chips based on the direction of optics and mechanical motion.

component. So, when comparing technology, one should also look at the effort needed to create a new system, because there is a huge amount of effort put and know-how created in many of these companies on how the MEMS can be made to reliable work in actual application conditions.

One interesting topic in various technologies is their future potential. What are the limits? Are they at their limit? Especially, if one is selecting a new platform, it is good to understand how this platform might look like in 5 years. The MEMS FPI technology, for instance, shows highest miniaturization potential because of its small chip size (> 3×3×0.5 mm$^3$) and extremely simple optical system layout (stacked on top of a single pixel detector).

More details on the various MEMS approaches can be found, for example, in [286]. There are also trials to make spatial, or nonmoving, FTIR devices using multipixel detectors, development on tunable NIR laser sources and other special approaches. In addition, there are additional approaches and actors when going through the NIR wavelengths to mid infrared (MIR), but these are not handled here.

## Comparing basic specifications

As the principles for the spectral measurements differ, the technical specifications cannot always be directly compared. This is especially tricky, if a person has been accustomed to using one type of technology, then he or she tries to transfer those specifications to a new type of device. There are two major practical differences to look at: the total measurement time and SNR, and instrument function.

## The total measurement time and SNR

The time you spend on getting your measurement result is usually the real specification you want, and not on the integration time, which is actually only one of the parameters contributing to the total measurement time and the SNR. Selecting the time, one can tune several parameters in the spectrometers to achieve as good an SNR within that time as possible. The main difference is that scanning devices need time for the mechanical scanning when spatial devices do not. Scanning devices do not have a parameter called integration time, because they use a single pixel detector and there, unlike in multipixel detectors, received NIR radiation is converted to electricity directly, without first accumulating and then reading out the result. A special ability exists in the scanning FPIs and micromirror-based grating spectrometers, as one can select the position and number of wavelengths, so one can also optimize the wavelength points for an application. Figure 6.63 shows a practical example of this when Spectral Engines FPI-based sensor has been compared with a grating spectrometer at 1,940 nm, using the same total measurement time of 200 ms, same fiber coupling, target and light source and changing the amount of spectral points to be measured. As the amount of spectral points in a grating spectrometer is fixed, there is no change in the signal level, but as the amount of points are decreased for the FPI, one can compensate this by added averaging and thus get a better signal. In many practical applications, such as moisture measurement, 10 points is well adequate, so SNR improvements of 2–5 are easily achievable with this optimization. The total time, therefore, consists of scanning time, integration time, number of averaging and data readout, transfer and analysis; hoowever, these parameters are relevant or governing depends on the spectrometer type.

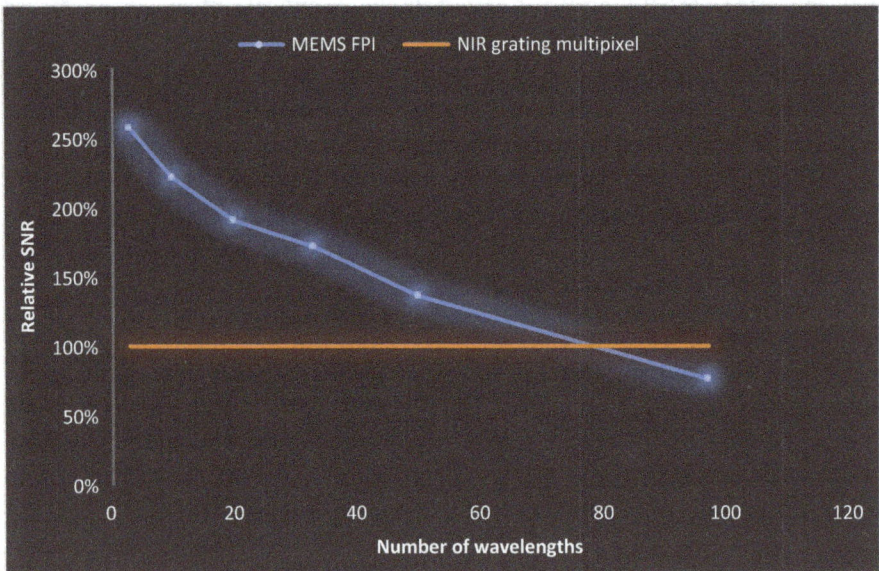

**Figure 6.63:** Comparing a grating spectrometer and a Spectral Engines FPI device. The total measurement time for both was 200 ms and the amount of wavelength points to be measured with the Spectral Engines' device was changed.

### Instrument function

All the different types of spectrometers (FTIR, FPI, grating and multichannel) have a different response to the spectrum. An NIR spectrometer never exactly replicates the real spectrum, but adds its own instrument shape to the results (this is called convolution). This makes it difficult to translate a database built with one type of device to be used by another type. When people are talking about the "resolution" of their device, this can therefore mean different things. The wavelength resolution for most devices are depicted by the Full Width at Half Maximum (FWHM) value, but even this is not directly comparable as the "sideslobes" around the main peak are of different shape, and different portion of the main energy fall within the FWHM, typically from 70% to more than 90%. The FTIR has a different definition all together: the resolution of an FTIR device is the width of the first zero crossings on both side of the main peak. In addition, to make it even more complex, the value can be affected by different kinds of preprocessing, called apodization. Figure 6.64 illustrates the difference between technology types.

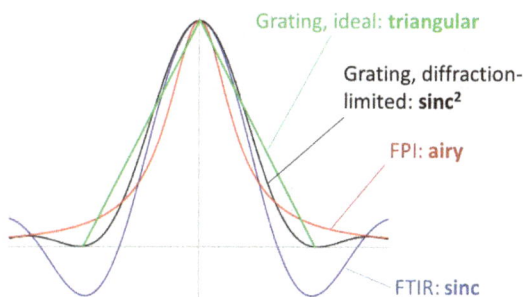

**Figure 6.64:** Various instrument functions for different spectrometer types.

Table 6.9 summarizes most of the spectrometer specifications for different types of devices. The values in the table can be held as indicative, as for example, there are many companies manufacturing grating spectrometers and therefore the size, weight and many other things can vary significant. One quickly sees that for some technologies the manufacturers do not for some reason give specifications and some specifications are not relevant to some of the technologies. One also needs to pay attention on the definition of specifications, as terms such as accuracy, repeatability and the resolution might mean different things.

## Practical comparisons

How can one at the end tell how a specification affects the final result? If possible, the first thing one should do is to make a measurement series using a high-end benchtop FT-NIR with high resolution and wide wavelength range. From that data, one can pin down the actually needed wavelength ranges mathematically. In NIR spectroscopy, one finally measures molecular vibrations: different materials react to light at different wavelengths depending on their molecular bonds. The same molecules react at different "overtone" ranges, that is, water can be measured at multiple wavelengths.

**Table 6.9:** Specification comparison table with notes for common types of spectrometers. Table 6.9 has been compiled by using data gathered from public sources in 2017. Spectral Engines does not take responsibility of the accuracy of this table but recommends always checking from the manufacturer.

| | Things to consider | Surface-micromachined MEMS FPI | MEMS FTIR | LVF | General grating | Monolithic grating | Grating with TI DLP | Multichannel detector |
|---|---|---|---|---|---|---|---|---|
| **Wavelength ranges** | These depend on technology and detector type used | 1.3–1.7 µm 1.5–2.0 µm 1.7–2.2 µm 1.9–2.5 µm | 1.25–1.7 µm 1.3–2.1 µm 1.35–2.5 µm | 0.95–1.65 1.15–2.15 | 0.9–1.7 µm 1.1–2.2 µm 1.1–2.5 µm (Ext-)InGaAs | 0.9–1.7 µm 1.1–2.2 µm (Ext-) | 0.9–1.7 µm | 0.6–0.9 µm 0.75–1.0 µm |
| **Detector** | Single pixel detector area is 100× bigger and cost 100× less. | Single-point (Extended) InGaAs | Single-point (Extended) InGaAs | (Ext-)InGaAs array | InGaAs linear array with 2 stage TE-cooling, 256 or 512 pixels | InGaAs array, 128 or 256 elements | Single-point InGaAs | Silicon image sensor |
| **Spectral separating element** | The basic technology type for separating the spectrum | MEMS Fabry–Perot | MEMS FTIR | LVF | Grating | Monolithic grating | MEMS DMD from Texas instruments | Fixed filter array on silicon array detector |
| **Optical interface** | Many devices have been only fiber-coupled | Open or fiber | Open or fiber | Open or fiber | Fiber | Fiber | Open or fiber | Open |
| **Entrance slit** | Relevant for grating spectrometers | Not relevant | Not relevant | Not relevant | 10–200 µm | 60µm × 300 µm | 1800×25 µm | Not relevant |
| **Resolution** | Commonly FWHM figures. | Typical <13–20 nm | 8 nm or 16 nm @ 1,550 nm, FWHM | Typical 10–20 nm | 4–90 nm (depends on grating and slit) | 10–16 nm | 10–12 nm | 20 nm |

| | Description | | | | | | | |
|---|---|---|---|---|---|---|---|---|
| **Wavelength accuracy** | Relates to absolute, traceable accuracy. Often already the calibrated references are ± 1.5 nm at least. | ±0.3 nm (low-pressure Ne of Kr lamp at single wavelength) | ± 1.5 nm, <40 °C | < 3 nm | ± 1.3 nm | Not given | Not given | Not given |
| **Wavelength repeatibility** | Measurement-to-measurement or device-to-device? Very rarely given by companies. Optical benches with discrete components experience hysteresis. | ±0.3 nm (device-to-device) | ± 0.1 nm | <1 nm | ± 0.3 nm | Not given | Not given | Not given |
| **Temperature induced drift** | | < 0.03 nm/°C | Not given | Not given | 0.05–0.1 nm/°C | <0.05 nm/°C (from VIS version) | Not given | Not given |
| **Mechanical scanning time** | Scanning modes can be different: step-wise or continuous. Different between | 1 ms between wavelengths | Not given | Not relevant | Not relevant | Not relevant | 15 ns | Not relevant |
| **Wavelength step size** | scanning (programmable) and fixed pixel devices | 0.01–400 nm | Not given | 6 nm | 2–10 nm | 4–8 nm | Not given, programmable | Not given |

(continued)

Table 6.9 (continued)

| | Things to consider | Surface-micromachined MEMS FPI | MEMS FTIR | LVF | General grating | Monolithic grating | Grating with TI DLP | Multichannel detector |
|---|---|---|---|---|---|---|---|---|
| **Signal-to-noise ratio** | Often announced for some measurement time or amount of averaging at ideal lighting conditions. | >10'000 @ 0.5 s | >3'000 @ 2 s | 23'000 with 100 averaging | 5'000 | 5'000 | 5'000 typical | Not given |
| **Integration time** | Integration time is fixed for single pixel devices. Signal quality there is increased only by averaging. | 30 µs per wavelength | 2.5 ms min | 10 µs min | 10 µs – 1,000 ms, depends on type | 2–40.000 ms | Not Given | Not Given |
| **Operating temperature range** | Very limited if active cooling is applied. MEMS range can often be tailored. | +10–+50 °C | –5–+40 °C | –20 to 50 °C, non-condensating | +5–50 °C C (+5–35 °C), non-condensating | 5 °C below ambient. 40 °C (cooled) | 0–50 °C | +5–50 °C, non-condensating |
| **Power requirement** | Questions: Lamps included, cooling included, idle condition or "full speed"? | <0.3W | <2.5 W | <2.5 W | <2.5 W, <40W (with detector cooling) | 7.5W (cooled) | Not given | <0.02W |

| | Notes | | | | | | | |
|---|---|---|---|---|---|---|---|---|
| **Size** | Envelope size. Some measures include light sources, some include bluetooth and so on extra boards. | ca. 25×25×20 mm, including light sources | 80x65x45 mm | 45 (diam) × 48 (height) mm, including light sources | 300 × 200 × 100 mm | 108 × 76,6 × 21,5mm | 82.2 mm (L) × 66 mm (W) × 45 mm (H) | 4.5 × 4.7 × 2.5 |
| **Weight** | Check what is included | < 15 g | 150 g | 60 g | 1–3 kg | 130 g | 136 g | Not given |
| **Shock** | Usually not given. If given, refers to certain standard, but not the same from tech to tech | IEC 60068-2-31 | Not given | Drop test ISTA 2A | Not given | Not given | Not given | Not given |
| **Vibration** | Usually not given. If given, refers to certain standard, but not the same from tech-to -tech | MIL-STD-810G | Not given | MIL-PRF-28800F Class 2 | Not given | Not given | Not given | Not given |

Therefore, if you need a moisture sensor, you do not need a certain range in the spectrum. The higher the wavelengths, the more sensitive and specific one gets and the lower, the more insensitive and ambiguous the spectral shapes get. This means that even though the FPI platform looks like it has a major restriction in the specifications in the above-mentioned table because of the wavelength range of a single device, the vast majority of NIR applications can be covered with this technology, as one very rarely needs to cover multiple overtone regions in a high-volume, commercial-sensing application. Figure 6.65 below illustrates the overtones and molecular bonds to be detected.

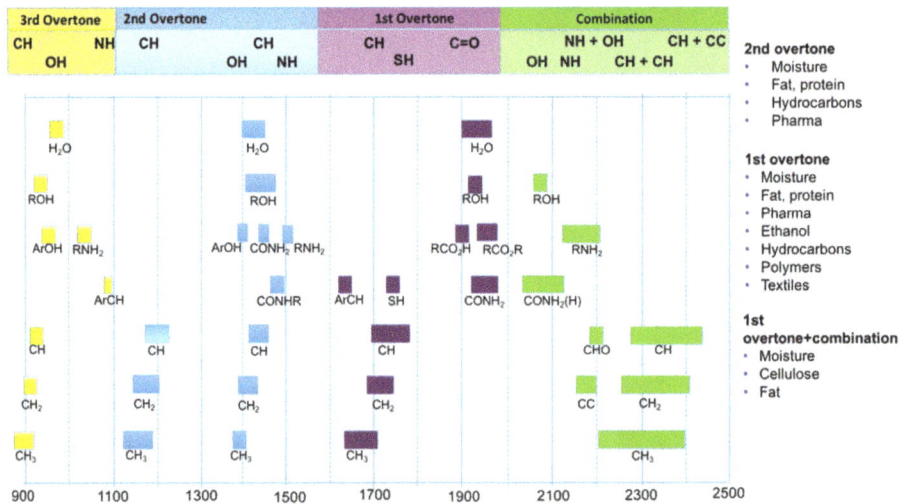

**Figure 6.65:** The NIR absorption bands and typical application areas.

Even after going through all specifications, it is still down to testing the application at hand to see how the devices perform. What does a narrower spectral range do to the final error, Is it better to have a good SNR or good resolution? VTT Technical Research Centre of Finland made practical comparison measurements between three different types of spectrometers in three applications: (1) polyethylene coating thickness on paper, (2) paper moisture content and (3) paracetamol, lactose and microcrystalline cellulose in pharmaceutical blends. To overcome the ambiguity of specifications, VTT set the total measurement time to be equal, and optimized other parameters within that restriction. In addition, optics was similar for all fiber-optical probes with external light source.

Table 6.10 and Table 6.11 summarize the results. Interestingly, the MEMS FPI (Spectral Engines' sensor) outperformed the wider range MEMS FTIR and grating instruments in moisture measurement and was equal to the grating spectrometer with cooled detector in polyethylene measurements. In both cases, the error was 20 times smaller than that of the MEMS FTIR. This can be mostly explained by the low throughput of the MEMS FTIR, as a result of the in-plane configuration similar to what is explained in the previous sections, leading to long measurement times or high noise. In the pharmaceutical blends, the MEMS FPI and the grating spectrometer produced very similar results, with linear array being slightly better. This demonstrates well that the end performance is not always straightforward to forecast from technology specifications only and even from one application to another!

**Table 6.10:** Error of achieved application results for polyethylene thickness and moisture content.

| Spectrometer | Polyethylene 2-σ error [g/m²] | Moisture 2-σ error [weight-%] |
|---|---|---|
| Micro FTIR | 0.44 | 0.100 |
| Micro FPI | 0.02 | 0.005 |
| Linear Array | 0.01 | 0.017 |

**Table 6.11:** Error of achieved application results for pharmaceutical blends.

| Spectrometer | Paracetamol RMSEP [weight-%] | Lactose RMSEP [weight-%] | Microcrystalline cellulose RMSEP [weight-%] |
|---|---|---|---|
| Micro FPI | 2.12 | 3.45 | 2.20 |
| Linear array | 2.10 | 2.69 | 1.79 |

Another practical example is the EU Horizon Foodscanner Prize of 800,000€. This prize was set up to motivate companies and communities to build a device that can measure the nutrition values in food, allergens and potential harmful residues. The idea was to get new tools to fight against obesity, diabetes and so on. All top 3 finalists represented the (V)NIR spectroscopy technology, showing the power of the method. The finalists were Spectral Engines (MEMS FPI), TellSpec (grating with Texas Instruments' DMD) and Consumer Physics (VNIR multichannel). Fifty unknown food substances were measured and a 15-person panel reviewed the technologies and the solutions. At the end, Spectral Engines was selected as the winner of the main prize, demonstrating once again the power of MEMS FPI-based sensor technology.

## Summary
There are plenty of technologies within NIR spectroscopy to choose from nowadays as development is going fast toward smaller and cheaper devices. It is generally believed that NIR-based material sensors will appear for various industrial and consumer applications within this decade. Selecting the right platform for the needs requires some background research, but with a few main points to consider the options are readily narrowed down.

The questions to answers are as follows:

Do you want to purchase components and invest in instrument and calibration development yourself?

Do you want to get a ready-made platform which is fitted to your application and gives you short time-to-market?

Do you aim at high-volume market, a few special devices or something in between?

Do you need a shock- and vibration-resistant technology?

Do you need to connect devices to cloud platforms to perform analysis?

After answering those top-level questions first, one can start getting into the details of comparing spectrometers. Finally, if one is selecting a platform, one needs to evaluate the *future* of the platform, too: Can this be made even smaller and cheaper or is this the limit?

### 6.3.7.3 Design of in-line probes for liquids

There is a great variety of different in-line probes (see for example the companies Hellma and Avantes) which cover every possible situation in process analysis [287, 340, 350]. Depending on the probe design, the techniques can be divided into flow through cells and immersion cells. Figure 6.66 shows a transmission immersion probe which is mainly used in the NIR and UV/Vis. The major advantage of this probe is that it can be used directly inside the liquid and no bypass is necessary. The transmission flow cell, which is also shown, can only be used in a by-pass. The advantage here is

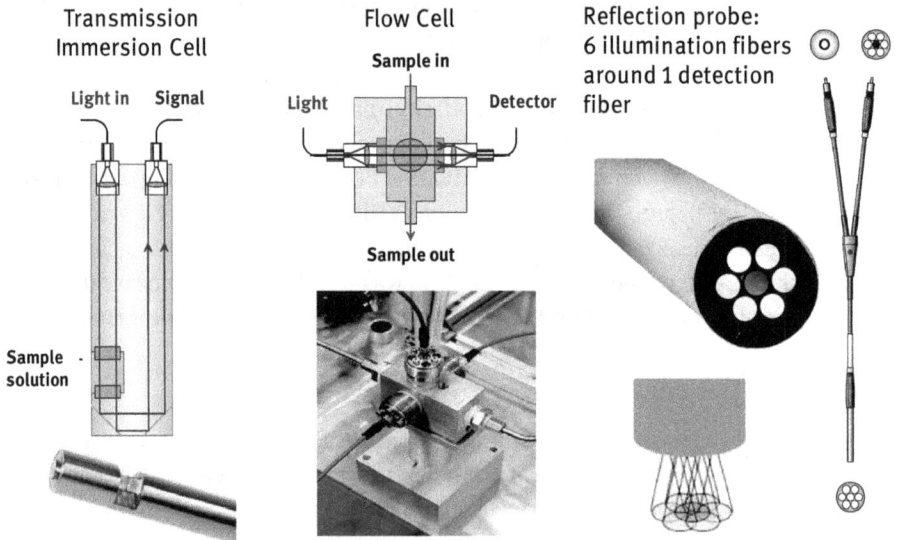

**Figure 6.66:** Probe technologies for UV/Vis and NIR spectroscopy. The Reflections probe is shown with the permission of Ocean Optics.

that it also has a viewing port at 90° to the direct transmission port. This feature allows the measurement of the fluorescence of an analyte, but can also be used to measure the scattered light of turbid liquids. The pathlengths are typically between 1 and 10 mm. Reflection probes are optimized for measuring diffuse and specular reflectance, color, fluorescence, and backscattering of solid, liquid, and powder samples.

The reflection probe shows that six fiber bundles are connected with the light source to illuminate the sample volume optimally and one fiber bundle is connected to the spectrometer.

Based on the tendency to absorb IR radiation for solvents, IR cells are normally much narrower (0.01 to 1 mm) in comparison to NIR. Therefore, often, high sample concentrations from 0.1% to 10% are required because the pathlength and the molar absorptivity are small. Liquid IR-cells use Teflon spacers to allow variation in the pathlength. A variety of window materials exists with a range of transparency and solubility in the solvent (Table 6.12).

**Table 6.12:** Typical IR materials for spectroscopy.

| Window material | Usable range (cm$^{-1}$) | Water solubility g/100 g Water, 20 °C |
|---|---|---|
| Zinc selenide | 10,000–500 | insoluble |
| Fused silica | 50,000–2,500 | insoluble |
| Thallium bromide-iodide, KRS-5 | 16,500–250 | 0.048 |
| Calcium fluoride | 50,000–1,100 | $1.5 \times 10^{-3}$ |
| Sodium chloride | 40,000–620 | 36.0 |

The attenuated total internal reflectance (ATR) technique enables samples to be examined directly without further preparation. It is mainly used in the NIR/IR region. In the principle of total internal reflection, a beam of infrared light is passed through a crystal in such a way that it reflects at least once off the internal surface in contact with the sample. This reflection results in an evanescent wave which escapes from the crystal surface. The angle at which total internal reflection occurs is called the critical angle ($\theta_c = sin^{-1}n_2/n_1$). When the sample is brought into contact with the prism surface, the evanescent wave will be attenuated in the region of spectrum at which the sample absorbs (Figure 6.67).

The depth of this penetration depends on the wavelength of the beam, the refractive index of the sample, and the crystal as well as the beam angle. The penetration depth can be calculated by the equation

$$d_p = \frac{\lambda_c}{2\pi \left[ sin^2\theta - \left(\frac{n_s}{n_c}\right)^2 \right]^2}$$

where $\lambda_c$ is the wavelength in the crystal ($\lambda/n_c$), $\theta$ is the angle of incidence, and $n_c$ and $n_s$ are the refractive indexes of the sample and crystal, respectively. The advantage of

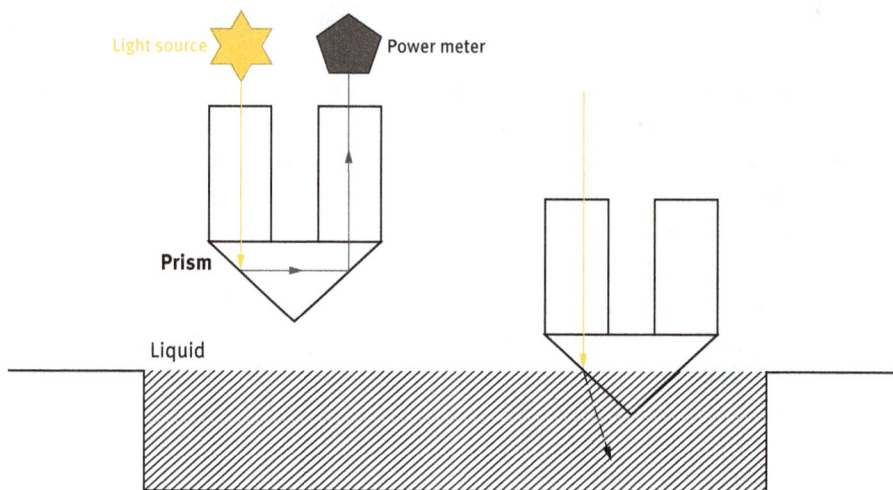

**Figure 6.67:** Schematic of the ATR-principle.

this technique is that the inert crystal materials such as diamond and sapphire can be used, which is important in harsh environments. Bubbles and particulate materials do not have a significant effect on measurements, but using high concentrations without sample dilution is not feasible with this method.

Figure 6.68 shows probe technologies for ATR measurement. Tunnel cells have been designed for continuous online analysis. The stainless steel construction, integrated permanently aligned optics, and integrated temperature control insure the high level of reliability needed for process applications. By proper adjustment of the incident angle, the radiation undergoes multiple internal reflections before passing to the detector. Both absorption and attenuation take place at each of these reflections. ATR probes are suitable for direct measurement in strongly absorbing solutions where standard transmission probes cannot be used.

The use of transmission cells is straightforward as these cells can directly quantify the resulting spectra with high linearity. The Lambert-Beer law will be valid as long as there is no scattering. Even then, the optical geometry is robust and highly reproducible. This is also true for the ATR immersion cell. However, there are some additional problems with quantification when an ATR tunnel cell is used. There are severe sources of errors in cylindrical cells which come from the optical construction of the probe; for example, the distribution of incident angles, the number of reflections, optical aberrations, skew ray effects (light coupling in and out), polarization, and refractive index variations. A feature which can solve almost all limitations is the integration of a conical reflector (Company AXIOM) at the input and output of the internal reflectance element. The input cone reflects each paraxial ray so that it strikes the conical end surface of the element at approximately normal incidence. The output cone redirects the emerging cone of rays to the paraxial direction. In addition, stops located at the

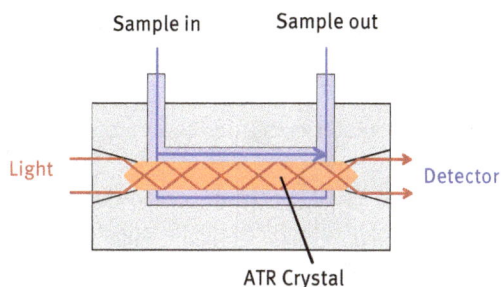

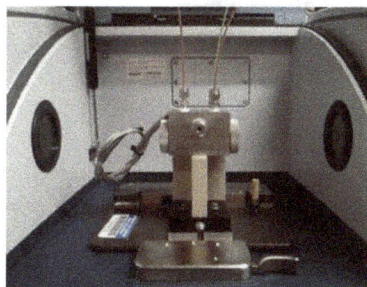

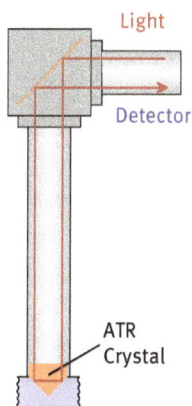

**Figure 6.68:** On-line ATR tunnel cell and ATR immersion probe [350].

large end of each reflecting cone block any rays which would otherwise enter the rod without first being reflected from the cone. This design results in three significant enhancements: an increase in absorbance by a factor of 2–3; as the calibration is a fixed factor for a given cell, so no adjustments are necessary; and there is minimization of optically induced non-linarites.

## Self-Assessment Questions

1. For a grating, the first-order diffraction line for $\lambda$ = 400 nm is observed at a reflection angle of 5° with the angle of incidence of 45°. How many lines per millimeter would be required?
2. A spectrometer uses a 25 µm slit, a 14 µm 2048 pixel detector, and a wavelength range from 350–1,050 nm. What will be the resolution of the system?
3. Calculate the frequency range of a modulated signal from a Michelson interferometer with a mirror velocity of 0.20 cm/s for visible radiation of 700 nm.
4. Why do qualitative and quantitative analyses often require different monochromators slit width?

5. A 10-cm interference wedge is to be built that has a linear dispersion from 400 to 700 nm. Describe details of its construction. Assume that a dielectric with a refractive index of 1.32 is to be used.

6. What are the advantages of an FT-IR spectrometer compared with a dispersive spectrometer?

7. For measuring trace component in a liquid, a FT-IR spectrum is recorded with a set of sixteen interferograms. The signal-to-noise ratio for a specific spectral band was about 5:1. How many interferograms would have to be collected and averaged if the goal is to obtain a S/N of 20:1?

8. An aqueous solution with a refractive index of 1.003 is measured with an ATR crystal with a refractive index of 2.8. The incidence angle is 45°. What is the effective penetration depth at 3,000 cm$^{-1}$, 2,000 cm$^{-1}$, and 1,000 cm$^{-1}$. Is absorption by the aqueous solvent as much of a problem in ATR as in normal IR absorption measurement? Why or why not?

## 6.3.8 Raman spectroscopy

In 1928, C.V. Raman and Krishnan discovered that a small fraction of visible radiation is scattered by certain molecules and the scattered light differs from the wavelength of the incident beam. Raman found that these shifts in wavelength depend on the chemical structure of the molecule that is responsible for the inelastic scattering. For this discovery, Raman was awarded a Nobel Prize in 1931 for his systematic investigations.

### 6.3.8.1 Theoretical background

Nearly all scattering occurs without loss of energy (Rayleigh scattering). However, 1 in $10^6$ scattering events may result in the gain or loss of energy. This is also the reason why Raman spectroscopy has gained more interest of late, as in the early 1930s, the technology was not available for such low detection limits. Today, Raman spectroscopy is used in liquids, solids, or gases [317]. Raman spectroscopy considers the fundamental vibrational information contained in the visible and NIR light scattered from a sample. Molecules in the ground vibrational state ($v = 0$) can absorb a photon of energy $h v_{ex}$ to a virtual level j which is far away from an absorption band and they subsequently reemit a photon of energy $h(v_{ex} - v_v)$ of lower energy. Molecules in a vibrationally excited state ($v = 1$) can scatter inelastically and return to the ground state, producing a Raman effect with energy $h(v_{ex} + v_v)$.

The lower frequency transition is the Stokes scattering transition and the higher frequency transition is the Anti-Stokes scattering (Figure 6.69). If the system is in thermal equilibrium, the equilibrium populations of the ground and the excited

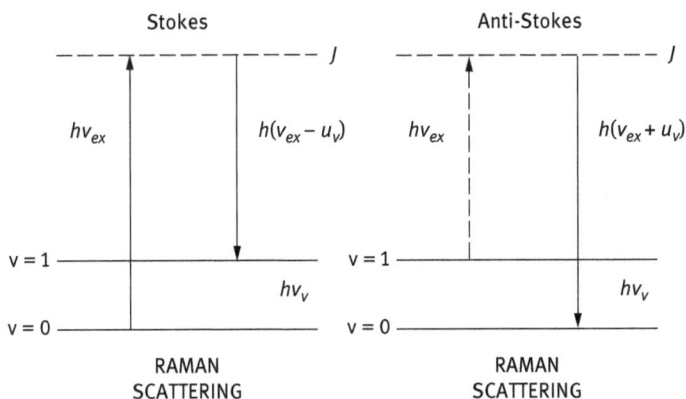

**Figure 6.69:** Schematic illustration of Raman scattering.

states follow a Boltzmann distribution. Stokes scattering is more intense than Anti-Stokes scattering as the ground state is more populated.

At room temperature, the thermal population of vibrational excited states is low; therefore, for the majority of molecules, the initial state is the ground state, and the scattered photon will have lower energy (longer wavelength) than the exciting photon. This Stokes shifted scatter is what is usually observed in Raman spectroscopy. The time spent in this virtual excited state, j, is very small ($10^{-14}$ s). In contrast, the lifetime of a true vibrational excited electronic state is about $10^{-8}$ s. In a Stokes Raman spectrum, energy is lost from the laser frequency. As the amount lost depends on the different Raman active vibrations, that can take place in the molecules. The spectrum produced is a series of sharp bands that exists at different wavelengths shifted from that of the initial laser wavelength. The wavelength difference is small and the spectrum is normally recorded as bands at certain wavenumbers shifted from the Rayleigh scattered laser line. This is convenient for comparisons with the spectral features/vibrations of a mid-IR spectrum. Although there can be striking similarities between Raman spectra and IR spectra, enough differences remain between the kinds of groups that are IR active and Raman active to make the techniques complementary rather than competitive. Figure 6.70 shows the Raman spectrum of cyclohexane.

The energy units normally used on the horizontal axis are in wavenumbers. Since a wavenumber in a Raman spectrum is a difference measurement, it is independent of the excitation frequency used to generate the spectrum. This unit is also the one used for infrared spectroscopy. The intensity of Raman scattered light is proportional to the number of molecules that produce the Raman scattered light. As a result, the intensity of Raman scattered light can be used to measure how much of a material is in a sample for quantitative analysis.

The intensity of the Raman scattered radiation is given by

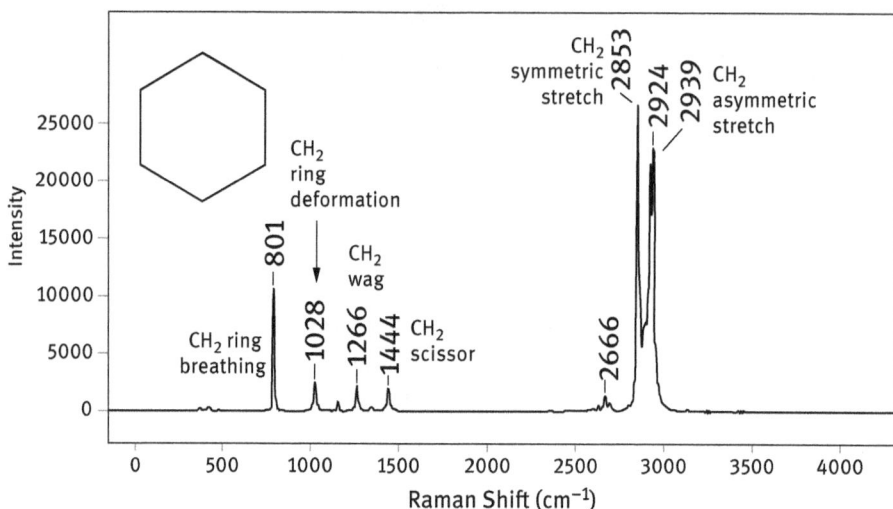

**Figure 6.70:** Raman spectrum of cyclohexane.

$$I \sim v^4 I_0 N \left( \frac{\partial \alpha}{\partial q} \right)^2$$

where v is the frequency of the exciting laser; $I_0$ is the intensity of the exciting laser; N is the number of scattering molecules, and $\partial \alpha / \partial q$ is the polarizability change.

The Raman selection rule of a change in polarizability during a vibration is analogous to the selection rule of an infrared-active vibration, which states that there must be a net change in the permanent dipole moment during the vibration. From group theory, it is straightforward to show that if a molecule has a centre of symmetry, vibrations that are Raman active will be silent in the infrared and vice versa. If a vibration does not greatly change the polarizability, then the intensity of the Raman band will be low. The symmetry of a molecule defines what vibrations are Raman and IR active. Generally, symmetric or in-phase vibrations and non-polar groups can be easily measured by Raman spectroscopy, while asymmetric and out-of-phase vibrations and polar groups are easily measured by IR. For example, the vibration of the highly polar group like O–H is usually Raman weak because an external electric field cannot induce a large change in the dipole moment, and stretching or bending the bond does not change this. An important advantage of Raman spectroscopy over IR lies in water being a useful solvent. Typically strong Raman scatterers are substances with distributed electron systems such as C=C bonds or the π-electron cloud of aromatic systems.

The shape of a Raman spectrum can be used to determine the types of molecular vibrations in a sample. This vibrational information can be used to identify materials in a sample (qualitative analysis). Many molecular vibrations are sensitive to their micro-environment. Differences in stress, temperature, crystal structure, micro-heterogeneity, etc., can, therefore, often be measured using Raman spectroscopy.

Raman scatter is partially polarized, even for molecules in a gas or liquid, where the individual molecules are randomly oriented. The effect is most easily seen with an exciting source that is plane polarized. In isotropic media, polarization arises because the induced electric dipole has components that vary spatially with respect to the coordinates of the molecule. Raman scatter from totally symmetric vibrations will be strongly polarized parallel to the plane of polarization of the incident light. The scattered intensity from partially symmetric vibrations is 3/4 times strong as in the plane perpendicular to the plane of polarization of the incident light as in the plane parallel to it.

### 6.3.8.2 Instrumentation

The main reason for the increase in use of Raman spectroscopy in on-line analysis is the development of more compact and reliable spectrometers. Advances in laser technology and detectors have also had beneficial effects. In addition, because signals are usually in the visible or near-IR region, glass or quartz cells can be employed avoiding the inconvenience of working with sodium chloride or other atmospherically unstable window materials.

A Raman spectrometer can be used with a number of probes for a variety of applications (Figure 6.71). Monochromatic light is coupled with an optical fiber and relayed to the universal probe head. The universal probe head acts as a small spectrograph to filter background (Silica Raman and luminescence) generated by the interaction of the laser light with the optical fiber. The laser light is then brought to a focus on the sample. The sampling optic can be any one of a wide variety of optical elements ranging from a high powered microscope objective to a process hardened immersion probe.

The backscattered Raman signal is then collected through the sampling optic that filters out the laser radiation on the probe head and relay on the Raman signal to a second optical fiber which is connected to the analyzer. The Raman signal passes through the pre-filter portion of the analyzer where any residual laser signal is removed. The Raman signal enters the spectrograph portion of the analyzer where the light is dispersed and imaged to the two-dimensional CCD detector. A PC is used to analyze and display the Raman spectrum.

The laser choice is important. Wavelengths typically used range from 700 to 1,200 nm, but can go down as far as 500 nm. As the wavelength decreases, scattering intensity increases. However, at low wavelengths, problems can be encountered with background fluorescence. Also, the transmission of laser light through quartz is more efficient as you approach 1,200 nm. Some of the available lasers with wavelengths are as follows (Table 6.13).

**Table 6.13:** Common Laser Sources for Raman spectroscopy

| Laser Type | Wavelength (nm) |
|---|---|
| Argon-Ion | 488 or 514.5 |
| Krypton-Ion | 530.9 or 647.1 |
| Helium-Neon | 632,8 |
| Diode | 785 or 830 |
| Nd-YAG | 1064 |

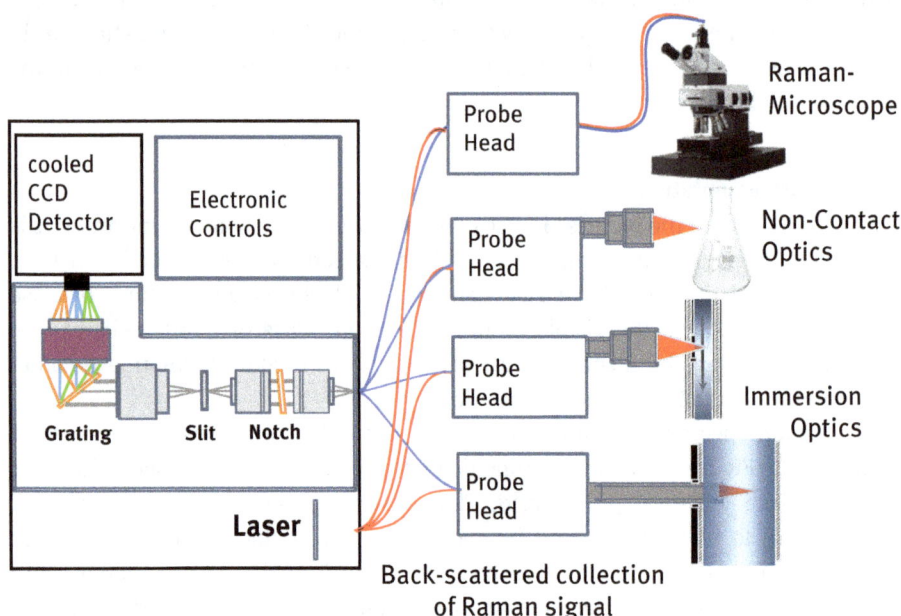

**Figure 6.71:** Fiber-optic Raman spectrometer with different probes. Figure modified according Kaiser Optical Systems (KOSI)

A dispersive Raman spectrometer uses a volume holographic transmission grating in combination with photographic lenses to achieve almost aberration-free imaging performance in a compact instrument. The system includes a pre-filter stage in which the first lens collimates the input signal, which then passes through the notch filter to attenuate laser light. The filtered signal, with laser light intensity, reduced by at least $10^6$ is focused on the spectrograph entrance slit with the second lens. The slit functions as a spatial filter to minimize transmission of stray light into the dispersion stage of the spectrograph. The diffracted light from the spectrograph stage is focused onto the surface of a CCD detector, which records the actual spectrum. When in-line Raman spectra are to be measured, the spectrometer is operated with a fiber optic attachment and probe head. Multimode optical fibers make excellent general-purpose sample delivery and collection devices.

The common uses of probes in Raman spectroscopy are the concentric unfiltered fiber bundle or "n-around-1" probe and the filtered probe. The fiber bundle is less expensive, while the filtered probe offers better signal-to-background ratio in certain applications.

Figure 6.72 shows a concentric fiber bundle probe in which the central excitation fiber delivers laser light to the sample and the surrounding fibers collect Raman scatter. For coupling to the spectrograph entrance slit, the output ends of the collection fibers are arranged in a line. The linearized bundle is placed directly against the entrance slit of the spectrograph. The optical fibers themselves are significant source of background signal. Laser light travels down a meter of fiber or more, generating an intense silica Raman spectrum. Fluorescence from the fiber cladding and from the cement holding the fiber bundle together are also generated. These signals emerge from the delivery fiber along with laser light. They are reflected from the surface of any liquid or solid and collected along with the Raman spectrum of the sample. Reflected laser light generates more background as it travels through the collection fibers. In most cases, the background can be subtracted adequately but it may obscure weak Raman signals. Photomultipliers, photodiodes, or charge-coupled devices (CCDs) are used to convert photons from the spectrometer into measurable electrical current.

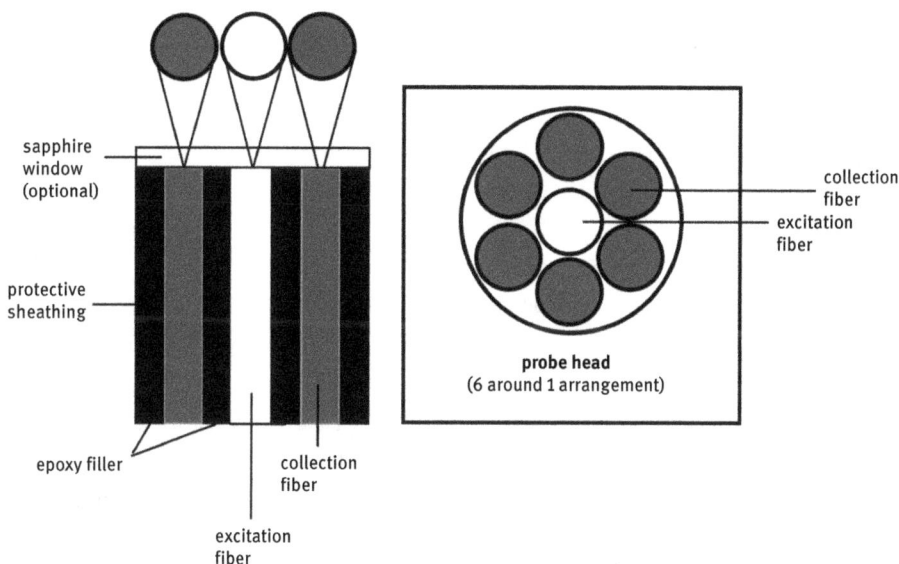

**Figure 6.72:** Representation of an unfiltered n-around-1 probe; in this case a 6-around-1 probe: (a) cross-sectional view and (b) end-on view.

The pseudo-color image from Figure 6.74 represents the signal when taking a spectrum of the cooled CCD detector. The image of an optical fiber illuminated by a white light source after the fiber image has passed through the spectrograph. The dispersed

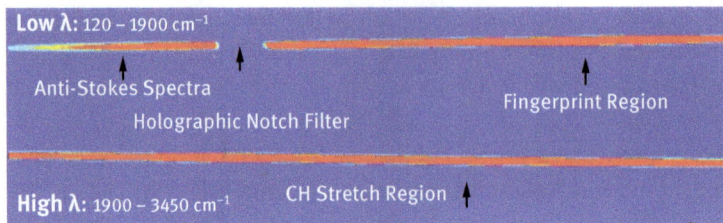

**Figure 6.73:** CCD image of a Raman spectrometer.

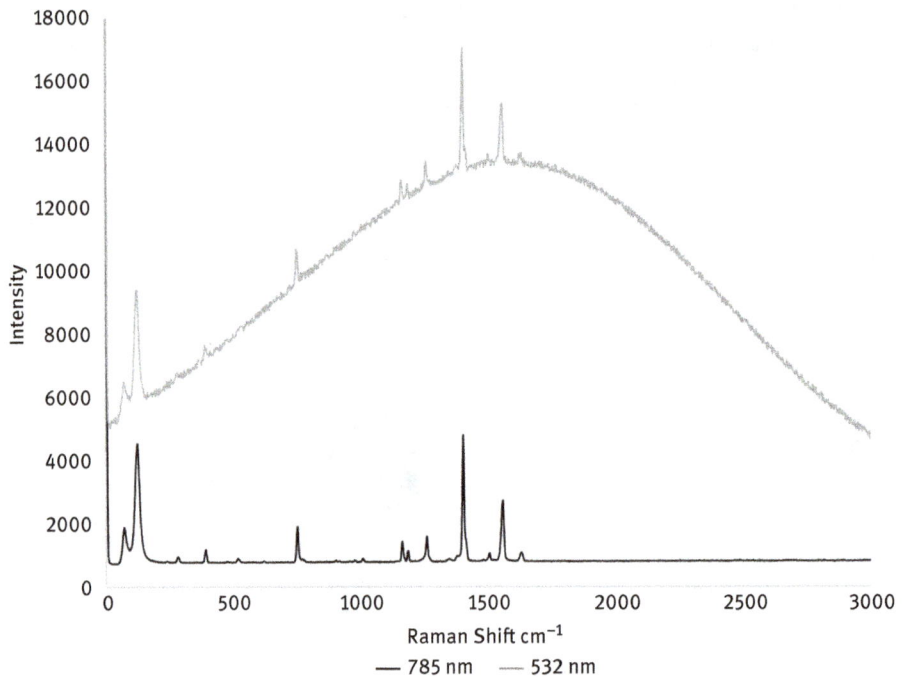

**Figure 6.74:** Raman spectra of anthracene taken with a 785 nm and a 532 nm laser source.

light appears as two streaks; one from each grating on the detector image. The upper streak represents the spectra from about 120 cm$^{-1}$ to 1,900 cm$^{-1}$. The lower streak represents the spectra collected from 1,900 cm$^{-1}$ to 3,450 cm$^{-1}$.

Raman spectroscopy is not a very sensitive technique. Compounds present at concentrations below 0.1% to 1% range are often not detected in typical samples. The limited sensitivity is a direct consequence of the low probability for the detection of a Raman scattered photon. Also, Raman photons are scattered in nearly all directions; so only a fraction of them can be collected by the Raman instrument. Optical processing of the collected Raman photons by traditional Raman instruments loses 80–97%

of the photons. Photomultipliers, the traditional light measuring devices in Raman instruments, detect only 1 photon in every 5. Reliable measurements require many detected Raman photons; so low analyte concentrations are difficult to measure using Raman spectrometry.

In addition to Raman scattered light, most samples produce other types of scattered light when illuminated [323]. The contribution from this "non-Raman light" can usually be subtracted from the Raman spectrum, but the noise from the non-Raman light cannot. The minimum noise in an optical intensity measurement is equal to the square root of the detected photons due to the statistical nature of light. This minimum noise is called "shot noise." When the shot noise from the background is too large compared to the intensity of the Raman signal, the Raman measurement cannot be made reliably. The most common source of background encountered in Raman measurements is fluorescence. Fluorescence is a much more efficient process than Raman scattering. A single incident photon can produce a fluorescence photon with a high probability. Consequently, impurities at the part-per-million level that fluoresce can obscure the Raman spectrum of an otherwise pure material. This situation is common and has seriously limited in the practical application of Raman spectroscopy. Spectra of anthracene taken with a Raman instrument with an argon-ion laser source at 514.5 nm (A) and Nd-YAG source at 1,064 nm (B).

There are several variations of the Raman technique available today. Some of the most common and used ones are:

- SERS: Surface-enhanced Raman spectroscopy [319, 329]
- TERS: Tip-enhanced Raman spectroscopy [335, 352]
- CARS: Coherent anti-Stokes Raman spectroscopy [292, 302]
- SRS: Spontaneous Raman scattering [342]

## Surface-Enhanced Raman Scattering (SERS)

Raman scattering from a molecule adsorbed on to a structured metal surface can be 100 times greater than in solution. This surface-enhanced Raman scattering effect is very strong on silver and gold but also observable on copper. SERS occurs via two mechanisms. In the first mechanism, an enhanced electromagnetic field is produced at the surface of the metal. When the wavelength of the incident light is close to the plasma wavelength of the metal, conduction electrons in the metal surface are excited into an extended surface electronic excited state which is called surface plasmon resonance. Molecules adsorbed onto or are close to the surface experience an exceptionally large electromagnetic field. Vibrational modes normal to the surface are the most strongly enhanced. The second mechanism for enhancement is by the formation of a charge-transfer complex between the surface and analyte molecule. The electronic transitions of many charge-transfer complexes are in the visible region; so, resonance enhancement can occur when the complex is excited with visible radiation in Raman spectroscopy. Molecules with lone-pair electrons or π-clouds show the strongest SERS.

The effect was first discovered with pyridine. Other aromatic nitrogen- or oxygen containing compounds, such as aromatic amines or phenols, are also strongly SERS active. The effect can also be seen with other electron-rich functional groups, such as carboxylic acids. The intensity of the surface plasmon resonance is dependent on many factors, including the wavelength of the incident light and the morphology of the metal surface. The wavelength should match the plasma wavelength of the metal. SERS is used to study monolayers of materials adsorbed on metals, including electrodes. However, many materials other than electrodes can be used; and the most popular include colloids, metal films on dielectric substrates, and, more recently, arrays of metal particles bound to metal or dielectric colloids through short linkages. Although SERS allows easy observation of Raman spectra from solution concentrations in the micromolar ($10^{-6}$ M) range, slow adsorption kinetics and competitive adsorption limit its application in analytical chemistry.

Raman spectroscopy, as it is applied today, is ideally suited to real-time reaction monitoring and to the characterization of industrial compounds in the polymer, chemical, and pharmaceutical industries. Polymorphs may have different properties, such as solubility, dissolution rate, stability, or bioavailability. Raman spectroscopy is able to discriminate between polymorphs because different crystal forms provide intensity and frequency changes in the Raman spectrum. The Raman technique can be applied without sample preparation and allows for non-destructive and in-situ measurements.

Handheld Raman spectrometers have gained particular attention recently, especially in the security areas where it is crucial to know on the spot, chemicals, with the most current drugs and hazardous materials, including explosives. Several examples are available on the market with increasing performance. Implementation of Raman spectroscopy in Optofluidic systems is of high interest to the Optofluidics community and current work is focused towards this goal. An example of such a development is a sensor based on the so-called waveguide-confined Raman spectroscopy (WCRS) presented in [355].

One of the smallest MEMS Raman spectrometers with integrated sample handling has been developed by Metrohm with a size of 13.0 cm (H) × 8.5 cm (B) × 4.0 cm (T). A photograph of the device is shown in Figure 6.75. The resolution of the MEMS Raman spectrometer is 14 cm$^{-1}$ and the spectral range is 400–2,300 cm$^{-1}$.

## Self-Assessment Questions

1.  Create a table to summarize characteristics of NIR, MIR, and Raman spectroscopy. Use the following terms: wavenumber, bonds, absorption due to, absorption strength, absorption bands, signal intensity, quantification, excitation conditions, selectivity, interferences, particle size, applicability for at-line, online, in-line, radiation source, and sample preparation.

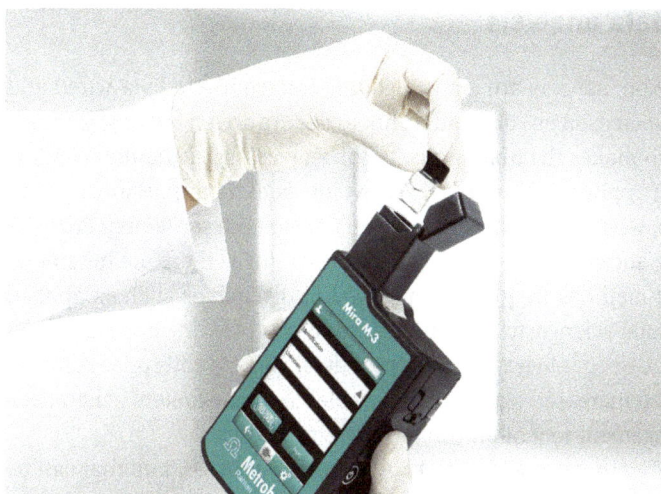

**Figure 6.75:** Mira M-3 Raman handheld spectrometer from Metrohm.

2.  At what wavelengths in nanometres would the Stokes and anti-Stokes Raman lines for carbon tetrachloride ($\Delta\bar{\nu}$ = 218, 314, 459, 762, and 790 cm$^{-1}$) appear if the source were (a) a helium-neon laser (632.8 nm) and (b) an argon-ion laser (488.0 nm)?

3.  Assume the excitation sources in Problem 2 have the same power. (a) Compare the relative intensities of the CCl, Raman lines when each of the two excitation sources is used. (b) If the intensities were recorded with a typical monochromator-photo-multier system, why would the measured intensity ratiodiffer from the ratio calculated in Part (a)?

4.  For vibrational states, the Boltzmann equation can be written as

$$\frac{N_1}{N_2} = e^{-\Delta E / kT}$$

where $N_O$ and $N_1$ are the populations of the lower and the higher energy states, respectively; $\Delta E$ is the energy difference between the states; $k$ is the Boltzmann constant, and $T$ is the temperature in kelvin. For temperatures of 20 °C and 40 °C, calculate the ratios of the intensities of the anti-Stokes and Stokes-lines for CCl$_4$ at (a) 218 cm$^{-1}$ and (b) 790 cm$^{-1}$.

5.  The following questions deal with laser sources in Raman spectroscopy. (a) Under what circumstances would a helium-neon laser be preferable to an argon-ion laser? (b) Under what circumstances would a diode laser be preferable to an argon-ion or helium-neon laser? (c) Why are ultraviolet emitting sources avoided?

## 6.4 Multivariate data analysis

In the age of digitalization, sensors are embedded and interconnected via wired and wireless networks. The increased use of cloud technologies are driven by the significant reduction in costs which makes the ubiquitous use of sensors and actuators possible. Internet of Things (IoT) applications process large volumes of data that flow to computers for data analysis, while all physical objects are able to both sense their environment and communicate autonomously among one another. IoT is therefore inherently interdisciplinary, using methods frequently employed in disciplines such as applied mathematics and computer science, to describe industrial processes in chemistry or life science applications. The deep integration of process data with the enterprise resource planning system and the increased transparency available for customers generates a highly integrated management tool of all core business processes.

There are many different types and variety of industrial process data that can be generated and used as real-time measurements (Figure 6.76). For example, temperatures, pressures, and flows are used for basic evaluation. Additional periodic lab measurements like viscosity of fluids or pH-values are integrated to describe quality. Moreover, array data from instruments like NIR or Raman spectrometers enhance the picture for process understanding, monitoring, and control.

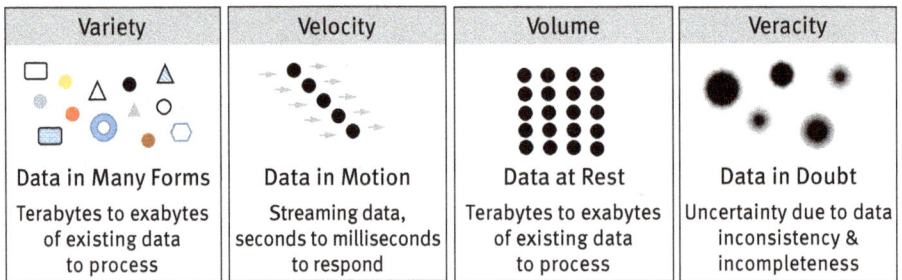

| Variety | Velocity | Volume | Veracity |
| --- | --- | --- | --- |
| Data in Many Forms | Data in Motion | Data at Rest | Data in Doubt |
| Terabytes to exabytes of existing data to process | Streaming data, seconds to milliseconds to respond | Terabytes to exabytes of existing data to process | Uncertainty due to data inconsistency & incompleteness |

**Figure 6.76:** The four dimensions of data.

We differentiate between structured data, which can be stored in database with rows and columns; and semi-structured data that has not been organized into a specialized repository, such as a database; but that nevertheless has associated information, such as metadata, that makes it more amenable to processing than raw data and unstructured data with text, individual notes, or images. Real-time monitoring of industrial process data implies a velocity that depends on the system dynamics. As production equipment becomes more highly instrumented and connected, there will be more data streams to be analysed. In Big Data terms, veracity means problems in data accuracy and integrity. Industrial process data has noise in the measurements and missing data values. Missing data happens because of data connectivity issues, sensor malfunctions, sporadic testing or maintenance work. In typical areas of Big Data, the volume is also crucial. During processing, hundreds of process sensors, raw material data, and QA lab measurements are involved. Traditional methods cannot handle diverse data blocks that must be combined for analysis. In addition, huge numbers of highly correlated measurements or simultaneous prediction of multiple y-variables require a different type of evaluation.

Wherever many measurements on many variables are performed, this data is ideal for multivariate data analysis (MVA). Multivariate models give us a means to find the relationships in the data, and provides tools to visualize the relationships between samples and variables. It can be used for both qualitative and quantitative analysis. The visualization of data is also very important. Making various plots of data helps us to see trends and relationships that are not evident when looking at lists or tables of numbers. An illustration of why this is important is shown in Figure 6.77, where the univariate plots of two variables, pH and temperature, are shown along with a multivariate plot of these two variables.

As there is a correlation between these univariate measurements, the univariate plots do not capture the fact that one sample is showing an anomalous relationship between the pH and temperature values, as shown by the red dot. In this context, control charts are often used for understanding of process variability. Depending on the number of process parameter to be monitored, there are two basic types of control charts. The first type referred to as a univariate control chart that is a graphical chart of one quality parameter, for example, temperature or pH-value. In general, the chart contains a centre line that represents the mean value for the in-control process. Two other horizontal lines, called the upper control limit (UCL) and the lower control limit (LCL), are also shown on the chart. These control limits are chosen so that almost all of the data points will fall within these limits as long as the process remains in control. A process control chart is based on past experience and it can be predicted how the process will vary (within limits) in future. When a process is stable and in control, it displays common cause variation that is inherent to the process. If a data point falls outside the control limits, we assume that the process is probably out of control and that an investigation is warranted to find and eliminate the cause or causes. In Figure 6.77(a, b), the graphs seems to be no "out of control" situation of the process. However, individual univariate charts cannot explain situations that are a result of some problems in the covariance or correlation between the variables. This is why a dispersion chart must also be used (Figure 6.77(c)).

Data on its own has no significance and no worth without any corresponding knowledge. Knowledge is the appropriate collection of information, in such a way that it is intended to be useful. Knowledge is a deterministic process. When sensors "memorize" information, then they have amassed knowledge. This knowledge has useful meaning to the process, but it does not provide for, in and of itself, an integration, such that it would infer further knowledge. The difference between understanding and knowledge is the difference between "learning" and "memorizing." Understanding is an analytical and probabilistic process. It is the process by which a sensor can take knowledge and synthesize new knowledge from the previously held knowledge.

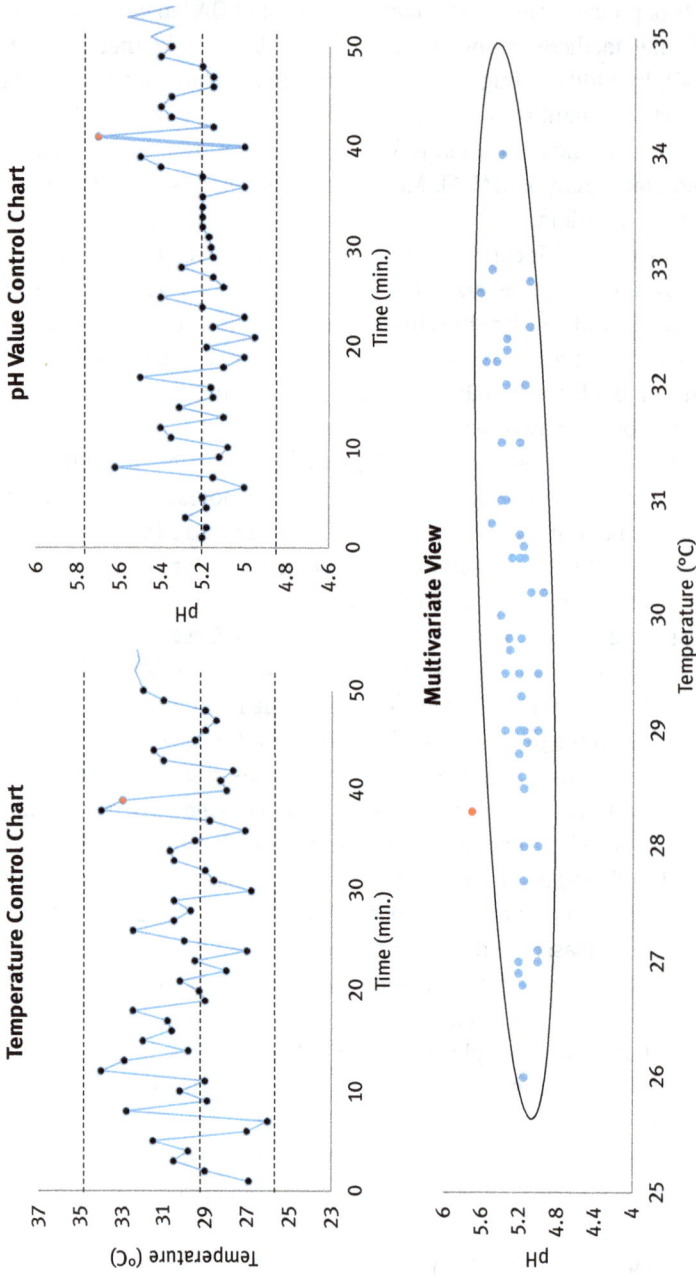

**Figure 6.77:** Two univariate control charts and a simple multivariate control chart with a 95% confidence interval (CI).

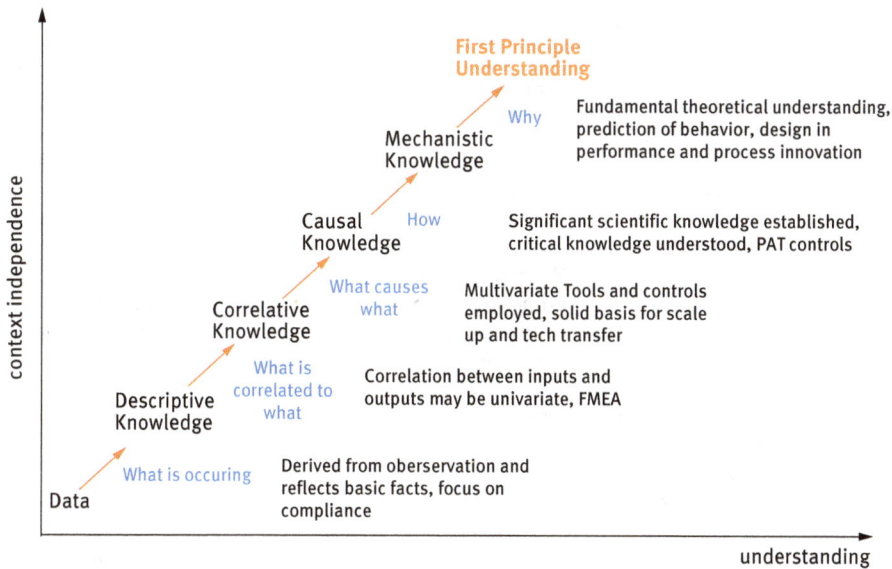

**Figure 6.78:** The difference between data and first principle understanding.

Important aspects for modeling of data:
- Causality and indirect correlation
- The difference between significance and relevance
- No prediction without interpretation
- No interpretation without a valid model
- First Principle vs. empirical modelling (deduction vs. induction)
- When do we need non-linear methods?

Multivariate techniques can be classified into *descriptive* data mining, also known as exploratory data analysis and *predictive* data mining. Descriptive data mining is used to search large data sets to discover locations of unexpected structures, patterns, clusters, trends, and relationships. In descriptive applications, properties of optofluidic systems are modeled with the intent of learning the underlying relationships and structure of the system (i.e., model understanding and identification). In predictive applications, properties of optofluidic systems are modeled with the intent of predicting new properties or behavior of interest. In both cases, the datasets can be small but are often very large and highly complex, involving hundreds to thousands of variables, and hundreds to thousands of cases or observations (Table 6.14).

Experimental design is a strategy to gather *empirical* knowledge based on the analysis of experimental data and not on the theoretical models. When collecting new data during optofluidic experiments, attention should be paid to efficiency to get more information from fewer experiments, and through focusing only on the information that is really needed will be collected. To make quantitative models, regression methods,

**Table 6.14:** Multivariate tools and their purposes.

| Tool | Purpose |
| --- | --- |
| **Design of Experiments (DoE):** Factorial Designs, Mixture Designs, Response Surface and Combined | Identifying relationships between cause and effect; providing an understanding of interactions among causative factors; determining the levels at which to set the controllable factors in order to optimize reliability; minimizing experimental error. |
| **Exploratory Data Analysis (EDA):** Principal Component Analysis (PCA), Cluster Analysis, Multivariate Curve Resolution (MCR) | Maximize insight into a data set; uncover underlying structure; extract important variables; detect outliers and anomalies; test underlying assumptions; develop parsimonious models and determine optimal factor settings. |
| **Classification & Discrimination:** PCA, Support Vector Machine Classification (SVMC), linear Discriminant Analysis (LDA) | Identify new species; define the state of an evolving chemical or biological system |
| **Regression and Prediction:** Multiple Linear Regression (MLR); Principle Component Regression (PCR), Partial Least Square Regression (PLS) | Identifying relationship between a dependent variable and one or more independent variables; Predict important attributes in new samples or find out which variables most influence the system. |

such as partial least-squares regression (PLSR), are used. Regression is used to develop a model using known samples and responses that can provide a predictive model. To identify or classify samples into groups with similar characteristics, there are various classification models such as Soft Independent Modelling by Class Analogy (SIMCA) or discriminant analysis. The following table gives an overview on multivariate tools.

One thing to note is that modeling is an iterative process. As we work through data, we may need to remove outliers, focus on just some of the measurements, and make other adjustments to optimize analysis and give a picture that is descriptive of the system. Depending on the data analysis objective, multivariate data can be used to understand and model numerous outcomes. It can be used for measuring data sets with many input variables or for investigating the trends in time series data, all of which provide a better understanding of a given issue and often result in resource and time savings.

## 6.4.1 The importance of data quality

Prior to any other analysis, you may use a few simple statistical measures to check your data. These analyses can be computed either on samples or on variables and include number of missing values, minimum, maximum, mean, and standard deviation. Checking these statistics is useful if you want to detect out-of-range values or pick out variables and samples that have too many missing values to be reliably included in a model.

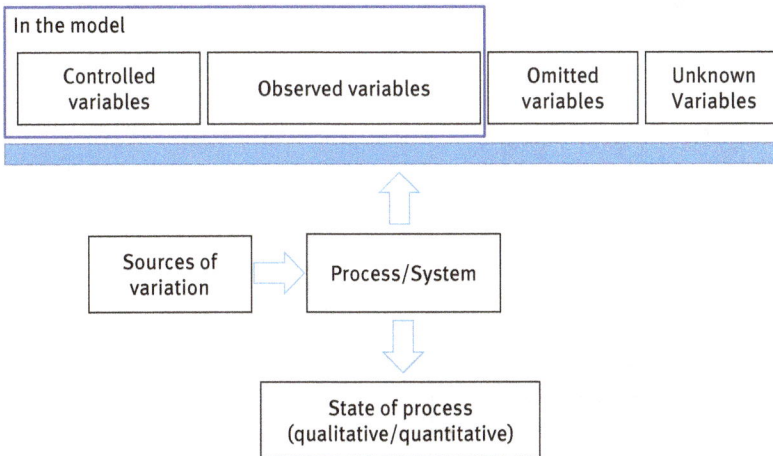

**Figure 6.79:** Which variables are present in a process?

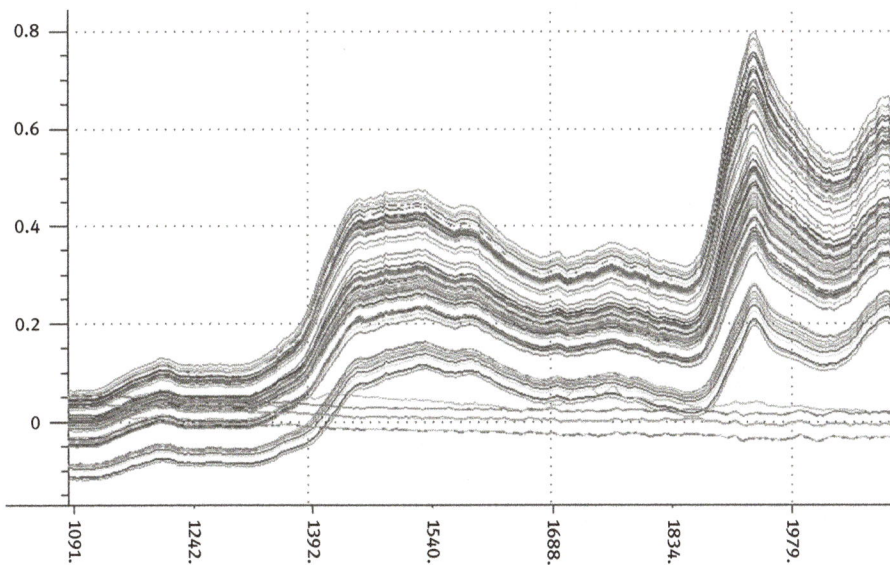

**Figure 6.80:** NIR Spectra of different solid samples.

## 6.4.2 Inspect the data

Check plausibility of the data; for example, see Figure 6.81.
- Look if spectra look like what they should. Instrument performance OK?
- Noise OK? Spectra everywhere in dynamic range?
- Identify real and potential outliers.
- Determine useful areas of the spectra, e.g., to eliminate noise.

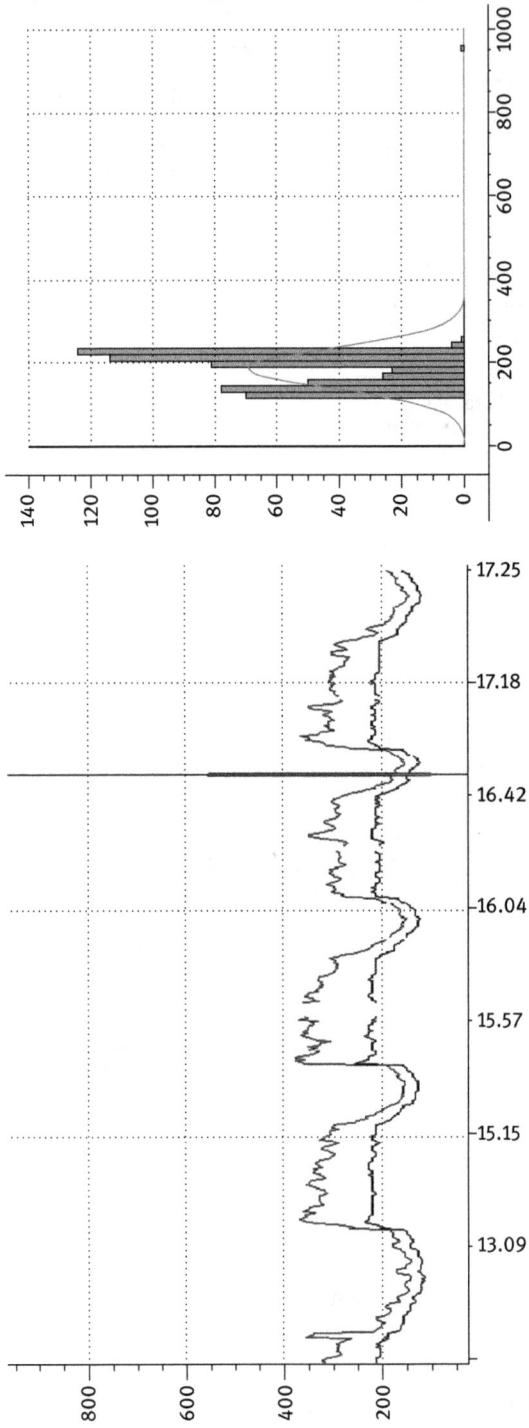

**Figure 6.81:** Process control chart with reference data.

### 6.4.3 Check the process or reference data

–   Look for problems in the quality of the process or reference data.
–   Examine distribution of the samples, e.g., uniform or groups.
–   Linear range? One calibration or several calibrations are necessary?

Look into simple summaries about the sample and about the observations that have been made. *Descriptive statistics* is a summary of the distribution of one or two variables at a time. It is not supposed to tell much about the structure of the data, but it is useful if you want to get a quick look at each separate variable before starting an analysis.

–   **One-way statistics** – mean, standard deviation, variance, median, minimum, maximum, lower and upper quartile – can be used to spot any out-of-range value, or to detect abnormal spread or asymmetry. You should check this before proceeding with any further analysis, and look into the raw data if they suggest anything suspect. A transformation might also be useful.
–   **Two-way statistics** – correlations – show how the variations of two different variables are linked in the data you are studying.

The methods of MVA enable, for example, the determination of the concentration of a substance from a recorded spectrum or determine an unknown substance out of a dataset of known substances. MVA is essential when spectrometers are used as sensors that deliver information, which is extracted from recorded spectral data.

To provide a hands-on approach into the field of MVA, the example from Section 6.3.3 on measuring the UV-Vis spectrum in order to determine the color of a liquid is used. Using UV-Vis spectroscopic measurement techniques not only provides information on the color of a liquid, but also as previously described on the concentration of the used substances. The goal is to use MVA models in liquids of different color to determine the concentrations of the known ingredients used to produce this specific color. A chemometrics model is needed which analyses the UV-Vis measured spectral data; and by a suitable calibration of the model, this type of model delivers the wanted results, the concentration of the ingredients of the measured substance.

Multivariate measurements rely in contrary to univariate measurements on the fact that several parameters are required for obtaining a result at the same time, as it is the case in spectroscopy. Here, a spectrum contains information on the wavelengths measured and their corresponding intensity. In the case of calculating the concentration, the entire spectrum is needed for the sample and is compared with the spectra of the known ingredients. The fact, which the chemometrics model needs to resolve, is that the spectra of the ingredients overlap with one another. As known from experiments and literature, absorption spectra of different substances add up to an "integrated" absorption spectrum when mixed, making it necessary to use multivariate data mining to determine the concentration of each used substance.

### 6.4.3.1 Principal component analysis

A widely used descriptive data mining method, not only used in the context of MVA, is the principal component analysis (PCA). Applying this method to spectral data with several spectral measurements from several samples results in a matrix of the form:

$$X = \begin{bmatrix} x_{11} & \cdots & x_{1n} \\ \cdots & \cdots & \cdots \\ x_{m1} & \cdots & x_{mn} \end{bmatrix}$$

Every row $m$ contains $n$ variables. One row contains, for example, a spectrum and the intensity values corresponding to a wavelength are in each corresponding column. As can be seen by this example, a matrix X can become quite large very fast, just by the number of measurements performed. The goal of the PCA is now to reduce this obtained data set into several principal components. These principal components are new variables, which are not directly assessable through measurements. They are formed by linear combinations of the measured variables using different weighing factors. The interesting fact is that the PCA reduces the data set, keeping all obtained and measured information. In order to calculate the principal components, a singular value decomposition of the covariance matrix of matrix X is derived, obtaining a sum of vectors:

$$X = TP^T + E$$

Matrix T contains the factor scores, matrix P contains the factor loadings, and matrix E is the so-called residual matrix. The residual matrix contains originally measured data, which cannot be described by principal components. Matrix E can be zero or very small if all principal components are used. The rows in matrix P contain the eigenvectors of matrix X, the so-called factor loadings. The number of rows of matrix P corresponds to the number of columns of matrix X. The factor loadings correspond to the amount of a variable contributes to the principal component. In our spectroscopy measurement example, the factor loading is the amount a specific wavelength contributes to the principal component. The number of components is equal to the rows of matrix P. The rows of matrix P form the new coordination system where the measurement data can be plotted. The new coordination system is chosen in such a way that only principal components, which have a major contribution to the total variance, are used. This in turn reduces the data and also defines the dimension of the coordination system. The maximum number of principal components, which can be calculated, is equal to the number of variables n of the matrix X.

Principle component analysis usually does only derived dependencies in the original data set and other sophisticated methods are needed to further interpret new measured data to find similar dependencies. Principle component analysis can therefore serve as the basis for further measurement data interpretations.

Another important descriptive data mining technique is neural networks, also referred to as artificial neural network. Neural networks are known for more than 50 years, but their implementation at that time was hindered due to lack of necessary computing power. This problem has been of course overcome and neural networks are applied widely for all kinds of problem solving, for example, in machine learning, cognitive science, neurobiology, philosophy, and of course mathematics. The question, what is a neural network, has been answered for example by one of the first inventors of a neurocomputer, Dr. Robert Hecht-Nielsen. He defines a neural network as "... a computing system made up of a number of simple, highly interconnected processing elements, which process information by their dynamic state response to external inputs" in "Neural Network Primer: Part I" by Maureen Caudill, AI Expert, Feb. 1989. The role model for neural networks is of course the human brain, which contains about 10 billion nerve cells, called neurons. Neurons are connected to one another by so-called synapses, forming a parallel computing system.

Neural networks are able to identify and analyze data by operating experience and expert knowledge that also enables them to interpret randomly structured databases, including nonlinear behavior. Neural networks are used where classical methods are limited and require extensive calculation power. When coming back to our original goal of determining the concentrations of known ingredients used to produce a specific color, the neural network calculation method is far too powerful and necessary as we are dealing here with a linear determination with a known dataset, leading to predictive data mining methods.

One of the most widely used predictive data mining techniques is the multivariate regression analysis. This method is used to determine certain wanted parameters, which are difficult to be obtained through measurements. Therefore, simple measurements with interdependency are fed into this calculation method. Regression analysis requires two data sets, one of independent variables and one of dependent variables. The set of independent variables contains the measurements, which can be readily made. The independent variables are the source for the predicted answer one wants to model. The dependent variables in turn are the answers, which one models.

Another example of predictive data mining is variance analysis, which is not explained further here as it only provides a true/false answer. A method which also comes in handy is the so called multivariate curve resolution (MCR), which is defined as (Source: www.camo.com) "a group of techniques which help resolve mixtures by determining the number of constituents, their response profiles (spectra, pH profiles, time profiles, elution profiles) and their estimated concentrations, when no prior information is available about the nature and composition of these mixtures." MCR would come in use for our case as well, but as there are no chemical reactions taking place, we will focus on the following paragraph on multivariate regression analysis. In order to develop a regression model, the matrix X is needed, as is the case for the descriptive calculation model. In addition, another matrix Y with reference values is required.

$$Y = \begin{bmatrix} y_{11} & \cdots & y_{1c} \\ \cdots & \cdots & \cdots \\ y_{m1} & \cdots & y_{mc} \end{bmatrix}$$

This second data set contains the reference values for matrix X. The number of columns c is equal to the number of wanted results. The number of rows correspond to the number of samples. The reference values provide more information on the samples; for example, the concentration of single reagents in the sample. The goal for predicting the wanted results from future measurements is now to form a mathematical relationship between the two matrices X and Y. The developed regression model needs a calibration and a validation before it can be used on real world data sets in order to calculate the uncertainty of the predicted results. There are several multivariate regression models to be used. The three most common ones are:

- multiple linear regression (MLR)
- principal component regression (PCR)
- partial least squares regression (PLSR)

The MLR method links the matrices X and Y with the help of a linear model:

$$y = b_0 + b_1 x_1 + b_2 x_2 + b_3 x_3 + \cdots + b_n x_n + e$$

where e is the random noise and $b_0$ is the intercept of the regression model. With the help of the regression model, it is now possible to determine the coefficients $b_i$ by conducting measurements of y by several conditions $x_i$. Every measurement can then be written as in the equation above resulting in a minimum of n measurements for n coefficients. The matrix can then be written as:

$$y = Xb + e$$

The random noise e can be eliminated if the number of regression coefficients n is equal to the number of independent linear equations. The resulting matrix is given by:

$$b = X^{-1} y$$

If there are more linear equations available than necessary for calculating the regression coefficients, then a solution is derived where the sum of the square of the random noise $e_i$ is minimized. Another vector $\hat{y}$ is introduced with an estimated vector b, leading to:

$$\hat{y} = Xb$$

giving the random noise $e_i$:

$$e_i = y_i - \hat{y}_i$$

The goal is of course to minimize the random noise $e_i$ ideally to zero. In order to achieve this, the so-called predicted residual sum of squares (PRESS) is calculated, which is given by:

$$PRESS = \sum_{i=1}^{n} (y_i - \hat{y}_i)^2 = \sum_{i=1}^{n} e_i^2$$

where $y_i$ is the actual value of y for object i, and $\hat{y}_i$ is the y-value for object i predicted with the model under evaluation, $e_i$ is the residual for object i (the difference between the predicted and the actual y-value), and n is the number of objects for which $\hat{y}$ is obtained by prediction. The regression coefficient b can then be calculated as follows:

$$b = (X^T X)^{-1} X^T y$$

leading to one of the solutions. If more solutions are required to be derived, then instead of using vector b, matrix B is used and matrix Y is used instead of vector y. In conclusion, multiple linear regression can be thought of as an extension of simple linear regression, where there are n variables, or simple linear regression can be regarded as a special case of multiple linear regression, where n = 1. The term "linear" is used because in multiple linear regression we assume that y is directly related to a linear combination of the variables.

Another multivariate regression method is the principal component regression (PCR). PCR is a multivariate method and in a first step a principle component analysis (PCA) is performed on matrix X. In a second step, a multiple linear regression is conducted between the scores obtained in the PCA step and the characteristics y to be modeled. An advantage of PCR over MLR is that the number of performed measurements does not need to be equal to the number of regression coefficients. Only principle components will be considered which contribute to the total variance. All other principle components will be neglected, leading as wanted to a reduction in data. The main issue is to choose the right amount of principle components for calculating the regression. After performing PCA on matrix X, the second step in PCR consists of the linear regression of the scores and the y property of interest. The linear model between y and T is of the form:

$$y = Tq + e$$

where the regression coefficients of the scores are contained in vector q. This equation does not contain a direct relationship of the values of matrix X and the corresponding y values. This is obtained using the following calculation. The matrix of the loadings P is orthonormal, which is defined as:

$$(P^T)^{-1} = P$$

leading to:

$$T = XP$$

inserting T in the previous equation:

$$y = XPq + e = Xb + e$$

As vector b contains the scores related regression coefficients q, it is possible to calculate the corresponding y values from a new data matrix X after the regression coefficients b have been derived. This means that, coming back to our original starting point: the use of chemometrics in liquids of different color to determine the concentrations of the known ingredients used to produce this specific color, we are able to determine the score values $t = x \cdot P$ from a measured absorption spectrum x and calculate the concentration $y = t \cdot q$ of the known ingredients.

The third multivariate regression model is the partial least squares regression method (PLS) or also known as projection to latent structures. PLS is a method for constructing predictive models when the factors are many and highly collinear (An Introduction to Partial Least Squares Regression, Randall D. Tobias, SAS Institute Inc., Cary, NC). PLS is an iterative calculation method. The principal components are called latent variables or latent factors. The general idea of PLS is to try to extract these latent factors, accounting for as much of the manifest factor variation as possible while modeling the responses well. As was in the previously described regression methods, the basis for calculating the reference value y is the data matrix X. Developing a PLS model enables to find only one value y or several values of y. The matrices X, T, and P are related as already previously described, given by:

$$X = TP^T + E$$

Matrix Y is calculated using the PCA method and is given by the loadings matrix U and the score matrix Q:

$$Y = UQ^T + F$$

The relationship between the X and Y data is derived with the help of matrix W. Calculating the vectors of every principle component derives the matrices. The first scores t are calculated with the help of the vector u with the largest absolute value, leading to the weight w:

$$w = \frac{X^T u}{u^T u}$$

the score t are then calculated to be:

$$t = Xw$$

the loadings p are then derived to be:

$$p = \frac{X^T t}{t^T t}$$

with the help of the loadings p, one is able to calculate the loadings q of the Y matrix:

$$q = \frac{t^T Y}{t^T t}$$

These steps are repeated until the convergence of t is satisfied. Usually this is seen by the fact that the previous iteration step is identical with the following one. The first convergence step is by comparing the vectors t and u. If the convergence relation is not satisfied, then u is determined with the following equation:

$$u = \frac{Yq}{q^T q}$$

If the convergence relation is satisfied, then:

$$Y = XB + E$$

with

$$B = W \left( P^T W \right)^{-1} Q^T$$

Using the example of determining the concentration of known ingredients, it is possible to calculate the concentrations by inserting new measurement data x if the regression coefficients are determined.

Multivariate data mining techniques have been briefly introduced and, as can be seen, are based on statistical methods. In order to choose the appropriate chemometrics model, several factors need to be taken into account and balanced against the favored method, which can be a descriptive data mining analysis or a predictive data mining analysis. The example of determining the concentration of a liquid could be solved with a multiple linear regression, if the effort is taken of measuring one spectrum per wanted concentration. A more promising method could be the principal component regression, which can also be applied more readily to newly measured data. On the other hand, comparing the PCR with the partial least squares regression method reveals an important advantage of the PLS regression method. The PLS regression method already takes into account the reference values during the determination of the principal components; the PCR method does this only in a second step. Therefore, for this example, the partial least squares regression method is the most promising method, also because it is mainly used for analyzing mixtures.

Being an important tool for optofluidic systems, chemometrics is essential for deriving the necessary control parameters from the performed measurements.

## Self-Assessment Questions

1.  What are the pitfalls of univariate analysis?
2.  Explain wanted and unwanted variability in relation to MVA.
3.  What is meant by the statement: The world is under indirect observation?
4.  What is a principal component?
5.  Why is each variable usually mean-centered before bi-linear modeling?
6.  What is the difference between the first and second PC?
7.  What is the difference between scores and loadings?

## Further reading

Martens H, Martens M. Multivariate analysis of quality. An introduction, 2001.
Martens H, Naes T. *Multivariate calibration*. John Wiley & Sons, 1992.
Kessler W. *Multivariate datenanalyse: für die pharma, bio-und Prozessanalytik*. John Wiley & Sons, 2007.
Gemperline P, editor. *Practical guide to chemometrics*. CRC press, 2006.
Dean A, Morris M, Stufken J, Bingham D, editors. *Handbook of design and analysis of experiments*, vol. 7. CRC Press, 2015.

## 6.5 Conclusion

Sensors are key components and enable the high versatility of optofluidic systems. Especially, their availability in miniaturized versions enables the possibility of integrating application-specific combinations, leading to novel devices and true innovations. In this respect, miniaturized spectrometers will be the future drivers of optofluidic systems as their use opens up a wide range of applications with only a limited amount of "sensor" elements needed. UV-Vis spectrometers can already be regarded as sensors, as several versions and sizes are commercially available.

The key issue for designing optofluidic systems is the creation of the whole measurement chain, taking into account the photonics, the fluidics, the necessary sensors, and the required software.

# 7 Bus technologies for optofluidic systems

This chapter is derived from the knowledge pool of Buerkert fluid control systems and used with permission. To combine intelligent components, it is essential beforehand to decide on the system architecture and thus the internal bus, linking each component to one another forming a real system to perform a dedicated solution for a specific application.

The "evolution" of network technology has essentially developed from the principle of centralization to distributed intelligence. Today, it is possible to decentralize intelligence as electronic hard- and software is readily available providing basically unlimited calculation power. Of course, this also necessitates components that comply with all aspects of the new "command structure". Maximum availability and minimum possible downtimes are only two key aspects of more efficient, that is, advanced, operation of an optofluidic system, which is based on future-proof field bus technology. Three common network architectures are as follows: centralized, distributed, and distributed networks (distribution of variables in the network).

From today's technology experience, it is not trivial to create complex systems. Each individual component needs to be tested with all the other components with which an interaction is performed to guarantee error-free operation. Solid component architecture can eliminate unwanted negative effects. Therefore, common device architectures in soft- and hardware are essential for robust optofluidic systems.

Various user associations track the ongoing development of individual bus system. Examples are as follows:
- AS-International Association: www.as-interface.net
- CANopen: www.can-cia.de
- DeviceNet: www.odva.org
- Ethernet: www.iaona-eu.com, www.ida-group.org, www.odva.org, www.profibus.com
- FOUNDATION Fieldbus: www.fieldbus.org
- HART Communication Foundation: www.hartcomm.org
- INTERBUS Club: www.interbusclub.com
- PROFIBUS International (PI): www.profibus.com

Field buses enable digital networking of open-loop control systems, sensors, and actuators enabling true optofluidic systems. Data are exchanged both horizontally between the devices of a level and vertically to the systems in the next hierarchy level. To achieve a practically oriented classification, communication structures are assigned in automation engineering to various application levels.

The coordinating level monitors higher level system control while the automation level controls the actual process. The focus is on a reliable transmission

https://doi.org/10.1515/9783110546156-007

of very long messages (file transfer). At the field level, data transmission of the measured values and manipulated variables is cyclic in many cases and necessitates efficiency as high as possible so as not to impair real-time characteristics of the open-loop control application. This connection is referred to as data-oriented communication. Moreover, field buses also support access to field units of the upper levels, for example, engineering stations from the automation or coordinating level. Process data and status information can be read out, parameters can be polled and set and, in some cases, software can even be downloaded and program routines started for configuration, operation, monitoring, and diagnosis by the user. This form of acyclic data exchange is referred to as message-oriented communication.

All field bus systems are based on the same idea of allowing addressable devices to use a common transmission medium. The network topology describes the spatial extent of a field bus network, but also the logical arrangement of the devices during communication.

The bus or linear structure is clearly arranged and features little complexity. This is where all users communicate via a common line. The devices are linked either with or without very short stubs. Occasionally, this leads to untidy cabling in practice.

The tree structure is similar to the linear structure, the only difference being that several bus branches may converge at the nodes. The tree structure enables large areas to be networked more easily and with more flexibility.

If a physical ring is constructed with several two-point connections, it is termed as a ring structure. A message to be transferred is passed on from one user to the next. Since the signal can be amplified each time it is passed on, it is possible to span very large distances.

A central station is linked to all users with two-point connections in a star structure. This central station may either be responsible for network control as the Master or as a "star coupler", may simply establish the connection between the current sender and receiver.

Complex network structures frequently consist of several independent sub-networks. Each of these sub-networks can work with differing topology and a different communication protocol.

Rules must be defined for all communication partners so that communication between various users and across the network hierarchy levels can occur effectively and without misunderstandings. This is achieved with the ISO/OSI model (Figure 7.1), which describes all elements required for communication, such as the cable type or physical mode of transmission of the messages.

The model features seven layers that build upon each other, each of which describes a specific task. The ISO/OSI model has also become established as a virtual standard representation when implementing communication services outside of field bus technology, since it fundamentally describes the communication sequence. If

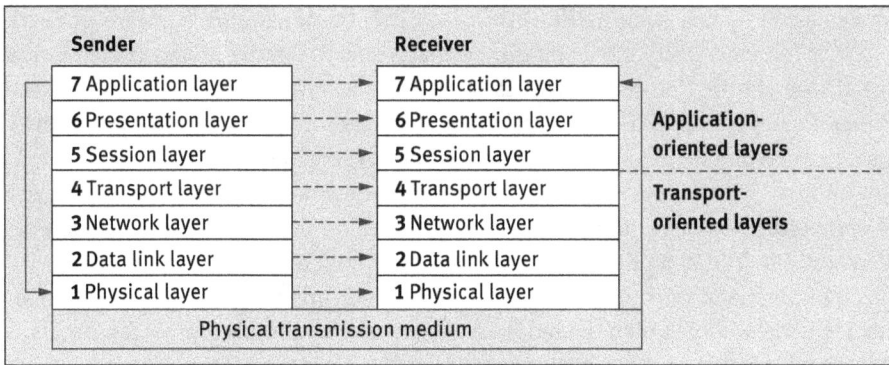

**Figure 7.1:** ISO/OSI model (Courtesy Buerkert fluid control systems).

specific services (layers) are no longer required within a communication system, these layers remain empty.

Generally, only layers one and two are fully defined when specifying field bus networks, while the application process itself or the subordinate layer seven handles all other services.

–   Layer one defines the way in which data transmission is performed physically, that is, electrically and mechanically. This includes, for example, the method of coding (e.g., NRZ) and the transmission standard used (e.g., RS-485).

–   Layer two has the task of providing integral, that is, error-free information. It must detect any errors which have occurred in layer one and remedy these errors via suitable error routines.

–   Layer seven forms the interface to the application program and contains all functions with which the user, generally a computer program, can access the communication functions.

If Ethernet is to be used in automation engineering instead of a classic field bus system, ("hard") real-time capability is particularly important. Basically, the term "real time" is a question of definition. Thus, real time in the case of synchronization of drives or actuators may amount to microseconds, while times in terms of seconds are adequate in process engineering applications. If we compare the various field bus systems and Ethernet with regard to efficiency of data transmission, Ethernet achieves a poorer value. This results from the carrier sense multiple access/collision detection (CSMA/CD) procedure used, which must operate with a long minimum telegram length due to unconditional and secure collision detection. However, this disadvantage is compensated for by the high transmission speeds of up to 100 Mbit/s (and growing). Such high transmission speeds can be implemented only by a point-to-point connection between the units, which, besides Ethernet, is only offered by the INTERBUS system. On systems with a variable transmission speed, such as PROFIBUS or CAN, the maximum possible network extent is reduced with increasing transmission speed, the higher the transmission speed, the shorter the line length is. This may lead to a situation in which the communication link to be laid is only a few meters long, which is not necessarily disadvantageous within closed systems or system sections.

Another important aspect of the implementation of field busses for the operation of optofluidic systems is the operational optimization by what it is known as asset management. The aim of asset management in optofluidic system automation is to effectively manage and optimize the use of equipment and systems. This includes, for instance, the ability to plan the required maintenance, minimization of faults occurring, enhancement of process diagnostics, and process monitoring as well as both identification and utilization of functional reserves. This requires complex information that must be obtained from the overall optofluidic automation system.

Asset management can only operate by the interplay of intelligent modules, a highly developed communication structure and corresponding operating software. Thus, for example, diagnostic information from each module in an optofluidic system is sent via the field bus to corresponding asset management modules where it is evaluated. System-oriented asset management is aimed not only at maintaining an existing system, but is already deployed in the engineering of components in optofluidic control engineering. It includes programming and configuring the modules and opens up access to the system documentation and to the operating environment of the optofluidic system installation.

Measured on the basis of the lifecycle of an optofluidic system installation, maintenance and, in particular, status detection of different modules and other system components are very important. Measured values which characterize the status of module and which are processed by the asset management system (AMS) on the basis of characteristics or models to form trend information or which combine these values centrally from other information systems are used for this purpose. As the decision-making basis for maintenance measures, AMS must also offer access to documents such as shift logs, system documentation, and computer-aided engineering (CAE) systems, in addition to current status information. With all asset management solutions, it is necessary to bear in mind the basic requirement that all technical operational support activities have to be performed from one central module.

If a field bus network is constructed on the basis of devices from a single institution (e.g., a company, university, or research center) and if the modules from this institution can all be operated in the same way, a single software package is sufficient for being the user interface. However, an optofluidic system can certainly cover several different field unit types from different institutions, which frequently also may involve several different operating programs for configuring and programming the modules. To at least partially ameliorate this situation, proprietary descriptions (languages) based on the standard device description languages (DDLs) have been developed. However, each of these languages is tailored to the specific communication systems for which it was originally developed. Virtually, every configuration tool and every field bus standard have implemented its own device description language or at least uses a "dialect" of the HART device description, which was developed at a very early point. The operating methods for PROFIBUS (GSD, EDD, DTM), HART devices (DD), DeviceNet (EDS), and FOUNDATION Fieldbus devices (DD, DTM) are examples that emerged from this.

Thanks to the creation of an open and standardized communication platform, at least referring to the relevant system, modules can be easily integrated in the given control and instrumentation system structure and operated centrally via a common engineering tool. This has to be considered and implemented when designing and creating an optofluidic system.

Advantages of using a common field bus technology are manifold.

Using a field bus drastically reduces the cabling expenses, effort, and complexity. Field bus technology enables integrating the same I/Os using one pair of conductors. Another tremendous advantage is the reduction in wiring effort and expense leading to the same reduction in documentation expenditures related to all electrical wiring diagrams and ladder diagrams.

A new field unit can be added to a field bus at any point without having to separately rewiring. Subsequent modifications and extensions thus pose no problem. This applies in particular to wiring with the two-wire system in which data and power are transferred in one cable.

Significant advantages can be anticipated in relation to the duration of the planning phase. State-of-the-art technologies mean faster integration of the modules (loop check and calibration) in the optofluidic control system. Wiring errors are prevented thanks to the simple cabling. If difficulties within the network structure occur nevertheless, these are rapidly diagnosed using bus testers and bus monitors.

All process data, device data, or business management data are available via a universal communication structure from all locations, and even outside of the system via the Internet. This allows central and distributed operation and engineering. Comprehensive and central data management forms the basis for operational optimization in every system.

Various factors are crucial to selecting the field bus system to be used. Essentially however, the field bus requirements necessitated by the application plays a crucial role. Since every system is able to meet specific requirements particularly well due to its technical characteristics, various field buses have a high share of the market in individual industries.

Production based on lot sizes and executions of repetitive, frequently mutually independent work steps, are characteristics of the production industry. The level of decentralization within a production plant is low. The requirements applicable to communication between programmable controller (PLC) and field units are very stringent. In many sectors, such as robotics, measuring technology and test and inspection technology, stringent real-time requirements arise with cycle times less than 20 ms. In many cases, equidistant data transmission is required in drive engineering, for example, for axis interpolation. The requirement regarding safety against system failure is only moderate. In many cases, halting production in the event of device failure costs less than designing the entire system fully redundantly. More stringent requirements related to fail-safe design exist in areas in which persons may be placed at risk, for example, burner controls, press controls, and lathes, etc.

Production is generally batch-oriented in the food industry, chemical industry, and pharmaceuticals industry as well as within the process industry and process engineering. Typically, the process industry utilizes very complex installations that are greatly decentralized and implemented over a very large area in the form of distributed systems. The volume of project data in such installations may comprise several hundred thousand data points. This means that the requirements for process control systems are very stringent with regard to the handling of the data volume. However, the time aspect is less critical and is within the range of seconds in many cases. One fundamentally important criterion in the process industry is high availability. The systems are not switched off due to the complex and long, drawn-out starting procedures for continuous processes, which often last several hours. Expensive redundancy concepts with hot standby prevent fault or error-related interruption in the process. The requirements in the areas of maintenance and commissioning are stringent. It must be possible to convert or extend the system during ongoing operation. Additional safety requirements apply to explosion-hazard environments, such as in the petrochemical and gas industries. The entire system in the field must ensure that the legally required safety regulations, such as ATEX directives, are complied with. Depending on the level of danger or hazard, there is a classification into zones from 0 to 2 which, in turn, allow only specific automation concepts, including the communication method to be used.

In summary, when in the concept phase of designing an optofluidic system, the network architecture of the communication structure should be well aligned with the application, bearing in mind the additional features required today and tomorrow.

The following chapter provides examples of platforms for creating true optofluidic systems.

## 7.1 Optofluidic systems

The book presented a deep ultraviolet (DUV) polymer technology platform for the simultaneous realization of photonic integrated devices, fluidic channels, and immobilization of living cells including neural cells. The introduced deep ultraviolet (DUV) process enables the fabrication of single-mode and multi-mode waveguides in methacrylate-based polymers for both visible and infrared wavelengths. In addition to the planar waveguide-based devices, ridge wave-guides and MMI couplers have been described, which have been fabricated by hot embossing and DUV modification of the used polymer for increasing the refractive index. The unique feature of the DUV process makes it possible to integrate fluidic channels with planar and ridged waveguides in a simple way. Due to the surface chemistry induced by DUV radiation, selective patterning of cells including neural cells is possible, thus demonstrating the integration capability of this process even with biology. The further integration of organic light sources and detectors lead to a true optofluidic platform. Therefore, this developed DUV platform can be the driving force for future optofluidic systems, as the technology has been proven and can be adapted to the application. The reader should also keep the DUV platform in mind when looking at the other examples provided in this chapter, as it is possible to merge and integrate the DUV platform with the other examples in a following step.

Most LoC systems require a platform with external supply and control units to be operated. This practical example, which is based on [27], describes the development of a modular optoelectronic microfluidic backplane, enabling the flexible interconnection, supply, and control of microfluidic and optofluidic devices ideally suited for optofluidic systems. The developed system is fabricated in polymers and consists of backplane modules that may be individually connected with each other. Each module holds one dedicated port on top for a device to be operated. In particular, an optical backplane module is described based on a novel opto-mechanical light switch to guide light to the device of choice within the system. This modular approach allows assembling an arbitrary number of different devices in three dimensions. In conclusion, the backplane provides a configurable platform for multiple optofluidic applications.

Modular system design is one of the most favorable means of advancing development in strongly expanding areas with complex applications such as microsystems for fluid analysis. Recently, various LoC systems with integrated optics and fluidics have been demonstrated, inherent to high functional integration and compact size [168, 201, 256] These systems are mostly designed for dedicated applications and intended to be fabricated by means of mass-production technologies. Their main advantages are disposability due to low cost, short sensor response times, low analyte consumption due to minimized volumes, and biocompatibility due to the use of adequate polymer materials. These characteristics are particularly interesting for portable analyzing systems. However, certain applications, for example, in biomedical, environmental, and process analytics, request a broad range of similar but individual configurations and device combinations. Additionally, since each LoC needs its own external operating units, an update to a deployed system requires the adaptation of these units. Therefore flexible and extendable interconnection platforms are required. Moreover, in addition to disposable LoC, the development of reusable micro-total analysis systems has been enforced, especially for continuous environmental and process monitoring, where various different analysis methods are combined in one system [167, 284]. Existing optical backplane systems are mostly used for communication networks in the near infrared, thus not adapted for visible light and optimized for unidirectional light propagation only [15]. Available microfluidic backplane systems are often composed of passive circuits, thus needing external valves for fluidic switching [218]. Typical microfluidic systems with integrated active elements are limited to low pressures or low flow rates [73]. Moreover, most microfluidic and optical backplanes are not designed to be combined in one system and not modularly extendable in three dimensions.

The developed system consists of backplane modules that are individually connected with each other. Every backplane module with lateral dimensions of 40 mm × 40 mm is fabricated in polymers and holds one slot on top for a microfluidic or optofluidic device. This approach allows combining, supplying, and operating as many devices as needed for the dedicated application.

Each module consists of three networking layers:
- an optical backplane with waveguides and optical switches to guide light to the devices,
- a microfluidic backplane with channels and microvalves for fluid control, and
- an electronic backplane for the electrical supply and control of the active elements in the system, for example, microvalves, optical switches, and the connected devices.

The optical backplane modules are designed to guide light from one neighboring module to each other connected module, thus enabling the supply of all connected optofluidic devices with light of an external or central optical light module. Two different concepts to achieve this goal are as follows: (1) all devices are supplied with light simultaneously, that is, only passive elements such as optical waveguides and fibers are needed and (2) light is guided to one device at a time only. Despite its higher complexity and need for integrated optical switches, the second concept is followed, since the light intensity reaching the supplied module is only reduced by guiding and coupling losses but not by dividing the light intensity at each device crossing, as this would be the case if the first concept was applied.

As switching element in the prototype of the backplane module, a linearly actuated mirror assembly is implemented (Figure 7.2). It permits the controlled guiding of light from one neighboring module horizontally to the three others and vertically to a mounted device or a second level of modules.

The switching element connects five 960 μm core, polymer optical fibers with a numerical aperture of 0.25 that are positioned perpendicularly to each other, four of which are in plane and the fifth out of plane guiding light to the mounted device. A lens focuses the light from each fiber, which is either deflected or transmitted by one element of the mirror assembly, focused again by another lens, and finally coupled

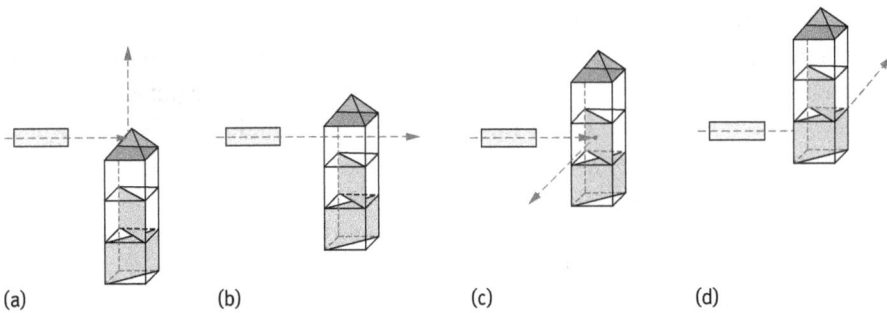

**Figure 7.2:** Switching positions of the opto-mechanical light switch. Chart (a), Vertical light deflection at the pyramid mirror. Chart (b), Light transmission through the transparent glass cube. Chart (c), (d), Horizontal light deflection at the prism mirrors (Courtesy Marko Brammer).

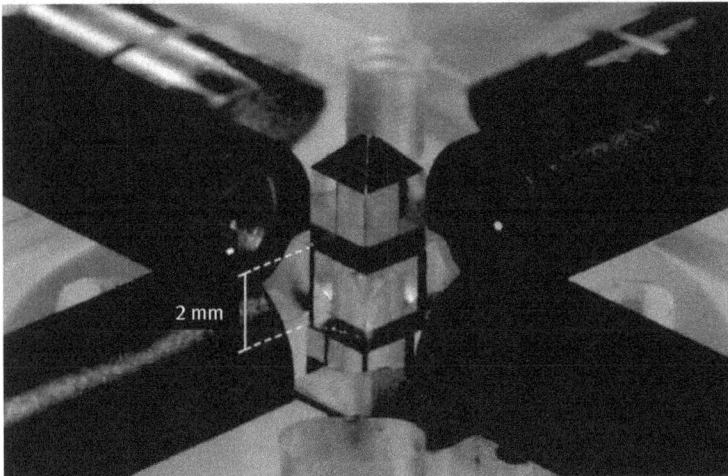

**Figure 7.3:** Prototype of the opto-mechanical light switch for horizontal coupling. The lens and fiber for vertical coupling are not depicted (Courtesy Marko Brammer).

into the selected other fiber. Figure 7.3 shows the prototype of the opto-mechanical light switch.

The horizontal deflection is achieved by 2 mm × 2 mm × 2 mm right-angle prisms that are coated with aluminum at the hypotenuse. Two prisms glued together at their mirror surfaces with epoxy adhesive (353ND, EPO-TEK®) build a double-sided mirror. The vertical deflection is realized by an aluminum coated 2 mm × 2 mm × 1 mm pyramid that is fabricated by micromachining on top of a 2 mm × 2 mm × 2 mm silica cube, which is used as transmitting element. The 150 nm aluminum mirror coating of the pyramid surfaces is deposited by ultra-high vacuum evaporation deposition at 150 K and $5.2 \cdot 10^{-7}$ Pa. The two perpendicularly oriented double-sided mirrors and the cube with the pyramid were glued on top of each other. This stack is used to guide the light to the desired output and is therefore moved vertically within the module. The actuation is implemented by a brushless micromotor with integrated 125:1 gear (smoovy 0308B, Faulhaber) that is turning a M1.6 × 0.2 spindle. The base on which the mirror assembly is mounted has a mating internal screw thread and is glued with epoxy adhesive (UHU plus 300) to a linear bearing (LWL1, IKO), thus transforming the rotation into a linear movement. The optical switch is integrated in an insert, which allows the active alignment in the housing relatively to the fibers during the mounting process. At the angle of highest coupling efficiency, the insert is fixed. The fiber and lens are inserted into a holder made of polyimide, assuring reliable alignment and defined distances.

The optical design of the lenses and the dimensions are optimized by ray-tracing simulations with ZEMAX.

The configuration of the optical elements in plane is symmetric. For a compact design, the optimal distances amount to 3.6 mm between fiber and lens and 3 mm between lens and prism. The chosen lenses are plano-convex with diameter, center thickness, and focal length of 3 mm × 0.8 mm × 3 mm. The minimum distance between pyramid top and upper lens is equal to the maximum stroke of the mirror assembly and amounts to 6 mm. The chosen distance between upper lens and upper fiber is 2.7 mm for a 3 mm × 2 mm × 3 mm double-convex lens. All lenses and prisms have been purchased through Edmund optics.

The main advantages of this concept are its simple actuation principle and the low positioning accuracy needed (± 0.5 mm), since the extension of the light beam, which depends on the light source, is less than 1 mm in diameter, while the area to transmit or deflect it horizontally is 2 mm × 2 mm. Since the pyramid mirror surfaces for each fiber are smaller and not centered with respect to the upper lens, the precise positioning of the pyramid is crucial for the coupling efficiency. To fulfill this requirement, this position is set by a stop at the bottom of the base. Hence, the position, in which light is transmitted to the opposite fiber, is implemented by a transparent cube and not through free space.

The coupling efficiency of the optical backplane prototype measured by a silicon optical power head connected to an optical multimeter (OMH-6703B and OMM-6810B, ILX Lightwave).

A red laserdiode ($\lambda$ = 655 nm, Popt = 5 mW, ADL65055TL, Laser Components) was used as source. The reproducibility is displayed by the standard deviation. It is higher for vertical deflection with the pyramid (0.6% in relation to the four different impact directions) than for horizontal deflection and transmission (0.2%). This is due to the smaller size of the pyramid surface compared to the prism mirror and the larger distance between the pyramid and the upper lens compared to the distance between prism mirror and lens. A means of achieving higher coupling efficiencies is using fibers with smaller diameter and lower numerical aperture, which, however, raises the need for a more precise fabrication and alignment. An alternative is using lenses with larger diameters and focal length. This would, however, increase the dimensions of the device. Additionally, using antireflection-coated optical elements would reduce losses.

In the presented system, each module has six fluidic connectors to its neighboring modules and two connectors to the device mounted on top of the module (Figure 7.4). The integration of 12 microvalves (refer Section 4.5 microvalves) [90] and a compatible fluid circuit in one module allows controlled guiding of fluids from the six neighboring modules to each of the other ones as well as through the supplied device [26]. Thus, this fluid circuit serves as a base for a three-dimensional system. Furthermore, two levels of fluid channels are integrated to permit independent crossing of two fluid flows through one module and thus a higher degree of interconnection variability.

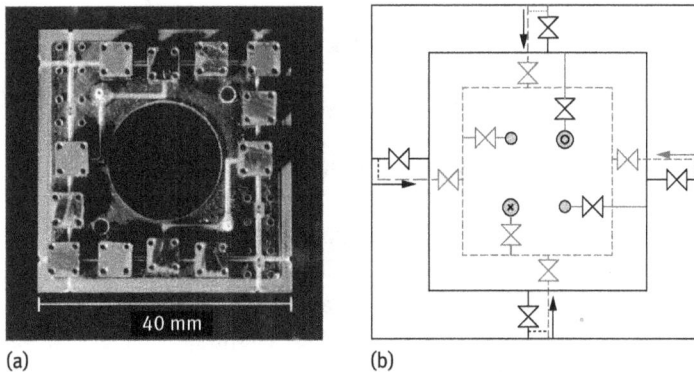

(a)                                        (b)

**Figure 7.4:** Microfluidic backplane. Chart (a), Prototype with mounted dummy valves. Chart (b), Logical fluidic circuiting. The input and output to the mounted device and the connectors to the neighboring modules above and below are depicted as circles, and the four connectors to the horizontally neighboring modules as arrows. The second level of fluid channels is depicted as dashed lines (Courtesy Marko Brammer).

The channels of the presented prototypes have a diameter of 0.8 mm and were fabricated out of cyclic-olefin copolymer (COC) by micromilling. For industrial production, replication technologies may be applied (refer also Chapter 3). The channels are covered by a plate made of COC doped with 3 vol.% carbon, which absorbs the light of the diode laser beam ($\lambda = 940$ nm) used for laser-beam welding.

The backplane modules are interconnected mechanically by magnets (Figure 7.5). This method allows a flexible configurability and reassembly of the system. Furthermore, it respects the need for a clearance fit, in order not to statically overdetermine the system. Fluidic sealing is assured by nitrile rubber O-rings with inner diameter and cross-section of 1 mm × 1 mm that are integrated in each interface.

**Figure 7.5:** Prototype of the magnetic fluid connector with integrated fluidic sealing element (Courtesy Marko Brammer).

Leak tests showed that the fluidic contacts are leak proof for fluid pressure differences of up to $24 \cdot 10^5$ Pa.

Optical connectors (Figure 7.6) assure self-alignment of the fibers between two neighboring modules with measured coupling losses less than 0.4 dB. The tolerance in misalignment amounts to $\pm$ 35 $\mu$m (measured by an opto-tactile sensor with an accuracy of 5 $\mu$m), which corresponds to a reduction of the coupling efficiency by 0.2 dB. Fabricating the devices by injection molding may reduce these losses. Using index-matching gel may reduce losses due to surface and interface imperfections.

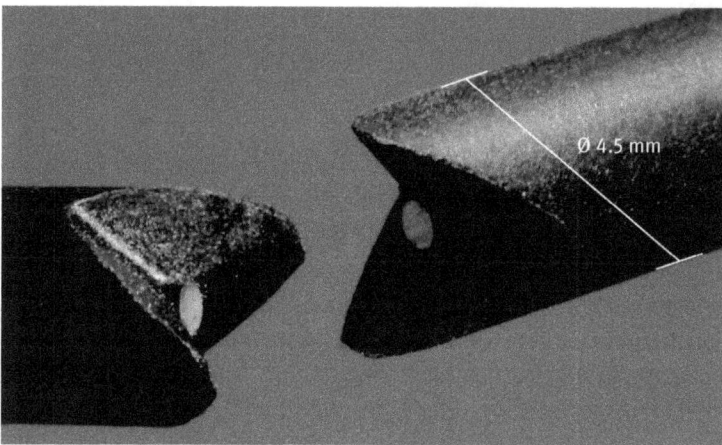

**Figure 7.6:** Prototype of the optical fiber connector for automatic alignment of optical fibers (Courtesy Marko Brammer).

The interface to the mounted device features an optical input, as well as two fluidic contacts as input and output to operate the device. The system is operated with the intended functionality separately for each backplane layer as well as for the assembled modules by interconnecting a miniature spectrometer (C10988MA, Hamamatsu) and a combined conductivity-temperature sensor module, and supplying them with multiple light sources and sample fluids. Figure 7.7 shows the measurement setup.

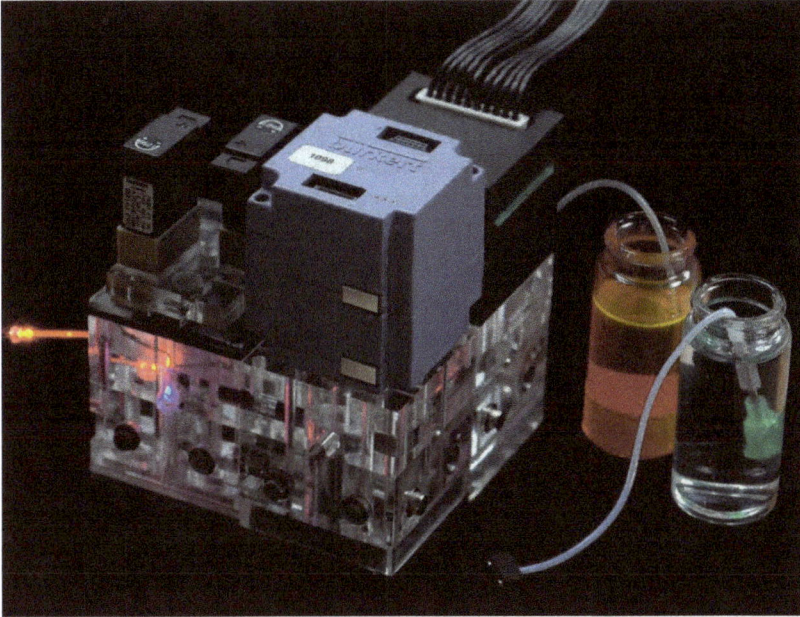

**Figure 7.7:** Measurement setup with backplane modules interconnecting a miniature spectrometer and a combined conductivity temperature sensor module. The modules are operated with light-emitting diodes, a micropump, and sample fluids with different dyes (Courtesy Marko Brammer).

The development of the described backplane modules made out of polymers enables configurable system design in optofluidic systems. This approach permits the interconnection in three dimensions of an arbitrary number of modules, each one mounted with a device. This system design is adaptable to multiple applications in optofluidic systems and may be considered as a major step toward custom-made true optofluidic systems, which includes the necessary electronic control and readout. A major advantage of this modular architecture is the fact that basically all kinds of LoC devices can be integrated and automized with this type of platform, leading to a tremendous development time reduction when heading for commercial applications.

Another example of a purely fluidic backplane is presented by [9], where a high density microchannel network with integrated valves and photodiodes is fabricated and analyzed.

Another example of an optofluidic system with the topic "BANSAI – A new approach for clinical chemistry analysis utilizing laser light" is provided with kind permission by Graciete M. Rodrigues Ribeiro, (1) Helena Buchmiller, (2) Horst Mayer, (3) Thomas Woggon, (2) Jens Brümmer, and (1) Christian Karnutsch (1. Institute for Optofluidics and Nanophotonics (IONAS), Karlsruhe University of Applied Sciences, Germany; 2. Zentralinstitut für Laboratoriumsmedizin, Mikrobiologie und Transfusionsmedizin, Städtisches Klinikum Karlsruhe, Germany; and 3. VISOLAS GmbH, Eggenstein-Leopoldshafen, Germany).

The IONAS is developing a novel technology approach for blood analysis in collaboration with the department of clinical laboratory medicine of the municipal hospital Karlsruhe.

The aim of the BANSAI project is to perform photometric blood analysis at the established wavelengths of routine clinical chemistry tests, but furthermore create the possibility to use additional wavelengths to form a basis for future clinical tests with new detection capabilities. With the use of a continuously tuneable organic semiconductor laser (Visolas do, VISOLAS GmbH, Eggenstein-Leopoldshafen, Germany) instead of a conventional discharge lamp, we can additionally use short light pulses. The use of these nanosecond pulses allows highly dynamic detection with a very narrow bandwidth. The authors in [271, 272] showed that a similar photometric system only needs a small number of components to match the resolution and sensitivity of a commercial spectrophotometer. The approach uses tuneable organic laser light sources, which exhibit a unique potential for miniaturization. A miniaturization of the current macroscopic platform would enhance the portability and enable mobile point of care applications even in remote areas and rough environments.

The photometric measurement is the most used analysis method in clinical chemistry. This method measures the light absorbance of molecules in the visible and ultraviolet spectral range. Photometric analysis is based on Lambert and Beer's law, which, for low concentrations, states that the absorbance of liquid samples changes linearly with the concentration of a dissolved absorbing material.

For our experiments, we started with the wavelength range of 500–650 nm and we used the albumin bromcresol green (BCG) kit from mti-diagnostics GmbH (Idstein, Germany) to measure human albumin (HA). The HA is a common parameter in clinical chemistry analysis. This manual method is simple, inexpensive, and specific enough for the selected wavelength of 628 nm. A 5-minute settling time was strictly adhered between mixing the sample with the BCG reagent and measuring the extinction. This slow kinetics is well suited for a detailed study of the BANSAI system.

Albumin is one of the proteins in human serum and it is the primary protein in terms of quantity, normally accounting for more than half of the total serum protein. The HA is formed exclusively in the liver and serves as a transport and binding protein for calcium, fatty acids, bilirubin, hormones, vitamins, trace elements, drugs, and other substances. It is a major contributor in maintaining the colloidal osmotic pressure. Diminished serum albumin concentrations occur in case of a severe impairment of the hepatic synthesis capacity (e.g., in hepatic cirrhosis, severe hepatitis, or chronic malnutrition) as well as in marked protein loss (nephrotic syndrome, severe burns, gastrointestinal loss). It has been demonstrated that the serum albumin concentration may be indicative as a prognostic marker for mortality in hospitalized patients [265].

### 7.1.1 Specimens and analytical control material

Twenty-one serum specimens of obviously healthy blood donors from the department of blood preservation of the municipal hospital of Karlsruhe were used for the pooled HA. From each sample, 3 ml were collected and mixed gently for about 30 min and then filtrated through a filter paper. The pooled HA standard was aliquoted and stored in capped aliquot vials at −20 °C. Pure HA (Sigma Aldrich, Taufkirchen, Germany) in the form of lyophilized powder was used as well. It was dissolved in cold physiologic salt solution without adding preservatives. The dissolved pure HA standard stock solution (80 g/L) was immediately diluted again (from 80 g/L to 5 g/L) with cold physiologic salt solution and then used for experiments.

Two commercial Bio-Rad (Munich, Germany) Liquid Assayed Multiqual controls were selected to monitor the precision of the method validation. We used Bio-Rad control level 1 with a low albumin concentration of 28 g/L and level 3 with a high albumin concentration of 48 g/L.

According to the Bio-Rad product guidelines, the controls were thawed, aliquoted, and then stored at 20 °C.

### 7.1.2 Chemicals and calibrators

A commercial kit for the quantitative in-vitro determination of HA in serum was purchased from mti-diagnostics GmbH (Idstein, Germany) based on the BCG reagent (0.15 mmol/L BCG reagent prepared in pH 4.2 succinat buffer (75 mmol/L)) and an HA standard solution (40 g/L) (kit standard was not used). The BCG reagent was stored at 8 °C.

In this survey, the N protein's standard SL calibrator (48.3 g/L) from Siemens Healthcare Diagnostics Products GmbH (Marburg, Germany) was used.

### 7.1.3 Apparatus

The BANSAI system uses a tuneable organic semiconductor laser VISOLAS (VISOLAS GmbH, Eggenstein-Leopoldshafen, Germany), which can be tuned continuously from

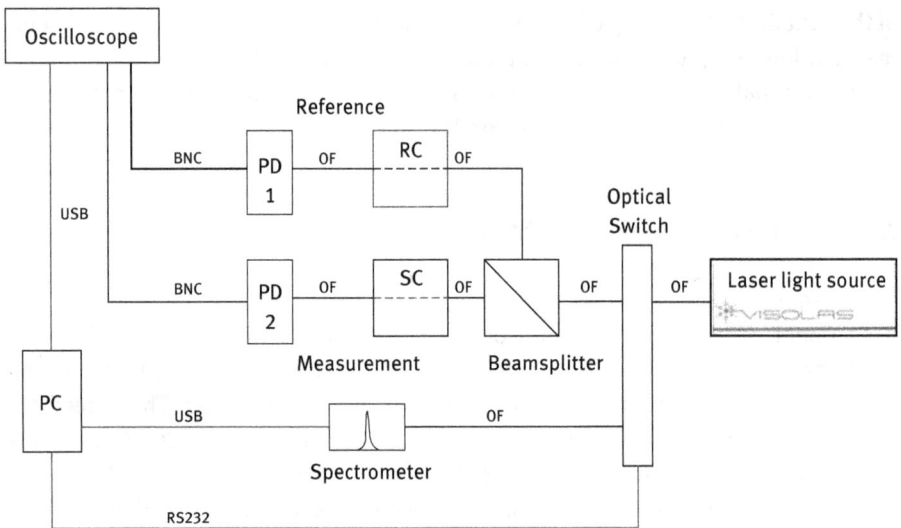

**Figure 7.8:** Schematic overview of the BANSAI setup. PD: Photodetector; OF: Optical fiber; RC: Reference cell; SC: Sample cell.

500 to 650 nm. A schematic representation and a photograph of the employed BANSAI implementation are shown in Figure 7.8.

The individual optical components are connected by optical fibers. The monochromatic laser output is split into two equal parts with equal physical properties and is directed to a measurement and a reference cell. After each cell, the transmitted radiation is measured by two individual photodiodes. The system performs 25 measurements in 1.2 s. A single measurement is performed in less than 50 ms.

The measurement cuvette contains the sample (i.e., serum). To compensate reflections and scattering losses, the reference cuvette is filled with pure BCG reagent. The concentration of the sample is calculated on the basis of the absorbance, using the Lambert–Beer Law [176]. The measurement is controlled with a self-developed software based on LabVIEW® (National Instruments, Austin, USA).

### 7.1.4 BANSAI photometric measurement with laser light

The manual BCG assay was carried out as described by [67] with a commercial reagent kit (mti-diagnostics GmbH, Idstein, Germany). This dye-binding endpoint method works at pH 4.2 and a wavelength of 628 nm. For the manual assay, we prepared a 1:101 dilution with the BCG reagent and the prepared sample series of albumin standards, calibrators, and controls. After mixing the sample with the BCG reagent, the mixture was incubated for 5 min at 37 °C until it was measured. We used several concentrations to measure the extinction of HA in 10-mm cuvettes (Hellma Analytics, Müllheim, Germany), ranging from 10 g/L to 80 g/L using lyophilized reference material or pooled HA.

### 7.1.5 Reference method

To compare the automated reference method for the determination of albumin with the BANSAI system, we used in vitro diagnostic assays for quantitative measurement of HA in serum, plasma, or cerebrospinal fluid on a Dimension Vista clinical analyser system (Siemens Healthcare Diagnostics products, Marburg, Germany) at the Zentrum für Labormedizin, Mikrobiologie und Transfusionsmedizin (ZLMT) Karlsruhe. This assay (specialty Albumin – sALB) is an immunonephelometric test, based on an immunochemical reaction with specific antibodies against human albumin. These form immune complexes, which scatter a beam of light (840 nm) passing through the sample cuvette. The intensity of the scattered light is proportional to the concentration of the investigated protein in the sample. We followed the calibration procedure provided by the manufacturer of the calibrator N Protein Standard SL (Siemens Healthcare Diagnostics Products GmbH, Marburg, Germany).

### 7.1.6 Imprecision

Within-run imprecision experiments were performed for both sera to determine repeatability with at least two different control concentrations with 22 samples per concentration. The control samples were analyzed consecutively within one run on different working days spanning a period of 10 days. On the basis of the individual data, inaccuracies were calculated as deviation of the respective target value. Between-days imprecision was analyzed daily in the same manner as the within-run imprecision. The results contribute an additional validation of the accuracy of the system.

### 7.1.7 Drift

To determine the drift, we investigated the stability of the measurement system during one calibration interval. The BANSAI system and reagents must be operated according to the known procedure using two control materials and the pooled HA sample. For a duration of 3 h, the control materials and the pooled HA serum were measured every 30 min.

### 7.1.8 Accuracy

Accuracy was calculated according to the recovery of control concentration. The recovery was evaluated over two days with Bio-Rad controls low and high. The calibration of the BANSAI system was carried out with the calibrator N Protein standard SL from Siemens.

### 7.1.9 Method comparison

The method comparison is commonly used to compare the method procedure with different analysers that measure the same parameter. In our case, we compared the BCG method on the BANSAI system with the immunonephelometric method (Specific Albumin – sAlb) on the Siemens Dimension Vista system. The immunonephelometric method is the standard method at the ZLMT. We wanted to examine whether the chemically measured results of the samples on the BANSAI are comparable to the standard method results.

   Specimens of 21 patients were used and selected to represent various HA concentrations between 10 g/L and 52 g/L. The fresh human specimens were measured first with the Siemens Dimension Vista system and then with the BANSAI system.

### 7.1.10 Linearity

Pure HA stock solution with a stock concentration of 80 g/L was diluted with physiologic salt solution to 16 concentration levels between 5 g/L and 80 g/L.

### 7.1.11 Limits of detection and quantification

The limit of detection is the lowest result of a measurement by a given procedure above which a blank value can be excluded with a stated confidence level; it is an exclusion limit for a blank value (blank exclusion limit), that is, the lowest result that can be considered being a signal caused by the quantity to be measured. The detection limit is mainly determined by the amount of noise. The imprecision of the detection limit is relatively high and intra-individual trends can hardly be detected. Therefore, some authors have proposed a definition for a lower limit of detection: the lower limit is the lowest result of a measurement that can be obtained by a stated measurement procedure and with a statement of measurement uncertainty (1). Several recommendations for this limit have been published and the common definition says that the lower limit of detection is the quantity at which a fixed coefficient of variation, for example, 15% is obtained [69].

As a result of these considerations, we prepared and measured samples with different dilutions of lyophilized HA with concentrations between 10 g/L and 0.162 g/L and evaluated the corresponding coefficients of variation.

### 7.1.12 Statistical analysis

Statistical analysis was performed using Sigma Plot 12.0 (SPSS, Munich, Germany).

Following the guidelines of [92], we started the evaluation with the serum component HA and dye reagent BCG.

In Figure 7.9, we compare the measured albumin concentration of low controls, high controls, and pooled HA over 10 days.

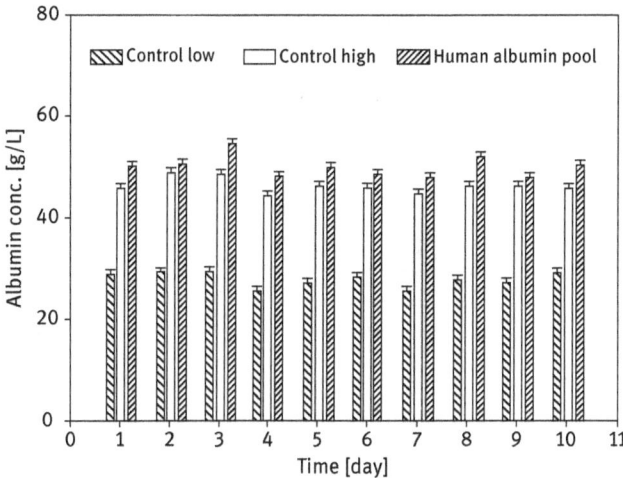

**Figure 7.9:** Imprecision experiments measured with the BANSAI system over a time period of 10 days.

Each control group was treated in the same manner within-days and between-days. It becomes apparent that the results for each group are stable. The low control showed a mean concentration of 27.2 g/L, the high control 45.3 g/L, and the pooled HA has an average concentration of 48.4 g/L. As the determined concentration, the coefficient of variation also remained constant throughout the whole experiment. We calculated the appendant coefficients of variation to compare the deviation of the three control groups. The coefficients of variation of the low controls were below 3.5%, of the high controls below 2.8%, and of the pooled HA below 3.6% over 10 days (Figure 7.10).

In the next experimental investigation, we examined the short time variation of the measurements (drift) (Figure 7.11).

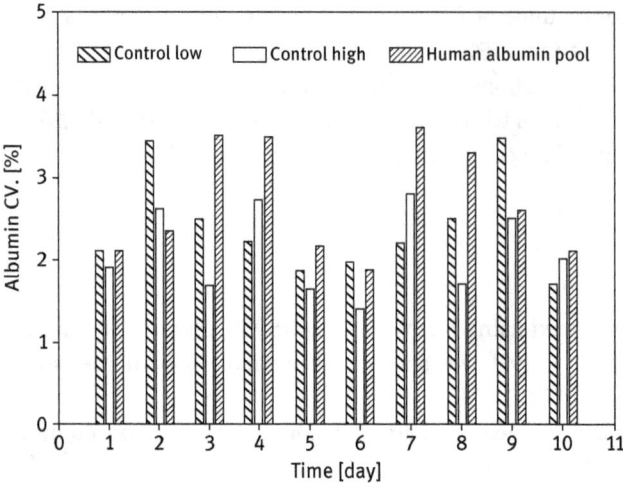

**Figure 7.10:** Coefficient of variation of the measured imprecision samples from Figure 7.20 over 10 days.

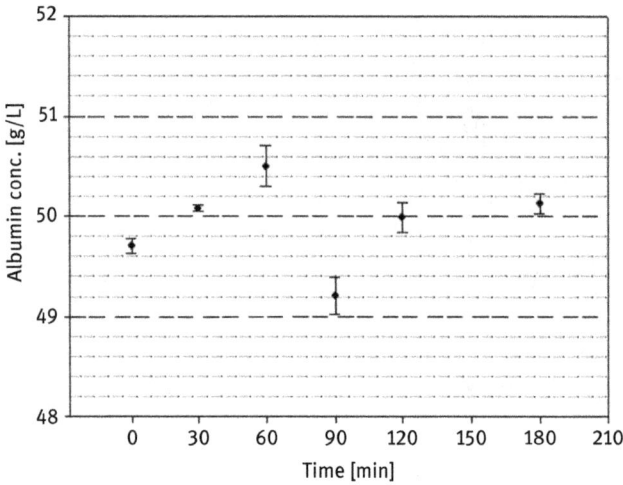

**Figure 7.11:** Drift measurement over 3 h with human serum pool with the BANSAI system.

We found that the measurement of pooled HA with the BCG method on the BANSAI system showed no significant drift effect. Each of the six measurements was delivered a result around 50 g/L over 3 h.

The accuracy and recovery of the controls were also analyzed using the BANSAI system. We investigated the recovery of the known control material concentrations of 28 g/L and 48 g/L (low and high control).

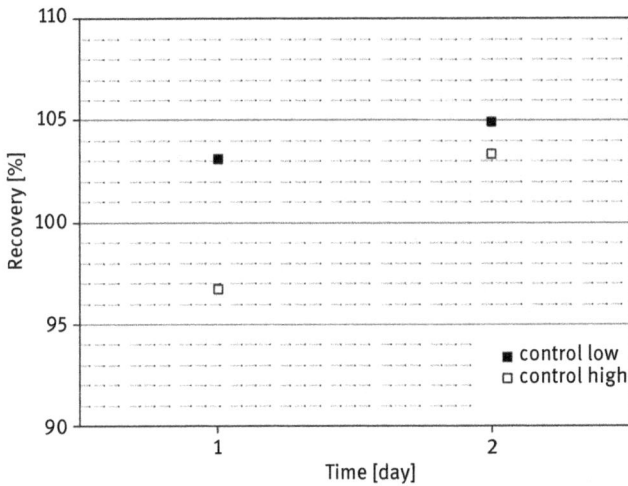

**Figure 7.12:** Accuracy and the associated recovery of controls.

Close to 100% recovery was found for the low and high controls (Figure 7.12).

Comparative studies of measurement procedures are widely used in laboratory medicine to assess an agreement between two procedures that measure the same quantity. One procedure is usually considered as the comparative (x-method) and the other method as the test procedure (y-method) [91]. In our case, the comparative method is the immunonephelometric method on the Siemens Dimension Vista (x-scale) and the test method is the BCG method on the BANSAI system (y-scale). The specimens of 21 patients (concentrations between 10 g/L and 52 g/L) were analyzed. The results are shown in Figure 7.13.

We observed an almost linear behavior and no errant values over the measurement range.

A good linearity could also be observed in the measurement of the linear dilution with a solution of pure HA (Figure 7.14).

In addition to linearity, we measured the limits of detection for HA. Although this was the first implementation of the BANSAI system, it already shows a remarkable low detection limit for albumin of 2.5 g/L (Figure 7.15).

The system was also able to produce results with coefficients of variations of 10% below the threshold of 1.25 g/L. These values, however, were not taken into consideration due to their inconsistency with the concentration values expected according to the linearity of the dilution series.

Serum albumin predicts mortality especially in patients with end-stage renal diseases undergoing a hemodialysis and is used to assess their health status and the quality of delivered care [216].

We demonstrated the applicability of the first implementation of the BANSAI concept for the clinical determination of the serum parameter HA. To evaluate the

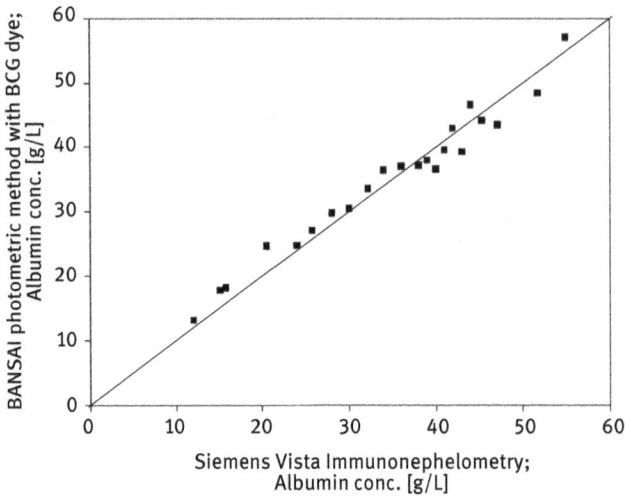

**Figure 7.13:** Method comparison between the BANSAI system and the standard method (immuno-nephelometry) with the Siemens Dimension Vista system.

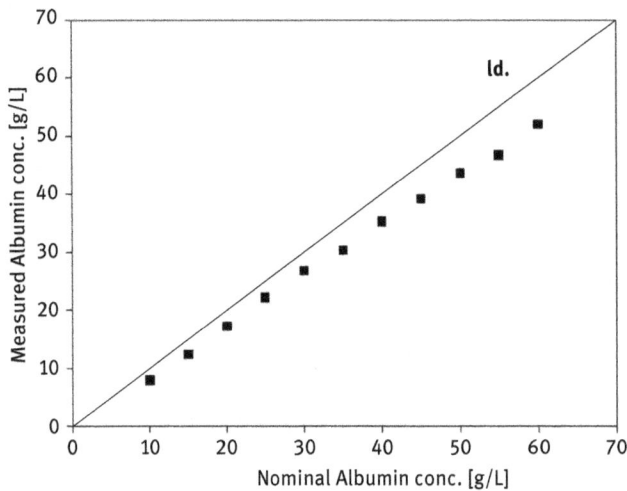

**Figure 7.14:** Linearity of diluted pure HA samples. Id. = Identity line.

measurement quality of our system, our results were compared with the results of a Siemens Dimension Vista system. We tested the within-day imprecision and between-days imprecision, which gives information about repeatability and reproducibility of the experiment. We were able to show that BANSAI reproduced constant results in each control group. Comparable with these results are the coefficients of variation.

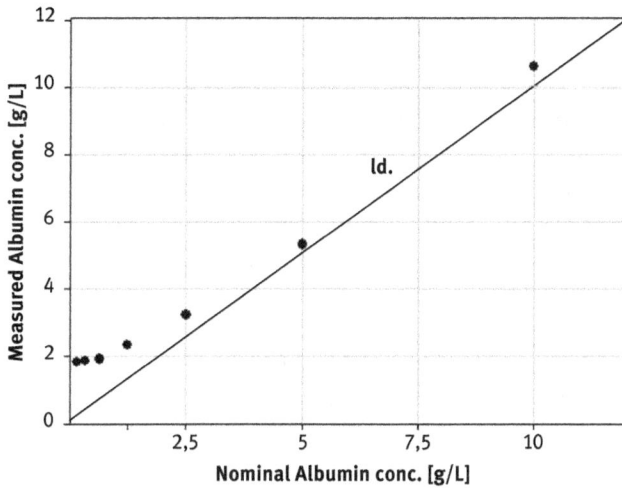

**Figure 7.15:** Lower limit of detection. Pure human albumin solution diluted until 0.162 g/L.
Id.= Identity line.

We also evaluated drift, accuracy and recovery, linearity, and we determined a limit of detection. Furthermore, we performed a method comparison with an immunonephelometric method. Calibration of the system was carried out at the beginning of the measurement, and we were able to show that the BANSAI system is stable over at least 3 h.

The accuracy or rather the recovery ratio of the control materials was close to 100%.

Additionally, the good recovery results confirmed the method comparison between BANSAI and Siemens Dimension Vista. The results were linear and well comparable to the immunonephelometric method.

The results of the measuring dilution of pure HA solution showed the expected linearity. We found that the clinically relevant range can be reliably reproduced and recovered. At a HA concentration of over 60 g/L, the graph indicates a small aberration from the ideal linearity. This can be neglected as such concentrations go far beyond the clinically relevant references for patients (35–53 g/L) [249].

Especially noteworthy are the results of the limit of detection/quantification.

We were able to measure concentrations as low as 2.5 g/L, while the coefficient of variation stayed well below 10%. These values match the concentration defined by the dilution series. In a classic detection limitation, the variances exceed 15%. This indicates that we may not have found the actual detection limit and that the system may be able to measure even lower concentrations. We have not fully understood the reasons for this phenomenon but we will investigate further. Succeeding in lowering the detection limit, the BANSAI system capabilities could be extended toward albumin measurements in urine samples (reference range for adults is 0.02–0.2 g/L [223]). At the moment, urinary albumin is measured routinely using the nephelometric method. The BANSAI system with the BCG method would be an interesting alternative.

The research aims to develop compact and portable alternatives to conventional clinical photometers. The BANSAI system requires only a limited number of components, which can be easily reduced in size. Therefore, a miniaturization of the current macroscopic platform is feasible, which would enhance the mobility and enable mobile point of care applications. Photometric analysis methods would then be possible everywhere while maintaining laboratory grade result quality.

This example already demonstrates the system aspect where every component is addressed and required for creating a true optofluidic system.

A related review on emerging optofluidic technologies for point-of-care genetic analysis systems is provided by [30].

The following two paragraphs provide examples of two different industrial approaches toward creating optofluidic systems.

The following paragraph has been provided with kind permission by Thomas Hahn, Inamullah Naveed (Software), Christopher Christie (Design), Maik Fuchs (Elektronics) – all Buerket Fluid Control Systems, Ingelfingen, Germany.

To run experiments in LoC applications, scientists and industrial developers utilize a variety of equipment such as syringe pumps, pressure controller, a temperature unit, a microscope, a personal computer, and many other devices. The Buerkert microfluidic demonstration device (MDD) shown in Figure 7.16 is capable of providing all the features in one device that are normally carried out by different units. At the moment, there are only a few systems on the market that provide similar flexibility but without integrating all features in one device.

**Figure 7.16:** Buerkerts microfluidic device demonstrator (Courtesy Thomas Hahn, Buerkert Fluid Control Systems).

To provide a standard tool that enables scientists and industrial developers to deal with liquids of small volumes, Buerkerts MDD is capable of helping out in any experimental design taking care about the fluidic actuation. The MDD can hold so-called LoC systems and controls the liquids within these chips with high accuracy. The user is free of using own chip designs that can be made from glass-PDMS hybrids or thermoplastic polymers. The MDD is a fluidic control system for pulsation-free handling of liquids in the range from nanoliters to microliters with high accuracy. In addition, it offers fully integrated optical analysis capabilities such as for, for example, fluorescence and spectrophotometry, thus being a truly optofluidic control system.

General lab applications are meant to be more a "lab around the chip" rather than a "Lab-on-a-Chip" system as the control of an experiment is distributed among many devices. The MDD combines the features of several devices such as a touchscreen, controlling syringe pumps, which are controlled by valves with continuous pressure readout.

The main housing of the MDD consists of standard elements such as a touch screen, three syringe pumps, and electronics. A mobile unit is attached to the main housing via tubes and cables, providing additional functionality. The tubes end at specific connection modules that hold the microfluidic chip via magnetic force. The mobile unit can, therefore, be even placed on an inverted (epi-fluorescence) microscope to monitor the reaction chamber of the microfluidic chip. The integrated magnets provide a flat chip surface and allow the observation of living cells or other objects with large magnification objectives and small working distances (>100 $\mu m$). The mobile unit is placed into a dedicated place on the main housing when not in use. The mobile unit contains several modules and functionalities. It holds connection modules that allow the transfer of the gas pressure produced by the syringe pumps onto the microfluidic chip. Driving the fluid in the microfluidic chip is pressure controlled with an accuracy of ± 2 mbar using 1 bar pressure transmitters. Using pressure transmitters up to 100 mbar can increase the accuracy. The syringe pumps can also be used without the pressure control functionality.

Each syringe pump is connected to a 3/2 valve (Buerkert Type 6624) that enables the system to recharge the syringe pump with air while the system is disconnected from the mobile unit. A 2/2 valve (Buerkert Type 6624) is added in series, which allows the pressure storage until it reaches a given value (e.g., 20 mbar). If the pressure is below the required value, the actuator moves the syringe pump accordingly. The fluidic layout is shown in Figure 7.17.

The mobile unit contains several modules:
- a temperature module to be able to control the temperature locally in the microfluidic chip
- an optical module which is able to provide light and a measurement functionality in the form of a photodiode for the microfluidic chip. The microfluidic chip needs to contain mirrors in the form of slanted angles to be able to handle the incoming light from underneath

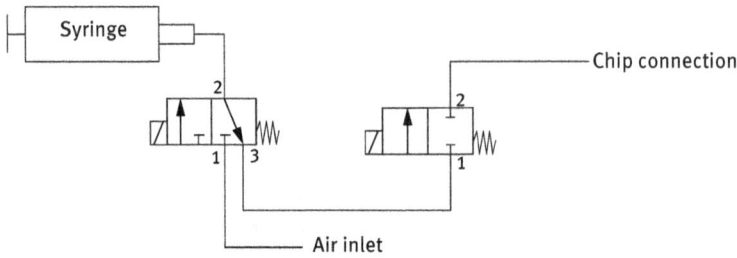

**Figure 7.17:** Fluidic layout of one syringe pump. The syringe pump is connected to a 3/2 valve and in nc-position connected to a 2/2 valve (type 6624). In nc position, the 2/2 valves stores the pressure generated by the syringe. Once the valve is open, it releases the pressure to the chip and the syringe pump regulates the system to keep the pressure constant at the given value (Courtesy Thomas Hahn, Buerkert Fluid Control Systems).

– connection modules for delivering the required air pressure to drive the fluid in the microfluidic chip

The optical module contains an LED emitting at 488 nm (OD 6 filter) to excite fluorescence in molecules like fluorescein or YOYO-1 that have an emission maximum around 520 nm (OD 6 filter). The LED is pulsed with a frequency of 228 Hz to separate the signal from ambient light. Besides fluorescence measurements, the module can be rearranged to measure absorbance through a microchannel.

This example of another true optofluidic system can be combined with both previously described platforms, the DUV, and/or the modular optofluidic backplane platform for solving numerous applications and providing unlimited functionality and solutions, again with the possibility of creating a true optofluidic product. Another example of a true optofluidic system within an industrial environment is described in the following paragraph.

The following paragraph has been provided with kind permission by Festo AG & Co. KG, Esslingen, Germany. The author would like to thank all members of the optofluidics project especially Florian Zieker and Elias Maria Knubben for inspiring discussions and results achieved.

## 7.1.13 Festo optofluidic

To demonstrate processes in the field of optofluidics, Festo has developed optofluidic, a bionic test medium (Figure 7.18). For this purpose, Festo is benefiting from its broad range of optical technology components such as cameras, sensors, LEDs, lasers, and photodiodes. These are combined with fluid engineering devices such as valves, pumps, and mixers, which control the flow of the media.

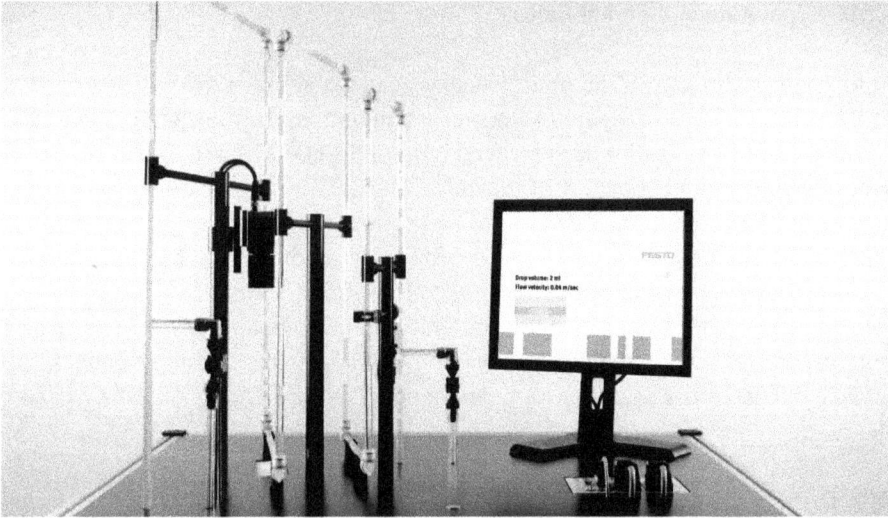

**Figure 7.18:** Optofluidic: Demonstration of optical analysis and separation processes (Courtesy Festo AG & Co. KG)

A transparent liquid is metered into a blue fluid by means of a valve. The two immiscible fluids flow through the lines of the display. The blue liquid is the intelligent medium that conveys the information required for optical diagnosis. Possible steps in optofluidic analysis and separation are demonstrated in the following three stations:

### 7.1.14 Real-time optical analysis

The first station incorporates an optical sensor SOEC and a compact camera SBOC, which analyze the fluid in real time in terms of its composition and characteristics and evaluates parameters such as the transparent drop's volume. For this, the in-house developed Software CheckKon and CheckOpti is used. A monitor connected to the camera system displays the collected data and allows continuous real-time monitoring of the processes.

### 7.1.15 Visualization of hidden information

At the second station, a blue LED with a wavelength of 490 nm renders a previously implemented fluorescent dye visible. This process shows how optofluidics can allow certain desired parameters and information to be registered and displayed in readable form by means of optical components.

### 7.1.16 Separation of different fluids

At the third station, an SOEC optical sensor detects the blue liquid. By means of a VODA valve, the system separates the blue medium from the transparent liquid.

The overall application is controlled by a CPX-CEC-C1 type CoDeSys front-end controller and a number of CPX I/O modules. The CoDeSys-capable compact camera is directly connected to the CPX control and communicates directly in real time via Ethernet. The color sensors are connected to the I/O modules and thus constitute an integrated closed chain of communication and control.

This chapter on true optofluidic systems has demonstrated the system aspect from small systems to medium to large optofluidic systems and their potential to provide the necessary solutions for different applications.

## 7.2 Conclusion

Optofluidics systems technology can be defined as the art of implementing individual intelligent components for creating a functional system based on optical principles. The terminology "system" reflects the necessity for providing a stable solution in the corresponding working environment.

For an optofluidic system to be successful, all system aspects have to be addressed in the designing phase, assuring component, and platform compatibility. This will be the task for future optofluidic system designers and suppliers (industrial and academic) to define a common standard on every system level, thus being able to merge photonics, fluidics, and biology for creating true optofluidic systems.

## Further reading

Brammer M. Modulare Optoelektronische Mikrofluidische Backplane. KIT Scientific Publishing, 2012
Fainman Y, Lee LP, Psaltis D, Yang C. Optofluidics: Fundamentals, Devices, and Applications. McGraw-Hill, 2010
Hawkins AR, Schmidt Holger. Handbook of Optofluidics. CRC Press, 2010
Love J. Process Automation Handbook. Springer, 2007

# 8 Outlook

Optofluidic systems are an emerging class of new solutions and the question is what will be the driving force for their future. Optofluidics is definitely a cross-discipline technology requiring the close interaction of all fields covered in this book and necessitates a common language of all disciplines, which is a challenge for today's and tomorrow's Optofluidics specialists. This also implies a broadening of the curriculum at educational institutions, which should provide a basic understanding of this emerging technology field.

The chapters in this book have touched the surface of different fundamental requirements for creating true Optofluidic systems:

**Materials**, with an overview on smart materials ranging from material modifications to actuators to smart fluids;

**Photonics and Biophotonics**, with an introduction to the basics of waveguides, their characterization and types of waveguides, the corresponding fabrication technology leading to planar and ridge waveguide based devices, active photonic devices ranging from LEDs to detectors to lasers;

**Fluidics and fluid control systems**, providing the basics on valves, fluid mechanics, control theory, microvalves and fluidic channels, sensors describing commonly used process analytical sensors, spectroscopy and miniature spectrometers;

**Optofluidic system technology**, presenting examples of integration of components.

Optofluidics in the context of process analytical technology and systems will be part of the next industrial revolution.

Process analytical technology (PAT), which is defined according to the USFDA (2009) as a system for designing, analyzing, and controlling manufacturing through timely measurements of critical quality and performance attributes of raw and in-process materials and processes is a perfect environment for optofluidics.

Miniaturization of optofluidic devices and systems enables numerous applications paired with robustness and easy-to-use emerging cloud-based interfaces, an ideal playground for what can be defined as PAT 4.0.

What is sure is that the Optofluidics revolution has just started and will continue like an avalanche effect once it will penetrate all industries including applications which have so far not been touched by it because of its price and miniaturization hurdles.

The digital transformation is desperately in need of smart optofluidic systems, especially IoT based applications, services, and products in the framework of Industry 4.0.

Industry 4.0 is the synonym of the fourth industrial revolution, whereas Industry 1.0: Mechanical industrialization, Industry 2.0: Mass production, and Industry 3.0: Automation.

https://doi.org/10.1515/9783110546156-008

Industry 4.0 is the melting pot of digitization and automation, where smart production machines communicate directly with one another by creating a highly flexible and automated production environment.

This communication is not only happening amongst smart production machines, but spans the entire value chain, from product design to production – worldwide, leading to smart factories. This value chain can even include the customer.

Whereas Industry 4.0 is an initiative driven by Germany and European countries (www.plattform-i40.de), the Industrial Internet Consortium (www.iiconsortium.org) initiative is driven by USA.

Going even a step further and looking into the future, the successful integration, fusion, and merger of optofluidic technologies and disciplines will lead to novel Optofluidic systems with the potential to create bio-artificial systems with so far unreached features and performances unable to be achieved purely by artificial systems.

This development could create, what could be defined as "cybernetic organisms," that is an organism which is a self-regulating integration of artificial and natural systems and components – Optofluidics 4.0. ☺

# 9 Glossary of optofluidics terms and definitions

This is a summary of abbreviations and terms, which one comes across when dealing with an interdisciplinary topic like optofluidic system technology.

It is a collection from various references without the guarantee for completeness.

**Abrasion:** A process by which a surface is worn away by friction.

**Absolute Pressure:** Absolute pressure is zero-referenced against a perfect vacuum, so it is equal to **Gauge Pressure** plus atmospheric pressure.

**Absorption:** Light is not only reflected at a surface but also absorbed or taken up. This energy is taken up by the molecules of the irradiated body, which in turn change this energy to kinetic energy, to movement. This movement produces heat.

**Activation Energy:** This is the energy required for a substance to perform a reaction. In the Arrhenius equation for calculating the speed of a chemical reaction, activation energy is used as an indicator for expressing how easily a chemical reaction occurs. The Arrhenius equation is used to calculate the service life of an LED, etc. The activation energy of LED degradation is obtained from the failure rate under several temperature conditions (Courtesy Hamamatsu).

**Actuator:** An actuator is a device that converts energy (air, electricity, or liquid) into motion or force. Actuators are used in manufacturing or industrial applications such as motors, pumps, switches, and valves.

**Adsorbent:** Selecting an adsorbent is to choose one that can retain by usually intermolecular forces onto its surface, a specific, or group of analytes for a specified sample volume. The adsorbent must also be able to release the analyte(s) during the desorption process.

**Adsorption:** A chemical process that takes place when a liquid or a gas accumulates on the surface of a solid forming a molecular or atomic film.

**Agglomeration:** Two particles, whether micro or nano, are submitted to attractive and repulsive forces. Agglomeration occurs when the attractive forces are predominant and the smaller the particle size is, the higher these forces are. This explains why agglomeration plays an important role for nanoparticles. According to ISO TS27687 2008, an agglomerate is a collection of loosely bound particles or aggregates or mixtures of the two where the resulting external surface area is similar to the sum of the surface areas of the individual components. The standard adds that the forces holding agglomerate together are weak forces, for example, Van der Waals forces, as well as simple physical entanglement. The terms agglomerate and aggregate are often assimilated, but as also explained in ISO TS27687 2008, an aggregate is made of particle comprising of strongly bonded fused particles. As a consequence, aggregation is often considered as a non-reversible phenomenon.

https://doi.org/10.1515/9783110546156-009

Agglomerates are brittle structures, which can be broken down and rebuilt depending on the strength of the external solicitations. They can be described by the number and the arrangement of the particles, which they contained [1].
More information can be found here: http://ewpa.group.shef.ac.uk/agglomeration.html

**Amorphous**: Non-crystalline state having no definite form. For example, when a liquid or gaseous semiconductor is cooled and solidified so rapidly that no crystals are formed, it becomes amorphous. In this state, the crystal structure has a short-distance order but does not have a long-distance order, and a tail level appears at the band gap edge, making the optical characteristics different from those of monocrystalline or polycrystalline materials (Courtesy Hamamatsu).

**Anisotropic Etching**: An etching process in which the etching speed in a particular direction is different from that in other directions. For example, when a (100) silicon substrate is alkaline-etched, V-grooves are formed due to the fact that the etching speed on the (100) plane is faster than that on the (111) plane. Etching in which the etching speed is the same in all directions is called isotropic etching (Courtesy Hamamatsu).

**Anodic Bonding**: When the flat surface of glass containing alkali metal is attached to the flat surface of silicon and heated while a voltage is being applied, an electrostatic attractive force is generated at the interface between the glass and silicon. Anodic bonding is the bonding technique that makes use of this phenomenon. During anodic bonding, the silicon side is used as the anode (Courtesy Hamamatsu).

**Arrhenius equation**: The following equation describes the temperature dependence of chemical/physical reaction speeds, proposed by S.A. Arrhenius (Sweden) in 1889. This equation is used to calculate the expected life of a component when a major cause of degradation of the component is probably temperature (Courtesy Hamamatsu).

$$K = A \cdot exp\left(\frac{-E_a}{kT}\right)$$

where K is the reaction speed, A is a constant, Ea is the activation energy in eV, k is the Boltzmann constant in eV/K, and T is the absolute temperature in Kelvin.

**ASE** (amplified spontaneous emission): An optical amplifier amplifies signal light by induction radiation. However, it emits energy little by little even under conditions where no signal light is input. This is optical amplifier spontaneous emission and is referred to as ASE. This ASE is noise and degrades characteristics (Courtesy Hamamatsu).

**Autocatalysis**: A single chemical reaction is said to have undergone autocatalysis, or be autocatalytic, if the reaction product is itself the catalyst for that reaction.

**Autoclaving:** Is the process run by an autoclave that is an apparatus used for sterilizing using steam under high pressure. Autoclaving can reach temperatures greater than 100 °C, which not only kills bacteria but also bacterial spores. It is used, for example, in Bio-laboratories for sterilizing equipment.

**Band Gap Energy:** In a semiconductor, insulator, or metal, electrons surrounding the nucleus are present in energy levels with a certain width. In semiconductors or insulators, among the energy bands where electrons exist, the highest energy band filled with electrons at absolute zero degrees is called the valence band, and the energy band with no electrons is called the conduction band. The energy range in the band gap (forbidden band) between the valence band and the conduction band is called the band gap energy. In metals, there is no band gap because the valence band and conduction band overlap with each other (Courtesy Hamamatsu).

**Bioenergetics:** Is a field of study focusing on how energy flows through living systems.

**Biofiltration:** Biofiltration is a pollution control technology. It is a technique for eliminating malodorous gas emissions and of low concentrations of volatile organic compounds (VOCs).

**Biopolymers:** Those are polymers produced by living organisms. Cellulose, starch and chitin, proteins and peptides, and DNA and RNA are all examples of biopolymers, in which the monomeric units, respectively, are sugars, amino acids, and nucleotides.

**Bioreactor:** Is a system in which a biological conversion is effected. Bioreactors differ from conventional chemical reactors in that they support and control biological entities. As such, bioreactor systems must be designed to provide a higher degree of control over process upsets and contaminations, since the organisms are more sensitive and less stable than chemicals [1].

**Biosensor:** As per definition of IUPAC, a biosensor – in contrast to biotic sensors or biotests – is a self-contained, integrated receptor-transducer device, which is capable of providing selective quantitative or semi-quantitative analytical information and which uses a biological recognition element (bio-receptor) and a transducer in direct spacial contact [1].

**Biotechnology:** Is technology based on biology – biotechnology harnesses cellular and biomolecular processes to develop technologies and products.

**Biotechnology, Green:** The use of biotechnology for processes and products in the agricultural sector.

**Biotechnology, Red:** Red biotechnology refers to the use of organisms for the improvement of medical processes and manufacturing of pharmaceutical products.

**Biotechnology, White:** Refers to the application of biotechnology in industrial processes. For example, the use of cells.

**Blooming:** A phenomenon in which the photoelectrically converted signal charge in an image sensor exceeds a certain level and spills over into adjacent pixels or transfer region other than photodiodes (in IT type CCDs). In CCDs, the spill-out charge appears in the image as a vertical stripe occurring from the light incident position the same as with "smear." To prevent blooming, some means for discharging excess charge should be implemented. In CCDs, this blooming is suppressed using a vertical/horizontal anti-blooming or clocking method (Courtesy Hamamatsu).

**Bragg diffraction**: A coherent, strong reflection that occurs at a particular angle at which the phases are matched by multiple surface reflections when monochromatic light strikes a light-scattering material with a cyclically arranged structure. This technique is utilized to fabricate resonators in semiconductor lasers (Courtesy Hamamatsu).

**Breakdown voltage**: As the reverse voltage applied to a PN junction is raised, an abrupt increase in reverse current occurs at a certain voltage. This voltage is called the breakdown voltage. As a guide for convenience when evaluating silicon avalanche photodiode (Si APD), the voltage that produces a reverse current of 100 µA is specified as the breakdown voltage.

**Bypass**: A flow path besides the main flow path, usually using a smaller diameter of the tube or flow path. Used, for example, to insert sensors, which would be unable to insert into the main flow path.

**Calorimetry**: Calorimetry is the quantitative measurement of heat. It measures the transfer of energy from one system to another caused by temperature differences. Applications of calorimetry include measurements of the specific heats of solids and liquids, the heats of vaporization and combustion, and the rate of heat generation (power) from radio-nuclides.

**Candela (cd)**: The unit candela (cd) is defined as the luminous flux radiated from $1/60$ $cm^2$ of a blackbody with temperature 2.042 K.

**Capillary Action**: Is defined as the ascension of liquids through slim tube, cylinder or permeable substance due to adhesive and cohesive forces interacting between the liquid and the surface.

**Catalyst**: Increases the rate of a chemical reaction.

**Catalysis**: Is the reaction taking place in the presence of a catalyst.

**Catalytic Cracking**: Catalytic cracking is a refinery process for converting heavy oils into more valuable gasoline and lighter products. Originally, cracking was accomplished thermally but the catalytic process has almost replaced thermal cracking because more gasoline having a higher octane and less heavy fuel oils and light gases are produced.

**Cavitation**: Cavitation may occur when the local static pressure in a fluid reaches a level below the vapor pressure of the liquid at the actual temperature. According to the Bernoulli Equation, this may happen when the fluid accelerates in a control valve or around a pump impeller. A liquid is said to cavitate when vapor bubbles form and grow as a consequence of pressure reduction.

**CFD (Computational fluid dynamics)**: "The physical aspects of any fluid flow are governed by the following three fundamental principles: (1) mass is conserved; (2) F = ma (Newton's second law); and (3) energy is conserved. These fundamental principles can be expressed in terms of mathematical equations, which in their most general form are usually partial differential equations. Computational fluid dynamics is, in part, the art of replacing the governing partial differential equations of fluid flow with numbers, and advancing these numbers in space and/or time to obtain a final numerical description of the complete flow field of interest." John F. Wendt (Ed.), Computational Fluid Dynamics, An Introduction, Springer 2009.

**Chemisorption:** It is a kind of adsorption, which involves a chemical reaction between the surface and the adsorbate. New chemical bonds are generated at the adsorbant surface.

**Chemometrics:** The science of extracting information from chemical systems by data-driven means. Refer Section 7.4.

**CIE:** The International Commission on Illumination – also known as the CIE from its French title, the Commission Internationale de l´Eclairage – is devoted to world-wide cooperation and the exchange of information on all matters relating to the science and art of light and lighting, color and vision, photobiology and image technology. With strong technical, scientific and cultural foundations, the CIE is an independent, non-profit organization that serves member countries on a voluntary basis. Since its inception in 1913, the CIE has become a professional organization and has been accepted as representing the best authority on the subject and as such is recognized by ISO as an international standardization body. More information can be found at: www.cie.co.at

**CIP:** Clean in place. To clean the product contact surfaces of vessels, equipment and pipework in place, that is, without dismantling.

**Clamp-on:** A method to place a sensor on a tube or pipe without the need for providing a fitting or extra insert. The mounting on the outside of the pipe ensures that there is no material incompatibility with the process being measured or the pressure in the tube or pipe.

**Clark-electrode/Clark-sensor:** Is an oxygen sensor named after the inventor, Dr Leland Clark. The Clark electrode consists of an anode and cathode in contact with an electrolyte solution. It is covered at the tip by a semi-permeable membrane, which is permeable to gases. The cathode is in a glass envelope in the body of the electrode. Reduction of oxygen occurs at the surface cathode, which is exposed at the tip of the electrode. Oxygen molecules diffuse through the semi-permeable membrane and combine with the electrolyte solution. The current produced is a result of the following reduction of oxygen at the cathode.

**Coagulant:** A substance or agent that causes or promotes a liquid to coagulate.

**Coagulation:** Is the clotting of blood in medicine.

**Colloid:** A colloid refers to a substance which consists of particles » atoms or molecules. Colloidal systems exist as dispersions of one substance in another, for example, smoke particles in air.

**COD:** Chemical oxygen demand (COD) is a measure of the capacity of water to consume oxygen during the decomposition of organic matter and the oxidation of inorganic chemicals.

**Condition Monitoring:** It is the data recording and interpretation of data from machines, systems, and their components, with the aim of conducting predictive maintenance programs based on the system condition.

**Conductivity:** Conductivity and resistivity are both measures of the ability of a fluid to conduct electrical current. Conductivity is the reciprocal of resistivity. Conductivity units are used, for example, when referring to water ranging from drinking water to sea water, while resistivity units are reserved for ultra pure water such as deionized or reverse-osmosis water. The unit of conductivity is Siemens (S).

**Conductivity, Conductive:** The measuring cell consists of two open electrodes to which an AC voltage is applied. The medium is in direct contact with the electrodes. The applied voltage generates a current dependent on the resistance of the medium (Ohm's law). The geometry of the measuring cell (area S and distance d) is defined by its quotients $K = d/S$. The conductivity of the solution is calculated on the basis of this known cell constant K and by measuring the current generated.

**Conductivity, Inductive:** An inductive conductivity cell consists of two coils: a transmitter coil and a receiver coil. The coils are integrated in a finger-shaped housing. A bore is routed through the finger and the coil integrated into it. The fluid encloses the finger and is also in the bore. A sinusoidal AC voltage is applied to the transmitter coil. This produces a current in the fluid, proportional to the conductivity. This current in turn generates a voltage in the receiver coil. By measuring this voltage and knowing the cell constant, it is possible to determine the conductivity. A temperature sensor is integrated for temperature compensation.

**COP:** Clean out of place.

**Continuity Equation:** One of the fundamental principles used in the analysis of uniform flow is known as the Continuity of Flow. It is derived from the fact that mass is conserved in fluid systems regardless of the dimension or shape of the pipe or direction of flow.

**Coriolis (flow meter):** Coriolis mass flow meters measure the force resulting from the acceleration caused by mass moving toward (or away from) a center of rotation. In a Coriolis mass flow meter, the "swinging" is generated by vibrating the tube in which the fluid flows. The amount of twist is proportional to the mass flow rate of fluid passing through the tube. Sensors and a Coriolis mass flow meter transmitter are used to measure the twist and generate a linear flow signal (www.flowmeters.com).

**Corrosion:** Corrosion can be defined as the degradation of a material due to a reaction with its environment. Degradation implies deterioration of physical properties of the material. This can be a weakening of the material due to a loss of cross-sectional area, it can be the shattering of a metal due to hydrogen embrittlement, or it can be the cracking of a polymer due to sunlight exposure. (Fundamentals of Corrosion and Corrosion Control, http://corrosion.ksc.nasa.gov/corr_fundamentals.htm)

**Cross-contamination**: Cross-contamination is the transfer of disease-causing microorganisms, such as bacteria and viruses, from one substance (e.g., food or chemicals) to another.

**Crossflow-Microfiltration (CFMF)**: The liquid in CFMF systems is pumped tangentially across the surface of a microfilter or filtration medium. Purified and clear liquid permeates through the filter and is collected below it.

**Crystallization**: Crystallization is a technique, which chemists use to purify solid compounds. It is one of the fundamental procedures each chemist must master to become proficient in the laboratory. Crystallization is based on the principles of solubility: compounds (solutes) tend to be more soluble in hot liquids (solvents) than they are in cold liquids. If a saturated hot solution is allowed to cool, the solute is no longer soluble in the solvent and forms crystals of pure compound. Impurities are excluded from the growing crystals and the pure solid crystals can be separated from the dissolved impurities by filtration (http://orgchem.colorado.edu/Technique/Procedures/Crystallization/Crystallization.html).

**Dark current**: A small current which flows when a reverse voltage is applied to a photodiode even in a dark state. This current is called the dark current. Noise resulting from dark current becomes dominant when a reverse voltage is applied to photodiodes (PIN photodiodes, etc.) (Courtesy Hamamatsu).

**DBR (distributed Bragg reflector)**: This is a reflector containing a diffraction grating having a cycle of $\lambda/2n$ ($\lambda$: wavelength in vacuum, n: refractive index of medium) formed outside the light-emission region in light-emitting devices such as LEDs and semiconductor lasers to selectively reflect the light of wavelength $\lambda$. In vertical cavity surface emitting lasers (VCSEL), forming DBR layers as the upper and lower layers of the light-emitting layer at an appropriate distance causes resonance only at a specific wavelength, so the laser beam can be emitted in the direction perpendicular to the surface. In some LEDs, a DBR layer is formed underneath the light-emitting layer to increase the light level (Courtesy Hamamatsu).

**Dedusting**: Dedusting (also known as aspirating) is a process that involves screening and other pneumatic means to remove fine impurities (e.g., dust).

**Degasification**: Is a water treatment method that removes dissolved gases from the water. This is especially important in optofluidic systems, as, for example, air bubbles can lead to false measurement when present in the optical path. Specially designed degassers need to be integrated which usually involve the use of membranes.

**Deliquescence:** Deliquescent compounds absorb moisture from the air which results in the fact that they dissolve. Examples are calcium chloride and sodium hydroxide.

**Desalination (freeze):** Desalination by freezing processes is based on the fact that ice crystals are made up of pure water. This process consists of three main steps: ice formation, ice cleaning, and ice melting.

**Desorption:** An adsorbed species present on a surface at low temperatures may remain almost indefinitely in that state. As the temperature of the substrate is increased, however, there will come a point at which the thermal energy of the adsorbed species is such that one of following points may occur:
– a molecular species may decompose to yield either gas phase products or other surface species.
– an atomic adsorbate may react with the substrate to yield a specific surface compound, or diffuse into the bulk of the underlying solid.
– the species may desorb from the surface and return into the gas phase.

The last of these options is the desorption process. In the absence of decomposition, the desorbing species will generally be the same as that originally adsorbed but this is not necessarily always the case (The Desorption Process, http://www.chem.qmul.ac.uk/surfaces/scc/scat2_6.htm).

**Destillation:** Distillation is a process, which can be used to separate a pure liquid from a mixture of liquids. A prominent example is the separation of ethanol from water.

**Differential Pressure:** Differential pressure is the difference in pressure between two points.

**Diffraction Grating:** An optical element designed to obtain a spectrum by making use of light diffraction. Reflective diffraction gratings usually have a great number of grooves formed in their surfaces and utilize diffraction images created by interference with light beams reflected from the grating surface (Courtesy Hamamatsu).

**Diffusion (chemical):** Diffusion (chemical diffusion) is the transport of a material or chemical by molecular motion.

**Dispersion (chromatic):** Chromatic dispersion is the change of index of refraction with wavelength.

**Disposables:** Disposables in optofluidics systems are usually devices, which can be discarded after they have been used without the intention of reusing them. Devices which are only used for a single measurement are also referred to as single use devices.

**Disruption, Cell:** Cell disruption is breaking apart of cell wall or cell membrane to affect the release of intracellular product. Cell disruption is required to extract biological products of interest that are not secreted from cells.

**DoE:** Design of experiment

**Downstream Processing**:

**D-value**: The time in minutes required for a one-log or 90% reduction of a specific microbial population under specified lethal conditions. For steam sterilization, it is determined at a constant temperature (Source: http://pharmaceuticalvalidation.blogspot.de).

**EBC**: European Brewery Convention

**Electrolyte**: Electrolytes are minerals in fluids that carry an electric charge and thus enable the fluid to conduct electricity.

**EHEDG**: European Hygienic Engineering and Design Group. EHEDG is a consortium of equipment manufacturers, food industries, research institutes as well as public health authorities and was founded in 1989 with the aim to promote hygiene during the processing and packing of food products. The principal goal of EHEDG is the promotion of safe food by improving hygienic engineering and design in all aspects of food manufacture. More information can be found here: http://www.ehedg.org/

**Elastomer**: Elastomers are flexible or 'rubbery' materials, which can readily be deformed, and return rapidly to almost their original shape and size once released from stress, thus making them able to form reliable seals. Natural and synthetic rubbers are common examples of elastomers (Polymer types, http://www.ami.ac.uk/courses/topics/0210_pt/).

**Electromagnetic compatibility (EMC)**: EMC is the interaction of electrical and electronic equipment with its electromagnetic environment, and with other equipment. The European standard can be found here: http://ec.europa.eu/enterprise/policies/european-standards/harmonised-standards/electromagnetic-compatibility/index_en.htm

**Electromagnetic interference (EMI)**: EMI is an unwanted signal at the signal receiver, and in general, methods are sought to reduce the level of this unwanted interference. For further information, refer, for example, EMI Electromagnetic Interference Basics, http://www.radio-electronics.com/info/circuits/emc-emi/electromagnetic-interference-basics-tutorial.php)

**Electro-osmosis**: When electric fields are applied across capillaries or microchannels, bulk fluid motion is observed. The velocity of this motion is linearly proportional to the applied electric field, and dependent on both:
- the material used to construct the microchannel and
- the solution in contact with the channel wall.

This motion is referred to as electro-osmosis. Further information can be found here: http://www.kirbyresearch.com/textbook

**Electrophoresis**: Electrophoresis is an analytical method, which is frequently used in molecular biology and medicine for the separation and characterization of proteins, nucleic acids, and subcellular-sized particles.

**Emulsion:** An emulsion is defined by combining two liquids that will maintain their distinct characteristics after they have been mixed.

**Energy Harvesting:** Energy harvesting is referred to as the conversion of ambient energy into electrical energy. More information can be found, for example, at: http://www.energyharvesting.net/

**Enzyme:** Enzymes are biological molecules that catalyze (i.e., increase the rates of) chemical reactions.

**EPDM:** Ethylene propylene diene monomer (M-class) rubber

**Electrostatic Discharge (ESD):** ESD is referred to as the transfer of charge between bodies at different electrical potentials. The ESD can destroy electrical circuits and when designing any electrical system needs, precautions have to be made to avoid ESD.

**eV (electron volt):** Energy acquired by an electron when it is accelerated through a potential difference of 1 V in a vacuum. This is generally used as a unit to express the energy of elementary particles, atomic nuclei, atoms, and molecules, etc., 1 eV = $1.602 \times 10^{-19}$ J

**Excitation:** In semiconductors, excitation refers to the process of raising electrons from a low-energy valence band to the higher-energy conduction band. If electrons are excited by heat, this process is called thermal excitation. If an electron is excited by light, then this process is called photoexcitation. Light absorption by a photodiode is photoexcitation (Courtesy Hamamatsu).

**Extraction:** Extraction is a laboratory or industrial procedure used when isolating or purifying a product. Organic chemistry employs solid–liquid, liquid–liquid, and acid–base extractions. More information can be found, for example, here: Extraction Theory and General Procedure, http://academics.wellesley.edu/Chemistry/chem211lab/Orgo_Lab_Manual/Appendix/Techniques/Extraction/extraction_n.html.

**FDA:** Food and Drug Administration

**Fed-Batch Process:** A fed-batch process means that nutrient substrate is added to the culture medium to increase growth and achieve a high cell density in the bioreactor. Generally, the feed solution is highly concentrated. Adding nutrient in a controlled manner has a positive effect on the culture's growth rate and production (Source: Novasep).

**Fermentation:** Is referred to a process in which an agent (e.g., bacteria or yeast) cause an organic substance to break down into simpler substances. Prominent examples are beer or wine making.

**Fermenter:** Either an apparatus for fermentation or an organism which causes fermentation.

**Filtration:** Filtration is a technique used either to remove solid impurities from an organic solution or to isolate an organic solid. The two types of filtration commonly used in organic chemistry laboratories are gravity filtration and vacuum or suction filtration. More information can be found here: Filtration,

http://orgchem.colorado.edu/Technique/Procedures/Filtration/Filtration.
html

**Flange:** A flange is a method of connecting pipes, valves, pumps, and other equipment to form a piping system, providing easy access for cleaning, inspection, or modification.

**Fluidization:** Fluidization is a process by which solids are made to behave like fluids by blowing gas or liquid upward through, for example, a solid-filled reactor.

**Formazine Turbidity Unit (FTU):** Is a standard unit for measuring turbidity.

**FWHM (full width at half maximum):** This is used to describe the width of a normal distribution (Gaussian distribution). FWHM is the full width at half (1/2) maximum of a normal distribution.

**Gas chromatography (GC):** Gas chromatography is a chromatographic technique that can be used to separate organic compounds that are volatile. A gas chromatograph consists of a flowing mobile phase, an injection port, a separation column containing the stationary phase, a detector, and a data recording system. The organic compounds are separated due to differences in their partitioning behavior between the mobile gas phase and the stationary phase in the column. (Source: http://www.files.chem.vt.edu/chem-ed/sep/gc/gc.html)

**Gauge Pressure:** Gauge pressure is zero-referenced against ambient air pressure, so it is equal to absolute pressure minus atmospheric pressure. Negative signs are usually omitted.

**Halo:** This is a circular light spot encircling the primary ray; the halo of the Moon, for example, results of light refraction at the ice crystals in the atmosphere.

**Halogenation:** The halogenation is a chemical reaction, which allows to introduce one or several halogen atoms on a molecule such as, for example, bromination reaction.

**Homogenization:** Homogenizing is a process that combines various substances to produce a uniformly consistent mixture. It is primarily used with components that are not dissolvable in each other and which are barely mixable or not mixable.

**Hydrogenation:** The act or process of subjecting a fluid to the action of hydrogen in the presence of a catalyst.

**Hygienic Design:** see EHEDG

**Immobilization:** Immobilization is a general term describing a variety of cell or particle attachment or entrapment onto a surface. It can be applied to basically all types of biocatalysts including enzymes, cellular organelles, animal, and plant cells.

**Incubator (biotechnology):** Is an apparatus in which environmental conditions can be set and controlled for culturing cells, bacteria, and other microorganisms.

**Ion Exchange:** Ion exchange is the process through which ions in solution are transferred to a solid matrix, which, in turn, releases ions of a different type but of the same polarity. The ions in solutions are replaced by different ions originally present in the solid.

**Ion Sensitive Field Effect Transistor (ISFET):** An ISFET is a miniature ion sensor manufactured using standard semiconductor manufacturing processes (e.g., CMOS compatible). The measurement principle is based on the field effect induced by ions across an insulated film acting as a capacitor. An ISFET operates in a similar way like a metal oxide semiconductor (MOS) FET. A reference electrode and the measured electrolyte solution replace the metal gate electrode. Adsorbed ions on the insulated gate modify the electronic current between the source and drain in the p-type silicon channel of the transistor.

**LASER:** **L**ight **A**mplification by **S**timulated **E**mission of **R**adiation

**Laser Ablation:** Is a process in which a laser beam is focused on a sample surface to remove material from the irradiated zone. Refer, for example, Laser ablation, http://www.appliedspectra.com/technology/laser-ablation.html.

**LED:** **L**ight **E**mitting **D**iode

**LEP:** **L**ight **E**mitting **P**olymer

**MBE (molecular beam epitaxy):** An epitaxial growth technology for forming a thin-film crystal on a heated substrate crystal by supplying molecular beams of crystal-constituent elements, which are created by evaporating each element from separate cells in an ultra-high vacuum (Courtesy Hamamatsu).

**MEMS (micro-electro-mechanical systems):** MEMS is a system for integrating electronic circuits with microactuators, micromechanisms, and microsensors, etc., fabricated using semiconductor-processing technologies.

**Metal-Oxide-Semiconductor Field Effect Transistor (MOSFET):** Metal-oxide semiconductor field-effect transistor; metal-oxide silicon field-effect transmitter. In a MOSFET, the conductive channel between the drain and source contacts is controlled by a metal gate separated from the channel by a very thin insulating layer of oxide. The gate voltage establishes a field that allows or blocks current flow.

**MOCVD (metal organic chemical vapor deposition):** An epitaxial growth technology for forming a thin-film crystal on a heated substrate crystal by thermally decomposing and chemically reacting an organic metal supplied in the form of vapor, which is solid or liquid at normal temperatures and pressures.

**MOEMS:** micro-opto-electro-mechanical systems

**Multimode fiber:** An optical fiber for transmitting light in multiple transverse mode (electromagnetic field distribution). Multimode fibers are not suitable for long-distance transmissions because the transmitted waveform is distorted due to differences in the signal light arrival time depending on the mode (modal dispersion). Compared to single-mode fibers, the core diameter is large so that connecting to a light emitter is easy. However, condensing a light beam onto a light receiver element for high-speed communication is difficult because the element must have a small light receiving area. (Courtesy Hamamatsu).

**NAMUR:** NAMUR is an international user association of automation technology in process industries. More information can be found here: www.namur.de

**National Institute of Standards and Technology (NIST):** NIST is the federal technology agency of the USA that works with industry to develop and apply technology, measurements, and standards. Refer also www.nist.gov

**Newtonian Fluid:** Newton's Law of Viscosity: a moving plate, which is separated from fixed plate by a fluid, is considered. For a "Newtonian fluid", the force required to move the plate is proportional to the velocity and area and inversely proportional to the distance between the plates. Examples of Newtonian fluids are water, oils, glycerin, air, and other gases at low to moderate shear rates.

**Nitrification:** Nitrification is the process by which ammonia is converted to nitrites ($NO^{2-}$) and then nitrates ($NO^{3-}$). Refer also, for example, www.nitrification.org

**Nephelometric Turbidity Unit (NTU):** A unit for measuring turbidity.

**Noise Equivalent Power (NEP):** Is the minimum incident power required on a photodiode to generate a photocurrent equal to the photodiode noise current. Since the photodiode light power-to-current conversion depends on the radiation wavelength, the NEP power is quoted at a particular wavelength. The NEP is nonlinear over the wavelength range, as is responsivity.

**Outgassing:** The release of volatile gases from a solid material as a part of aging, decomposition, or curing. The rate of outgassing of a material is expressed in torr liter per second per square centimeter.

**Organic Light Emitting Diode (OLED):** An LED made with organic materials.

**Osmosis:** In biology, Osmosis is defined as the diffusion of water across a cell membrane.

**Ozonization:** Ozonation is a water treatment process that destroys bacteria and other microorganisms through an infusion of ozone, a gas produced by subjecting oxygen molecules to high electrical voltage.

**Passive alignment:** Precise mechanical positioning for coupling alignment between two or more optical elements during optical module assembly. Since the positioning of optical elements usually requires accuracy ranging from submicrons to several microns, highly precise metalization patterns and V-grooves formed by semiconductor process technology are utilized as the positioning reference. In contrast to passive alignment, active alignment performs the positioning of optical modules or fibers while making the optical modules emit light by simulating actual operation (for instance, operating a laser diode to emit light) and monitoring the emitted light to obtain the required characteristics of the optical modules. (Courtesy Hamamatsu)

**Pervaporation:** Is a membrane process used for dehydrating organics, such as ethanol and isopropanol. It is the only membrane process primarily used to purify chemicals.

**Photodiode Responsivity:** Responsivity is the ratio between the photocurrent output in ampères and radiant power (in watts) incident on the photodiode. It is expressed in A/W

**Photoconduction:** Those photoinduced electrons and holes recombine radiatively (photoluminescence) and non-radiatively. If an electrical field is applied to the semiconductor, some of the induced carriers take part in electric conduction and this leads to a decrease in electrical resistance of the semiconductor.

**Photo Sensitivity:** The ratio of photocurrent expressed in amperes (A) or output voltage expressed in volts (V) to the incident light level expressed in watts (W). Photo sensitivity is represented as an absolute sensitivity (A/W or V/W) or as a relative sensitivity (%) to the peak wavelength sensitivity normalized to 100. We usually define the spectral response range as the range in which the relative sensitivity is higher than 5% or 10% of the peak sensitivity. (Courtesy Hamamatsu)

**Photoelectric effect:** A phenomenon in which a substance absorbs light and generates free electrons.

**Photovoltaic effect:** If there is a pn-junction in the illuminated area, the electrons and the holes are separated by the electrical field at the pn-junction without any electric bias, and an electromotive force between the p- and n-side semiconductors is generated. This is called the photovoltaic effect, and with regard to the effect there is basically no difference between a pn-junction and a pn-heterojunction. Based on the phenomena described above, light power can be converted into electrical power in photodiodes.

– When a photodiode is reverse-biased (photoconductive mode), a small current flow even in the absence of incident light: the so-called dark current.
– The dark current increases noise at the output of the receiver, reducing the signal-to-noise ratio.
– Typical values of dark current span from tens to hundreds of nAmpères.
– Dark current is temperature dependent; the higher the temperature, the higher the dark current.

**Phototransistor:** It is in essence nothing more than a normal bipolar transistor that is encased in a transparent case so that light can reach the base-collector diode. The phototransistor works like a photodiode, but with a much higher sensitivity to light, because the electrons that tunnel through the base-collector diode are amplified by the transistor function.

**Process Analysis:** A step-by-step breakdown of the phases of a process, used to convey the inputs, outputs, and operations that take place during each phase. A process analysis can be used to improve understanding of how the process operates and to determine potential targets for process improvement through removing waste and increasing efficiency.
(Source: http://www.businessdictionary.com/)

**Polymerase Chain Reaction (PCR):** PCR is a revolutionary method developed by Kary Mullis in the 1980s. The PCR is based on using the ability of DNA polymerase to synthesize new strand of DNA complementary to the offered template strand. More information can be found here: PCR, http://www.ncbi.nlm.nih.gov/genome/probe/doc/TechPCR.shtml

**Process Analytical Technologies (PAT):** A system for designing, analyzing, and controlling manufacturing through timely measurements (i.e., during processing) of critical quality and performance attributes of raw and in-process materials and processes with the goal of ensuring final product quality. It is important to note that the term analytical in PAT is viewed broadly to include chemical, physical, microbiological, mathematical, and risk analysis conducted in an integrated manner. More information can be found for example here: Guidance for Industry, FDA, http://www.fda.gov/downloads/Drugs/Guidances/ucm070305.pdf.

**Quantum efficiency:** This is the number of electrons or holes that can be extracted as photocurrent divided by the number of incident photons. It is commonly expressed in percent (%). The quantum efficiency ($QE$) and photo sensitivity ($S$) (unit: A/W) have the following relationship at a given wavelength (unit: nm).

$$QE = \frac{S \cdot 1240}{\lambda} \cdot 100 \ [\%]$$

**Registration, Evaluation and Authorization of Chemicals (REACH):** REACH is the Regulation on Registration, Evaluation, Authorization and Restriction of Chemicals. It entered into force on 1 June 2007. It streamlines and improves the former legislative framework on chemicals of the European Union (EU). The main aims of REACH are to ensure a high level of protection of human health and the environment from the risks that can be posed by chemicals, the promotion of alternative test methods, the free circulation of substances on the internal market, and enhancing competitiveness and innovation. The REACH makes industry responsible for assessing and managing the risks posed by chemicals and providing appropriate safety information to their users. In parallel, the EU can take additional measures on highly dangerous substances, where there is a need for complementing action at EU level. More information can be found here: http://ec.europa.eu/enterprise/sectors/chemicals/reach/index_en.htm

**Reynolds Number:** In fluid mechanics, the Reynolds number Re is a dimensionless number that gives a measure of the ratio of inertial forces to viscous forces and consequently quantifies the relative importance of these two types of forces for given flow conditions:

$$Re = VL_c/\nu$$

where $\nu$ is the kinematic viscosity, V is the mean velocity of the fluid, and $L_c$ is the characteristic length of the geometry. The higher the Reynolds number is, the more turbulent the flow will be.

(Source: http://energy.concord.org/energy2d/reynolds.html)

**Rheology:** Rheology is the science of deformation and flow.

**Restriction of Certain Hazardous Substances (RoHS):** The RoHS directive aims to restrict certain dangerous substances commonly used in electronic and electronic equipment. See also: http://ec.europa.eu/enterprise/policies/european-standards/harmonised-standards/restriction-of-hazardous-substances/index_en.htm

**Risetime (tr):** This is the measure of the photodiode response speed to a stepped light input signal. It is the time required for the photodiode to increase its output from 10% to 90% of final output level.

**Separation:** A chemical separation usually involves changing the chemical composition of one of the ingredients of a solution, so as to achieve a separation of the ingredients.

**Shot Noise:** The shot noise current is related to the intrinsic uncertainty in the process of generation of electrons from quanta of light. It is also called Quantum Noise and can be expressed as a current $I_S$ by the following shot noise equation:

$$I_S = \sqrt{\left(I_d + I_p\right) \cdot q \cdot 2B}$$

where $q = 1.6 \times 10^{-19}$ C, $I_p$ is the photogenerated current (A), $I_d$ is the dark current (A) and $B$ is the bandwidth (Hz).

**Single-mode fiber:** An optical fiber designed to transmit light in the single transverse mode (electromagnetic field distribution). Single-mode fibers have low transmission loss and are not affected by modal dispersion, making them suitable for long-distance transmission. However, they require precise core alignment when connecting to a light emitter since their core diameter is small (Courtesy Hamamatsu).

**SIP:** Sterilize in Place. To ensure product contact, surfaces are sufficiently sterile to minimize product infection.

**SOA (semiconductor optical amplifier):** An optical amplifier using a semiconductor. The structure is very similar to a Fabry-Perot laser diode but is designed not to cause reflection at the edge. An SOA enables amplification over a wide spectral range and requires fewer components than an erbium-doped fiber amplifier (EDFA), which makes the amplifier device smaller and reduces power consumption (Courtesy Hamamatsu).

**Spectral response:** The relation (photoelectric sensitivity) between the incident light level and resulting photocurrent differs depending on the wavelength of the incident light. This relation between the photoelectric sensitivity and wavelength is referred to as the spectral response characteristic and is expressed in terms of photo sensitivity or quantum efficiency. (Courtesy Hamamatsu)

**Sterilization:** "The act or process, physical or chemical, that destroys or eliminates all viable microbes including resistant bacterial spores from a fluid or a solid."

Examples of sterilization methods are as follows: steam treatment at 121 °C, dry heat at 230 °C, flushing with a sterilizing solution such as hydrogen peroxide ($H_2O_2$) or ozone ($O_3$), irradiation, and filtration. (Source: http://pharmaceuticalvalidation.blogspot.de)

**Thermistor**: A thermally sensitive resistor that greatly changes its electrical resistance as the temperature changes. Thermistors are used for temperature sensing (Courtesy Hamamatsu).

**Thermoelectric cooling element**: When an electric current flows through the junction of two dissimilar electric conductors, heat absorption (or heat generation) occurs on one side while heat generation (or heat absorption) occurs on the other side. Thermoelectric cooling elements make use of this effect (known as Peltier effect). Reversing the direction of the electric current reverses the relation between the heat absorption and generation (Courtesy Hamamatsu).

**Translucent, ideal**: A directional ray is partly absorbed and diffuse scattered; it exits the medium with lower intensity.

**Translucent diaphanous**: The primary ray in the optical medium is surrounded by a halo. The primary ray is more attenuated than in the ideal translucent case.

**Translucent gleaming**: The energy of the primary ray is spread out in an intense halo. There is virtually no primary ray remaining at the second boundary surface.

**Turbidity**: Turbidity is the phenomena where by a specific portion of a light beam passing through a liquid medium is deflected from undissolved particles.

**Underwriters Laboratories (UL)**: UL is a global independent safety science company with more than a century of expertise innovating safety solutions from the public adoption of electricity to new breakthroughs in sustainability, renewable energy, and nanotechnology. Dedicated to promoting safe living and working environments, UL helps safeguard people, products, and places in important ways, facilitating trade and providing peace of mind. More information can be found here: www.ul.com

**Vapor Pressure**: The vapor pressure of a liquid is the equilibrium pressure of a vapor above its liquid (or solid), that is, the pressure of the vapor resulting from evaporation of a liquid (or solid) above a sample of the liquid (or solid) in a closed container. Refer also, for example, Vapor pressure, http://www.chem.purdue.edu/gchelp/liquids/vpress.html

**VCSEL (vertical cavity surface emitting laser)**: Semiconductor lasers usually resonate light in a direction parallel to the substrate surface and emit the light in that direction. In VCSEL however, light resonates in a direction perpendicular to the substrate surface and is emitted perpendicular to the substrate. Therefore, the resonant cavity can be formed and laser characteristics can be tested without cleaving the substrate during the manufacturing process. This fact makes VCSEL suitable for high volume production.

Compared to edge-emitting lasers, VCSEL can be manufactured at a lower cost and can be easily arrayed in two dimensions. Other features include low threshold current, high-speed modulation at low current, and small temperature dependence. (Courtesy Hamamatsu)

**Vial**: A small vessel usually made out of glass for use in laboratories or hospitals.

**Voltage (Maximum Reverse Vr)**: Applying excessive reverse voltage to photodiodes may cause breakdown and severe degradation of device performance. Any reverse voltage applied must be kept lower than the maximum rated vale, (Vrmax).

**Water for Injection**: Water for Injection (WFI) by definition is water, that is, intended for use in the manufacture of medicines for parenteral administration whose solvent is water (WFI in bulk), or water that is used to dissolve or dilute substances or preparations for parenteral administration (heat-sterilized WFI).

Biopharmaceutical production methods often require large quantities of "pharmaceutical-grade" ultra pure water. However, this water is only approved as WFI for the manufacture (direct or indirect) of medicines if it is of reproducible quality and satisfies the relevant specifications. Depending on local water quality, several processing steps are required to manufacture pharmaceutical quality water from the raw material "drinking water". The microbioloical, chemical, and physical parameters to be observed are specified in the European Pharmacopoeia (Ph. Eur.) and elsewhere. Validation of WFI manufacture, storage, and shipping is a legal requirement.

(Source: www.biochrom.de)

# 10 Chemical resistance properties of materials

The following data are based on the Bürkert Chemical Resistance chart and are intended to serve as a materials guide for selecting the appropriate material for a wide range of media without liability. Recently also available as app for download. App store: https://itunes.apple.com/app/apple-store/id1337027781?pt=118942767&ct=News%2009-2018%20Banner&mt=8 Google play store: https://play.google.com/store/apps/details?id=com.buerkert.resistapp&referrer=utm_source%3DNews%252009-2018%26utm_medium%3DBanner%26utm_content%3DGoogle%2520Play%2520Store%26utm_campaign%3DresistApp

Since corrosion performance is influenced by several factors, the information contained in this chapter should be treated as a guide only and is not valid for all operating conditions. Increased temperatures, higher concentrations, and the inadvertent ingress of water in originally pure chemicals can all lead to accelerate corrosion. Dependent on the purity of the fluid as well as the compounding and nature of vulcanization of the gasket materials, deviations can result that affects the suitability and durability of plastics and elastomers.

The information quoted in this chapter does not consider the effect of mechanical loading, which may also have a bearing on the material performance in the fluid. In case of doubt, it is recommend testing samples of material combinations, in order to establish and check their suitability under real-world operating conditions of the application. Where liquid food products are involved, plastics and elastomers employed must normally conform to the local food and hygiene regulations

The suffix "pure" in the following tables refers to the technical pureness of the fluid, which in most cases exceeds 95% purity. As a rule, organic fluidic or gaseous media have this supplement. Acetic acid – pure means, for example, a 98% acetic acid. The suffix "aqueous" is mostly used for water miscible substances (such as Ethanol) but also for aqueous solutions of inorganic salts. Due to the great number of possible concentrations, an average concentration is always assumed. Saturated aqueous solutions are described only if explicitly noted and the reference temperature is room temperature if nothing else is stated. To note is that at higher temperatures, a reduced chemical resistance has to be considered. (Tables 10.1 and 10.2)

Terms and symbols used:
+    material is not affected or is slightly affected by the chemical:
     suitable
o    various attack level depending on prevailing conditions:
     limited suitability
–    material exhibits severe attack:
     unsuitable

https://doi.org/10.1515/9783110546156-010

**Table 10.1:** Chemical resistance properties gasket and housing materials.

| Material | Designation | Chemical resistance | Permissible temperatures | | |
|---|---|---|---|---|---|
| | | | Neutral fluids long-term °C(°F) | Neutral fluids short-term °C(°F) | Aggressive fluids long-term °C(°F) |
| Epoxy Resin | EP | Resistant to nearly all chemicals. Unsuitable for short-chain organic acids of high concentration and for strong oxidizing substances. | −20 (−4) to +150 (+302) | | |
| Ethylene propylene diene rubber | EPDM | Good resistance to Ozone and weathering. Particularly suitable for aggressive chemicals. Unsatisfactory for oils and fats. | −30 (−22) to +130 (+266) | | Dependent on aggressiveness of fluid and on mechanical load. |
| Fluorine rubber | FKM | Chemical properties superior to all other elastomers | 0 (+32) to +150 (+302) | 0 (+32) to +200 (+392) | |
| Nitrile rubber | NBR | Fairly resistant to oil and petrol. Unsatisfactory with oxidizing fluids | −10 (+14) to +90 (+194) | −10 (+14) to +120 (+248) | |
| Chloroprene rubber | CR | The chemical properties are very similar to those of PVC and are between those of NBR and EPDM | −10 (+14) to +100 (+212) | −10 (+14) to +110 (+230) | |
| Perfluorinated elastomers | FFKM | Similar to PTFE | 5 (+41) to +230 (+446) | 5 (+41) to +230 (+446) | |
| Steel | 1.4112 | | −20 (−4) to +450 (+842) | | −20 (−4) to +150 (+302) |
| Grey cast iron | GG 25 | For neutral fluids | −20 (−4) to +180 (+356) | | |

| Material | Abbreviation | Properties | | | |
|---|---|---|---|---|---|
| S.G. cast iron | GGG 40.3 | For neutral fluids | −20 (−4) to +400 (+752) | | |
| Cast steel | GS-C, C22, C25 | For neutral fluids | −20 (−4) to +400 (+752) | | |
| Rigid polyvinyl | PVC | Resistant to most acids and bases, salt solutions, and water miscible, organic solvents. | 0 (+32) to +60 (+140) | 0 (+32) to +60 (+140) | 0 (+32) to +40 (+104) |
| Rigid polyvinyl | PVC-HT | Non-resistant to aromatic and chlorinated hydrocarbons. | 0 (+32) to +90 (+194) | 0 (+32) to +110 (+230) | 0 (+32) to +40 (+104) |
| Poly propylene Poly ethylene | PPPE | Resistant to organic solvents, aqueous solutions of acids, bases, and salts. Unsuitable for concentrated, oxidizing acids. | 0 (+32) to +100 (+212) | | 0 (+32) to +60 (+140) |
| Poly amide | PA | Resistant to fats, oils, waxes, fuels, weak bases, aliphatic, and aromatic hydrocarbons. | 0 (+32) to +100 (+212) | | 0 (+32) to +60 (+140) |
| Poly tetrafluoroethylene | PTFE | Resistant to nearly all chemicals. Unsuitable for liquid sodium and fluorine compounds. | −20 (−4) to +200 (+392) | −20 (−4) to +260 (+500) | −20 (−4) to +150 (+302) |
| Poly vinylidene fluoride | PVDF | Unsuitable for hot solvents as well as for ketones, esters, and strong bases. | −20 (−4) to +100 (+212) | | |
| Poly phenylene sulfide | PPS | Resistant to dilute mineral acids, bases, aliphatic, and aromatic hydrocarbons, ketones, alcohols, chlorinated hydrocarbons, oils, fats, water, and to hydrolysis. | to +200 (+392) | to +260 (+500) | |
| Poly ether ketone | PEEK | Resistant to most chemicals. Unsuitable for concentrated sulfuric and nitric acid and certain chlorohydrocarbons. | to +220 (+428) | to +280 (+536) | |

**Table 10.2:** Resistance in basic chemicals

**A**

| Chemical | Formula | 1.4305/1.4105 | 1.4401/1.4571 | 66, GS | RG | MS | PEEK | PPS | PVDF | PA | PP | PVC | PTFE | CR | FFKM | FKM | EPDM | NBR |
|---|---|---|---|---|---|---|---|---|---|---|---|---|---|---|---|---|---|---|
| Acetaldehyde – pure | $CH_3CHO$ | + | + | – | + | + | + | o | + | o | o | – | + | – | + | o | + | – |
| Acetic acid ethyl ester (ethyl acetate) – pure | $CH_3CO_2CH_2CH_3$ | + | + | + | + | – | + | + | o | o | – | – | + | – | o | – | o | – |
| Acetic acid – pure | $CH_3COOH$ | – | + | – | – | – | + | + | + | o | – | o | + | – | o | – | o | – |
| Acetic anhydride – pure | $CH_3COCOCH_3$ | o | o | o | o | – | | + | – | – | – | – | + | – | o | – | o | – |
| Acetoacetic ester (acid free, pure) | $CH_3COCH_2CO_2C_2H_5$ | + | + | o | o | o | | + | – | + | – | – | + | – | + | – | – | – |
| Acetone – pure | $CH_3COCH_3$ | + | + | + | + | + | + | + | – | + | + | – | + | o | + | – | + | – |
| Acetophenone – pure | $C_6H_5COCH_3$ | + | + | + | + | + | | o | o | + | – | – | + | – | + | – | – | – |
| Acetylacetone – pure | $CH_3COCH_2COCH_3$ | + | + | + | – | – | | + | – | + | – | – | + | – | + | – | – | – |
| Acetyl chloride – pure | $CH_3COCl$ | o | o | o | o | o | | + | – | – | o | o | + | – | + | – | – | – |
| Acetylene – technical grade | HCCH | + | + | + | – | ²+ | + | + | + | + | o | o | + | –¹ | +¹ | –¹ | +¹ | –¹ |
| Acrylic acid ethyl ester – pure | $CH_2CHCOOC_2H_5$ | + | + | + | + | + | | + | o | + | + | – | + | – | + | – | o | – |
| Acrylonitrile – pure | $CH_2CHCN$ | + | + | + | + | + | | o | – | o | + | – | + | – | + | – | – | – |
| Adipic acid – aqueous (saturated) | $HO_2C((CH_2)_4CO_2H$ | + | + | o | o | o | + | + | + | + | + | + | + | + | + | + | + | + |
| Albumin solutions | | o | o | o | o | o | | + | + | + | + | + | + | + | + | + | + | + |
| Alum (potassiumaluminiumsulphate)–aqueous (saturated) $KAl(SO_4)_2 \cdot 12\ H_2O$ | | + | + | o | o | o | + | + | + | + | + | + | + | + | + | + | + | + |
| Albumin – pure | | o | + | – | o | o | + | + | + | + | + | + | + | + | + | + | + | + |
| Allyl alcohol – pure | $CH_2CHCH_2OH$ | + | + | – | + | + | | + | + | + | + | – | + | – | + | – | + | + |
| Aluminium acetate – aqueous (saturated) | $Al(OOCCH_3)_3$ | + | + | – | o | o | | + | + | + | + | o | + | + | + | + | + | o |

| Chemical | Formula | | | | | | | | | | | | | | | | | | |
|---|---|---|---|---|---|---|---|---|---|---|---|---|---|---|---|---|---|---|---|
| Aluminium chloride – aqueous (saturated) | AlCl₃ | o | o | o | o | o | + | + | + | o | + | + | + | + | + | + | + | + | + |
| Aluminium fluoride – aqueous (saturated) | AlF₃ | − | − | o | + | + | | + | + | + | + | + | + | + | + | + | + | + | + |
| Aluminium sulfate – aqueous (saturated) | Al₂(SO₄)₃ | o | o | − | − | − | + | + | + | o | + | + | + | + | + | + | + | + | + |
| Aminoacetic acid (glycocoll) | NH₂CH₂COOH | + | + | o | o | o | | + | + | o | + | + | + | + | − | + | + | o | | |
| Ammonia – anhydrous (liquid) – pure | NH₃ | + | + | + | o | o | | o | − | − | − | − | + | − | + | − | 3o | − | − | |
| Ammonia (gas) – pure | NH₃ | + | + | + | − | − | + | o | − | − | − | − | + | + | + | − | + | + | + | |
| Ammonium hydroxide solution | NH₄OH | + | + | + | − | − | + | o | + | − | + | + | + | + | + | − | + | − | − | |
| Ammonium acetate – aqueous | CH₃COONH₄ | + | + | o | o | o | | + | + | | + | + | + | + | | + | − | | | |
| Ammonium carbonate – aqueous | (NH₄)₂CO₃ | + | + | o | − | − | + | + | + | + | + | + | + | + | + | + | + | + | + |
| Ammonium chlorid – aqueous | NH₄Cl | o | o | o | o | o | | + | + | + | + | + | + | + | + | + | + | + | + |
| Ammonium citrate – aqueous | | + | + | o | o | o | + | + | + | o | + | + | + | + | + | + | + | | − |
| Ammonium fluoride – aqueous | NH₄F | o | o | o | o | o | | + | − | − | + | + | + | + | + | + | + | + | + |
| Ammonium fluorsilicate – aqueous | | + | + | o | o | o | | + | + | o | + | + | + | o | o | + | + | + | + |
| Ammonium formiate – aqueous | HNCOONH₄ | + | + | o | o | o | | + | − | + | + | + | + | + | + | + | + | + | + |
| Ammonium nitrate – aqueous (saturated) | NH₄NO₃ | + | + | o | − | − | + | + | + | + | + | + | + | + | + | + | + | + | + |
| Ammonium oxalate – aqueous | NH₄O₂CCO₂NH₄ | + | + | − | o | o | | − | | o | + | + | + | + | + | + | + | + | + |
| Ammonium persulfate – aqueous | (NH₄)S₂O₈ | o | o | + | o | o | | + | + | − | o | o | + | o | + | + | | + | + |
| Ammonium phosphate – aqueous | (NH₄)₂HPO₄ | + | + | o | o | o | + | + | | + | | + | + | + | + | + | + | + | + |
| Ammonium thiocyanate – aqueous | NH₄NCS | o | o | o | − | − | | + | + | + | + | + | + | + | + | + | + | + | + |
| Ammonium sulfate – aqueous | (NH₄)₂SO₄ | + | + | o | − | − | + | + | + | o | + | + | + | + | + | + | + | + | + |
| Ammonium sulfide – aqueous | (NH₄)₂S | o | o | o | − | − | | + | + | + | + | + | + | + | o | o | + | + | + |
| Ammonium sulfite – aqueous | (NH₄)₂SO₃ | + | + | o | + | + | | + | + | + | + | + | + | + | + | + | + | + | + |
| Amyl acetate – pure | CH₃COO(CH₂)₄CH₃ | + | + | o | + | + | + | + | − | + | o | − | + | − | + | − | o | − | − |

(continued)

**Table 10.2:** (Continued)

| | | NBR | EPDM | FKM | FFKM | CR | PTFE | PVC | PP | PA | PVDF | PPS | PEEK | SW | RG | GG, GS | 1,4401/1,4571 | 1,4305/1,4105 |
|---|---|---|---|---|---|---|---|---|---|---|---|---|---|---|---|---|---|---|
| Amyl alcohols – pure | $H_3C(CH_2)_4OH$ | + | o | + | + | + | + | + | + | + | + | + | | + | + | o | + | + |
| Aniline – pure | $C_6H_5NH_2$ | – | + | o | o | – | + | – | o | – | + | o | + | – | – | + | + | + |
| Aniline hydrochloride – aqueous | $C_6H_5NH_3Cl$ | o | + | o | + | o | + | o | o | – | + | | | – | – | – | – | – |
| Anisole – pure | $C_6H_5OCH_3$ | o | o | – | + | – | + | – | – | + | | + | | + | + | + | + | + |
| Anone (cyclohexanone) – pure | $C_6H_{10}O$ | – | – | – | + | – | + | – | – | + | + | + | + | o | o | o | + | + |
| Anthracene oil – pure | | | – | – | + | – | + | – | – | + | | | | + | + | + | + | + |
| Anthraquinone sulfonic acid – aqueous | $C_6H_4COCOC_6H_4SO_3H$ | o | + | + | + | + | + | + | + | o | | | | o | o | o | o | o |
| Antimony chloride – aqueous | $SbCl_3$ | o | + | + | + | + | + | + | + | – | + | + | + | o | o | o | – | – |
| Apple acid – aqueous | $(HO)CH(COOH)CH_2COOH$ | + | + | + | + | + | + | + | + | – | + | + | + | – | – | – | + | + |
| Aqua regia | $HNO_3 + HCl$ | – | – | – | + | – | + | o | – | – | – | – | – | – | – | – | – | – |
| Arabic acid – aqueous | | + | + | + | + | + | + | + | + | + | + | + | – | – | – | – | + | + |
| Argon – pure | Ar | + | + | + | + | + | + | + | + | + | + | + | + | + | + | + | + | + |
| Arsenic acid – aqueous | $H_3AsO_4$ | + | + | + | + | + | + | + | + | o | + | + | | – | o | – | + | + |
| Arsenic trichloride – aqueous | $AsCl_3$ | + | + | + | + | + | + | + | + | – | + | + | | – | – | o | o | o |
| Arsenious acid – aqueous | $H_3AsO_3 (As_2O_3+H_2O)$ | + | + | + | + | + | + | + | + | – | + | + | | o | o | – | + | + |
| Aryl silicates – aqueous | | o | o | o | + | o | + | + | + | | | | | o | o | + | + | + |
| Ascorbic acid – aqueous | | + | + | + | + | + | + | + | + | + | | + | | – | – | – | + | + |
| Aspartic acid – aqueous | $(HOOC)CH(NH_2)CH_2COOH$ | + | + | + | + | + | + | + | + | + | | + | | – | – | o | + | + |

| Substance | Formula | 1 | 2 | 3 | 4 | 5 | 6 | 7 | 8 | 9 | 10 | 11 | 12 | 13 | 14 | 15 | 16 | 17 | 18 |
|---|---|---|---|---|---|---|---|---|---|---|---|---|---|---|---|---|---|---|---|
| Barium chlorate – aqueous | Ba(ClO$_3$)$_2$ | + | + | o | + | + |  | + |  | – | + | + | + | + | + | + | + | + | + |
| Barium chloride – aqueous | BaCl$_2$ | o | + | o | + | + | + | + | + | + | + | + | + | + | + | + | + | + | + |
| Barium hydroxide – aqueous | Ba(OH)$_2$ | + | + | + | + | + | + | + | o | + | o | + | + | + | + | + | + | + | + |
| Barium sulfide and polysulfide – aqueous | BaS | + | + | o | o | o | + | o | o | + | + | o | + | + | + | + | + | + | + |
| Benzaldehyde – aqueous | C$_6$H$_5$CHO | o |  | – | o | o |  | + | – | o |  | – | + | – | – | + | + | o | o |
| Benzene – pure | C$_6$H$_6$ | – | + | – | – | – | + | + | + | + | + | + | – | – | – | – | – | – | + |
| Benzene sulfonic acid – aqueous | C$_6$H$_5$SO$_3$H | + | + | + | + | + | + | + | + | + | + | + | + | + | + | + | + | + | + |
| Benzidine sulfonic acid – aqueous | (NH$_2$)C$_6$H$_4$C$_6$H$_3$(SO$_3$H)(NH$_2$) | + | + | o | + | + | + | + | + | + | + | + | + | + | + | + | + | + | + |
| Benzoic acid – aqueous (saturated) | C$_6$H$_5$COOH | + | + | + | + | + | – | + | + | + | + | + | + | + | + | + | + | + | + |
| Benzyl alcohol – pure | C$_6$H$_5$CH$_2$OH | – | + | o | o | o | o | + | o | + | + | + | o | + | + | + | + | + | + |
| Benzyl butyl phtalate – aqueous | | o | + | o | o | o | o | + | o | + | + | + | + | + | + | + | + | + | + |
| Bergamot essence | | + | + | + | + | + | + | + | o | + | + | + | + | + | + | + | + | + | + |
| Bisulfite (sodium bisulfite) – aqueous | NaHSO$_3$ | o | o | – | o | o | – | + | + | + | + | + | o | + | o | o | + | + | + |
| Borax – aqueous | Na$_2$B$_4$O$_7$ | + | + | + | + | + | + | + | + | + | + | + | + | + | + | + | + | + | + |
| Borfluoric acid | HBF$_4$ | + | + | + | + | + | + | + | + | o | o | o | + | + | + | + | + | o | + |
| Boric acid – aqueous | H$_3$BO$_3$ | + | + | + | + | + | + | + | + | + | + | + | + | + | + | + | + | + | + |
| Brines | | + | – | – | – | – | o | – | – | i | – | – | – | i | i | o | o | – | o |
| Bromine (liquid) – pure | Br$_2$ | o | o | o | o | o | + | i | i | o | o | o | o | o | o | o | o | o | o |
| Butadiene (gas) – pure | CH$_2$CHCHCH$_2$ | o | o | o | o | + | + | + | + | + | + | + | + | + | + | + | + | + | + |
| Butane (gas and liquid) | C$_4$H$_{10}$ | + | + | + | + | + | + | + | + | + | + | + | + | + | + | + | + | + | + |
| Butanediol – aqueous (10 %) | HO(CH$_2$)$_4$OH | + | + | + | + | + | + | + | + | + | + | + | + | + | + | + | + | + | + |
| Butanol (butylalcohol) – aqueous | C$_4$H$_9$OH | o | + | + | + | + | + | + | + | + | + | + | + | + | + | + | + | + | + |

(continued)

**Table 10.2:** (Continued)

| | | 1.4305/1.4305 | 1.4401/1.4571 | GG, GS | RG | WS | PEEK | PPS | PVDF | PA | PP | PVC | PTFE | CR | FFKM | FKM | EPDM | NBR |
|---|---|---|---|---|---|---|---|---|---|---|---|---|---|---|---|---|---|---|
| Butinediol – pure | HOCH₂C₂CH₂OH | + | o | + | + | + | + | + | | + | + | o | + | o | | o | o | o |
| Butoxyl (methoxybutyl acetate) – pure | CH₃OC₄H₄O₂CCH₃ | + | + | o | o | o | | | | | + | – | + | + | + | o | o | + |
| Butyl acetate – pure | CH₃(CH₂)₃O₂CCH₃ | + | + | o | + | o | + | + | + | + | – | – | + | – | + | – | + | – |
| Butyl alcohol (butanol) – pure | CH₃(CH₂)₃OH | + | + | + | + | + | + | + | + | + | + | + | + | + | + | + | + | o |
| Butylene (liquid) – pure | H₃CCH₂CHCH₂ | + | + | o | + | + | | + | + | + | + | + | + | + | + | + | o | + |
| Butyl phtalate – pure | C₆H₄(CO)₂(OCH₂CH₂CH₂CH₃)₂ | + | + | o | + | + | | + | | + | o | – | + | – | + | – | – | – |
| Butyric acid – aqueous | H₃CCH₂CH₂COOH | o | + | – | o | o | + | + | + | o | – | o | + | o | o | o | o | o |
| **C** | | | | | | | | | | | | | | | | | | |
| Calcium bisulfite – aqueous | Ca(HSO₃)₂ | o | + | – | – | – | + | + | + | – | + | + | + | + | + | + | + | + |
| Calcium chloride – aqueous | CaCl₂ | o | o | o | – | – | + | + | + | o | + | o | + | + | + | + | + | + |
| Calcium hydroxide – aqueous | Ca(OH)₂ | + | + | + | – | – | + | + | + | + | o | + | + | + | + | + | + | + |
| Calcium hypochlorite – aqueous | Ca(OCl)₂ | o | o | o | – | – | + | – | + | – | + | o | + | o | + | o | + | – |
| Calcium nitrate – aqueous | Ca(NO₃)₂ | o | o | o | o | o | + | + | | + | + | + | + | + | + | + | + | + |
| Camphor oil – pure | | + | + | o | o | o | | o | | | – | + | + | – | o | + | – | + |
| Car–battery fluid (20 % sulphuric acid, aqueous) | H₂SO₄ | o | + | – | – | – | o | + | + | – | + | + | + | o | + | + | + | o |
| Carbolic acid (phenol) – aqueous | C₆H₅OH | + | + | o | o | o | + | + | + | – | + | + | + | o | + | o | o | o |
| Carbolineum | | + | + | + | + | + | | | | + | + | + | + | o | + | o | o | o |
| Carbon dioxide – dry | CO₂ | + | + | + | + | + | + | + | + | + | + | + | + | o | + | + | o | + |

| Medium | Formula |
|---|---|
| Carbon dioxide – wet | $CO_2$ |
| Carbon disulfide – pure | $CS_2$ |
| Carbon monoxide | $CO$ |
| Carbon tetrachloride – pure | $CCl_4$ |
| Carbonic acid – aqueous | $H_2CO_3$ |
| Caro's acid – aqueous | $H_2SO_5$ |
| Caustic potash (potassium hydroxide) – aqueous | $KOH$ |
| Caustic soda (sodium hydroxide) – aqueous | $NaOH$ |
| Cellosolve (glycol ethyl ether) – pure | $HO(CH_2)_2OCH_2CH_3$ |
| Chloral hydrate (chloral) – aqueous | $CCl_3CH(OH)_2$ |
| Chlorbenzenes – pure | $C_6H_5Cl$ |
| Chloric acid – aqueous | $HClO_3$ |
| Chlorine (gas) – dry | $Cl_2$ |
| Chlorine (gas) – wet (chlorinated water) | $Cl_2$ |
| Chlorine (liquid) – pure | $Cl_2$ |
| Chlorine dioxide – aqueous | $ClO_2$ |
| Chloroacetic acid – aqueous | $ClCH_2COOH$ |
| Chloroethanol (ethylene chlorhydrine) – pure | $ClCH_2CH_2OH$ |
| Chloride of lime (calcium hypochlorite) – aqueous | $Ca(OCl)_2$ |
| Chlorinated water (chlorine gas – wet) | $Cl_2$ |
| Chlormethane (methyl chloride) – pure | $ClCH_3$ |
| Chloronaphtalene – pure | $C_{10}H_7Cl$ |
| Chloroform (trichloromethane) – pure | $CHCl_3$ |

(continued)

**Table 10.2:** (Continued)

| Chemical | Formula | 1,4305/1,4105 | 1,4401/1,4571 | GG, GS | RG | WS | PEEK | PPS | PVDF | PA | PP | PVC | PTFE | CR | FFKM | FKM | EPDM | NBR |
|---|---|---|---|---|---|---|---|---|---|---|---|---|---|---|---|---|---|---|
| Chlorophenol – pure | $C_6H_4(OH)(Cl)$ | + | + | o | + | + |  | o |  |  | + | o | + | – | + | – | – | – |
| Chlorophenoxyacetic acid | $(OC_6H_5)(Cl)CHCOOH$ | + | + | o |  |  | – | – | o | – | – | + | + | + |  | + | + | + |
| Chlorsulfonic acid – pure | $ClSO_3H$ | o | o | o | o | o |  |  |  |  |  | o | + | – | o | – | – | – |
| Chlorxylenol – pure | $C_6H_2(OH)(CH_3)_2(Cl)$ | + | + | o | + | + |  |  |  |  |  | o | + | – | + | – | – | – |
| Choline chloride – aqueous | $[HOCH_2CH_2N(CH_3)_3]Cl$ |  |  | o | – | – |  |  |  |  | o | o | + | + |  | + | + | + |
| Chromic acid – aqueous | $H_2CrO_4$ | o | o | o | – | – | o | – | + | – | o | + | + | o | + | + | o | – |
| Chromium alum – aqueous | $KCr(SO_4)_2 \cdot 12H_2O$ | o | o | o | o | o |  |  | + | o | + | o | + | + | + | + | + | + |
| Chromium sulfate – aqueous | $Cr_2(SO_4)_3$ | o | o | – | o | o |  | + |  | o | + | + | + | + | + | + | + | + |
| Citral (citronella oil) – pure |  | + | + | – | + | + |  | + |  | + | – |  | + | – |  | – | – | – |
| Citric acid – aqueous |  | o | + | o | + | o | + | + | + | + | + | + | + | + | + | + | + | + |
| Cresol – aqueous (lysol) | $C_6H_4(OH)(CH_3)$ | o | + | – | + | + | + | + | o | – | o | o | + | – | + | o | – | – |
| Common salt (sodium chloride) – aqueous | $NaCl$ | o | o | o | o | – | + | + | + | + | + | + | + | + | + | + | + | + |
| Copper acetate – aqueous | $Cu(CH_3COO)_2$ | + | + | – | – | o | + | + | + | o | + | + | + | + | + | + | + | o |
| Copper chloride – aqueous | $CuCl_2$ | – | – | o | o | o | + | + | + | o | + | + | + | + | + | + | + | + |
| Copper sulfate – aqueous | $CuSO_4$ | o | o | o | o | o | + | + | + | o | + | + | + | + | + | + | + | + |
| Cyclohexane – pure | $C_6H_{12}$ | + | + | o | + | + | + | + | + | + | – | + | + | – | + | o | – | – |
| Cyclohexanol – pure | $C_6H_{11}OH$ | + | + | + | + | + | + | + | + | + | + | o | + | + | + | + | – | – |
| Cyclohexanone (anone) – pure | $C_6H_{10}O$ | + | + | + | o | o | + | + | o | + | – | – | + | – | + | – | – | – |
| Cymene – pure | $C_6H_4(CH_3)[CH(CH_3)_2]$ | + | + | + | + | + | + | o | + | + | – | – | + | – | + | – | – | – |

| Chemical | Formula | | | | | | | | | | | | | | | | | |
|---|---|---|---|---|---|---|---|---|---|---|---|---|---|---|---|---|---|---|
| **D** | | | | | | | | | | | | | | | | | | |
| Decahydronaphtalene (decalin) – pure | C₁₀H₁₈ | + | + | + | + | + | – | + | + | + | + | + | + | | + | + | + | + |
| Dextrose – aqueous | C₆H₁₂O₆ | + | + | + | + | + | + | + | + | + | + | + | + | | + | + | + | + |
| Diacetone alcohol – anhydrous | (CH₃)₂C(OH)CH₂COCH₃ | + | + | o | + | + | – | o | – | o | + | o | o | | + | + | + | + |
| Dibutyl phthalate – pure | C₆H₄(COOC₄H₉)₂ | + | + | + | + | + | – | + | + | o | + | o | o | | + | + | + | + |
| Dibutyl sebacate – pure | (C₄H₉COO)(CH₂)₈(OOC₄H₉) | + | + | o | + | + | – | + | + | o | + | o | o | | + | + | + | + |
| Dichlorethane (ethylene chloride) – pure | ClCH₂CH₂Cl | | + | + | + | + | o | o | o | | + | o | o | + | + | + | + | + |
| Dichlorethylene – pure | Cl₂CHCH₃ | o | + | + | + | + | o | o | o | | + | o | o | + | + | + | + | + |
| Dichlormethane (methylene chloride) – pure | CH₂Cl₂ | | | | | | – | – | + | + | – | + | – | – | – | o | – | o |
| Dicyclohexyl –ammonium nitrite – pure | [(C₆H₁₁)₂NH₂]NO₂ | + | + | o | + | + | + | + | – | + | – | – | – | o | – | + | + | + |
| Diethyl ether – pure | CH₃CH₂OCH₂CH₃ | o | + | | o | + | – | – | – | – | o | + | | + | – | – | o | – |
| Dimethyl amine – pure | (CH₃)₂NH | + | + | | o | – | – | – | – | – | – | – | | – | – | – | – | – |
| Dimethyl formamide – pure | HCON(CH₃)₂ | + | + | + | + | + | + | + | + | + | + | + | + | + | + | + | + | + |
| Dimethyl sulfoxide – pure | (CH₃)₂SO | – | + | o | – | – | – | – | – | + | – | – | | – | – | – | – | – |
| Dioctyl phtalate – pure | C₆H₄(COOC₈H₁₇)₂ | + | + | + | + | + | + | + | + | + | + | + | + | + | + | + | + | + |
| Dioxane – pure | C₄H₈O₂ | – | + | o | – | – | – | – | – | – | – | o | + | o | – | + | – | + |
| Diphenyl + diphenyloxide | | + | + | + | + | + | + | + | + | + | + | + | + | + | + | + | + | + |
| Dissousgas (acetylene + acetone) | | – | + | o | + | + | o | o | + | + | – | o | – | – | + | + | + | + |
| **E** | | | | | | | | | | | | | | | | | | |
| Ethane – pure | CH₃CH₃ | + | + | + | + | + | + | + | + | – | + | – | + | + | + | + | + | + |
| Ethanol (ethyl alcohol) – pure | CH₃CH₂OH | + | + | + | + | + | + | + | + | + | + | + | + | o | + | o | + | + |
| Ethanolamine – pure | NH₂CH₂CH₂OH | + | + | + | + | o | o | + | + | o | o | o | + | o | – | – | o | o |
| Ether (diethyl ether) – pure | CH₃CH₂OCH₂CH₃ | + | + | + | + | + | + | + | + | – | + | – | + | + | – | + | – | – |

(continued)

**Table 10.2:** (Continued)

| | Formula | 1.4305/1.4105 | 1.4401/1.4571 | GG, GS | RG | MS | PEEK | PPS | PVDF | PA | PP | PVC | PTFE | CR | FFKM | FKM | EPDM | NBR |
|---|---|---|---|---|---|---|---|---|---|---|---|---|---|---|---|---|---|---|
| Ethyl acetate (acetic acid ethyl ester) – pure | CH₃CO₂CH₂CH₃ | + | + | + | + | – | + | + | o | o | – | – | + | – | o | – | o | – |
| Ethyl alcohol (ethanol) – pure | CH₃CH₂OH | + | + | + | + | + | + | + | + | o | + | o | + | + | + | – | + | o |
| Ethyl alcohol + acetic acid | CH₃CH₂OH + CH₃COOH | + | + | o | o | o | + | + | + | – | + | o | + | o | + | o | + | o |
| Ethyl alcohol – fermentation mash | | + | + | o | + | + | + | + | + | o | + | + | + | + | + | + | + | + |
| Ethyl alcohol – denaturated (depending on denaturant) | | + | + | + | o | o | + | + | | o | + | + | + | o | + | o | o | o |
| Ethyl benzene – pure | C₆H₅CH₂CH₃ | + | + | + | + | + | | o | + | + | – | – | + | – | + | o | – | – |
| Ethyl chloride – pure | CH₃CH₂Cl | + | + | – | – | – | | o | + | + | – | – | + | + | + | + | + | + |
| Ethyl formiate | HCOOCH₂CH₃ | + | + | o | + | + | + | + | + | + | o | – | + | – | + | – | – | – |
| Ethylene – pure | CH₂CH₂ | + | + | + | + | + | o | + | + | o | + | + | + | – | + | + | + | + |
| Ethylene chlorhydrine (chloroethanol) – pure | ClCH₂CH₂OH | + | + | + | + | + | | o | + | o | + | – | + | – | + | o | o | – |
| Ethylene diamine – pure | NH₂CH₂CH₂NH₂ | + | + | o | – | – | | o | + | o | + | – | + | + | o | o | – | o |
| Ethylene dibromide – anhydrous | CH₂CHBr | o | + | o | + | + | – | o | + | + | – | – | + | – | + | – | – | – |
| Ethylene dichloride (dichloroethane) – pure | ClCH₂CH₂Cl | + | + | – | – | – | + | o | + | + | – | – | + | – | + | – | + | – |
| Ethylene glycol (glycol) – pure | HOCH₂CH₂OH | – | + | o | o | o | + | + | + | o | + | + | + | + | + | – | – | + |
| Ethylene oxide (liquid) – pure | CH₂CH₂O | + | + | – | – | – | | | + | – | – | – | + | – | o | – | – | – |
| Essential oils | | + | + | o | o | o | | o | | – | – | – | + | – | + | | – | – |
| Fat alcohols | | o | + | o | + | + | | + | + | + | o | + | + | + | + | + | o | + |
| Fat alcohol sulfates – aqueous | | + | + | o | o | o | | | + | o | + | + | + | + | + | + | o | + |

| Chemical | Formula | | | | | | | | | | | | | | | | | | |
|---|---|---|---|---|---|---|---|---|---|---|---|---|---|---|---|---|---|---|---|
| Ferrous/ferric chloride – aqueous (saturated) | FeCl₃ | – | – | – | – | – | + | + | + | + | + | + | + | + | + | + | + | + | + |
| Ferrous/ferric sulfate – aqueous | FeSO₄ | + | + | – | o | o | + | + | + | + | + | + | + | + | + | + | + | + | + |
| Fluorine (dry) – pure | F₂ | + | + | – | o | o | – | o | o | + | –* | + | o | + | – | + | + | – | – |
| Fluorine (wet) – pure | F₂ | o | o | – | – | – | – | – | o | o | – | – | – | – | – | – | – | – | – |
| Fluoboric acid (borofluoric acid) | HBF₄ | – | – | – | – | – | + | o | + | + | + | + | + | + | + | + | + | + | + |
| Fluorocarbons (see freon) | | | | | | | | | | | | | | | | | | | |
| Fluosilicic acid – aqueous | | o | o | – | – | – | + | | o | + | + | o | o | + | | | o | | o |
| Formaldehyde – aqueous | CH₂O | + | + | – | + | + | + | o | + | + | + | o | + | + | + | + | + | + | + |
| Formaldehyde – pure | CH₂O | + | + | + | + | + | + | + | + | + | + | + | + | + | + | + | + | + | + |
| Formamide – pure | HCONH₂ | o | + | o | o | o | + | o | + | + | + | o | + | + | o | + | + | + | + |
| Freon 12 B1 | CBrClF₂ | + | + | o | + | + | + | o | + | + | + | o | o | o | + | + | + | + | + |
| Freon 13 | CClF₃ | + | + | + | + | + | + | + | + | + | o | o | o | o | o | o | + | + | + |
| Freon 13 B1 (Halon 1301) | CBrF₃ | + | + | o | + | + | + | o | + | + | + | + | o | + | o | o | + | + | + |
| Freon 22 | CHClF₂ | + | + | + | + | + | + | + | + | + | o | + | | + | | o | | o | + |
| Freon 23 | CHF₃ | + | + | o | + | + | + | + | + | + | – | | o | | | o | | | + |
| Freon 502 | C₆H₁₂O | + | + | + | + | + | + | + | + | + | + | o | | + | | – | | o | o |
| Freon substitute HFCKW 123 | | + | + | + | o | + | + | + | o | + | – | | | | | | | | + |
| Freon substitute HFCKW 134a | | + | + | + | + | + | + | + | + | + | – | | o | | | | | | + |
| Freon TF (Freon 113) | Cl₃FCCClF₃ | o | + | + | + | + | + | + | + | o | – | – | – | – | – | – | – | o | + |
| Formic acid – aqueous | HCO₂H | o | + | – | – | – | + | + | + | + | – | – | – | – | – | – | – | – | + |
| Formic acid – pure | HCO₂H | – | + | – | – | – | o | o | o | – | o | o | – | o | o | – | o | o | + |

(continued)

* Resistant only with specific formulations

**Table 10.2:** (Continued)

| Chemical | Formula | 1,4305/1,4105 | 1,4401/1,4571 | GG, GS | RG | MS | PEEK | PPS | PVDF | PA | PP | PVC | PTFE | CR | FFKM | FKM | EPDM | NBR |
|---|---|---|---|---|---|---|---|---|---|---|---|---|---|---|---|---|---|---|
| **G** | | | | | | | | | | | | | | | | | | |
| Gas liquor | | + | + | o | - | - | | | | | | o | + | - | | o | - | + |
| Glucose – aqueous | C₆H₁₂O₆ | + | + | + | + | + | + | + | + | + | + | + | + | + | + | + | + | + |
| Glycerin – aqueous | HHCO₂CH(OH)CH₂OH | o | + | o | o | o | + | + | + | + | o | o | + | + | + | + | + | + |
| Glycerin – pure | HOCH₂CH(OH)CH₂OH | o | + | o | o | o | + | + | + | + | o | o | + | o | + | + | + | o |
| Glycocoll (aminoacetic acid) – aqueous | NH₂CH₃CHCO₂H | + | + | o | o | o | | + | + | o | + | + | + | + | | + | o | o |
| Glycol – aqueous | HOCH₂CH₂OH | + | + | o | o | o | + | + | + | o | + | + | + | + | + | + | + | + |
| Glycol ethyl ether (cellosolve) | HO(CH₂)₂OCH₂CH₃ | + | + | + | + | + | | + | + | + | - | - | + | - | + | - | - | - |
| Glycolic acid – aqueous | HOCH₂COOH | o | o | o | o | o | | + | + | - | + | + | + | + | + | + | + | + |
| **H** | | | | | | | | | | | | | | | | | | |
| Helium | He | + | + | o | o | o | + | + | + | + | + | + | + | + | + | + | + | + |
| Heptane, Hexane (petrol) – pure | | + | + | + | + | + | + | + | + | + | o | + | + | + | + | + | - | o |
| Hexamethylene tetramine – aqueous | | + | + | o | o | o | | o | | + | + | + | + | + | + | + | + | + |
| Humic acids | | + | + | o | + | + | | | | - | + | + | + | + | | + | + | + |
| Hydrazine hydrate – aqueous | NH₂NH₂*2 H₂O | o | - | - | - | - | + | | o | | - | + | + | - | + | + | + | - |
| Hydrobromic acid – aqueous | HBr | - | - | o | - | - | - | - | + | - | + | + | + | o | + | + | + | - |
| Hydrochloric acid (gas) – pure | HCl | o | + | - | - | - | + | - | + | - | + | + | + | o | + | +[4] | o | o |
| Hydrochloric acid – aqueous (36%) | HCl | o | o | - | - | - | o | - | + | - | + | + | + | - | + | +[4] | o | - |

| Substance | Formula | 1 | 2 | 3 | 4 | 5 | 6 | 7 | 8 | 9 | 10 | 11 | 12 | 13 | 14 | 15 | 16 | 17 |
|---|---|---|---|---|---|---|---|---|---|---|---|---|---|---|---|---|---|---|
| Hydrocyanic acid – aqueous | HCN | o | + | o | + | + | | + | + | − | + | + | + | + | + | + | o | o |
| Hydrofluoric acid – aqueous | HF | − | o | − | − | − | + | − | + | | − | o | o | o | o | − | − | + |
| Hydrogen peroxide 0,5% | $H_2O_2$ | o | + | − | − | − | + | + | + | + | − | − | − | + | + | + | − | + |
| Hydrogen peroxide 30% | $H_2O_2$ | − | + | − | − | − | + | + | o | o | − | − | − | + | + | + | − | + |
| 5Hydrogen – pure | $H_2$ | + | o | − | − | − | + | + | + | − | − | − | + | + | + | + | + | + |
| Hydrogen sulfide – aqueous | $H_2S$ | + | + | + | + | + | | o | − | o | + | o | o | o | o | + | + | + |
| Hydroquinone – aqueous | $C_6H_4(OH)_2$ | + | + | o | o | + | + | + | + | o | + | + | + | + | + | o | o | + |
| Hydroxylamine sulphate – aqueous | $(NH_3OH)_2SO_4$ | + | + | + | + | + | + | | | − | + | + | o | + | + | + | | + |
| Inert gases | | + | o | + | o | o | + | + | + | + | + | + | + | + | + | + | + | + |
| Illuminating gas | | + | + | + | + | + | + | + | + | + | + | + | + | + | + | + | + | + |
| Iodine + Potassium iodide – aqueous | $I_2 + KI$ | o | o | o | − | − | o | o | + | − | o | o | o | o | o | − | o | + |
| Isobutyl alcohol – pure | $(CH_3)_2CHCH_2OH$ | o | + | + | + | + | + | + | + | + | + | + | + | + | + | − | o | + |
| Isooctane – pure | $CH_3C(CH_3)_2CH_2CH(CH_3)CH_3$ | − | + | + | + | + | + | + | + | + | + | + | + | + | + | + | o | + |
| Isopropanol (propanol) – pure | $CH_3CH(OH)CH_3$ | o | + | + | + | + | + | + | o | o | + | + | + | + | + | + | + | + |
| Kerosene (petroleum spirit, benzine) | | + | + | o | + | + | + | + | + | o | o | o | + | + | + | + | − | + |
| Lactic acid – aqueous | | o | o | o | o | o | + | + | + | o | + | o | o | o | + | o | o | o |
| Laughing gas (nitrous oxide) | $N_2O$ | + | + | + | + | + | + | + | + | + | + | + | + | + | + | + | + | + |

(continued)

**Table 10.2:** (Continued)

| | | 1.4305/1.4105 | 1.4401/1.4571 | GG, GS | RG | MS | PEEK | PPS | PVDF | PA | PP | PVC | PTFE | CR | FFKM | FKM | EPDM | NBR |
|---|---|---|---|---|---|---|---|---|---|---|---|---|---|---|---|---|---|---|
| Lead acetate – aqueous | Pb(CH$_3$COO)$_2$ | + | + | – | o | o | + | + | + | + | + | + | + | + | + | + | + | o |
| Lead nitrate – aqueous | Pb(NO$_3$)$_2$ | + | + | o | – | – | | + | | | + | + | + | + | + | + | + | + |
| Lead tetraethyl (tetraethyl lead) – pure | Pb(CH$_2$CH$_3$)$_4$ | + | + | + | o | o | | | + | + | + | + | + | o | + | + | o | o |
| Linoleic acid | | o | + | o | o | o | | + | + | | – | + | + | – | + | o | – | o |
| Lithium chloride – aqueous | LiCl | o | o | o | o | o | | + | + | o | + | + | + | o | + | + | + | + |
| **M** | | | | | | | | | | | | | | | | | | |
| Magnesium chloride – aqueous | MgCl$_2$ | o | o | o | o | o | + | + | + | o | + | + | + | + | + | + | + | + |
| Magnesium sulfate – aqueous | MgSO$_4$ | + | + | – | + | + | + | + | + | o | o | o | + | + | + | + | + | + |
| Maleic acid – aqueous | | o | + | o | o | o | + | + | + | o | + | + | + | + | + | + | + | + |
| Manganese chloride – aqueous | MnCl$_2$ | o | o | o | o | o | | + | | + | + | + | + | + | + | + | + | + |
| Manganese sulfate – aqueous | MnSO$_4$ | o | + | o | + | o | | + | + | + | + | + | + | + | + | + | + | + |
| Mercaptanes | | + | + | – | o | o | | | o | + | | + | + | – | + | o | – | – |
| Methane (marsh gas) – pure | CH$_4$ | + | + | + | + | + | + | + | o | + | o | o | + | – | + | + | – | + |
| Methanol (methyl alcohol) | CH$_3$OH | o | + | o | o | o | + | + | o | o | o | + | + | + | + | – | + | – |
| Methoxybutanol – pure | CH$_3$O(CH$_2$)$_3$CH$_2$OH | + | + | + | + | + | | + | o | | + | + | + | o | + | + | + | + |
| Methylacetate – pure | CH$_3$COOCH$_3$ | o | o | o | + | o | + | + | o | + | + | – | + | – | + | – | o | – |
| Methyl alcohol (methanol) – pure | CH$_3$OH | o | + | o | o | o | + | + | o | o | o | + | + | + | + | – | + | – |
| Methyl amine – aqueous | CH$_3$NH$_2$ | o | o | o | – | – | + | o | – | o | + | o | + | o | – | o | o | – |

| Substance | Formula | M1 | M2 | M3 | M4 | M5 | M6 | M7 | M8 | M9 | M10 | M11 | M12 | M13 | M14 | M15 | M16 |
|---|---|---|---|---|---|---|---|---|---|---|---|---|---|---|---|---|---|
| Methyl chloride (chloromethane) – pure | $CH_3Cl$ | + | + | – | + | + | + | o | o | – | – | – | + | – | + | + | – |
| Methylene chloride (dichloromethane) – pure | $CH_2Cl_2$ | + | + | – | + | + | o | o | – | – | – | – | + | – | + | o | – |
| Methyl ethyl ketone – pure | $CH_3COCH_2CH_3$ | + | + | o | + | + | o | o | – | o | – | – | + | – | + | – | o |
| Mercury | $Hg$ | + | o | + | – | – | + | + | + | + | + | + | + | + | + | + | + |
| Mercury chloride | $HgCl_2$ | o | o | + | – | – | + | + | + | + | + | o | + | + | + | + | + |
| Mercury salts – aqueous | | + | + | – | – | – | + | + | + | – | + | o | + | + | + | + | + |
| Morpholine – pure | | + | + | + | + | + | – | – | + | + | – | – | + | o | – | o | – |

**N**

| Substance | Formula | M1 | M2 | M3 | M4 | M5 | M6 | M7 | M8 | M9 | M10 | M11 | M12 | M13 | M14 | M15 | M16 |
|---|---|---|---|---|---|---|---|---|---|---|---|---|---|---|---|---|---|
| Natural gas | | + | + | o | o | o | + | + | + | o | o | + | o | – | | | |
| Nickel sulfate – aqueous | $Ni(SO4)_2$ | o | o | – | o | – | + | + | + | + | o | + | + | + | + | + | + |
| Nitric acid – aqueous (40%) | $HNO_3$ | – | + | – | – | – | o | – | + | o | o | + | – | + | +[4] | – | – |
| Nitrobenzoic acids – aqueous | | + | + | – | + | + | + | + | + | + | + | + | + | + | + | + | + |
| Nitrobenzenes – pure | $C_6H_5NO_2$ | + | + | o | + | + | o | o | o | – | o | – | + | + | o | + | + |
| Nitrogen oxides – gaseous, wet and dry | $(NO, NO_2, N_2O_4)$ | + | + | o | + | + | + | o | – | o | – | + | – | o | – | – | o |
| Nitrotoluenes (o-, m-, p-) – pure | $C_6H_4(NO_3)(CH_3)$ | + | + | – | o | o | o | o | – | o | – | – | + | – | o | – | – |
| Nitrogen oxides | | – | o | – | – | – | + | o | o | + | o | – | + | – | o | – | o |
| Nitrogen | $N_2$ | + | + | + | + | + | + | + | + | + | + | + | + | + | + | + | + |
| Nitrous oxide | $N_2O$ | + | + | + | + | + | + | + | + | + | + | + | + | + | + | + | + |

**O**

| Substance | Formula | M1 | M2 | M3 | M4 | M5 | M6 | M7 | M8 | M9 | M10 | M11 | M12 | M13 | M14 | M15 | M16 |
|---|---|---|---|---|---|---|---|---|---|---|---|---|---|---|---|---|---|
| Oleum (fuming sulfuric acid) | $H_2SO_4$ | o | + | o | – | – | – | – | o | – | o | – | + | – | + | o | – |
| Oxalic acid – aqueous (saturated) | $HOOCCOOH$ | o | + | – | – | – | + | + | – | + | – | + | + | + | + | + | o |

(continued)

**Table 10.2:** (Continued)

| | 1.4305/1.4105 | 1.4401/1.4571 | GG, GS | RG | MS | PEEK | PPS | PVDF | PA | PP | PVC | PTFE | CR | FFKM | FKM | EPDM | NBR |
|---|---|---|---|---|---|---|---|---|---|---|---|---|---|---|---|---|---|
| [6] Ozon – wet and dry $O_3$ | | + | o | o | o | o | – | + | – | – | + | + | – | o | o | o | – |
| +Oxygen $O_2$ | + | + | – | + | + | + | +[5] | – | + | – | – | + | o | + | +[7] | o | o |
| Paraffin oil | + | + | + | + | + | + | + | + | + | + | o | + | o | + | + | – | + |
| Perchloroethylene (tetrachlorethylene) – pure $Cl_2CCl_2$ | + | + | o | o | o | + | o | + | o | – | – | + | – | o | o | – | – |
| Peracetic acid – aqueous (6%) $CH_3CO_3H$ | + | + | – | – | – | | – | o | o | – | + | + | – | o | o | o | – |
| Petrol (gasoline) – pure $C_6H_{14}$ | + | + | + | + | + | + | + | + | + | o | + | + | + | + | + | – | – |
| Petrolether | + | + | o | + | + | + | + | + | + | o | + | + | + | + | + | – | o |
| Phenol – aqueous (saturated) $C_6H_5OH$ | + | + | o | o | o | o | + | o | – | + | + | + | o | + | o | o | + |
| Phosgene (liquid) – pure $COCl_2$ | + | + | + | + | + | | | | o | o | o | + | – | + | o | – | o |
| Phosgene (gaseous) – pure $COCl_2$ | + | + | + | + | + | | | + | o | – | + | + | – | + | + | – | |
| Phosphor chloride – pure | o | o | o | o | | + | + | + | – | + | – | + | – | + | o | – | – |
| Phosphoric acid – aqueous $H_3PO_4$ | – | + | – | – | – | + | | + | – | + | + | + | – | + | + | o | o |
| Picric acid (trinitrophenol) – pure $C_6H_2(OH)(NO_2)_3$ | + | + | + | + | + | + | + | + | | + | – | + | – | + | o | – | o |
| Pinene (turpentine oil) – pure | + | + | + | o | o | + | + | + | + | – | o | + | – | + | o | – | o |
| Potash (potassium carbonate) – aqueous $K_2CO_3$ | + | + | o | o | o | + | + | + | o | + | + | + | o | + | + | + | + |
| Potassium aluminium sulfate (alum) – aqueous $KAl(SO_4)_2 \cdot 12\,H_2O$ | o | + | – | – | – | + | + | – | + | + | + | + | + | | + | + | + |
| Potassium bifluoride – aqueous $KHF_2$ | + | + | o | o | o | + | + | + | – | + | + | + | + | + | + | + | + |
| Potassium bromate – aqueous $KBrO_3$ | o | + | o | o | – | | – | + | | + | + | + | + | + | + | + | + |

| | C1 | C2 | C3 | C4 | C5 | C6 | C7 | C8 | C9 | C10 | C11 | C12 | C13 | C14 | C15 | C16 |
|---|---|---|---|---|---|---|---|---|---|---|---|---|---|---|---|---|
| Potassium bromide – aqueous (KBr) | o | o | o | + | + | + | + | + | + | + | + | + | + | + | o | o |
| Potassium carbonate (potash) – aqueous (K₂CO₃) | + | + | o | o | o | + | - | o | + | + | + | o | + | + | + | + |
| Potassium chlorate – aqueous (KClO₃) | o | o | o | o | o | + | o | o | + | + | o | o | + | + | o | - |
| Potassium chloride – aqueous (KCl) | o | o | o | o | + | + | + | + | + | + | + | + | + | + | + | + |
| Potassium chromate – aqueous (K₂CrO₄) | o | o | + | + | + | + | + | o | + | + | o | o | + | + | + | o |
| Potassium cyanide – aqueous (KCN) | + | + | - | - | + | + | + | + | + | + | + | + | + | + | + | + |
| Potassium dichromate – aqueous (K₂Cr₂O₇) | + | + | o | o | - | + | - | + | + | + | o | o | + | + | o | o |
| Potassium ferricyanide (potassium cyano ferrat III) – (red potassium prussiate) – aqueous (KFeCN₄) | + | + | o | - | o | + | + | + | + | + | + | + | + | + | + | + |
| Potassium ferrocyanide (potassium cyano ferrat II) – (yellow potassium prussiate) – aqueous (KFeCN₃) | - | o | + | + | + | + | + | + | + | + | - | o | + | + | + | + |
| Potassium hydroxide (caustic potash) – aqueous (KOH) | + | o | + | + | + | + | + | + | + | + | + | - | + | + | - | - |
| Potassium hypochlorite – aqueous (KOCl) | + | + | - | - | + | o | o | o | + | + | o | o | + | + | o | o |
| Potassium iodide – aqueous (KI) | o | o | o | o | + | o | + | + | + | + | + | + | + | + | + | + |
| Potassium nitrate – aqueous (KNO₃) | o | o | o | o | + | + | + | o | + | + | o | o | + | + | + | + |
| Potassium nitrite – aqueous (KNO₂) | o | o | o | o | + | + | + | + | + | + | + | + | + | + | + | + |
| Potassium permanganate – aqueous (KMnO₄) | + | + | + | + | + | o | + | o | + | + | o | - | + | + | - | - |
| Potassium peroxide – aqueous (K₂O₂) | o | + | o | o | + | + | o | o | o | + | - | - | + | + | + | - |
| Potassium persulfate – aqueous (K₂S₂O₈) | + | + | o | - | + | + | o | + | + | + | o | - | + | + | - | - |
| +Potassium phosphate – aqueous | + | + | - | - | + | o | o | o | + | + | o | o | + | + | o | o |
| Potassium sulfate – aqueous (K₂SO₄) | + | + | o | o | + | + | o | + | + | + | + | + | + | + | + | + |
| Potassium sulfide – aqueous (K₂S) | + | + | o | + | + | + | o | + | + | + | + | + | + | + | + | + |
| Potassium sulfite – aqueous (K₂SO₃) | o | + | o | o | + | + | + | o | + | o | + | + | + | + | + | + |

(continued)

**Table 10.2:** (Continued)

| Substance | Formula | 1.4305/1.4105 | 1.4401/1.4571 | GG, GS | RG | MS | PEEK | PPS | PVDF | PA | PP | PVC | PTFE | CR | FFKM | FKM | EPDM | NBR |
|---|---|---|---|---|---|---|---|---|---|---|---|---|---|---|---|---|---|---|
| Propane (liquid and gas) – pure | $C_3H_8$ | + | + | + | + | + | + | + | + | + | + | + | + | + | + | + | – | + |
| Propanol (isopropanol) – pure | $CH_3CH(OH)CH_3$ | + | + | + | + | + | + | + | + | o | + | + | + | + | + | + | + | – |
| Propylenglycol – pure | $HOCH_2CH_2CH_2OH$ | + | + | + | o | + | + | + | + | o | + | + | + | + | + | + | + | + |
| Pyridine – pure | $C_5H_5N$ | o | + | + | + | + | + | o | o | + | o | – | + | – | + | – | – | – |
| **S** | | | | | | | | | | | | | | | | | | |
| Silicon oil | | + | + | + | + | + | + | + | + | + | + | + | + | + | + | + | + | + |
| Silver nitrate – aqueous | $AgNO_3$ | + | + | – | – | – | + | + | + | + | + | o | + | + | + | + | + | o |
| Sodium arsenate, sodium arsenite – pure | $Na_3AsO_4$ u. $Na_3AsO_3$ | + | + | + | + | + | | | | + | + | + | + | + | + | + | + | + |
| Sodium benzoate – aqueous | COONa | + | + | + | + | + | + | | + | + | + | + | + | + | + | + | + | + |
| Sodium bicarbonate – aqueous | $NaHCO_3$ | + | + | o | + | o | | + | + | + | + | + | + | + | + | + | + | + |
| Sodium bisulfate – aqueous | $NaHSO_4$ | o | o | o | o | o | | | + | + | + | + | + | + | + | + | + | + |
| Sodium bisulfite – aqueous (bisulfite) | $NaHSO_3$ | o | + | – | o | o | + | | + | + | + | + | + | + | + | + | + | o |
| Sodium bromate – aqueous | $NaBrO_3$ | o | + | o | o | – | + | – | + | o | + | + | + | o | + | + | + | + |
| Sodium bromide – aqueous | NaBr | o | o | o | o | o | + | + | + | – | + | + | + | + | + | + | + | + |
| Sodium carbonate (soda) – aqueous | $Na_2CO_3$ | + | + | o | o | o | | + | + | + | + | + | + | o | + | + | + | + |
| Sodium chloroacetates | | + | + | o | o | o | | + | | | + | + | + | + | + | + | + | + |
| Sodium chlorate – aqueous | $NaClO_3$ | o | o | o | o | o | + | – | + | o | + | + | + | o | + | o | o | o |
| Sodium chloride (common salt) – aqueous | NaCl | o | o | – | o | – | + | + | + | + | + | + | + | + | + | + | + | + |
| Sodium chlorite – aqueous | $NaClO_2$ | – | o | – | o | o | | + | + | – | o | o | + | – | + | o | o | – |

| Chemical | Formula | | | | | | | | | | | | | | | | | | |
|---|---|---|---|---|---|---|---|---|---|---|---|---|---|---|---|---|---|---|---|
| Sodium chromate – aqueous | NaCrO$_4$ | o | o | o | + | + | o | + | + | – | + | + | + | o | o | o | + | o |
| Sodium cyanide – aqueous | NaCN | + | + | o | – | – | + |  | + | + | + | + | + | + | + | + | + | + |
| Sodium dodecylbenzene sulfonate – aqueous | | + | + | o | o | o |  |  |  | + | o | + | + | + | + | + | + | + |
| Sodium fluoride – aqueous | NaF | o | + | o | + | + |  |  | + | + | + | + | + | + | + | + | + | + |
| Sodium glutamate – aqueous | | + | + | o |  |  |  |  |  |  | + | + | + | + | + | + | + | + |
| Sodium hydroxide (caustic soda) – aqueous | NaOH | + | + | o | – | – | + | o | – | + | + | + | + | + | + | – | + | – |
| Sodium hypochlorite (chlorine bleach) | NaOCl | o | o | o | o | o | + | – | o | – | o | o | + | – | + | + | o | o |
| Sodium iodide – aqueous | NaI | o | o | o | o | o |  | o | + |  | + | + | + | + | + | + | + | + |
| Sodium mercaptobenzothiazol – pure | | + | + | + | + |  |  |  |  |  | + | o | o | o | + | + | + | o |
| Sodium nitrate – aqueous | NaNO$_3$ | – | + | – | – | – | + | + | + | + | o | + | + | + | + | + | + | + |
| Sodium nitrite – aqueous | NaNO$_2$ | + | + | + | + | + | + |  | + | + | + | + | + | + | + | + | + | + |
| Sodium pentachlorphenolate – pure | C$_6$Cl$_5$ONa | + | + | o | + | + |  |  |  | + | + | + | + | + | – | + | o | o |
| Sodium perborate – aqueous | NaBO$_3$ | + | + | o | o | o |  | – | + |  | + | + | + | + | + | + | + | o |
| Sodium persulfate – aqueous | K$_2$S$_2$O$_8$ | o | + | – | – | – |  | – | + | – | + | + | + | + | + | + | + | + |
| Sodium phosphate – aqueous | Na$_3$PO$_4$ | o | o | o | o | o |  |  | + | + | + | + | + | + | + | + | + | + |
| Sodium propionate – aqueous | CH$_3$CH$_2$COONa | + | + | + | + | + |  |  | + | + | + | + | + | + | + | + | + | o |
| Sodium pyrosulfite – aqueous | Na$_2$S$_2$O$_5$ | o | + | – | o | o | + | + |  | + | + | + | + | + | + | + | + | + |
| Sodium silicate – aqueous | | + | + | + | o | o |  |  | + | + | + | + | + | + | + | + | + | + |
| Sodium stannate – aqueous | Na$_2$SnO$_3$ | + | + | + | o | o | + | + | o | o | + | + | + | + | + | + | + | + |
| Sodium sulfate – aqueous | Na$_2$SO$_4$ | + | + | – | – | – | + | + | + | + | + | + | + | + | + | + | + | + |
| Sodium sulfide – aqueous | Na$_2$S | + | + | o | – | o | + |  |  | + | + | o | – | + | + | + | + | o |
| Sodium sulfite – aqueous | Na$_2$SO$_3$ | o | + | o | – | o | + |  |  | + | + | + | + | + | + | + | + | + |
| Sodium tartrate – aqueous | | + | + | o | + | + |  |  |  | + | + | + | + | + | + | + | + | + |

(Continued)

**Table 10.2:** (Continued)

| Chemical | Formula | NBR | EPDM | FKM | FFKM | CR | PTFE | PVC | PP | PA | PVDF | PPS | PEEK | WS | RG | GG, GS | 1.4401/1.4571 | 1.4305/1.4105 |
|---|---|---|---|---|---|---|---|---|---|---|---|---|---|---|---|---|---|---|
| Sodium thiosulfate – aqueous | $Na_2S_2O_3$ | + | + | + | + | + | + | + | + | + | + | + |  | o | – | o | o | o |
| Sodium zincate – aqueous | $Na_2[Zn(OH)_4]$ | o | + | + |  | + | + | o | o |  | + |  | + | + | + | + | + | + |
| Solvent naphtha (Shellsol D 60 and D 70) |  | o | – | o | + | o | + | + | + | + | + | + | + | o | o | + | + | + |
| Starch solutions – aqueous |  | + | + | + | + | + | + | – | – | – | + | + | + | o | + | o | + | + |
| Steam (rubber seals up to 130 °C) | $H_2O$ | o | + | + | + | o | + | + | + | – | + | o |  | o | + | + | + | + |
| Stearic acid | $C_{18}H_{37}COOH$ | + | + | + | + | + | + | – | o | + | + | + | + | o | o | – | + | + |
| Styrene | $C_6H_5CHCH_2$ | – | – | o | + | – | + | + | + | + | + |  |  | + | + | o | + | + |
| Succinic acid – aqueous | $HOOCCH_2CH_2COOH$ | + | + | + | + | + | + | – | – | – | + | + | + | o | o | o | + | + |
| Sulfur chlorides and oxychlorides – pure |  | – | – | + | + | – | + | o | o | o | + |  | + | – | – | o | + | – |
| Sulfur dioxide (gas, wet) | $SO_2$ | – | + | + | + | – | + | o | o | o | + | o | + | o | o | – | + | o |
| Sulfur dioxide (gas, dry) – pure | $SO_2$ | – | + | + | + | – | + | o | o | o | + | + | + | + | + | o | + | o |
| Sulfur dioxide (liquid) – pure | $SO_2$ | – | + | + | + | – | + | – | – | – | + | + | + | + | + | + | + | + |
| Sulfur hexafluoride – pure | $SF_6$ | + | + | o | o | + | + | + | + |  | + |  | o | – | – | + | + | + |
| Sulfuric acid – aqueous (30 %) | $H_2SO_4$ | o | + | + | + | o | + | + | + | – | + | o | – | – | – | – | – | – |
| Sulfuric acid – concentrated (96 %) | $H_2SO_4$ | – | – | o | + | – | + | – | – | – | + | o | + | – | – | – | – | – |
| Sulfurous acid – aqueous | $H_2SO_3$ | – | + | + | + | + | + | + | + | – | + | o |  |  |  | – | + | – |
| Tall oil |  | o | o | o |  | o | + | + | + | + | + |  |  | – | – | – | + | o |
| Tannic acid |  | + | + | + | + | + | + | + | + | + | + | + |  | o | o | o | + | + |

| Chemical | Formula | Resistance (materials, left→right) |
|---|---|---|
| Tannin (tannic acid) | | + + o o o + + + o + + + + + + |
| Tartaric acid – aqueous | | + + – – – + + + – + + + + + + |
| Tar oil (carbolineum) | | + + + + + o o o o o o o o o o |
| Tetrachloroethylene (perchloroethylene) | Cl₂CCCl₂ | + + o o o + o + o – o – o o o |
| Tetraethyl lead | Pb(CH₂CH₃)₄ | + + + o o + o + o – o – + + + |
| Tetrahydrofuran – pure | C₄H₈O | + + + o o + + + + + + + + + + |
| Tetrahydronaphtalene (tetralin) – pure | C₁₀H₁₂ | + + + + + o – + – – – – + + + |
| Thiophene – pure | C₄H₄S | + + + o – o o + – o – – + + + |
| Tin chlorides (stannous and stannic chlorides) – aqueous | | – o + – – + + + + + + + + + + |
| Toluene – pure | C₆H₅CH₃ | + + o + + + + + o – o – + + + |
| Tributyl phosphate – pure (phosphoric acid tributylester) | PO(OC₄H₉)₃ | + + o + + + o o + – + – + + + |
| Trichloroacetic acid – aqueous | (Cl)₃CCOOH | – – + – – + + – – – – – – – – |
| Trichloroethylene – pure | Cl₂CCCl | + + – o o + – o – – o – + + + |
| Trichloromethane (chloroform) | CHCl₃ | + + – – – + – o – – – – + + + |
| Tricresyl phosphate – pure | | + + o + + + o + + + + – + + + |
| Triethanolamine – pure | N(CH₂CH₂OH)₃ | + + o o o + + + o + + – + + + |
| Uranium hexafluoride – pure | UF₆ | o + o o o + + + – + + + o + + |
| Urea – aqueous | NH₂CONH₂ | o o o o o + + + + + + + o + + |
| Vinyl acetat – pure | CH₂CHOOCH₂CH₃ | + + o o o + + o + – + – + + + |
| Vinyl chloride – pure | CH₂CHCl | o o o – – + + + o – o – + – o |

(Continued)

**Table 10.2:** (Continued)

| | NBR | EPDM | FKM | FFKM | CR | PTFE | PVC | PP | PA | PVDF | PPS | PEEK | MS | RG | GG, GS | 1.4401/1.4571 | 1.4305/1.4105 |
|---|---|---|---|---|---|---|---|---|---|---|---|---|---|---|---|---|---|
| **W** | | | | | | | | | | | | | | | | | |
| Waste gases – with carbon dioxide | + | + | + | + | + | + | + | + | + | + | + | + | + | + | o | + | o |
| Waste gases – with carbon monoxide | + | + | + | + | + | + | + | + | + | + | + | + | + | + | + | + | + |
| Waste gases – with hydrochloric acid | + | + | + | + | + | + | + | + | − | + | − | o | o | o | − | o | − |
| Waste gases – with hydrogen fluoride | + | + | + | + | + | + | + | + | o | + | − | − | o | o | o | o | o |
| Waste gases – with nitrous gases | o | + | + | + | + | + | + | + | − | + | | + | − | − | o | + | + |
| Waste gases – with sulfur dioxide (dry) | o | + | + | + | + | + | + | + | o | + | + | + | + | + | + | + | + |
| Waste gases – with sulfuric acid – (sulfur trioxide wet) | o | + | + | + | + | + | + | + | − | + | o | − | − | − | − | + | o |
| Waste gases – with sulfur trioxide (dry) | o | + | + | + | + | + | + | + | + | + | | + | o | o | o | + | + |
| Water – distilled $H_2O$ | + | + | + | + | + | + | + | + | + | + | o | + | o | + | − | + | o |
| Water – seawater $H_2O$ | + | + | + | + | + | + | + | + | + | + | + | + | o | o | o | o | o |
| Wood tar, Wood oil (impregnating oils) | − | − | − | + | − | + | o | − | | | + | | + | + | o | + | + |
| **X** | | | | | | | | | | | | | | | | | |
| Xenon $Xe$ | + | + | + | + | + | + | + | + | + | + | + | + | + | + | + | + | + |
| Xylene – pure $C_6H_4(CH_3)_2$ | − | − | + | + | − | + | − | − | + | o | o | + | + | + | + | + | + |
| **Y** | | | | | | | | | | | | | | | | | |
| Yeast – aqueous | + | + | + | + | + | + | + | + | + | + | | + | o | o | o | + | + |

| | | | | | | | | | | | | | | | | | |
|---|---|---|---|---|---|---|---|---|---|---|---|---|---|---|---|---|---|
| Zinc chloride – aqueous | ZnCl$_2$ | + | + | + | + | + | + | + | + | − | + | + | + | − | − | − | O | − |
| Zinc sulfate – aqueous | ZnSO$_4$ | + | + | + | + | + | + | + | + | + | + | + | + | − | − | − | + | − |

---

**1** Technical acetylene contains solvents like alkanes, dimethyl formamide or acetone.
Bürkert generally does not know what solvent lack is used in the gas suppliers acetylene.
The chemical resistance of the gasket materials has to be proved according to the german specification DIN 9539.

**2** brass with up to 58% Cu

**3** diffuses through EPDM; attacks epoxy materials

**4** acid resistant FKM compound

**5** Hydrogen can lead to an embrittlement of metals.

**6** Most of the polymer materials get damaged by ozone. Therefore the resistances have to be put into perspective.

**7** under pressure permitted according to the BAM (Federal Institute for Materials Research and Testing)

# International standards and regulations

The following section is reprinted from the application note: "Test methods for color measurement," with kind permission from HunterLab, www.hunterlab.com.

More information on color measurement methods and regulations can be found at http://www.hunterlab.com/application-notes.html

The world governing body for color and appearance measurement is the CIE, Commission Internationale de l'Eclairage, or the International Commission on Illumination. They can be found online at www.cie.co.at. Their fundamental publication is Publication CIE 15:2004, Colorimetry, which provides detailed information on standard illuminants and observers, as well as the calculations for color scales. The primary standards organizations for the colorimetric industry are as follows:

- NIST, National Institute of Standards and Technology, www.nist.gov.
- NPL, National Physical Laboratory, www.npl.co.uk.
- NRC, National Research Council Canada, www.nrc-cnrc.gc.ca.
- BAM, Bundesanstalt für Materialforschung und -prüfung, www.bam.de

In addition, many test methods have been published by organizational bodies describing how color measurement should be performed in various industries. Brief descriptions of some of them (primarily addressing instrumental – as opposed to visual – assessments) are provided below. This list is not intended to be exhaustive.


## AATCC Evaluation Procedures – Textiles

American Association of Textile Chemists and Colorists (AATCC) evaluation procedures may be ordered through www.aatcc.org.

AATCC Evaluation Procedure 1, "Gray Scale for Color Change," describes the use of a gray scale for visually assessing color changes that occur during color fastness tests.

AATCC Evaluation Procedure 2, "Gray Scale for Staining," describes the use of a gray scale for visually evaluating staining of adjacent fabrics during color fastness tests.

AATCC Evaluation Procedure 6, "Instrumental Color Measurement," is a reference document covering instrumental reflectance measurement, transmittance measurement, and related calculations. It also contains an appendix describing sample handling techniques.

AATCC Evaluation Procedure 7, "Instrumental Assessment of the Change in Color of a Test Specimen," describes how to instrumentally assess color changes that occur during color fastness tests. This procedure is intended as an alternative to Evaluation Procedure 1's visual method.

https://doi.org/10.1515/9783110546156-011

AATCC Test Method 110, "Whiteness of Textiles," lists procedures for instrumentally measuring and calculating whiteness and tint of fabrics.

AATCC Test Method 173, "CMC: Calculation of Small Color Differences for Acceptability," describes how to calculate and use the dE CMC color difference scale.

AATCC Test Method 182, "Relative Color Strength of Dyes in Solution," describes determination of color strength of a dye spectrophotometrically by comparing its transmission measurements to those of a reference dye.

## AS/NZS Australian/New Zealand Standards

Australian Standard (AS)/New Zealand Standard (NZS) standards may be obtained through Standards Australia at www.standards.com.au.

AS/NZS 1301.455s, "Colour Measurement with a Diff/0° Geometry Tristimulus Reflectometer," describes color measurement of non-fluorescent paper and board using a diffuse/0° colorimeter.

AS/NZS 1580.601.2, "Paints and related materials – Methods of test: Colour – Principles of color measurement," provides an overview of colorimetric terms, illuminants, etc.

AS/NZS 1580.601.3, "Paints and related materials – Methods of test: Colour – Methods of colour measurement," describes the parameters of instrumental measurement, including geometries and illuminants.

AS/NZS 1580.601.4, "Paints and related materials – Methods of test: Colour – Calculation of colour differences," describes calculation of color differences using instrumental measurements.

AS/NZS 1580.601.5, "Paints and related materials – Methods of test: Colour – Calculation of small colour differences using the CMC equation," describes the use of the CMC equations for color tolerancing.

AS 2001.4.A02, "Methods of test for textiles – Colourfastness tests – Grey scale for assessing change in colour," adopts the ISO 105-A02 method described below for assessing gray scale during colorfastness testing.

AS 2001.4.A03, "Methods of test for textiles – Colourfastness tests – Grey scale for assessing staining," adopts the ISO 105-A03 method described below for assessing staining of adjacent fabrics during colorfastness testing.

AS 2001.4.A04, "Methods of test for textiles – Colourfastness tests – Instrumental assessment of the degree of staining of adjacent fabrics," allows for gray stain values to be assigned using instrumental color measurements.

AS 2001.4.A05, "Method of test for textiles – Colourfastness tests – Instrumental assessment of the change in colour for determination of grey scale rating," allows for gray scale values to be assigned using instrumental color measurements.

AS/NZS 2111.19.1, "Textile floor coverings – Tests and measurements: Colourfastness tests – Rubbing," describes how to evaluate visually or instrumentally the staining of cloths rubbed on floor coverings.

AS/NZS 2633, "Guide to the specification of colours," provides guidance on specifying colors, selecting a color measurement method, and establishing tolerances.

AS 4459.16, "Methods of sampling and testing ceramic tiles – Determination of small colour differences," provides a method for quantifying instrumentally color differences between glazed ceramic tiles.

## ASBC – Beer

American Society of Brewing Chemists (ASBC) methods may be ordered through www.asbcnet.org.

ASBC Beer-10, "Color," describes how to measure beer using a spectrophotometer and defines the ASBC Beer and Turbidity scales.

## ASTM Standard Specifications

American Society for Testing and Materials (ASTM) standards may be ordered through www.astm.org.

ASTM C1510, "Standard Test Method for Color and Color Difference of Whitewares by Abridged Spectrophotometry," describes how to measure reflectance and color of ceramics using a sphere spectrophotometer.

ASTM D156, "Standard Test Method for Saybolt Color of Petroleum Products (Saybolt Chromometer Method)," describes evaluating the color of refined oils visually using Saybolt Color Units.

ASTM D985, "Standard Test Method for Brightness of Pulp, Paper, and Paperboard (Directional Reflectance at 457 nm)," describes how to determine brightness of white paper products (even those that include fluorescent whitening agents) using a 45°/0° instrument.

ASTM D1003, "Standard Test Method for Haze and Luminous Transmittance of Transparent Plastics," provides a procedure for the instrumental measurement of luminous transmittance and haze of materials with haze values up to 30%.

ASTM D1209, "Standard Test Method for Color of Clear Liquids (Platinum-Cobalt Scale)," describes visual evaluation of light, yellow-colored liquids using the PtCo (APHA) scale.

ASTM D1500, "Standard Test Method for ASTM Color of Petroleum Products (ASTM Color Scale)," describes visual evaluation of petroleum products using the ASTM Color scale.

ASTM D1544, "Standard Test Method for Color of Transparent Liquids (Gardner Color Scale)," describes visual evaluation of oils and varnishes using the Gardner Color Scale.

ASTM D1746, "Standard Test Method for Transparency of Plastic Sheeting," covers instrumental measurement of regular transmittance of clear and colorless thin sheeting.

ASTM D2066, "Standard Test Method for Relative Tinting Strength of Paste-Type Printing Ink Dispersions," describes reflectance instrumental or visual evaluation of the relative tinting strength of ink dispersions.

ASTM D2244, "Standard Practice for Calculation of Color Tolerances and Color Differences from Instrumentally Measured Color Coordinates," describes the calculation of rectangular and elliptical color differences for opaque samples.

ASTM D2616, "Standard Test Method for Evaluation of Visual Color Difference With a Gray Scale," describes visual evaluation of small color differences and is based on the AATCC method that discusses the Gray Scale for Color Change.

ASTM D2745, "Standard Test Method for Relative Tinting Strength of White Pigments by Reflectance Measurements," explains how to determine the relative tinting strength of white pigments by using their reflectance measurements.

ASTM D3134, "Standard Practice for Establishing Color and Gloss Tolerances," describes a means of establishing color and gloss tolerances from standards for non-fluorescent opaque samples.

ASTM D3210, "Standard Test Method for Comparing Colors of Films from Water-Emulsion Floor Polishes," describes characterization of luminous reflectance for films treated with floor polishes using a tristimulus colorimeter.

ASTM D3964, "Standard Practice for Selection of Coating Specimens for Appearance Measurements," discusses selection of appropriate samples for appearance measurement and methods for their preparation and presentation.

ASTM D4086, "Standard Practice for Visual Evaluation of Metamerism," describes visual methods for detecting metamerism and for establishing the magnitude of such color differences.

ASTM D4265, "Standard Guide for Evaluating Stain Removal Performance in Home Laundering," describes how to evaluate fabrics that have been involved in laundry tests either visually or instrumentally.

ASTM D4303, "Standard Test Methods for Lightfastness of Colorants Used in Artists' Materials," explains how artists' colorants can be evaluated using a spectrophotometer or colorimeter that can exclude specular reflectance.

ASTM D4838, "Standard Test Method for Determining the Relative Tinting Strength of Chromatic Paints," explains how to determine absorption tinting strength relative to a standard paint using a spectrophotometer or colorimeter.

ASTM D4877, "Standard Test Method for Polyurethane Raw Materials: Determination of APHA Color in Isocyanates," provides a test method for measuring the color of clear, yellowish liquids.

ASTM D4960, "Standard Test Method for Evaluation of Color for Thermoplastic Traffic Marking Materials," describes using a 45°/0° colorimeter or spectrophotometer to measure these materials.

ASTM D5215, "Standard Test Method for Instrumental Evaluation of Staining of Vinyl Flooring by Adhesives," describes how to instrumentally measure the degree of staining that occurs in vinyl flooring as a result of its contact with adhesives. Color differences are expressed in CIELAB units.

ASTM D5326, "Standard Test Method for Color Development in Tinted Latex Paints," describes evaluation of paints by comparing a drawdown and a brushout of the same color using a colorimeter or spectrophotometer standardized in reflectance, specular excluded mode.

ASTM D5386, "Standard Test Method for Color of Liquids Using Tristimulus Colorimetry," gives an instrumental method for obtaining APHA values for near-clear liquid samples.

ASTM D5531, "Standard Guide for Preparation, Maintenance, and Distribution of Physical Product Standards for Color and Geometric Appearance of Coatings," discusses the types of physical product standards used in the coatings industry and techniques for creating them.

ASTM D6166, "Standard Test Method for Color of Naval Stores and Related Products (Instrumental Determination of Gardner Color)," describes color measurement of clear, yellow/brown liquids using an instrument and reporting of the results in the Gardner color scale.

ASTM D6290, "Standard Test Method for Color Determination of Plastic Pellets," explains how to measure the degree of yellowness of plastic pellets using a spectrophotometer or colorimeter.

ASTM E179, "Standard Guide for Selection of Geometric Conditions for Measurement of Reflection and Transmission Properties of Materials," suggests appropriate instrument types and measurement scales for evaluating appearance character-istics such as color, glossiness, and opacity.

ASTM E259, "Standard Practice for Preparation of Pressed Powder White Reflectance Factor Transfer Standards for Hemispherical and Bi-directional Geometries," defines preparation of highly-reflective transfer standards for both directional and diffuse instruments.

ASTM E275, "Standard Practice for Describing and Measuring Performance of Ultraviolet, Visible, and Near-Infrared Spectrophotometers," provides the require-ments of spectrophotometric performance for ASTM methods and explains how to test an instrument.

ASTM E284, "Standard Terminology of Appearance," defines terms used in describing appearance, such as color and opacity.

ASTM E308, "Standard Practice for Computing the Colors of Objects by Using the CIE System," explains how to calculate CIE XYZ and other color scales from spectral reflectance and transmittance values and defines the CIE illuminants and standard observers.

ASTM E313, "Standard Practice for Calculating Yellowness and Whiteness Indices from Instrumentally Measured Color Coordinates," defines the whiteness and yellowness indices.

ASTM E805, "Standard Practice for Identification of Instrumental Methods of Color or Color-Difference Measurement of Materials," describes how to effectively communicate color values and all the parameters that affect them.

ASTM E808, "Standard Practice for Describing Retroreflection," describes how to measure retroreflective samples and present the results.

ASTM E809, "Standard Practice for Measuring Photometric Characteristics of Retroreflectors," explains how to measure the photometric characteristics of retroreflectors.

ASTM E991, "Standard Practice for Color Measurement of Fluorescent Specimens," describes measurement of fluorescent samples and presenting numerically how they appear in daylight.

ASTM E1164, "Standard Practice for Obtaining Spectrophotometric Data for Object-Color Evaluation," covers instrumental measurement requirements for spectrophotometers and colorimeters.

ASTM E1247, "Standard Practice for Detecting Fluorescence in Object-Color Specimens by Spectrophotometry," describes how to use a spectrophotometer to determine whether samples are fluorescent.

ASTM E1331, "Standard Test Method for Reflectance Factor and Color by Spectrophotometry Using Hemispherical Geometry," defines colorimetric measurement using a spectrophotometer with a sphere (diffuse) geometry.

ASTM E1345, "Standard Practice for Reducing the Effect of Variability of Color Measurement by Use of Multiple Measurements," describes averaging as a technique for minimizing sample variation.

ASTM E1347, "Standard Test Method for Color and Color-Difference Measurement by Tristimulus (Filter) Colorimetry," describes color measurement using tristimulus colorimeters of either 45°/0° or diffuse geometry.

ASTM E1348, "Standard Test Method for Transmittance and Color by Spectrophotometry Using Hemispherical Geometry," describes color measurement of transparent and translucent samples using a spectrophotometer with a sphere (diffuse) geometry standardized in a transmittance mode.

ASTM E1349, "Standard Test Method for Reflectance Factor and Color by Spectrophotometry Using Bidirectional Geometry," describes color measurement of opaque samples using a spectrophotometer with a 45°/0° or 0°/ 45° geometry.

ASTM E1477, "Standard Test Method for Luminous Reflectance Factor of Acoustical Materials by Use of Integrating-Sphere Reflectometers," explains how to measure luminous reflectance (Y) of acoustical materials using a sphere instrument.

ASTM E1767, "Practice for Specifying the Geometries of Observation and Measurement to Characterize the Appearance of Materials," describes the different geometries of instruments that measure appearance.

ASTM E2022, "Standard Practice for Calculation of Weighting Factors for Tristimulus Integration," explains how to establish tables of weighting factors that are to be used in calculating tristimulus values from spectral reflectance or transmittance data.

ASTM E2214, "Standard Practice for Specifying and Verifying the Performance of Color-Measuring Instruments," explains how to characterize the performance of colorimeters and spectrophotometers used for color measurement.

ASTM E2301, "Standard Test Method for Daytime Colorimetric Properties of Fluorescent Retroreflective Sheeting and Marking Materials for High Visibility Traffic Control and Personal Safety Applications Using 45°:Normal Geometry," describes how to use a 45°/0° instrument to measure the color of fluorescent-retroreflective sheeting such as used for traffic control and safety applications.

ASTM F2109, "Standard Test Method to Determine Color Change and Staining Caused by Aircraft Maintenance Chemicals Upon Aircraft Cabin Interior Hard Surfaces," Explains how to describe, using the AATCC Gray Scale for Color Change and the AATCC Gray Color Scale for Staining, the effect of chemical cleaning on the color of aircraft interiors.

# DIN

Deutsches Institut für Normung e.V. (DIN) [German Institute for Standardization] industrial test methods may be ordered through www.en.din.de.

DIN 5033, "Colorimetry," has nine parts and covers visual and instrumental color analysis, as well as basic color concepts and terminology.

DIN 6174, "Colorimetric Evaluation of Colour Differences of Surface Colours According to the CIELAB Formula," describes measuring dL*, da*, and db* using an instrument.

DIN 6176, "Colorimetric evaluation of colour differences of surface colours according to DIN99-formula," explains how to measure using the elliptical DIN99 formula.

## ICUMSA Sugar Methods

International Commission for Uniform Methods of Sugar Analysis (ICUMSA) methods may be ordered through www.bartens.com/shop.html.

ICUMSA Method GS 2/3-9, "The Determination of Sugar Solution Colour at pH 7.0," describes the preparation and measurement of white sugar solution and the formula for calculating the ICUMSA 420 Sugar Score.

ICUMSA Method GS 2/3-10, "The Determination of White Sugar Solution Color," describes the preparation and measurement of white sugar solution and the formula for calculating the ICUMSA 420 Sugar Score.

# ISO

International Organization for Standardization (ISO) methods may be ordered through www.iso.org.

ISO 105-A01, "Textiles – Tests for colour fastness – Part A01: General principles of testing," provides general information about the various ISO methods for testing color fastness of textiles and dyes. This method may help you select an ISO test method for textiles if none of the ones listed below (not a complete list) is applicable.

ISO 105-A02, "Textiles – Tests for colour fastness – Part A02: Grey scale for assessing change in colour," describes the gray scale for color changes that occur during color fastness tests. The method includes a colorimetric specification for the scale.

ISO 105-A03, "Textiles – Tests for colour fastness – Part A03: Grey scale for assessing staining," describes the gray scale for determining staining of adjacent fabrics during color fastness tests. The method includes a colorimetric specification for the scale.

ISO 105-A04, "Textiles – Tests for colour fastness – Part A04: Method for the instrumental assessment of the degree of staining of adjacent fabrics," describes an instrumental method for assessing color changes that occur during color fastness tests. This method includes the calculations required to convert the instrumental values into a gray scale rating.

ISO 105-A05, "Textiles – Tests for colour fastness – Part A05: Instrumental assessment of change in colour for determination of grey scale rating," describes an instrumental method for assessing staining of adjacent fabrics during color fastness tests. This method includes the calculations required to convert the instrumental values into a gray scale rating.

ISO 105-J01, "Textiles – Tests for colour fastness – Part J01: General principles for measurement of surface colour," is a reference document describing general concepts and problems of reflectance color measurement.

ISO 105-J02, "Textiles – Tests for colour fastness – Part J02: Instrumental assessment of relative whiteness," explains how to quantify whiteness and tint of textiles, including fluorescent textiles.

ISO 105-J03, "Textiles – Tests for colour fastness – Part J03: Calculation of colour differences," explains how to calculate the color difference between two similar samples and allows for the specification of tolerances.

ISO 2211, "Liquid chemical products – Measurement of colour in Hazen units (platinum-cobalt scale)," describes visual and instrumental means of applying platinum-cobalt (APHA) values to clear, slightly-colored liquids.

ISO 4630-1, "Clear liquids – Estimation of colour by the Gardner colour scale – Part 1: Visual method," describes how to estimate the color of clear, yellow-brown liquids visually using the Gardner scale.

ISO 4630-2, "Clear liquids – Estimation of colour by the Gardner colour scale – Part 2: Spectrophotometric method," describes how to estimate the color of clear, yellow-brown liquids instrumentally using the Gardner scale.

ISO 5631, "Paper and board – Determination of colour (C/2 degrees) – Diffuse reflectance method," describes measurement of paper and paper board using a diffuse spectrophotometer or colorimeter.

ISO 6271-1, "Clear liquids – Estimation of colour by the platinum-cobalt scale – Part 1: Visual method," defines how to visually assign Pt-Co (APHA) values to clear liquids.

ISO 6271-2, "Clear liquids – Estimation of colour by the platinum-cobalt scale – Part 2: Spectrophotometric method," defines how to determine the color of clear liquids instrumentally using the Pt-Co (APHA) scale.

ISO 7724-1, "Paints and varnishes – Colorimetry – Part 1: Principles," describes color scales and requirements for determining color values for paint films (including specifying instrument geometry).

ISO 7724-2, "Paints and varnishes – Colorimetry – Part 2: Colour measurement," describes how to determine color values for uniform paint films using a spectrophotometer or colorimeter.

ISO 7724-3, "Paints and varnishes – Colorimetry – Part 3: Calculation of colour differences," explains how to quantify small color differences between paint films using a color measuring instrument.

ISO 8112, "Caprolactam for industrial use – Determination of colour of 50% aqueous caprolactam solution, expressed in Hazen units (platinum-cobalt scale) – Spectrometric method," describes assignment of Hazen (APHA) values to specific water solutions after measuring their wavelengths at 390 nm with a path length of 50 mm.

ISO/CIE 10526, "CIE standard illuminants for colorimetry," defines the CIE standard illuminants.

ISO/CIE 10527, "CIE standard colorimetric observers," defines the CIE standard observers.

ISO 11475, "Paper and board – Determination of CIE whiteness, D65/10 degrees (outdoor daylight)," explains how to instrumentally determine the whiteness of papers and boards with or without fluorescent whitening agents under outdoor lighting conditions.

ISO 11476, "Paper and board – Determination of CIE-whiteness, C/2 degree (indoor illumination conditions)," explains how to instrumentally determine the whiteness of papers and boards with or without fluorescent whitening agents under indoor lighting conditions.

ISO 13468-2, "Plastics – Determination of the total luminous transmittance of transparent materials – Part 2: Double-beam instrument," describes measurement in transmittance of transparent and colorless plastics using a double-beam spectrophotometer. It does not apply to plastics that contain fluorescent components.

## JIS

Japanese Industrial Standards may be obtained through www.jsa.or.jp.

JIS K 5101-2-2:2004, "Test methods for pigments – Part2: Comparison of colour – Section 2: Colorimetric method," describes how to compare the colors of pigments using an instrument.

JIS K 5600-4-4:1999, "Testing methods for paints – Part 4: Visual characteristics of film – Section 4: Colorimetry (Principles)," defines terms used in color measurement of paint films and related materials.

JIS K 5600-4-5:1999, "Testing methods for paints – Part 4: Visual characteristics of film – Section 5: Colorimetry (Measurement)," describes color measurement of opaque monochromatic (nonmetallic) paint films.

JIS K 5600-4-6:1999, "Testing methods for paints – Part 4: Visual characteristics of film – Section 6: Colorimetry (Calculation of colour differences)," explains how to instrumentally evaluate color differences between paint films.

JIS K 7105:1981, "Testing methods for optical properties of plastics," describes how to measure attributes of plastics such as light transmittance, color, and color difference.

JIS L 0809:2001, "Instrumental determination of colour fastness – Change in colour and staining," describes how to determine gray scale values instrumentally.

JIS P 8148:2001, "Paper, board and pulps – Measurement of diffuse blue reflectance factor (ISO brightness)," gives a method for instrumentally determining ISO brightness of pulps, paper, and boards.

JIS P 8150:2004, "Paper and board – Determination of colour (C/2 degree) – Diffuse reflectance method," explains how to instrumentally measure the color of white paper and board.

JIS Z 8105:2000, "Glossary of colour terms," defines terms used in communicating color.

JIS Z 8701:1999, "Colour specification – The CIE 1931 standard colorimetric system and the

CIE 1964 supplementary standard colorimetric system," defines the CIE 2-degree and 10-degree standard observers.

JIS Z 8720:2000, "Standard illuminants and sources for Colorimetry," defines standard illuminants.

JIS Z 8722:2000, "Methods of colour measurement – Reflecting and transmitting objects," gives specifications for measuring color using spectrophotometers and colorimeters.

JIS Z 8729:2004, "Colour specification – CIELAB and CIELUV colour spaces," defines the CIELAB and CIELUV color scales.

JIS Z 8781:1999, "CIE standard colorimetric illuminants," defines the CIE standard illuminants.

JIS Z 8782:1999, "CIE standard colorimetric observers," defines the CIE standard observers.

## SAE J1545

This Society of Automotive Engineers (SAE) publication may be obtained through www.sae.org.

SAE J1545, "Instrumental Color Difference Measurement for Exterior Finishes, Textiles, and Colored Trim," specifies how to measure and communicate color difference information for opaque materials used in the manufacture of vehicles.

## TAPPI Test Methods – Pulp and Paper

Technical Association of the Pulp and Paper Industry (TAPPI) test methods may be obtained through www.tappi.org.

TAPPI T452, "Brightness of Pulp, Paper, and Paperboard (Directional Reflectance at 457 nm)," describes a method for determining, using a 45°/0° instrument, the brightness of white, near-white, and naturally-colored pulp, paper, and paperboard.

TAPPI T515, "Visual Grading and Color Matching of Paper," describes the conditions to be used when visually evaluating paper, including those that contain fluorescent whitening agents.

TAPPI T524, "Color of Paper and Paperboard (45/0, C/2)," describes how to measure the color of paper and paperboard using a colorimeter or spectrophotometer with 45°/0° geometry.

TAPPI T527, "Color of Paper and Paperboard (d/0, C/2)," describes how to measure the color of paper and paperboard using a colorimeter or spectrophotometer with diffuse geometry.

TAPPI T646, "Brightness of Clay and Other Mineral Pigments (45°/0°)," gives a procedure for measuring, using a 45°/0° instrument, the brightness of clay and pigments compacted into a plaque.

TAPPI T1213, "Optical Measurements Terminology (Related to Appearance Evaluation of Paper)," defines terms such as brightness, whiteness, color, gloss, and opacity.

TAPPI T1215, "The Determination of Instrumental Color Differences," introduces color differences and color difference formulas.

## USP – Pharmaceuticals

U.S. Pharmacopeia (USP) monographs may be purchased through www.usp.org.

USP Monograph 631, "Color and Achromicity," defines color and colorlessness, the visual observing situation, and how color should be evaluated visually.

USP Monograph 1061, "Color – Instrumental Measurement," gives a synopsis of color measurement, provides formulas for the calculation of various color scales, and describes how to instrumentally measure the color of opaque solids and transparent liquids.

# References

[1] Ahn CH, Choi JW, Beaucage G, Nevin JH, Lee JB, Puntambekar A, Lee JY. Disposable Smart Lab on a Chip for Point-of-Care Clinical Diagnostics. Proceedings of the IEEE 2004, 92.

[2] Allen JW, Hassanein T, Bhatia SN. Advances in bioartificial liver devices. Hepatology 2001, 34(3), 447–55.

[3] Amendola V, Pilot R, Frasconi M, Maragò O M, Iatì M A. Surface plasmon resonance in gold nanoparticles: a review. J. Phys. Condens. Matter 2017, 29, 203002.

[4] Andreas B, Breunig I, Buse, K. Modeling of X-ray-Induced Refractive Index Changes in Poly (methyl methacrylate). ChemPhysChem 2005, 6, 1544–53.

[5] Ashfold MNR, Claeyssens F, Fuge GM, Henley SJ. Pulsed laser ablation and deposition of thin films. Chem. Soc. Rev. 2004, 33, 23–31

[6] Ashok PC, Praveen BB, Dholakia K. Optofluidic Raman sensor for simultaneous detection of the toxicity and quality of alcoholic beverages. J. Raman Spectrosc. 2013, DOI 10.1002/jrs.4301.

[7] Assaïd I, Hardy I, Bosc D. Controlled refractive index of photosensitive polymer: towards photo-induced waveguide for near infrared wavelengths. Optics Communications 2002, 214, 171–5.

[8] Aziz MJ. Film growth mechanisms in pulsed laser deposition. Applied Physics A: Materials Science & Processing 2008, 93, 3, 579–87.

[9] Baechi D, Buser R, Dual J. A high density microchannel network with integrated valves and photodiodes. Sensors and Actuators 2002, A 95, 77–83.

[10] Baehr-Jones TW, Hochberg MJ. Polymer Silicon Hybrid Systems: A Platform for Practical Nonlinear Optics. J. Phys. Chem. C 2008, 112, 8085–90.

[11] Baker HJ, Lindsey JR, Weisbroth SH. The Laboratory Rat: Volume 1: Biology and Diseases, chapter Historical Foundations. Academic Press, 1979.

[12] Baldo MA, Holmes RJ, Forrest SR. Prospects for electrically pumped organic lasers. Physical Review B 2002, 66, 035321.

[13] Balslev S, Jorgensen AM, Bilenberg B, Mogensen KB, Snakenborg D, Geschke O, Kutter JP, Kristensen A. Lab-on-a-chip with integrated optical transducers. Lab Chip 2006, 6, 213–7.

[14] Banker G, Goslin K. Culturing Nerve Cells, 2nd Edition, chapter Rat Hippocampal Neurons in Low-Density Culture. MIT Press, 1998.

[15] Bargiel S, Rabenorosoa K, Clévy C, Gorecki C., Lutz P. Towards micro-assembly of hybrid MOEMS components on a reconfigurable silicon free-space micro-optical bench. J. Micromech. Microeng. 2010, 20(4), 045012–24.

[16] Barker GM. Molluscs as Crop Pests, chapter Integrated Management of Cantareus asperses (Muller) (Helicidae) as a Pest of Citrus in California, CABI, 2002.

[17] Bathia SN, Ingber DE. Microfluidic organs-on-chips. Nature Biotechnology 2014, 32, 760–72.

[18] Becker R. Theorie der Elektrizitaet. Teubner, Stuttgart 1969.

[19] Bergveld P. ISFET, Theory and Practice. IEEE Sensor Conference 2003.

[20] Bergveld P. Thirty years of ISFETOLOGY, What happened in the past 30 years and what may happen in the next 30 years. Sensors and Actuators 2003, B 88, 1–20.

[21] Bhatia SN, Balis UJ, Yarmush ML, Toner M. Effect of cell– cell interactions in preservation of cellular phenotype: cocultivation of hepatocytes and nonparenchymal cells. FASEB 1999, 13, 1883.

[22] Boot BL. Polymers for optical interconnects. Proc. OFC San Diego, California, 1995, 252–75.

[23] Bosc D, Grosso P, Hardy I, Assaïd I, Batte T, Haesaert S, Vinouze B. High refractive index contrast in a photosensitive polymer and waveguide photo-printing demonstration. Optics Communications 2004, 235, 281–4.

[24] Boustheena A, Homburga FGA, Dietzel A. A modular microvalve suitable for lab on a foil: Manufacturing and assembly concepts. http://dx.doi.org/10.1016/j.mee.2012.07.057

https://doi.org/10.1515/9783110546156-012

[25]   Boyd RW, Heebner JE. Sensitive disk resonator photonic biosensor. Applied Optics 2001, 40, 5742–7.

[26]   Brammer M, Megnin C, Parvanta T, Siegfarth M, Mappes T, Rabus DG. A modular microfluidic backplane for control and interconnection of Optofluidic devices. IEEE PS Winter Topicals 2011, 101–2.

[27]   Brammer M, Megnin C, Siegfarth M, Sobich, S, Rabus DG, Mappes T. Optofluidic backplane as a platform for modular system design. Proc. SPIE 8251 2012, Microfluidics, BioMEMS, and Medical Microsystems X, 825100.

[28]   Branch DW, Corey JM, Weyhenmeyer JA, Brewer GJ, Wheeler BC. Microstamp patterns of biomolecules for high-resolution neuronal networks. Med. Biol. Eng. Comput. 1998, 36, 135.

[29]   Brandenburg A, Krauter R, Kuenzel C, Stefan M, Schulte H. Interferometric sensor for detection of surface-bound bioreactions. Applied Optics 2000, 39, 6396–405.

[30]   Brennan D, Justice J, Corbett B, McCarthy T, Galvin P. Emerging Optofluidic technologies for point-of-care genetic analysis systems: a review, Anal Bioanal Chem, DOI 10.1007/s00216-009-2826-5

[31]   Brewer, GJ, Prince PJ. Viable cultured neurons in ambient carbon dioxide and hibernation storage for a month. Neuroreport 1996, 7, 1509–12.

[32]   Bruendel M, Henzi P, Rabus DG, Ichihashi Y, Mohr J. Herstellung integrierter polymerer Wellenleiter durch UV-induzierte Brechzahländerung. Invited Paper, Workshop "Optik in der Rechentechnik", 27 Oktober 2006, Siegen, Germany.

[33]   Bruendel M, Henzi P, Rabus DG, Worgull M. Hot embossing of optical waveguides. Proc. 50. Internationales Wissenschaftliches Kolloquium 2005, Mechanical Engineering from Macro to Nano, Ilmenau, Germany, 19. - 23. September 2005, ISBN 3-932633-98-9, 351–2.

[34]   Bruendel M, Henzi P, Rabus DG. Replication of Optical Rib Waveguide Structures using Nickel Shims. Proc. SPIE 2006, 6185, 618508.

[35]   Bruendel M, Ichihashi Y, Mohr J, Punke M, Rabus DG, Worgull M, Saile V. Photonic Integrated Circuits fabricated by Deep UV and Hot Embossing. Invited Paper, IEEE LEOS Summer Topicals, 23–5 July 2007, Portland, USA.

[36]   Bruendel M, Rabus DG. 1 x 2 and 1 x 3 Multimode Interference Couplers Fabricated by Hot Embossing and DUV-induced Modification of Polymers. 19th IEEE/LEOS Annual Meeting, 29 October - 2 November 2006, Montreal, QC, Canada, Paper TuS 5.

[37]   Bruggeman DAG. Berechnung verschiedener physikalischer Konstanten von hetarogenen Substanzen. Ann. Phys. Leipzig 1935, 24, 636–79.

[38]   Brynda E, Houska M, Brandenburg A, Wikerstal A. Optical biosensors for real-time measurement of analytes in blood plasma. Biosensors and Bioelectronics 2002, 17, 665–75.

[39]   Buck J. Fundamentals of optical fibers. Wiley, 2004.

[40]   Buestrich R, Kahlenberg F, Popall FM, Dannberg P, Müller-Fiedler R, Rösch O. ORMOCER®s for Optical Interconnection Technology. J Sol-Gel Sci. Tech. 2001, 20, 181–6.

[41]   Burns MA, et. al. An Integrated Nanoliter DNA Analysis Device. Science 1998, 282, 484–7.

[42]   Chandross EA, Pryde CA, Tomlinson WJ, Weber HP. Photolocking – a technique for fabricating optical waveguide circuits. Applied Physics Letters 1974, 24, 72–4.

[43]   Chang JC, Brewer GJ, Wheeler BC. Micro-electrode array recordings of patterned hippocampal neurons for four weeks. Biomed. Microdev. 2000, 2, 245–53.

[44]   Chao CY, Guo LJ. Biochemical sensors based on polymer microrings with sharp asymmetrical resonance. Applied Physics Letters 2003, 83, 1527–9.

[45]   Charlesby A. Use of High Energy Radiation for Crosslinking and Degradation, Radiation Physics and Chemistry 1977, 9, 17–29.

[46]   Chen D, Fetterman HR, Chen A, Steier WH, Dalton LR, Wang W, Shi Y. Demonstration of 110 GHz electro-optic polymer modulators. Applied Physical Letters 1997, 70, 3335–7.

[47]  Chia-Wen, Tsao. Polymer Microfluidics: Simple, Low-Cost Fabrication Process Bridging Academic Lab Research to Commercialized Production Micromachines 2016, 7, 225–36.

[48]  Cho SY, Seo SW, Jokerst NM. Planar lightwave circuit balanced optical fanout using a thin film photodetector array embedded in a polymer MMI coupler. Lasers and Electro-Optics Society 2003, 1, 65–6.

[49]  Choi JO, Moore JA, Corelli JC, Silverman JP. Degradation of poly(methylmethacrylate by deep ultraviolet, x-ray, electron beam, and proton beam irradiation. J. Vac. Sci. Technol. B 1988, 6, 2286–9.

[50]  Chon CH, Li D. Quantum Dot. In: Li D. (eds) Encyclopedia of Microfluidics and Nanofluidics. Springer, Boston, MA 2008. doi.org/10.1007/978-0-387-48998-8.

[51]  Choy KL. Chemical vapour deposition of coatings. Progress in Materials Science 2003, 48, 57–170.

[52]  Choudhury B, Shinar R, Shinar J. Glucose biosensors based on organic light-emitting devices structurally integrated with a luminescent sensing element. Journal of Applied Physics 2004, 96, 5, 2949–54.

[53]  Clemens J T. Silicon Microelectronics Technology. Bell Labs Technical Journal 1997, 2, 76–102.

[54]  Colin S, Pardo F, Teissier R, Pelouard JL. Efficient light absorption in metal–semiconductor–metal nanostructures. Applied Physics Letters 2004, 85, 194.

[55]  Craighead HG, Turner SW, Davis RC, James C, Perez AM, St. John PM, Isaacson MS, Kam L, Shain W, Turner JN. Chemical and topographical surface modification for the control of central nervous system cell adhesion. Biomedical Microdevices 1998, 1, 49–64.

[56]  Cristea D, Craciunoiu F, Modreanu M, Caldararu M, Cernica I. Photonic circuits integrated with CMOS compatible photodetectors. Optical Materials 2001, 17, 201–05.

[57]  Curtis ASG, Forrester JV, McInnes C, Lawrie F. Adhesion of Cells to Polystyrene Surfaces. J. Cell Biology 1983, 97, 1500–6.

[58]  Dablé PJMR, Dakoury GG, Yao B, Gossan A, Ezouah CA. Theoretical Thermodynamic Study of Crude Extraction by Isentropic Relaxation of Natural Gas. The Open Chemical Engineering Journal, 2008, 2, 125–32.

[59]  Dalton L, Harper A, Ren A, Wang F, Todorova G, Chen J, Zhang C, Lee M. Polymeric Electro-optic Modulators: From Chromophore Design to Integration with Semiconductor Very Large Scale Integration Electronics and Silica Fiber Optics. Ind. Eng. Chem. Res. 1999, 38, 8–33.

[60]  Danks AE, Hall SR, Schnepp Z. The evolution of 'sol–gel' chemistry as a technique for materials synthesis, Mater. Horiz. 2016, 3, 91.

[61]  Darraud-Taupiac C, Decossas JL, Vareille JC. Study of free radicals in poly(diethyleneglycol bis (allyl carbonate)) irradiated by rays. Polymer 1995, 36, 3251–4.

[62]  Del Campo A, Arzt E. Fabrication Approaches for Generating Complex Micro- and Nanopatterns on Polymeric Surfaces. Chem. Rev. 2008, 108, 911–45.

[63]  DeLouise LA, Miller BL. Trends in Porous Silicon Biomedical Devices - Tuning Microstructure and Performance Trade-offs in Optical Biosensors. Proc. SPIE 2004, 5357, 111–25.

[64]  Doremus RH. Diffusion in Noncrystalline Silicates in Modern Aspects of the Vitreous State, Vol. 2, by J. D. Mackenzie, ed., Butterworth, Washington, D.C. (1962).

[65]  Doremus RH. Exchange and diffusion of ions in glass. Jour. Phys. Chem 1964, 68, 8, 2212–8.

[66]  Detrait E, Lhoest JB, Knoops B, Bertrand P, van den Bosch de Aguilar P. Orientation of cell adhesion and growth on patterned heterogeneous polystyrene surface. J. Neurosci. Methods 1998, 84, 193–204.

[67]  Doumas BT, Watson WA, Biggs HG. Albumin standards and the measurement of serum albumin with bromcresol green. Clin. Chem. 1971 Acta 31, 87–96.

[68]  Eddington DT, Beebe DJ. Flow control with hydrogels. Advanced Drug Delivery Reviews 2004, 56, 199–210.

[69]  Ekins RP. The precision profile control: its use in assays design, assessment and quality control. In Immunoassays for clinical chemistry. 1983, 2. Edition, Churchill Livingstone, Edingburgh 76–105.

[70]  Eldada L, Shacklette LW. Advances in Polymer Integrated Optics. IEEE J. Select. Topics Quantum Electron. 2000, 6, 54–68.

[71]  Eldada L, Yin S, Poga C, Glass C, Blomquist R, Nonvood RA. Integrated multichannel OADMs using polymer Bragg grating MZIs. IEEE Photonics Technology Letters 1998, 10, 1416–8.

[72]  Eldada L. Advances in polymeric integrated optical componentry. Proc. IPR ITuH1, 2001.

[73]  Fair RB. Digital microfluidics: Is a true lab-on-a-chip possible?. Microfluid. Nanofluid. 2007, 3(3), 245–81.

[74]  Famento I. The History of Contact Lenses. 2008. (Accessed October 1, 2011 at http://www.xtimeline.com/timeline/The-History-of-Contact-Lenses)

[75]  Fan X, Yun SH. The potential of Optofluidic biolasers. Nature Methods 2014, 11, 141–47.

[76]  Fang Y, Ferrie AM, Fontaine NH, Mauro J, Balakrishnan J. Resonant waveguide grating biosensor for living cell sensing. Biophysical Journal 2006, 91, 1925–40.

[77]  Feuchter T, Thirstrup C. High precision planar waveguide propagation loss measurement technique using a Fabry-Perot cavity. IEEE Photon. Technol. Lett. 1994, 6, 1244–7.

[78]  Fiorini GS, Chiu DT. Disposable microfluidic devices: fabrication, function, and application. BioTechniques 2005, 38, 429–46.

[79]  Fleger M, Siepe D, Neyer A. Microfabricated polymer analysis chip for optical detection. IEE Proc.-Nanobiotechnol. 2004, 151, 159–61.

[80]  Fleger M, Neyer A. PDMS microfluidic chip with integrated waveguides for optical detection. Microelectronic Engineering 2006, 83, 1291–3.

[81]  Fraden J. Handbook of modern sensors: Physics, Designs, and Applications. Springer, 2004.

[82]  Frank WFX, Schösser A, Stelmaszyk A, Schulz J. Ionizing Radiation for Fabrication of optical Waveguides in Polymers. SPIE Critical Review Conference 1996, 63, 65–83.

[83]  Gardeniers JGE, van den Berg A. Lab-on-a-chip systems for biomedical and environmental monitoring. Analytical and Bioanalytical Chemistry 2004, 378, 1700–3.

[84]  Ge J, Lee H, He L, Kim J, Lu Z, Kim H, Goebl J, Kwon S, Yin Y. Magnetochromatic Microspheres: Rotating Photonic Crystals. Journal of the American Chemical Society 2009, 131, 43, 15687–94.

[85]  Gee KR, Brown KA, Chen WNU, Bishop-Stewart J, Gray D, Johnson I. Chemical and physiological characterization of fluo-4 Ca2+-indicator dyes. Cell Calcium 2000, 27, 97–106.

[86]  Gibas I, Janik H. Review: Synthetic Polymer Hydrogels for Biomedical Applications. Chemistry and Chemical Technology 2010, 4, 4, 297–304.

[87]  Grassie N. Polymer degradation and stabilization. Cambridge University Press, 1988.

[88]  Green M A, Keevers M J. Optical properties of Intrinsic silicon at 300K. Progress in Photovoltaics: research ad applications 1995 3, 189–92.

[89]  Green M A. Self-consistent optical parameters of intrinsic silicon at 300 K including tempera-ture coefficients. Solar Energy Materials & Solar Cells 2008, 92, 1305–10.

[90]  Grund T, Megnin C, Barth J, Kohl M. Batch Fabrication of Shape Memory Actuated Polymer Microvalves by Transfer Bonding Techniques. J. Micro. Elect. Pack. 2007, 6(4), 219–27.

[91]  Haeckel R, Sonntag O. Validation of quantitative analytical procedures in laboratory medicine. J Lab Med 2012, 36 (2): 111–8

[92]  Haeckel R. Evaluation Methods in Laboratory Medicine. VCH 1993, 243–6.

[93]  Han J, Seo BJ, Han Y, Jalali B, Fetterman HR. Reduction of fiber chromatic dispersion effects in fiber-wireless and photonic time-stretching system using polymer modulators. IEEE Lightwave Technology 2003, 21, 1504–9.

[94]  Heckele M, Schomburg WK. Review on micro molding of thermoplastic polymers. Journal of Micromech. and Microeng. 2004, 14, R1–R14.

[95]   Heliotis G, et al. Emission Characteristics and Performance Comparison of Polyfluorene Lasers with One and Two-Dimensional Distributed Feedback. Advanced Functional Materials 2004, 14(1), 91–7.

[96]   Heliotis G, et al. Two-dimensional distributed feedback lasers using a broadband, red polyfluorene gain medium. Journal of Applied Physics 2004, 96, 6959–65.

[97]   Helmers H, Greco P, Rustad R, Kherrat R, Bouvier G, Benech P. Performance of a compact, hybrid optical evanescent-wave sensor for chemical and biological applications. Applied Optics 1996, 35, 676–80.

[98]   Henry F, Marchal P, Bouillard J, Vignes A, Dufaud O, Perrin L. The Effect of Agglomeration on the Emission of Particles from Nanopowders Flow. Chemical Engineering Transactions 2013, 31.

[99]   Henzi P, Bade K, Rabus DG, Mohr J. Modification of Polymethylmethacrylate by Deep UV Radiation and Bromination for Photonic Applications. J. Vac. Sci. Technol. B 2006, 24, 1755–61.

[100]  Henzi P, Rabus DG, Ichihashi Y, Bruendel M, Mohr J. Photonic Integrated Polymer Components and Circuits by UV-Induced Refractive Index Modification. Proc. SPIE 2006, 6185, 618502.

[101]  Henzi P, Rabus DG, Wallrabe U, Mohr J. Fabrication of Photonic Integrated Circuits by DUV-induced Modification of Polymers. Proc. SPIE 2004, 5451, 24–31.

[102]  Henzi P, Rabus DG, Wallrabe U, Mohr J. Low Cost Singlemode Waveguide Fabrication Allowing Passive Fibre Coupling using LIGA and UV flood exposure. Proc. SPIE 2004, 5454, 64–74.

[103]  Henzi P, Rabus DG, Wallrabe U, Mohr J. UV-induced Modification of Dielectric Properties of Polymers for Fabrication of Passive Planar Ligthwave Circuits. 17th IEEE/LEOS Annual Meeting, 7–11 November, 2004, Rio Grande, Puerto Rico, 792–3.

[104]  Hikita M, Satoru T, Enbutsu K, Ooba N, Yoshida R, Usai M, Yoshida T, Imamura S. Polymeric Optical Waveguide Films for Short-Distance Optical Interconnects. IEEE J. Select. Top. Quantum Electron. 1999, 5, 1237–42.

[105]  Hillborg H, Ankner JF, Gedde UW, Smith GD, Yasuda HK, Wikström K. Crosslinked polydimethylsiloxane exposed to oxygen plasma studied by neutron reflectometry and other surface specific techniques. Polymer 2000, 41, 6851–63.

[106]  Hofmann O, Wang X, Demello JC, Bradley DD, Demello AJ. Towards microalbuminuria determination on a dispoable diagnostic microchip with integrated fluorescence detection based on thin-film organic light emitting diodes. Lab on a Chip 2005, 5, 863–8.

[107]  Hong W, Woo HJ, Choi HW, Kim YS, Kim GD. Optical property modification of PMMA by ion beam implantation. Applied Surface Science 2001, 169–170, 428–32.

[108]  Höök F, Kasemo B, Nylander T, Fant C, Sott K, Elwing H. Variations in coupled water, viscoelastic properties, and film thickness of a Mefp-1 protein film during adsorption and cross-linking: a quartz crystal microbalance with dissipation monitoring, ellipsometry, and surface plasmon resonance study. Anal Chem. 2001, 15, 73, 24, 5796–804.

[109]  Hornak LA, Wedman TW, Kwock EW. Polyalkylsilyne photodefined thin-film optical waveguides. J. Appl. Phys. 1990, 67, 2235–9.

[110]  Horvath R, Lindvold LR, Larsen NB. Reverse-symmetry waveguides: theory and fabrication. Appl. Phys. B 2002, 74, 383–93.

[111]  Horvath R, Pedersen HC, Skivesen N, Selmeczi D, Larsen NB. Monitoring of living cell attachment and spreading using reverse symmetry waveguide sensing. Applied Physics Letters 2005, 86, 071101.

[112]  Horvath R, Pedersen HC, Skivesen N, Svanberg C, Larsen NB. Fabrication of reverse symmetry polymer waveguide sensor chips on nanoporous substrates using dip-floating. Micromech. Microeng. 2005, 15, 1260–4.

[113]  Hou X, Zhang Y S, De Santiago G T, Alvarez M M, Ribas J, Jonas S J, Weiss P S, Andrews A M, Aizenberg J and Khademhosseini A. Interplay between materials and microfluidics. Nature Reviews Materials 2017, 2, 17016.

[114] http://www.engineeringtoolbox.com/nozzles-d_1041.html

[115] http://www.fsl.orst.edu/geowater/FX3/help/8_Hydraulic_Reference/
Froude_Number_and_Flow_States.htm

[116] Huang JZ, Hu MH, Fujita J, Scarmozzino R, Osgood RM. High-performance metal-clad multimode interference devices for low-index-contrast material systems. IEEE Photonics Technology Letters 1998, 10, 561–3.

[117] Hunsperger RG. Integrated Optics, Springer 2009, ISBN: 978-0-387-89774-5

[118] Hyun J, Barletta P, Koh K, Yoo S, Oh J, Aspnes DE, Cuomo JJ. Effect of Ar+ Ion Beam in the Process of Plasma Surface Modification of PET Films. Journal of Applied Polymer Science 2000, 77, 1679–83.

[119] Ichihashi Y, Cristea D, Kusko M, Rabus DG, Mappes T, Mohr J. Integrated Silicon Photodiodes with Polymer (PMMA) Waveguides for Optical Interconnections and Sensing Applications. EOS Topical Meeting on Micro-Optics, Diffractive Optics and Optical MEMS - TOM 4, October 16th 2006 - October 19th, Paris, France.

[120] Ichihashi Y, Henzi P, Bruendel M, Mohr J, Rabus DG. Polymer Waveguides from Alicyclic Methacrylate Copolymer Fabricated by Deep-UV Exposure. Optics Letters 2007, 32, 379–381.

[121] Ichihashi Y, Henzi P, Bruendel M, Rabus DG, Mohr J. Influence of the Exposure Spectra on the Optical Properties of Alicyclic Methacrylate Copolymers Waveguides fabricated by Deep UV Exposure. Jpn. J. Appl. Phys. 2006, 45, 6654–7.

[122] Ichihashi Y, Henzi P, Bruendel M, Rabus DG, Mohr J. Material Investigation of Alicyclic Methacrylate Copolymer for Polymer Waveguide Fabrication. Jpn. J. Appl. Phys. (JJAP) 2006, 45, 2572–5.

[123] Ichihashi Y, Henzi P, Bruendel M, Rabus DG, Welle A, Mohr J. Novel Polymer Waveguides Consisting of an Alicyclic Methacrylate Copolymer. Proc. SPIE 2006, 6183, 618307.

[124] Ito Y. Regulation of cell functions by micropattern-immobilized biosignal molecules. Nanotechnol. 1998, 9, 200–4.

[125] IUPAC. A definition of biosensors. Electrochemical biosensors: proposed definitions and classification. Biosens. Bioelectron. 1996, 11 (4), i.

[126] Izawa T, Sudo S. Optical Fibers. Springer, 1986.

[127] James CD, Davis R, Meyer M, Turner A, Turner S, Withers G, Kam L, Banker G, Craighead H, Isaacson MS, Turner J, Shain W. Aligned microcontact printing of micrometer-scale poly-L-lysine structures for controlled growth of cultured neurons on planar microelectrode arrays. IEEE Transactions on Biomedical Engineering 2000, 47, 17–21.

[128] Janahmadi M, Malmierca MS, Hearne PG, Green GG, Sanders DJ. Morphological and electro-physiological features of F76 and D1 neurones of the sub-oesophageal ganglia of Helix aspersa in vitro and in culture. Anatomy and Embryology, 1999, 199(6),563–72.

[129] Jellinek HHG. Aspects of Degradation and Stabilization of Polymers. Elsevier Scientific Publishing Company, Amsterdam, 1978.

[130] Jimenez-Jorquera C, Orozco J, Baldi A. ISFET Based Microsensors for Environmental Monitoring. Sensors 2010, 10, 61–83, doi:10.3390/s100100061

[131] Kaibara M, Iwata H, Wada H, Kawamoto Y, Iwaki M, Suzuki Y. Promotion and control of selective adhesion and proliferation of endothelial cells on polymer surface by carbon deposition. J. Biomed. Mat. Res. 1996, 31, 429–35.

[132] Kamila S. Introduction, classification and applications of smart materials: An overview. Am. J. Applied Sci. 2013, 10, 876–80.

[133] Kaminow IP, Stulz LW. Loss in cleaved Ti-diffused LiNbO3 waveguides. Appl. Phys. Lett. 1978, 33, 62–4.

[134] Karpukhina N, Hilla R G and Law R V. Crystallisation in oxide glasses – a tutorial review. Chem. Soc. Rev. 2014, 43, 2174.

[135] Kawamura Y, et al. Ultraviolet amplified spontaneous emission from thin films of 4,4'-bis (9-carbazolyl)-2,2'-biphenyl and the derivatives. Applied Physics Letters 2004, 84(15), 2724–6.

[136] Keil N, Strebel B, Yao HH, Zawadzki C, Hwang WY. Optical polymer waveguide devices and their applications to integrated optics and optical signal processing. SPIE 1774, Nonconducting Photopolymers and Applications 1993, 130–41.

[137] Keil N, Yao HH, Zawadzki C, Bauer J, Bauer M, Dreyer C, Schneider J. Athermal all-polymer arrayed-waveguide grating multiplexer. Electronics Letters 2001, 37, 579–58.

[138] Keil N, Yao HH, Zawadzki C. 2 x 2 digital optical switch realized by polymer waveguide technology. Electron. Lett 1996, 32, 1470–1.

[139] Keller BK, DeGrandpre MD, Palmer CP. Waveguiding properties of fiber-optic capillaries for chemical sensing applications. Sensors and Actuators B 2007, 125, 360–71.

[140] Kelly PJ, Arnell RD. Magnetron sputtering: a review of recent developments and applications. Vacuum 2000, 56, 159–72.

[141] Kim JH, Dudley BW, Moyer PJ. Experimental Demonstration of Replicated Multimode Interferometer Power Splitter in Zr-Doped Sol–Gel. IEEE J. Lightwave Technol. 2006, 24, 612–6.

[142] Klein GA. Industrial Color Physics. Springer, 2010.

[143] Ko JS, Yoon HC, Yang H, Pyo HB, Chung KH, Kima SJ, Kim YT. A polymer-based microfluidic device for immunosensing biochips. Lab Chip 2003, 3, 106–13.

[144] Kobayashi J, Matsuura T, Sasaki S. Singlemode optical waveguides fabricated from fluorinated polymides. Appl. Optics 1974, 37, 1032–7.

[145] Komarov FF, Leontyev AV, Grigoryev VV, Kamishan MA. Ion implantation for local change of the optical constants of polymer films. Nuclear Instruments and Methods in Physics Research B 2002, 191, 728–32

[146] Koo J, Smith P, Williams R, Grossel M, Whitcombe M. Synthesis and characterization of methacrylate-based copolymers for integrated optical applications. Chemistry of Materials 2002, 14, 5030–6.

[147] Koo J, Williams R, Gawith C, Watts S, Emmerson G, Albanis V, Smith P, Grossel M. UV written waveguide devices using crosslinkable PMMA-based copolymers. Elec. Lett. 2003, 39, 394–5.

[148] Kozlov VG, et al. Structures for Organic Diode Lasers and Optical Properties of Organic Semiconductors Under Intense Optical and Electrical Excitations. IEEE Journal of Quantum Electronics 2000, 36, 18–26.

[149] Kragl H, Hohmann R, Marheine C, Pott W, Pompe G. Low cost monomode, integrated optics polymeric components with passive fiber-chip coupling. Electronics Letters 1997, 33, 2036–7.

[150] Krevelen DW. Properties of Polymers. Elsevier-Verlag, Amsterdam-Oxford-New York-Tokyo, 1990.

[151] Krioukov E, Klunder DJW, Driessen A, Greve J, Otto C. Sensor based on an integrated optical microcavity. Optics Letters 2002, 27, 512–4.

[152] Ksendzov A, Homer ML, Manfreda AM. Integrated optics ring-resonator chemical sensor with polymer transduction layer. Electronics Letters 2004, 40, 63–5.

[153] Kulish JR, Franke H, Singh A, Lessard RA, Knystautas EJ. Ion implantation, a method for fabricating light guides in polymers. J. Appl. Phys. 1988, 63, 8, 2517–21.

[154] Kulish JR, Franke H. Fabrication of Lightguides in in-diffused bulk PMMA. Appl. Phys. A 1990, 50, 425–30.

[155] Kumar A, Srivastava A, Galaev IY, Mattiasson B. Smart polymers: Physical forms and bioengineering applications. Prog. Polym. Sci. 2007, 32, 1205–37.

[156] Lambert CM. Chromogenic Smart Materials. Materials Today 2004, 7, 3, 28–35.

[157] Lampert M. Chromogenic smart materials. Materials Today 2004, 7, 3, 28–35, ISSN 1369-7021, http://dx.doi.org/10.1016/S1369-7021(04)00123-3.

[158] Langer G, Kavc T, Kern W, Kranzelbinder G, Toussaere E. Refractive Index Changes in Polymers Induced by Deep UV Irradiation and Subsequent Gas Phase Modification. Macromol. Chem. Phys. 2001, 202, 2359–3467.

[159] Last A, Mohr J. Fehllicht in LIGA-Mikrospektrometern. Forschungszentrum Karlsruhe GmbH, Wissenschaftlicher Bericht Nr. FZKA-6585, ISSN 0947–8620, 2003.

[160] Lendlein A, Kelch S. Shape memory polymers. Angewandte Chemie International Edition 2002, 41, 12, 2034–57.

[161] Li L, Lee L. Photopolymerization of HEMA/DEGDMA hydrogels in solution. Polymer 2005, 46, 11540–7.

[162] Lien V, Berdichevsky Y, Lo YH. A prealigned process of integrating optical waveguides with microfluidic devices. IEEE Photonics Technology Letters 2004, 16, 1525–7.

[163] Lu JG. Fundamental Properties of Zinc Oxide Nanowires, in Encyclopedia of Nanotechnology 2012, 919–27.

[164] Lucas P, Wilhelm AA, Riley MR, DeRosab DL, Collier JM. Cell-based bio-optical sensors using chalcogenide fibers. Proc. of SPIE 2006, 6083, 60830W.

[165] Lundin D, Pedersen H. High power pulsed plasma enhanced chemical vapor deposition: a brief overview of general concepts and early results, Physics Procedia 2013, 46, 3–11.

[166] Ma H, Jen AKY, Dalton LR. Polymer based optical waveguides: materials, processing, and devices. Adv. Mater. 2002, 14, 1339–65.

[167] Manz A, Harrison D, Fettinger JC, Verpoorte EM, Lüdi H, Widmer HM. Integrated electroosmotic pumps and flow manifolds for total chemical analysis systems. Transducers 1991, 939–41.

[168] Mappes T, Vannahme C, Klinkhammer S, Bog U, Schelb M, Grossmann T, Hauser M, Kalt H, Lemmer U. Integrated photonic lab-on-chip systems for biomedical applications. Proc. SPIE 2010, 7716, 77160R1–77160R10.

[169] Martınez-Manez R, Soto J, Garcıa-Breijo E, Gil L, Ibanez J, Gade E. A multisensor in thick-film technology for water quality control. Sensors and Actuators A 2005, 120, 589–95.

[170] Martinez, AW, Phillips ST, Butte MJ, Whitesides GM. Patterned Paper as a Platform for Inexpensive, Low-Volume, Portable Bioassays. Angew. Chem. Int. Ed. 2007, 46, 1318–20.

[171] Martinez AW, Phillips ST, Whitesides GM. Three-Dimensional Microfluidic Devices Fabricated in Layered Paper and Tape. Proc. Natl. Acad. Sci. USA. 2008, 105, 19606–11.

[172] Marx KA, Zhou T, Montrone A, Schulze H, Braunhut SJ. A quartz crystal microbalance cell biosensor: detection of microtubule alterations in living cells at nM nocodazole concentrations. Biosens. & Bioelectron. 2001, 16, 773–82.

[173] Marx KA. Quartz Crystal Microbalance: A Useful Tool for Studying Thin Polymer Films and Complex Biomolecular Systems at the Solution–Surface Interface. Biomacromolecules 2003, 4, 1099–120.

[174] Matsuda T, Sugawara T. Control of cell adhesion, migration, and orientation on photochemically microprocessed surfaces. J. Biomed. Mat. Res. 1996, 32, 165–73.

[175] Matsuzawa M, Kobayashia K, Sugiokab K, Knolla W. A Biocompatible Interface for the Geometrical Guidance of Central Neuronsin Vitro. Colloid Interface Sci. 1998, 202, 213–21.

[176] McPherson RA, Matthew PR. Henry's clinical Diagnosis and Management by Laboratory Method, Saunders Elsevier, 2011, 21. Edition, 252ff.

[177] Megnin C, Barth J, Kohl M. A bistable shape memory microvalve for three-way control. Solid-State Sensors, Actuators and Microsystems Conference (TRANSDUCERS), 2011, 1308–11.

[178] Micoulaut M. Relaxation and physical aging in network glasses: a review. Rep. Prog. Phys. 2016, 79 066504.

[179] Minzioni P, Osellame R, Sada C, et al. Roadmap on optofluidics. Journal of Optics 2017, 19, 9, 093003.

[180] Mogensen KB, El-Ali J, Wolff A, Kutter JP. Integration of polymer waveguides for optical detection in microfabricated chemical analysis systems. Applied Optics 2003, 42, 4072–79.

[181] Moore JA, Choi JO. Degradation of Poly(methylmethacrylate) by Deep UV, X-ray, Electron-Beam, and Proton-Beam Irradiation. Radiation Effects on Polymers. ACS Symposium Series Nr. 475 edited by R. L. Clough and S. W. Shalaby, Chapter 11, 156–192, American Chemical Society, Washington DC, 1991.

[182] Mrksich M. Model organic surfaces for mechanistic studies of cell adhesion. Chem. Soc. Rev. 2000, 29, 267–73.

[183] Mule AV, Joseph PJ, Allen SAB, Kohl PA, Gaylord TK, Meindl JD. Polymer optical interconnect technologies for polylithic gigascale integration. European Solid-State Device Research, 2003, 119–22.

[184] Mutschler T, Kieser B, Frank R, Gauglitz G. Characterization of thin polymer and biopolymer layers by ellipsometry and evanescent field technology. Anal. Bioanal. Chem. 2002, 374, 658–64.

[185] Nakanotani H, et al. Injection and Transport of High Current Density over 1000 A/cm2 in Organic Light Emitting Diodes under Pulse Excitation. Japanese Journal of Applied Physics 2005, 44(6A), 3659–62.

[186] Nesnidal RC, Walker TG. Multilayer dielectric structure for enhancement of evanescent waves. Appl. Opt. 1996, 35, 2226–9.

[187] Nespoli A, Besseghini S, Pittaccio S, Villa E, Viscuso S. The high potential of shape memory alloys in developing miniature mechanical devices: A review on shape memory alloy mini-actuators. Sensors and Actuators A 2010, 158, 149–60.

[188] Oh KW, Ahn CH. A review of microvalves. J. Micromech. Microeng. 2006, 16, R13–R39.

[189] Ohmori Y, Kajii H, Kaneko M, Katsumi Y, Masanori O, Akihiko F, Hikita M, Hisataka T, Taneda T. Realization of polymeric optical integrated devices utilizing organic light-emitting diodes and photodetectors fabricated on a polymeric waveguide. IEEE J. Select. Top. Quantum Electron. 2004, 10, 70–8.

[190] Olabi AG, Grunwald A. Design and application of magnetostrictive materials. Materials & Design 2008, 29, 2, 469–83.

[191] Oliveri RL, Sciuto A, Libertino S, D'Arrigo G, Arnone C. Fabrication and characterization of polymeric optical waveguides using standard silicon processing technology. IEEE/LEOS Workshop on Fibres and Optical Passive Components 2005, 265–70.

[192] Olivia AA, James CD, Kingman CE, Craighead HG, Banker GA. Patterning axonal guidance molecules using a novel strategy for microcontact printing. Neurochemical Research 2003, 28 (11), 1639–48.

[193] Park JU, Lee JH, Paik U, Lu Y, Rogers JA. Nanoscale patterns of oligonucleotides formed by electrohydrodynamic jet printing with applications in biosensing and nanomaterials assembly. Nano Letters 2008, 8(12), 4210–6.

[194] Park OH, Kim SJ, Bae BS. Photochemical reactions in fluorinated sol–gel hybrid materials doped with a photolocking agent for direct micropatterning. J. Mater. Chem. 2004, 14, 1749–53.

[195] Parriaux O, Veldhuis GJ. Normalized analysis for the sensitivity optimization of integrated optical evanescent-wave sensors. IEEE Journal of Lightwave Technology 1998, 16, 573–82.

[196] Peng GD, Xiong Z, Chu PL. Photosensitivity and grating in dye-doped polymer optical fibers. Opt. Fiber Technol. 1999, 5, 242–51.

[197] Petronis S, Stangegaard M, Christensen CBV, Dufva M. Transparent polymeric cell culture chip with integrated temperature control and uniform media perfusion. BioTechniques 2006, 40, 368–76.

[198] Pflumm C, et al. Parametric study of modal gain and threshold power density in electrically pumped single layer organic optical amplifier and laser diode structures. IEEE Journal of Quantum Electronics 2004, 1–29.

[199] Pimpin A, Srituravanich W. Review on Micro - and Nanolithography Techniques and their Applications, Engineering Journal 2012, 16, 1, 1–19

[200] Prokop A. Bioartificial organs in the twenty-first century: nanobiological devices. Ann NY Acad Sci 2001, 944, 472–90.

[201] Psaltis D, Quake SR, Yang C. Developing Optofluidic technology through the fusion of micro-fluidics and optics. Nature 2006, 442(27), 381–6.

[202] Punke M, Mozer S, Stroisch M, Bastian G, Gerken M, Lemmer U, Rabus DG, Henzi P. Organic devices for micro-optical applications. SPIE 2006, 6185, 618505.

[203] Punke M, Mozer S, Stroisch M, Lemmer U, Henzi P, Rabus DG. Coupling of organic semicon-ductor amplified spontaneous emission into polymeric single-mode waveguides patterned by deep-UV irradiation. IEEE Photon. Technol. Lett. 2007, 2, 61–3.

[204] Punke M, Stroisch M, Woggon T, Pütz A, Heinrich MP, Gerken M, Lemmer U, Bruendel M, Rabus DG, Wang J, Weimann T. Fabrication and characterization of organic solid-state lasers using imprint technologies. IEEE LEOS Summer Topicals, 2007.

[205] Punke M, Woggon T, Stroisch M, Heinrich MP, Karnutsch C, Mozer S, Lemmer U, Bruendel M, Rabus DG, Weimann T. All-organic waveguide coupled solid-state distributed feedback laser. CLEO/Europe-IQEC, 2007.

[206] Purves D, Augustine GJ, Fitzpatrick D, Katz LC, LaMantia AS, McNamara JO, Williams SM, editors. Neuroscience, 2nd Edition, chapter The Organization of the Nervous System, Sinauer Associates, Inc., 2001.

[207] Qing DK, Chen G. Enhancement of evanescent waves in waveguides using metamaterials of negative permittivity and permeability. Appl. Phys. Lett. 2004, 84, 669–71.

[208] Quigley GR, Harris RD, Wilkinson JS. Sensitivity enhancement of integrated optical sensors by use of thin high-index films. Appl. Opt. 1999, 38, 6036–9.

[209] Quinti L, Weissleder R, Tung CH. A fluorescent nanosensor for apoptotic cells. Nano Lett. 2006, 6, 488–90.

[210] Rabus DG, Bruendel M, Ichihashi Y, Welle A, Seger RA, Isaacson M. A Bio-Fluidic-Photonic Platform based on Deep UV Modification of Polymers, IEEE J. Select. Topics Quantum Electron. 2007, 13, 214–22.

[211] Rabus DG, DeLouise L, Ichihashi Y. Enhancement of the evanescent field using polymer waveguides fabricated by deep UV exposure on mesoporous silicon. Optics Letters 2007, 32, 2843–5.

[212] Rabus DG, Hamacher M. MMI-Coupled Ring Resonators in GaInAsP-InP. IEEE Photon. Technol. Lett. 2001, 13, 812–4.

[213] Rabus DG, Henzi P, Bruendel M, Hein H, Ichihashi Y, Rogge T, Welle A. Polymer Photonic Integrated Circuits by DUV-induced Modification. SPIE 2006, 6123, 61230I

[214] Rabus DG, Henzi P, Mohr J. Photonic Integrated Circuits by DUV-induced Modification of Polymers. IEEE Photon. Technol. Lett. 2005, 3, 591–3.

[215] Rabus, DG, Welle A, Seger RA, Ichihashi Y, Bruendel M, Hieb J, Isaacson M. Determination of living cell characteristics and behavior using biophotonic methods. SPIE 2006, 6329, 63290H.

[216] Rajnish M, Uyen D, Sirin J, Csaba PK, John M, Joel DK, Kamyar KZ. Serum Albumin as a Predictor of Mortality in Peritoneal Dialysis: Comparisons with Hemodialysis. AJKD Elsevier 2011, 58, 418–28.

[217] Ramsden JJ, Li SY, Heinzle E, Prinosil JE. Optical method, for measurement of number and shape of attached cells in real time. Cytometry 1995, 19, 97–102.

[218] Rapp BE, Carneiro L, Länge K, Rapp M. An indirect microfluidic flow injection analysis (FIA) system allowing diffusion free pumping of liquids by using tetradecane as intermediary liquid. Lab Chip 2009, 9(2), 354–6.

[219] Regener R, Sohler W. Loss in low finesse Ti: LiNbO3 optical waveguide resonators. Appl. Phys. B 1985, 36, 143–7.

[220] Ren K, Zhou J and Wu H. Materials for Microfluidic Chip Fabrication. Accounts Of Chemical Research 2013, 46, 2396–406.

[221] Reufer M, et al. Amplified spontaneous emission in an organic semiconductor multilayer waveguide structure including a highly conductive transparent electrode. Applied Physics Letters 2005, 86, 2211021–3.

[222] Riechel S, et al. Very compact tunable solid-state laser utilizing a thin-film organic semiconductor. Opt. Lett. 2001, 26, 593.

[223] Risler T, Kühn K. Facharzt Nephrologie. URBAN&FISCHER, 2008, 94.

[224] Rissanen A, Akujärvi A, Antila J, Blomberg M, Saari H. Moems miniature spectrometers using tuneable fabry-perot interferometers. J. Micro/Nanolith. MEMS MOEMS. 0001;11(2):023003-1-023003-6. doi:10.1117/1.JMM.11.2.023003.

[225] Rosenberger M, Hartlaub N, Koller G, Belle S, Schmauss B, Hellmann R. Polymer planar Bragg grating for sensing applications. Proc. SPIE. 2013, 8774, 87741P.

[226] Rozga J, Holzman MD, Ro MS, Griffin DW, Neuzil DF, Giorgio T, Moscioni AD, Demetriou AA. Development of a hybrid bioartificial liver. Annals of surgery 1993, 217, 502–9.

[227] Rück DM, Brunner S, Tinscher K, Frank WFX. Production of buried waveguides in PMMA by high energy ion implantation. Nuclear Instruments and Methods in Physics Research B 1995, 106, 447–51.

[228] Sada C, Argiolas N, Bazzan M, Mazzoldi P. Model of the erbium ion exchange process in lithium niobate crystals. Physical Review B 2004, 69, 144120

[229] Mazzoldi P, Carturan S, Quaranta A, Sada C, Sglavo VM, Ion exchange process: History, evolution and applications. Rivista Del Nuovo Cimento 2013, 36, 9

[230] Scherf U, et al. Conjugated polymers: lasing and stimulated emission. Current Opinion in Solid State & Materials Science 2001, 5, 143–54.

[231] Schneider D, et al. Deep blue widely tunable organic solid-state laser based on a spirobi-fluorene derivative. Applied Physics Letters 2004, 84(23),4693–5.

[232] Schneider F, Draheim J, Kamberger R, Wallrabe U. Process and material properties of poly-dimethylsiloxane (PDMS) for Optical MEMS. Sensors and Actuators A: Physical 2009, 151, 95–9

[233] Schösser A, Knoedler B, Tschudi TT, Frank WFX, Stelmaszyk A, Muschert D, Rueck DM, Brunner S, Pozzi F, Morasca S, De Bernardi CS. Optical Components in Polymers. SPIE Proceedings 1995, 2540, 110.

[234] Schösser A, Tschudi TT, Frank WFX, Pozzi F. Spectroscopic Study of Surface Effects in Polymer Waveguides Generated by Ionizing Radiation Related to Guiding Properties. SPIE Proceedings 1996, 2851.

[235] Schösser A, Tschudi TT, Frank WFX. Fiber chip coupling and refractive index profile of UV-generated waveguides in PMMA. SPIE Proceedings 1997, 3135, 144–51.

[236] Schvartzman M, Mathur A, Hone J, Jahnes C, Wind SJ. Plasma fluorination of carbon-based materials for imprint and molding lithographic applications. Applied Physics Letters 2000, 93, 153105-1-153105-3.

[237] Schwartzkroin PA, Prince DA. Changes in excitatory and inhibitory synaptic potentials leading to epileptic activity. Brain Research. 1980, 183(1), 61–76.

[238] Seeboth D, Loetzsch R, Ruhmann I. Piezochromic Polymer Materials Displaying Pressure Changes in Bar-Ranges, American Journal of Materials Science 2011, 1(2), 139–42.

[239] Seyler J. Assignment of the Glass Transition, Edizione 1249,

[240] Shapley JDL, Barrow DA. Novel patterning method for the electrochemical production of etched silicon. Thin Solid Films 2001, 388, 134–7.

[241] Shelby JE. Introduction to Glass Science and Technology: Edition 2. 2005 ED RSC http://dx.doi.org/10.1039/9781847551160.

[242] Shin J, Braun PV, Lee W. Fast response photonic crystal pH sensor based on template photo-polymerized hydrogel inverse opal. Sensors and Actuators B 2010, 150, 183–90.

[243] Spehr T, et al. Organic solid-state ultraviolet-laser based on spiro-terphenyl. Applied Physics Letters 2005, 87, 1611031–3.

[244] Strain AJ, Neuberger JM. A Bioartificial Liver – State of the Art. Science 2002, 295, 1005.

[245] Stankova NE. Optical properties of polydimethylsiloxane (PDMS). Applied surface Science 2016, 374, 96–103.

[246] Stjernstrom M, Roeraade J. Method for fabrication of microfluidic systems in glass. J. Micromech. Microeng. 1998, 8, 33–8.

[247] Suh KY, Langer R, Lahann J. Fabrication of elastomeric stamps with polymer-reinforced side-walls via chemically selective vapor deposition polymerization of poly(p-xylylene). Applied Physics Letters 2003, 83, 4250–2.

[248] Tarsa PB, Wist AD, Rabinowitz P, Lehmann KK. Single-cell detection by cavity ring-down spectroscopy. Applied Physics Letters 2004, 85, 4523–5.

[249] Thomas, Lothar. Labor und Diagnose. Dade Behring, 1998, 667–9.

[250] Tiwari SK, Woodruff ML. Helix aspersa neurons maintain vigorous electrical activity when co-cultured with intact H. aspersa ganglia. Comparative Biochemistryand Physiology. C Comparative Pharmacology and Toxicology, 1992, 101(1), 163–74.

[251] Thomson D, et al. Roadmap on silicon photonics. Journal of Optics 2016, 18, 073003

[252] Truckenmueller R, Henzi P, Herrmann D, Saile V, Schomburg WK. Bonding of polymer micro-structures by UV irradiation and welding at low temperatures. Microsystem Technologies 2004, 10, 372–4.

[253] Turcu F, Tratsk-Nitz K, Thanos S, Schuhmann W, Heiduschka P. Ink-jet printing for micropattern generation of laminin for neuronal adhesion. Journal of Neuroscience Methods 2003, 131, 141–48.

[254] Uhlig S, Robertsson M. Limitations to and solutions for optical loss in optical backplanes. Journal of Lightwave Technology 2006, 24(4), 1710–24.

[255] Ulbricht MJ, Matuschewskia H, Oechela A, Hicke HG, Photo-induced graft polymerization sur-face modifications for the preparation of hydrophilic and low-proten-adsorbing ultrafiltration membranes. J. Membr. Sci. 1996, 115, 31–47.

[256] Van Overmeire S, Ottevaere H, Desmet G, Thienpont H. Miniaturized detection system for fluorescence and absorbance measurements in chromatographic applications. IEEE J. Sel. Top. Quantum Electron. 2008, 14(1), 140–50.

[257] Vogt AK, Lauer L, Knoll W, Offenhausser A. Micropatterned substrates for the growth of functional neuronal networks of defined geometry. Biotechnology Progress 2003, 19, 1562–8.

[258] Vonau W, Guth U. pH Monitoring: a review. J Solid State Electrochem. 2006, 10, 746–52.

[259] Voros J, Graf R, Kenausis GL, Bruinink A, Mayer J, Textor M, Wintermantel E, Spencer ND. Feasibility study of an online toxicological sensor based on the optical waveguide technique. Biosensors & Bioelectronics 2000, 15, 423–9.

[260] Waali EE, Scott JD, Klopf JM. One- and Two-Dimensional Nuclear Magnetic Resonance Spectra of X-ray-Degraded Poly(methyl)methacrylate. Macromolecules 1997, 30, 2386–90.

[261] Wang DY, Huang YC, Chiang H, Wo AM, Huang YY. Microcontact printing of laminin on oxygen plasma activated substrates for the alignment and growth of Schwann cells.

Journal of Biomedical Materials Research Part B: Applied Biomaterials 2006, 80(2), 447–53.

[262] Welle A, Gottwald E. UV-based patterning of polymeric substrates for cell culture applications. Biomedical Microdevices. 2002, 4, 33–41.

[263] Welle A, Horn S, Schimmelpfeng J, Kalka D. Photo-chemically patterned polymer surfaces for controlled PC-12 adhesion and neurite guidance. Journal of Neuroscience Methods 2005, 142, 243–50.

[264] Wencel D, Abel T, McDonagh C. Optical Chemical pH Sensors, Analytical Chemistry 2014 86 (1), 15–29.

[265] Whicher J, Spence C. When is serum worth measuring? Ann Clin Biochem 1987, 24, 572–80.

[266] White JM, Heidrich PF. Optical waveguide refractive index profiles determined from measurement of mode indices: a simple analysis. Applied Optics 1976, 15, 151–5.

[267] White V, Ghodssi R, Fish G, Herdey C, Liu H, Denton DD, McCaughan L. A New Method for Producing Graded Index PMMA Waveguides. IEEE Photonics Technology Letters 1995, 7, 772–3.

[268] Wilbur JL, Kumar A, Kim E, Whitesides GM. Microfabrication by microcontact printing of self-assembled monolayers. Advanced Materials 1994, 6, 600–4.

[269] Williams JA, Keys to Bioreactor Selections, CEP March 2002 www.cepmagazine.org.

[270] Wittmann B, Johnck M, Neyer A, Mederer F, King R, Michalzik R, POF-based interconnects for intracomputer applications. IEEE J. Select. Top. Quantum Electron. 1999, 5, 1243–8.

[271] Woggon T, Klinkhammer S, Lemmer U. Compact spectroscopy system based on a tunable organic semiconductor lasers. Appl Phys B 2010, 99, 47–51.

[272] Woggon T, Punke M, Stroisch M, Gerken M, Lemmer U, Bruendel M, Rabus DG, Weimann T. All-organic waveguide coupled solid-state laser as integrated light source for lab-on-a-chip systems. Nanobioeurope, 2007.

[273] Wolffenbuttel RF. MEMS-based optical mini- and microspectrometers for the visible and infrared spectral range, J. Micromech. Microeng. 2005, 15, 145–52.

[274] Wu H, Odom T W, Chiu D T and Whitesides G M. Fabrication of Complex Three-Dimensional Microchannel Systems in PDMS. J. AM. CHEM. SOC. 2003, 125, 554–9

[275] Xia R, et al. Polyfluorene distributed feedback lasers operating in the green-yellow spectral region. Applied Physics Letters 2005, 87, 1–3.

[276] Xia Y, Whitesides GM. Soft lithography. Angew. Chem. 1998, 7, 550–75.

[277] Yalcın A, Popat KC, Aldridge JC, Desai TA, Hryniewicz J, Chbouki N, Little BE, King O, Van V, Chu ST, Gill D, Anthes-Washburn M, Ünlü MS, Goldberg BB. Optical sensing of biomolecules using microring resonators. IEEE Journal of Selected Topics in Quantum Electronics 2006, 12, 148–55.

[278] Yamamoto H, et al. Extremely-high-density carrier injection and transport over 12000 A/cm$^2$ into organic thin films. Applied Physics Letters 2005, 86, 0835021.

[279] Yanez F, Amoza J, Mararinos B. Hydrogels porosity and bacteria penetration: Where is the pore size threshold ?. Journal of Membrane Science 2010, 365, 248–55.

[280] Yao B, et al. A microfluidic device using a green organic light emitting diode as an integrated excitation source. Lab on a chip 2005, 5, 1041–7.

[281] Ye NW, Xiong Y. Review of silicon photonics: history and recent advances. Journal of modern Physics 2013, 60, 1299–320

[282] Yun KS, Yoon E. Fabrication of complex multilevel microchannels in PDMS by using threedimensional photoresist masters. Lab Chip 2008, 8, 245–50.

[283] Zhang C, Xing D, Li Y. Micropumps, microvalves, and micromixers within PCR microfluidic chips: Advances and trends. Biotechnology Advances 2007, 25, 483–514.

[284] Zhao Y, Zeng H. Fabricating non-photodefinable polymer microstructures for micro-total-analysis. Sens. Actuator B-Chem. 2009, 139(2), 673–81.

[285] Zhao YG, Lu WK, Ma Y, Kim SS, Ho ST, Marks TJ. Polymer waveguides useful over a very wide wavelength range from the ultraviolet to infrared. Appl. Phys. Lett. 2000, 77, 2961–3.

[286] Antila, J., Tuohiniemi, M., Rissanen, A., Kantojärvi, U., Lahti, M., Viherkanto, K., Kaarre, M. and Malinen, J. MEMS- and MOEMS-Based Near-Infrared Spectrometers. Enc Anal Chem 2014: 1–36.

[287] Archibong E, Stewart J, Pyayt A. Optofluidic spectroscopy integrated on optical fiber platform. Sens Bio-Sens Res 2015;3:,1–6.

[288] Ashok PC, Praveen BB, Dholakia K. Optofluidic Raman sensor for simultaneous detection of the toxicity and quality of alcoholic beverages. J Raman Spectrosc 2013;44(6):795–7.

[289] Bakeev KA, editor. Process analytical technology: spectroscopic tools and implementation strategies for the chemical and pharmaceutical industries. John Wiley & Sons. (2010).

[290] Barsan N, Gauglitz G, Oprea, IA, Ostertag E, Proll G, Rebner K, Schierbaum K, Schleifenbaum F, Weimar U. Chemical and biochemical Sensors, 1. Fundamentals. Ullmann's encyclopedia of industrial chemistry, 2016.

[291] Barsan N, Gauglitz G, Oprea, IA, Ostertag E, Proll G, Rebner K, Schierbaum K, Schleifenbaum F, Weimar U. Chemical and biochemical sensors, 2. Applications. Ullmann's encyclopedia of industrial chemistry, 2016.

[292] Baxter, GW, Johnson, MJ, Haub, JG, Orr, BJ. OPO CARS: coherent anti-Stokes Raman spectroscopy using tunable optical parametric oscillators injection-seeded by external-cavity diode lasers. Chem Phys Lett 1996;251(3–4):211–18.

[293] Bergveld P. Thirty years of ISFETOLOGY: what happened in the past 30 years and what may happen in the next 30 years. Sens Actuators B: Chem 2003;88(1):1–20.

[294] Bergveld P. ISFET, theory and practice. In IEEE Sensor Conference, Toronto, vol. 328, October 2003.

[295] Bogomolov A, Dietrich S, Boldrini B, Kessler, RW. Quantitative determination of fat and total protein in milk based on visible light scatter. Food Chem 2012;134(1):412–8.

[296] Boldrini B, Kessler W, Rebner K, Kessler, RW. Hyperspectral imaging: a review of best practice, performance and pitfalls for in-line and on-line applications. J Near Infrared Spectrosc 2012; 20 (5):483–508.

[297] Brown SW, Johnson BC, Feinholz ME, Yarbrough MA, Flora SJ, Lykke KR, Clark DK. Stray-light correction algorithm for spectrographs. Metrologia 2003;40(1):S81.

[298] Buijs K, Maurice, MJ. Some considerations on apparent deviations from lambert-beer's law. Anal Chim Acta 1969;47(3):469–74.

[299] Büning-Pfaue H. Analysis of water in food by near infrared spectroscopy. Food Chem 2003;82 (1):107–15.

[300] Burns DA, Ciurczak EW, editors. Handbook of near-infrared analysis. CRC press, 2007.

[301] Chung H, Arnold, MA, Rhiel M, Murhammer, DW. Simultaneous measurements of glucose, glutamine, ammonia, lactate, and glutamate in aqueous solutions by near-infrared spectroscopy. Appl Spectrosc 1996;50(2):270–6.

[302] Deckert-Gaudig T, Taguchi A, Kawata S, Deckert V. Tip-enhanced Raman spectroscopy–from early developments to recent advances. Chem Soc Rev 2017;46(13):4077–110.

[303] Ehrfeld W, Hessel V, Haverkamp V. Microreactors. Wiley-VCH Verlag GmbH & Co. KgaA, 2000.

[304] Farahani MA, Gogolla T. Spontaneous Raman scattering in optical fibers with modulated probe light for distributed temperature Raman remote sensing. J Lightwave Technol 1999;17(8): 1379–91.

[305] Hantelmann K, Kollecker M, Hüll D, Hitzmann B, Scheper T. Two-dimensional fluorescence spectroscopy: a novel approach for controlling fed-batch cultivations. J Biotechnol 2006; 121 (3):410–7.

[306] Hergert W, Wriedt T editors. The Mie theory: basics and applications, vol. 169. Springer, 2012.

[307] Hieftje GM, Haugen GR. Correction of quenching errors in analytical fluorimetry through use of time resolution. Anal Chim Acta 1981;123:255–61.

[308] Ho CK, Itamura MT, Kelley M, Hughes RC. Review of chemical sensors for in-situ monitoring of volatile contaminants. Sandia Report SAND2001-0643, Sandia National Laboratories, 2001:1–27.

[309] https://www.bruker.com/products/infrared-near-infrared-and-raman-spectroscopy/opus-spectroscopy-software/downloads/opus-downloads.html;

[310] Jian RS, Huang YS, Lai SL, Sung LY, Lu CJ. Compact instrumentation of a μ-GC for real time analysis of sub-ppb VOC mixtures. Microchem J 2013;108:161–167.

[311] Jimenez-Jorquera C, Orozco J, Baldi A. ISFET based microsensors for environmental monitoring. Sensors 2009;10(1): 61–83.

[312] Kandelbauer A, Kessler W, Kessler, RW. Online UV–visible spectroscopy and multivariate curve resolution as powerful tool for model-free investigation of laccase-catalysed oxidation. Anal Bioanal Chem 2008;390(5):1303–15

[313] Kandelbauer A, Rahe M, Kessler RW. Industrial perspectives. In: Handbook of Biophotonics, 2012.

[314] Kauppinen J, Partanen J. Fourier transforms in spectroscopy. John Wiley & Sons, 2011.

[315] Kessler RW,Perspectives in process analysis. J Chemom 2013;27(11):369–378.

[316] Kessler RW, editor. Prozessanalytik: Strategien und Fallbeispiele aus der industriellen Praxis. John Wiley & Sons, 2012.

[317] Kessler RW, editor. Prozessanalytik: Strategien und Fallbeispiele aus der industriellen Praxis. John Wiley & Sons, 2012:289–311.

[318] Kessler W, Oelkrug D, Kessler R. Using scattering and absorption spectra as MCR-hard model constraints for diffuse reflectance measurements of tablets. Analytica chimica acta 2009; 642 (1):127–34.

[319] Kiefer W. Surface enhanced Raman spectroscopy: analytical, biophysical and life science applications. John Wiley & Sons, 2011.

[320] Klein, GA, Meyrath T. Industrial color physics, vol. 7. New York: Springer, 2010.

[321] Larkin P. Infrared and Raman spectroscopy: principles and spectral interpretation. Elsevier, 2017.

[322] Last A. Fehllicht in LIGA-Mikrospektrometern. Doctoral dissertation, Karlsruhe, Univ, Diss, 2002, 2003.

[323] Lewis IR, Edwards H. Handbook of Raman spectroscopy: from the research laboratory to the process line. CRC Press, 2001.

[324] Marose S, Lindemann C, Scheper T. Two-dimensional fluorescence spectroscopy: a new tool for on-line bioprocess monitoring. Biotechnol Progr 1998;14(1):63–74.

[325] Martínez-Máñez R, Soto J, García-Breijo E, Gil L, Ibáñez J, Gadea E. A multisensor in thick-film technology for water quality control. Sens Actuators A: Phys 2005;120(2):589–95.

[326] Meng X, Wang S, Zhu M. Quinoline-based fluorescence sensors. In Molecular photochemistryvarious aspects. InTech, 2012.

[327] Merz T, Kessler RW. On-line Prozesskontrolle mittels 2D-Fluoreszenzspektroskopie. Process 2007;9:44–5.

[328] Metzger M, Riede D, Fleischmann A, Konrad A, Blendinger F, Meixner AJ, Bucher J, Brecht M. low-cost GRIN-lens based nephelometric turbidity sensing in the range of 0.1-1000 NTU (in preparation), 2018.

[329] Oelkrug D, Boldrini B, Rebner K. Comparative Raman study of transparent and turbid materials: models and experiments in the remote sensing mode. Anal Bioanal Chem 2017;409(3):673–81.

[330] Oelkrug D, Brun M, Hubner P, Rebner K, Boldrini B, Kessler R. Penetration of light into multiple scattering media: model calculations and reflectance experiments. Part II: the radial transfer. Appl Spectrosc 2013;67(4):385–95.

[331] Oelkrug D, Brun M, Rebner K, Boldrini B, Kessler R. Penetration of light into multiple scattering media: model calculations and reflectance experiments. Part I: the axial transfer. Applied Spectroscopy 2012;66(8):934–943.

[332] Opto-fluidic microsystem and method, EP 2833136 A1, 4. Feb. 2015

[333] Ou S, et al. A sugar-quinoline fluorescent chemosensor for selective detection of Hg2+ ion in natural water. Chemical Communications 2006;42:4392–4.

[334] Paaso et al., Microspectrometers for paper, plastic and pharmaceutical applications, 2017

[335] Pahlow S, et al. Bioanalytical application of surface-and tip-enhanced Raman spectroscopy. Eng Life Sci 2012;12(2):131–43.

[336] Persichetti G, Testa G, Bernini R. High sensitivity UV fluorescence spectroscopy based on an optofluidic jet waveguide. Opt Express 2013;21(20):24219–30.

[337] Prozess-Sensoren, TR. VDI/VDE-Gesellschaft Mess-und Automatisierungstechnik (GMA), 2009.

[338] Psaltis D, Quake, SR, Yang C. Developing optofluidic technology through the fusion of micro-fluidics and optics. Nature 2006;442(7101):381–6.

[339] Rathore AS, Winkle H. Quality by design for biopharmaceuticals. Nat Biotechnol 2009; 27 (1):26–34.

[340] Rissanen A, Akujärvi A, Antila J, Blomberg M, Saari H. MOEMS miniature spectrometers using tuneable Fabry-Perot interferometers. J Micro/Nanolithogr MEMS MOEMS 2012; 11(2):023003–1.

[341] SádeCká J, TóThoVá J. Fluorescence spectroscopy and chemometrics in the food classification review. Czech J Food Sci 2007;25(4):159–73.

[342] Schafer D, Squier, JA., Maarseveen, JV., Bonn D, Bonn M, Müller M. In situ quantitative measurement of concentration profiles in a microreactor with submicron resolution using multiplex CARS microscopy. Journal of the American Chemical Society 2008;130(35):11592–3.

[343] Schuler, LP., Milne, JS, Dell, JM, Faraone L. MEMS-based microspectrometer technologies for NIR and MIR wavelengths. J Phys D: Appl Phys 2009;42(13):133001.

[344] Shopova SI, White IM, Sun Y, Zhu H, Fan X, Frye-Mason G, et al. On-column micro gas chromatography detection with capillary-based optical ring resonators. Anal Chem 2008; 80 (6):2232–2238.

[345] Simon LL, et al. Assessment of recent process analytical technology (PAT) trends: a multiauthor review. Org Process Res Dev 2015;19(1):3–62.

[346] Socrates G. Infrared and Raman characteristic group frequencies: tables and charts. John Wiley & Sons, 2001.

[347] Technologie-Roadmap „Prozesssensoren 4.0 "VDI/VDE-Gesellschaft Mess-und Automatisierungstechnik (GMA), 2015.

[348] TEMATYS: Market and technology report "Miniature and Micro Spectrometers: End-user needs, Markets and Trends"

[349] Trefz P, Boldrini B, Kessler, RW, Löbbecke S. Online-Analyse von Mikroreaktionsprozessen mittels Pushbroom Imaging. Chemie Ingenieur Technik 2010;82(4):525–30.

[350] Udd E, Spillman Jr, WB, editors. Fiber optic sensors: an introduction for engineers and scientists. John Wiley & Sons, 2011.

[351] Vonau W, Guth U. pH monitoring: a review. J Solid State Electrochem 2006;10(9):746–752.

[352] Wang M, Jing N, Chou, IH, Cote, GL, Kameoka, J. (2007). An optofluidic device for surface enhanced Raman spectroscopy. Lab on a Chip, 7(5),630-632.

[353] Waters T. Industrial sampling systems. Solon, OH: Swagelok Company, 2013.

[354] Wencel D, Abel T, McDonagh C. Optical chemical pH sensors. Analytical Chemistry 2013; 86 (1):15–29.

[355] Wolffenbuttel, RF. MEMS-based optical mini-and microspectrometers for the visible and infrared spectral range. J Micromech Microeng 2005;15(7):S145.

[356] Zhu H, Nidetz R, Zhou M, Lee J, Buggaveeti S, Kurabayashi K, Fan X. Flow-through microfluidic photoionization detectors for rapid and highly sensitive vapor detection. Lab Chip 1 2015; 5 (14):3021–3029.

# Index

https://doi.org/10.1515/9783110546156-013